Publisher: Military History Group, London, United Kingdom.

E-Mail: milhisgroup@gmail.com

Print: Lulu Press, Inc., Lulu Press, Inc. 627 Davis Drive Suite 300 Morrisville, NC 27560, USA. Massachusetts, US; Wisconsin, US; Ontario, Canada; Île-de-France, France; Wielkopolska, Poland; Cambridgeshire, United Kingdom; Victoria, Australia.

Text © 2024 Ryan A. Then. Foreword © 2024 Paul J. Hazell. All Rights reserved.

Cover design © by Julien Lepelletier.

Photographs & illustrations © as individually credited.

Any total or partial reproduction, copying, representation, adaptation, modification, or transfer of this book, its content or visuals, whether for commercial or non-commercial purposes, is forbidden without express, prior and written consent by the authors.

Quotations of the book with proper citation (authors, title, page number) are of course allowed.

978-1-915453-21-1

http://militaryhistorygroup.com

Foreword	7
Preface	9
Acknowledgements	9
Introduction	10
Crewing the T-72	21
Blending Form and Function	21
Designing for Paper People	25
Soviet Tanker Anthropometry	28
Small, Not Cramped	32
Short Term Habitation	35
Long Term Habitation	36
Self-Defense	38
Ventilation	40
Crew Stations	45
Periscopes and Periscope Design	46
Commander's Station	52
Overview	52
Visibility	60
TKN-3	67
Radio and Intercom System	72
R-123M	75
R-173 and R-173P	76
R-130M	77
Gunner's Station	80
Sighting Systems	83
Day Sight	83
Sight Markings	84
TPD-2-49	86
TPD-K1	94
Day Sight Design	100

- Night Fighting ... 112
  - Features of Night Sights ... 114
  - TPN-1-49 ... 118
  - TPN-3-49 ... 119
  - 1K13-49 ... 121
  - Driver's Station ... 123
  - Driving at Night ... 132
  - Batteries ... 135
  - Inertial Navigation ... 138
- CBRN Protection ... 140
  - The Nuclear and Radiological Threat ... 140
  - Nuclear, Radiological Threat ... 149
  - Anti-Radiation Shielding ... 153
- Protection ... 165
  - Concealment and Signature Management ... 165
  - Tucha-2 Smoke Grenade System ... 173
- Firefighting Systems ... 175
  - ZETs11-3 ... 178
  - ZETs13-1 *Iney* ... 179
- Armor ... 182
  - Structural Overview ... 182
- Composite Armor Design ... 186
  - Side Screens ... 189
  - Upper Glacis ... 194
  - Three-layer STB armor (T-72) ... 198
  - Improved three-layer STB armor (T-72A) ... 202
  - Spaced Armor (T-72B) ... 204
  - Integral ERA (Kontakt-5) ... 208
  - Turret ... 213
  - Monolithic Turret ... 218

- Sand Bar Turret 221
- Reflective Plate turret 226
- Kontakt-1 ERA 229
- Kontakt-5 ERA 233
- Armament 235
  - Control & Aiming 235
  - 2E28M *Siren* 240
  - 2E42 *Zhasmin* 243
  - Autoloader 245
- Control System 256
- D-81 Main Gun 265
  - Designing a Big Gun for a Small Turret 270
  - Trunnion 278
  - Breech Assembly 282
  - Interior Ballistics 284
  - Power and Pressure 289
  - Barrel Replacement Process 296
  - Smoothbore and Barrel Wear 297
  - Bore Evacuator and Thermal Sleeve 302
- Ammunition 305
  - Overview 305
  - Powder Charges 308
  - Propellant Powders for 125 mm Charges 313
  - High-Nitrogen Nitrocellulose 316
  - APTs-235P 319
  - Charge Composition 321
  - Additives 322
  - APFSDS 324
  - Ring Sabot 326
  - Single-ramp Sabot 327

- 3BM9, 3BM17 329
- 3BM12, 3BM15 330
- 3BM22 331
- 3BM26, 3BM29 332
- HEAT 335
- 3BK12 337
- 3BK14, 3BK18 338
- HE-Frag 340

Machine guns 344
- PKT Coaxial Machine Gun 344
- NSV-12.7 Anti-Aircraft Machine Gun 352
- 12.7 mm Penetration Power 354
- ZU-72 355

Mobility 358
- Performance 358
- Drivetrain Design 362
- Water Obstacles 371
- Engine 376
- Engine Design 388
- Engine Air Cleaner 392
- Cooling System 394
- Transmission 399
- BKP Gearboxes 401
- BKP Design 404
- Steering 412
- Pneumatics 417
- Suspension 419
- Wheel Design 432
- Tracks 434

Gallery 440

Bibliography 446
    English Language Sources 446
    Russian Language Sources 454
    German Language Sources 482
    Other Language Sources 483

Foreword

Military history is awash with many main battle tanks and some of these stand out not just for their technological prowess, but for the indelible mark they leave on the landscape of warfare. Many armies around the world have used the T-72 and it has been widely tested in conflict. Entering service in 1973, it was used as early as 1980 in the Iran-Iraq war and is still being used in battles as I write 44 years later. The T-72 tank holds a place of undeniable significance. From the battlefields of the Cold War to conflicts on different continents, the T-72 has been an enduring influence, a symbol of Soviet pride, strength, innovation, and adaptability. From the deserts of the Middle East to the frozen expanses of Eastern Europe, the T-72 has proved its tenacity time and again, and in some ways, shaped the outcomes of conflicts and influenced the strategies of war-planners. And even today, it remains a steadfast presence in the largest conflict of its time, whilst also experiencing a revolution in land warfare due to the massive employment of drones on and beyond the front line.

Ryan A. Then's assessment of the T-72 is both useful and timely as it is still a troubling adversary despite its vintage. The fact that is still used today is a testament to the Soviet engineers of the 1960s not least because it went on subsequently to be the basis for the development of the future T-90. Not many of us would use a 1970's vehicle as our 'go-to' workhorse to do the shopping or travel to see family. And yet, time and again, this tank makes an appearance on the battlefield despite that it was largely based on the much earlier T-64. Of course, we should note that the fundamentals of the T-72 are arguably rather dated and as a system, it is of little worry to its modern counterparts, such as the Challenger 3 and the M1A2 Abrams. And more recently it has been known to fall victim to improvised drone attacks! Nevertheless, as a military war machine, it has become ubiquitous.

This book delves into the heart of the T-72 unravelling its technical specifications. Ryan also takes us through the development of the T-72, including a fascinating narrative around the deployment of different armour recipes, for example. Through meticulous research, readers will journey through the detail of this war machine whilst being able to look up over one thousand endnotes and references that include both Soviet (Russian) and Western literature.

Whether you are a military enthusiast, a history buff, or simply curious about the machines that shape our world, this book offers a comprehensive exploration of one of the most iconic tanks of the modern era. It is a testament to the enduring legacy of the T-72 and a tribute to the men and women who forged its place in history.

So, join me as we embark on a journey through the technical developments of the T-72 and uncover the detail that made it such a prolific war machine.

Professor Paul J. Hazell

Author of: Armour: Materials, Theory, and Design

UNSW, Canberra, Australia

June 2024

Preface

The T-72 was conceived as a major advancement over the previous generation of Soviet medium tanks in technical capability and usability. Taking the U.S. Army's M60 medium tank as the reference threat, the T-72 was armed with a potent 125 mm gun and shielded with armor proof to return fire from the M60's 105 mm gun. Mobility, though harder to rate qualitatively, was improved to the industry's limit. Particular attention was given to nuclear protection, a critical concern shared by the United States and West Germany, exhibited most famously in the deep nuclear hardening emphasized in the MBT-70 main battle tank project. Unlike the MBT-70, which ultimately failed to reach production, the T-72 achieved its ambitious goals and became a cornerstone of the Soviet tank forces.

This book is an examination of all of the technologies applied to the T-72, from its well-known autoloader and large gun to the glass in its periscopes and the pigments of its paint, and how the sum of all its design choices resulted in a tank that not only met but exceeded its intended objectives.

Tank design, like any technical endeavor, is a multidisciplinary field built on the expertise of specialists with the support of innumerable engineers across an immense industry supply chain. Rather than simply describing the T-72, we will focus on the hows and whys of its design and delve into the science behind it. Keeping in mind that these technologies exist in service to the tank's purpose as an instrument of military might, we cannot forget that a tank is first and foremost a vehicle for a crew of soldiers. It is in recognition of the human element that we begin with an examination of the fundamental design choices that shaped the layout of the T-72, and how the crew fit within it.

Acknowledgements

I would like to express my gratitude to Ivan Dashkov, Peter Kefallinos and Sebastian Stauber for their photography, and to the countless enthusiasts who have disseminated esoteric literature into the public sphere to spread knowledge for its own sake.

Special thanks to the Schweizerisches Militärmuseum for their generosity and photographic contributions, and to Andrew Gianelli and Paul A. Anderson for their unwavering assistance in reviewing and refining the book.

Introduction

On land, there are few tools more dependable than the modern tank to project one's will. The core belief placed on the tank as both a geopolitical tool and the backbone of mechanized warfare was true of every large army in the 20th century, but few have emphasized it so prominently as the Soviet Army.

The T-72 was a product of the desire to maintain a dominating position at the forefront of modern warfare. By combining new concepts and new technologies, the Soviet military leadership sought to hone an armored spearhead able to weather nuclear strikes and traverse hundreds of kilometers in titanic thrusts across Western Europe.

Like any other combat vehicle, tanks are developed by succession. Once any combat vehicle appears to be set for production, work on its successor begins immediately. The T-72 was an indirect successor to the T-54, born out of a long and rocky gestation period drawn out over two decades. The T-54 was meant to be replaced by the titular "New Medium Tank", the design of which was initially a two-way competition between two highly experienced tank factories, the KB-60 design bureau from Factory No. 75 in Kharkov, headed by veteran Chief Designer Aleksander Morozov, designer of the T-54, and the VNII-100 research institute associated with Factory No. 100 in Leningrad, headed by the legendary Iosif Kotin, father of the KV and IS heavy tanks of WWII fame. In a surprise upset, OKB-520 from Factory No. 183 in Nizhny Tagil was introduced as a third competitor and the VNII-100 proposal was eliminated.

The design requirements did not push for a major leap in capability, and as a result, both the Object 140 prototype from OKB-520 and the Object 430 from KB-60 were technically conservative designs, closely resembling the T-54 that they were meant to replace. This became the undoing of the entire project. The Object 140 was soon withdrawn from the competition due to irreconcilable issues baked into its design, but most crucially, Object 430 failed to win by merit.

It had transpired that the essential improvements outlined in the "New Medium Tank" program were achievable by simply upgrading the T-54 to the T-55, and Object 430 in its final form lacked a convincing armor and firepower advantage over new NATO tanks appearing in the late 1950's. Object 430 was armed with a powerful 100 mm rifled gun, yet it was not powerful enough to overmatch the new American M60 medium tank. It had an extremely compact drivetrain with a novel opposed-piston diesel engine at its center, but the tank was no faster than the T-55. Its armor, though thicker, mattered little to shaped charge shells fired from infantry recoilless rifles and most tank guns.

The Object 432 was created to rectify these shortcomings, armed with a 115 mm smoothbore gun as a stopgap solution against the M60 and equipped with composite armor capable of stopping 105 mm HEAT shells. To stay within the strict 36-tonne weight limit and lower the tank's silhouette, the loader was replaced by an autoloader, reducing the crew to three. An advanced fire control system with a stabilized optical rangefinder greatly reduced engagement times and improved the tank's long-range accuracy. The Object 432 entered mass production in 1963 and would formally enter service as the T-64 in 1966, but even this technological marvel of a tank was a stopgap to something better. A more powerful gun was desired to fight the M60, and the Object 434, armed with a 125 mm smoothbore gun, entered service in 1967 as the T-64A.

Under a resolution issued by the Central Committee of the CPSU and the Council of Ministers of the USSR on August 15, 1967, Factory No. 183 was ordered to prepare for the commencement of T-64A production in 1970. Not content with the idea of producing a competing product, OKB-520 campaigned to modify the T-64A into a "homegrown" tank. As nefarious and short-sighted as this plot might seem, its legitimacy was, at least, undeniably supported by the general dissatisfaction expressed by the Army during T-64A field exercises.

Under the auspices of resolving faults in the T-64A identified by the Army, OKB-520 successfully lobbied to "modernize" the T-64A into the Object 172. By order of the Minister of Defense Industry on 5 January 1968, OKB-520 formally started work on installing a cabinless autoloader, a four-stroke diesel engine and a new sight in the T-64A. The Object 172 was then evolved into the Object 172M at the design bureau's initiative. In May 1970, the government officially released Factory No. 183 of its obligation to produce the T-64A, and directed the factory to work on the Object 172M in the 1970-1971 period. In 1973, the T-72 *Ural* entered service, and full-scale production began in 1974.

The organization of T-72 production took place within the framework of the ninth Soviet five-year plan, marked for the 1971-1975 period. The ground forces were scheduled for a large-scale rearmament cycle, which involved the replacement of legacy tanks like the T-55 and T-62 by the T-64A, now joined by the T-72. After the T-72 entered production, it received a continuous stream of upgrades. Each design change meaningfully improved the tank in one way or another without significant disruption to the production line. The number of changes would accumulate until the degree of improvement to the tank's tactical-technical characteristics was major enough to warrant designating the tank under a new model name.

The T-72 was categorized under three official models, beginning its life with the T-72 *Ural*, followed by the T-72A in 1979, and the T-72B in 1985. The T-72 in its many modifications saw service in dozens of nations well after the fall of the USSR and will remain in service for the foreseeable future.

Table 1: Timeline of T-72 production and modernization

Year	Event
1973	T-72 *Ural* enters service on 7 August 1973.
1974	Beginning of T-72 *Ural* mass production.
1975	The first modernization of the T-72 was accepted into service, informally named T-72 *Ural-1*.
1976	T-72 *Ural-1* begins mass production.
1977	A new turret with "sand bar" composite armor replaces the original turret in mass production.
1978	The optical rangefinder for the gunner's sight is replaced by a laser rangefinder.
1979	The T-72A enters service on 22 July 1979 together with the T-72AK command tank.
1981	Development of the Object 184 begins on 5 July 1981. Testing of Kontakt-1 explosive reactive armor (ERA) on the T-72A begins in late 1981.
1983	T-72M1 is authorized for export.
1985	Kontakt-1 explosive reactive armor is accepted for service and is fitted to the T-72A, creating the T-72AV, and to the new T-72B as standard. Production of the T-72B and the simplified T-72B1 begins.
1987	Kontakt-5 explosive reactive armor is fitted to the T-72B for the first time.

The upgrades received by the T-72 during this period were shared with the T-64A. New side skirts, smoke grenade launchers, the 1A40 sighting system, a UVBU target leading unit, and other changes[1].

Throughout the 1970's, the welding of T-72 hulls on the UVZ production line was largely automated, but inside the structure itself, most of the smaller details were welded manually or semi-automatically inside on atop the finished hull. There were more than 700 welded parts and assemblies in total, and 97% of them were installed and welded inside the enclosed hull. As the tank evolved, its manufacturing methodology was progressively refined to improve working

---

[1] Чобиток, В., Саенко, М., Тарасенко, А., Чернышев, В. (2016). Основной танк Т-64 - 50 лет в строю. Москва: Яуза-каталог. p. 55.

conditions, reduce production time, improve the quality of welding, and reduce material consumption. The labor intensity of welding alone decreased by 2.8 times through the 1980's[2].

An extensive and expensive retooling took place in 1973 to switch the T-62 production line at *Uralvagonzavod* to the T-72. In spite of its familial resemblance to the T-62, the T-72 was a new tank in every sense of the word. Its continuity with the T-62 production line was very low[3]. Only 32 out of 231 casting blanks (14%) and 51 out of 709 forging blanks (7%) were carried over to the T-72 from the T-55 and T-62, and the degree of commonality was no higher in the final assembly parts. The T-72 borrowed only 284 out of 1,530 parts (18%) subjected to final heat treatment, and only 361 out of 5,240 machined parts (7%)[4].

The organization of T-72 production took place within the framework of the ninth Soviet five-year plan, marked for the 1971-1975 period. The ground forces were scheduled for a large-scale rearmament cycle, which included the replacement of legacy tanks like the T-55 and T-62 by the T-64A. However, despite entering service in 1967, the T-64A did not have the breakneck pace of mass production to support this rearmament plan.

A great deal of industrial cooperation was involved in the design of each tank. It is somewhat inaccurate for design bureaus like KB-60 or UKBTM to extend ownership over products. TsNII-73 was responsible for gun stabilizers. TsKB-393 developed tank sights with integral rangefinders in conjunction with KMZ[5]. TsNII-48 developed the manufacturing technology for the hull. The BKP transmission was designed by KB-60 in cooperation with VNII-100 and Department 15 of the Military Academy named after Stalin. The drivetrain of the T-72 involved VNII *Transmash*.

---

[2] Жирнов, А. Я., Левин, С. М., Шумилин, В. Г. (1987). 'Повышение Технологичности Изготовления Корпуса Танка', *Вестник Бронетанковой Техники* 1987, сборник 1. pp. 56-58.
[3] Ibid. p. 61.
[4] Лурье, В. И. (1981). 'О Влиянии Технологичности Конструкции На Сроки Подготовки И Мобильность Производства', *Вестник Бронетанковой Техники* 1975, сборник 1. p. 60.
[5] Красногорский механический завод (КМЗ)

Figure 1: T-72 platoon on an assault during *Zapad*-81 exercises. (A. I. Skrylnika, 1982)

Table 2: Basic tactical-technical characteristics of the three basic T-72 models

Model	T-72 *Ural*	T-72A	T-72B
Year of introduction	1973	1979	1984
Mass (metric tonnes)	41.0	41.5	44.5
Crew size	3		
Self-defense	1 x AKMS or AKS-74U, 3 x PM pistol		
Max. hull length (mm)	6,670	6,860	
Hull width to tracks (mm)	3,370		
Max. hull width (mm)	3,460	3,590	3,590
Max. height (mm)	2,230	2,230	2,230
Max. length, gun forward (rearward) (mm)	9,530 (9,320)		
Main gun	2A26M2	2A46	2A46M
Gun type	125 mm smoothbore, high-velocity		
Main gun barrel length (in mm) (in calibers)	6,000 (48)		
Gun stabilizer	2E28M		2E42-1
Elevation range	-6°13' to +13°47'		
Ammunition types	APFSDS, HEAT, HE-Frag		+ Gun-Launched ATGM
Standard combat load, in autoloader and in reserve	22 (39)	22 (44)	22 (45)
Coaxial machine gun	7.62 mm, PKT with 2,000 rounds		
Anti-aircraft machine gun	12.7 mm, NSV-12.7 with 300 rounds		
Elevation range	-5° to +75°		
Day sight, rangefinder type	TPD-2-49, optical rangefinder	TPD-K1, laser rangefinder	Improved TPD-K1
Max. direct fire range (m)	4,000		

Night sight	TPN-1-49	TPN-3-49	1K13-49
Night vision type	Active IR	Active IR + Passive	
Max. viewing range (passive) (m)	800	1,300 (500)	1,200 (500)
Hull armor composition	Glass textolite composite		Spaced high-hardness steel
Turret armor composition	Homogeneous Steel	Composite with sintered casting core filler	Spaced high-hardness steel, with bulging plates
Basic protection standard	105 mm APDS, HEAT	105 mm APFSDS, HEAT	105-120 mm APFSDS, HEAT
Smoke concealment system	Exhaust smokescreening	81 mm Smoke grenades + Exhaust smokescreening	
Firefighting system	Fully automatic with crew and engine compartment protection. Portable fire extinguishers as reserve.		
System model	ZETs11-3		ZETs11-3, ZETs13-1 (late)
Fire extinguishing agent	Composition 3.5	Halon 2402	Halon 1301
CBRN protection	Automatic overpressure protection system against nuclear detonation, nuclear fallout, sarin, soman and V-series nerve agents.		
Engine	V-46-4	V-46-6	V-84-1
Engine power (hp)	780		840
Power-to-weight (hp/tonne)	19.0	18.8	18.8
Transmission	Mechanical, twin-gearbox with geared steering		
Transmission gearing	7 forward, 1 reverse		
Top speed (technical max.) (km/h)	60 (68)		
Average speed on dirt road (km/h)	35-45		
Driving range (with fuel drums)	500 (650)		500 (700)
Fording depth (wading) (m)	1.2 (1.8)		
Deep wading depth, with snorkel (m)	5.0		
Navigation	GPK-59 gyrocompass, inertial navigation by waypoints		

Export T-72 models were divided between those intended for export within the Warsaw Pact, and to friendly states beyond the Warsaw Pact. The only meaningful distinction between these two types was the CBRN protection system. Tanks exported outside the Warsaw Pact lacked the GO-27 nuclear-chemical protection system, which contained a small sliver of weapons-grade plutonium (plutonium-239) in its chemical detection system as an alpha radiation emitter. Instead, these tanks were equipped with a simplified atomic defense system similar to the type available in the T-55 and T-62. This only changed at the end of the 1980's, when the T-72S "Shilden" was offered for export to Iran.

All export T-72 models lacked the cable anchor mounting points for the KMT-7 lane-clearing mine plow-roller, limiting it to the KMT-5M lane-clearing mine roller. The KMT-6 or KMT-4M mine plow could be fitted freely. Apart from mine roller compatibility and the CBRN protection system, export models broadly corresponded to their Soviet counterparts in terms of technical capability, but did not always match any specific model.

All export models were denoted by a special index as an administrative differentiator from T-72 models intended for the Soviet Army. Export T-72 models were divided by name into four generations. From 1976 until 1978, export models were known simply as the T-72. It was replaced by the T-72M, which in turn was superseded by the T-72M1 in 1983. The T-72M1 was analogous to the T-72A. The final export modification offered before the collapse of the USSR was the T-72S, based on the T-72B.

Table 3: T-72 export models and their definitive features

Two-page spread. Continued in next page

Model	T-72		T-72M	
Index	172M-E	172M-E1	172M-E2	172M-1-E3
Service year	1976	1978	1978	1980
Customer	WP	Other	WP	WP
Country in service	ČSSR, Poland, Yugoslavia (1978)	Libya, Algeria, Syria, Iraq, India	ČSSR, Poland	ČSSR, Poland, Bulgaria, DDR
Gun	2A46			
Ammunition load	39			44
Day sight	TPD-2-49		TPD-K1	
Night sight	TPN-1-49			
Stabilizer	2E28M			
Hull armor	Corresponding to T-72 *Ural*			
Hull side armor	"Gill" armor panels			
Turret armor	Corresponding to T-72 Ural			
CBRN defense	Atomic, Chemical	Atomic	Atomic, Chemical	
CBRN defense model	GO-27, FPT-100M filter	GD-1M, centrifugal ventilator	GO-27, FPT-100M filter	
Smokescreen system	TDA			
Engine	V-46-4 (780 hp)			
Transmission	BKP			
Suspension	Corresponding to T-72 *Ural*			

Поликарпов, В. В. (1997). 'Экспортные модификации танка Т-72', Невский Бастион. pp. 14-16.

T-72M	T-72M1		T-72S
172M-1-E4	172M-1-E5	172M-1-E6	172M-8
1982	1983	1983	1990
Other	WP	Other	Open
Libya, Algeria, Iraq, India, Finland	ČSSR, Poland, Yugoslavia,	India	Iran
2A46 with thermal sleeve			2A46M
44			45
TPD-K1			1A40-1
TPN-1-49-23			1K13
2E28M			2E42-2
	Corresponding to T-72A		Corresponding to T-72B (1985)
Full-length side skirts			Full-length side skirts with ERA
	Corresponding to T-72A		Corresponding to T-72B (1985)
Atomic	Atomic, chemical	Atomic	Atomic, chemical
GD-1M, centrifugal ventilator	GO-27, FPT-100M filter	GD-1M, centrifugal ventilator	GO-27, FPT-100M filter
TDA and *Tucha* 81 mm smoke grenade system			
V-46-6 (780 hp)			V-84MS (840 hp)
BKP with pneumatic brake			Reinforced BKP with pneumatic brake
	Improved shock absorbers		Improved torsion bars

Figure 2: Disembarking T-72 during *Zapad-81* exercises. (A. I. *Skrylnika*, 1982)

CHAPTER I
Crewing the T-72

By omitting a human loader - the only member of a tank crew obligated to stand while working - it was possible to lower the tank's silhouette, decrease the total internal volume and simultaneously allocate more space to the remaining crew members[6]. The reduction of the silhouette height was viewed as a critical component of a tank's protection, not only in the USSR, but abroad. On a battlefield expected to be saturated with shaped charge weapons, guided missiles and anti-tank artillery of ever-increasing lethality, lowering the silhouette of a tank reduced the likelihood of being seen. If seen, it reduced the likelihood of being hit.

There were a number of far more radical solutions to the issue of reducing silhouette height, explored both within the USSR and internationally. These included omitting the turret entirely, placing all crew members in a large turret, placing all crew members in the hull with an unmanned turret, and low-profile turrets on otherwise conventional tanks. An autoloader was the cornerstone to most of these radical designs.

Most of the attempts at implementing such novelties never left the prototyping stage. Of the few that did, very few were successful enough to enter service. Even so, suffice it to say, the merits of a low silhouette were understood and widely appreciated internationally, despite the unfortunate fate that radical solutions seemed to entail.

Blending Form and Function

The core qualitative metrics of firepower, protection and mobility bear heavily upon any tank designer. In the Soviet Union, tank designers worked not only to satisfy lofty technical requirements, but also their own sense of competition against the tank builders of the so-called "hypothetical enemy". The challenge was formidable. A fully enclosed, low-profile, lightweight, heavily armed and heavily armored tank was universally desired, but creating one was a different matter entirely.

By the late 1950's, the fact that combat performance was inextricably linked to the capabilities of the crew was well understood. For this reason, the fundamental principle that underlied postwar Soviet tank design was to achieve

---

[6] Попков В. Ф. (1984). 'Танк Т-64А (к 20-летию серийного производства)', *Вестник Бронетанковой Техники* 1984, сборник 4. р. 52.

progress in all aspects relevant to practical performance, rather than to approach the challenge as a zero-sum game. This idea was expressed quite succinctly in the 1958 design textbook *Fundamentals of designing artillery weapons for tanks and self-propelled artillery installations*[7]:

> The possibility of effective use of a vehicle's armament, and consequently, its firepower, largely depends on the correctness of the layout of its fighting compartment. When laying out the fighting compartment, the following basic requirements must be observed:
>
> 1) the size of the fighting compartment should be as small as possible, but sufficient to accommodate the armament and the main supply of ammunition, as well as the convenient location of crew members related to the handling of the armament

Studies carried out by the Soviet Military Academy of the Tank Forces[8] showed that in a duel between two tanks, one with a fatigued crew and one without, the tank with the unfatigued crew would emerge victorious 1.5-2 times more often[9].

The basic premise of the approach taken in the T-72's design[10] was to accept that the bodily dimensions of the crew in their working postures dictate the basic form of the internal volume of the vehicle, but any volume in excess of the minimum required by the crew carries with it a worsening of other combat characteristics. The most obvious source of extraneous volume was the loader. Replacing him with an automatic loader freed up room to expand the workstations of the other crew members, increase the ammunition capacity, and lower the tank[11] below the height of its predecessors, the T-54/55 and T-62. At the same time, the hull was widened slightly, and the crew compartment was lengthened. The difference in the total crew compartment volume did not decrease much from the T-54/55, and in fact, it is easy to see from this why the actual crew station volumes increased from the T-54/55 [**Table 4**].

---

[7] Самусенко М. Ф., Емелин М. И. (1958). *Основы проектирования артиллерийского вооружения танков и самоходно-артиллерийских установок.* Москва: Типография Артиллерийской Инженерной Академии. p. 19.
[8] Военная академия бронетанковых войск имени Р. Я. Малиновского (ВАБТВ).
[9] Заславский, Е. И., Романов, М. И. (1982). 'Эргономическое Обеспечение Проектирования ВГМ', *Вопросы Оборонной Техники,* 20(103). p. 5.
[10] This trait was inherited from the T-64A, which was an exemplary specimen of the second postwar generation of Soviet tank design.
[11] Due to no longer needing to allocate standing room for the loader.

Table 4: Total internal volumes of various tanks

Space	T-72[a]	T-54	Leopard 2A4[b]	M48[c]	M1 Abrams[c]	M60A1[c]
Crew compartment (m^3)	7.9	8.05	10.1	10.48	10.9	11.17
Engine compartment (m^3)	3.1	3.25	6.38	7.22	4.3[d]	7.24

a   Старовойтов, В. С. (ed.) (1990). *Военные гусеничные машины*. Том 1: Устройство, Книга 1. Москва: Издательство МГТУ им. Н. Э. Баумана. p. 49.
b   Hilmes R. (2007). *Kampfpanzer heute und morgen: Konzepte - System - Technologien*. Auflage: Motorbuch Verlag. p. 130.
c   Hilmes R. (1988). *Kampfpanzer. Die Entwicklungen der Nachkriegszeit*. Bonn: Mittler Report Verlag GmbH. p. 95.
d   Hilmes R. (1988). *Kampfpanzer. Die Entwicklungen der Nachkriegszeit*. Bonn: Mittler Report Verlag GmbH. pp. 52, 53, 56.

With that said, even in the absence of directly comparable figures, it is obvious that the T-72 was externally and internally smaller than almost all of its Western counterparts. The main factor behind this was, ultimately, a stringent initial weight limit of 36 tonnes [12] imposed on the T-64 by the tactical-technical requirements of the 1953 Soviet "New Medium Tank" program. The T-64, created as a continuation of the Object 430 tank developed for this program, shared this limit. A great deal of effort was expended on weight optimization to that end [13], and these efforts were inherited in the design of the T-72.

The total volume occupied by the autoloader was 1.1 m², excluding the cassette elevator and ramming mechanism. In total, the autoloader and full complement of reserve ammunition (43 rounds) in the T-72A occupied 1.6 m² of volume, which was 0.9-1.0 m³ (1.5-1.6 times) smaller than the T-55 and T-62 loading system inclusive of the space occupied by the human loader [14]. Considering that the total ammunition count was the same as the T-55 and the caliber difference between the T-72's 125 mm gun and the T-55's 100 mm gun, the volume savings were remarkable.

---

[12] Павлов М. В., Павлов, И. В. (2021). *Отечественные Бронированные Машины 1945-1965 гг. - Часть I: Легкие, средние и тяжелые танки*. Кемерово: ООО "Принт". p. 411.
[13] Чернышев В. Л. (ed.) (2006). *Танк и люди: Дневник главного конструктора Александра Александровича Морозова*. Available at: http://btvt.info/4ourarticles/morozov/all.htm (Accessed: 15 August 2023).
[14] Андреев, В. П., Изосимов, Н. Г., Кулемин, С. А. (1981). 'Влияние автоматизации заряжания пушки на общие свойства танка', *Вестник Бронетанковой Техники* 1981, сборник 3. p. 10.

Table 5: Crew spaces in different tanks (m^3) [a, b]

CREW	T-72	T-54/55*	REF. 1 [c]	REF. 2 [d]
Commander	0.615	0.828 (0.4)	0.44	0.5
Gunner	0.495	0.395 (0.5)	0.44	0.5
Driver	0.864	0.621 (0.8)	0.66	0.8
Loader	-	1.36 (1.0)	0.88	1.0

a   Тихонов М. Н., Кудрин Н. Д. (1991). 'Человеческий фактор и научно-технический прогресс в танкостроении', *Вестник Бронетанковой Техники* 1991, сборник 7. p. 34.
b   Попков В. Ф. (1984). 'Танк Т-64А (к 20-летию серийного производства)', *Вестник Бронетанковой Техники* 1984, сборник 4. p. 51.
c   Halbert G. A. (1983). 'Elements of Tank Design', *ARMOR*, (November-December).
d   Старовойтов, В. С. (ed.) (1990). *Военные гусеничные машины*. Том 1: Устройство, Книга 1. Москва: Издательство МГТУ им. Н. Э. Баумана. p. 49.
*   The commander's knees straddled the gunner's back due to the small seat clearance, so the values in parenthesis are more applicable. The same is true for the driver and loader.

However, despite the reduction in turret size achieved by automatic loading, the turret ring still had to be quite large to accommodate its two occupants. The outer diameter of the outer race was quite large at 2,283 mm. This was slightly larger than the outer race of the Leopard 2 turret ring (2,200 mm)[15]. The internal turret ring diameter was 1,934 mm.

The turret sat atop the turret ring along its front half, so the reduced internal diameter due to the internal flange was inconsequential, but along the rear half of the turret, it cut into the crew spaces. In the T-64A, the autoloader carousel had its own rotating ring that rode inside the inner turret ring, taking up most of the excess radial space. Inside the carousel, the crew cabin was only 1,590 mm in diameter - barely wider than the 1,575 mm fighting compartment of the T-34-85. The introduction of the cabinless autoloader in the T-72 freed up space at a comparatively small cost to headroom by placing the carousel below the turret, acting as the floor for the crew.

Each design bureau followed a set of practices typical in industrial design, but they had the freedom to approach the task of crew station design as they saw fit, mirroring the freedom traditionally enjoyed by automotive designers. And similar to the automotive industry at the time, there were relatively few governing standards for ergonomics that designers were obligated to follow, and little in the way of initial oversight through official channels. Ergonomics development in Soviet military industries was carried out with great reliance on

---

[15] Hilmes R. (2007). *Kampfpanzer heute und morgen: Konzepte - System - Technologien*. Auflage: Motorbuch Verlag. p. 132.

institutional knowledge, occasionally shared within a framework of peer-to-peer scientific cooperation between design organizations and research institutes[16].

Naturally, the final outcome of this type of development process depended a great deal on the capabilities of the chief designer, who had the unpleasant task of coordinating the marriage of a large number of mutually contradictory requirements. The fighting compartment of the T-72 was anomalous in this regard because it was created from the refit of an existing hull and turret with a new autoloader, without prior harmonization. Despite this, a significant overall improvement in the quality of habitation was achieved[17].

Designing for Paper People

For most of the Cold War, the practice of using an "average" tanker during the preliminary design stage was a working policy in the Soviet tank industry[18]. This was a 50th percentile model, A 50th percentile man refers to an individual whose characteristics, such as height and weight, fall exactly in the middle of his demographic, with half of the demographic having measurements lower and the other half higher. The "average man" model was used in the creation of the T-64, and the resulting form factor was subsequently inherited by the T-72.

During the initial drafting process for a tank design, the basic dimensions of the crew stations were assigned using reference figures based on tanks evaluated in the past. Then, preliminary design studies for the crew stations were carried out by applying model drawings of mannequins to set the working postures of each crew member in longitudinal, transverse and plan views. The same process was later repeated for the T-72. This mannequin [**Fig. 3**] was proportioned like an "average man", who was determined from Red Army surveys to be a slim, broad-shouldered fellow with a height of 170 cm, boots and helmet included. Design guidelines were then used for the finer details, e.g. the size of seats, the maximum acceptable force on pedals, the maximum acceptable effort on the gun elevation wheel, and so on.

As the draft design progressed into full-scale prototyping, the crew stations were subjected to military testing under the GBTU, invariably followed by a few rounds of design changes based on the comments compiled from the testing commission. Once all comments were resolved, the tank could enter service.

---

[16] Безъязычная С. et al. (eds.) (2000). *Эргономика: человекоориентированное проектирование техники, программных средств и среды.* Москва: Логос. pp. 138-142.
[17] Исаков, П. П. p. 181.
[18] Самусенко М. Ф., Емелин М. И. (1958). *Основы проектирования артиллерийского вооружения танков и самоходно-артиллерийских установок.* Москва: Типография Артиллерийской Инженерной Академии. p. 21.
Also stated in: Талу К. А. (1963). *Конструкция и расчет танков.* Москва: Военная академия бронетанковых войск. p. 29.

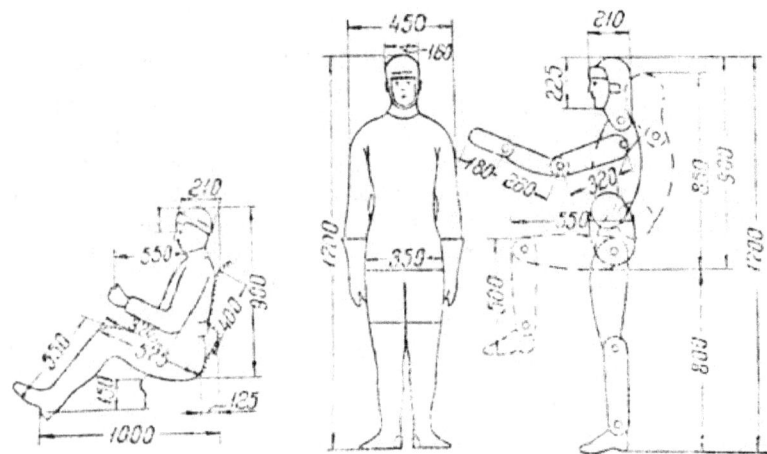

Figure 3: Dimensions of a flat mannequin drawing, superimposed into cross-sectional drawings of tank drafts to determine if an average person would fit. (M. F. Samusenko & M. I. Emelin, 1958)

During military service, further fine-tuning would be carried out based on feedback collected from tankers. Of course, not all complaints could be, or would be resolved, especially if it involved a major redesign effort. In actual service, it often transpired that tanks and other AFVs which were satisfactory in testing would still have notable deficiencies. For the most part, this can be attributed to the fundamental issues of designing for an "average man".

The fact that human anthropometry follows a normal distribution ostensibly implies that designing a workstation using a 50th percentile model will be "close enough" for most people. In practice, such accommodations would only be suitable for a fraction of the general population, and optimal to almost no one. The probability of an actual person meeting the 50th percentile in all anthropometric indicators is effectively negligible (≤0.05). Even if a man was indeed 50th percentile in height, he almost certainly wasn't in arm length, or torso height, forearm length, or a myriad of other indicators.

Quite understandably, the use of an "average man" model in the design of workstations and man-machine interfaces was considered bad practice by leading industrial designers even by the early 1950's[19]. In the U.S. military, designing according to 5th and 95th percentile models gained a foothold as a working policy not long after WW2, which contributed to the enormous size of

---

[19] Dreyfuss, H. (2012). *Designing for people*. New York: Allworth Press.

the M48 Patton medium tank. Rather than basing the dimensions of crew stations on a single reference, the design had to be large enough fit a 95th percentile tanker, but also adaptable enough for a 5th percentile tanker and everyone in between.

The U.S. Army Ordnance Corps was a pioneer in this field, establishing the Human Engineering Laboratory under the U.S. Army Materiel Command in 1951[20] to better integrate anthropometry and other forms of human-oriented design into military products. This not only allowed policies such as the 5th-95th percentile design concept to be enforced, it also enabled the U.S. Army to carry out ergonomics research focused on its own material needs. There was no equivalent in the Soviet tank industry - in fact, there was no organization dedicated to the study of ergonomics at all in the USSR until 1962, when the VNIITE[21] research institute was founded. By this point, the basic configuration of the T-64 had already been laid down[22], and the T-72 would later follow its template.

Nevertheless, even with the limited body of formal research available in the country at the time, the military and the tank industry understood that the operational requirements associated with modern nuclear warfare, namely the prolonged periods crews had to spend inside their tanks, had important implications on tank ergonomics[23]. In 1961, a special ergonomics research group was formed out of the radiation and chemical reconnaissance laboratory at the VNII-100 research institute to study these issues[24]. In 1966, it was transformed into the Department of Habitability with an expanded scope of study, covering issues important to tank design like volume allocation, crew size, and control automation. Soviet military ergonomists noted that the increased attention to ergonomics in the 1960's was technically justified by three primary factors[25]:

1. On a nuclear battlefield, up to 50% of the terrain crossed by a mechanized unit was expected to be contaminated.

2. Modern tanks were capable of covering ever-increasing distances on a march with few halts for maintenance and refueling.

3. With the growing sophistication of tank technologies, the scope

---

[20] *U.S. Army Human Engineering Laboratory* (1992). Aberdeen: Aberdeen Proving Ground.
[21] All-Union Scientific Research Institute of Technical Aesthetics, featuring the country's first scientific department dedicated to ergonomics research and development
[22] Чобиток В., Саенко М., Тарасенко А., Чернышев В. (2016). *Основной боевой танк Т-64: 50 лет в строю*. Москва: Яуза. p. 30.
[23] Агеев, Г. Н. (ed.) (1970). *Военная Инженерная Психология*. Москва: Воениздат.
[24] Потемкин, Э.К. (ed.) (1999). *ВНИИтрансмаш - страницы истории*. Санкт-Петербург: Издательство «Петровский фонд». p. 86.
[25] Тихонов М. Н., Кудрин Н. Д. (1991). 'Человеческий фактор и научно-технический прогресс в танкостроении', *Вестник бронетанковой техники* 1991, сборник 7. pp. 33-35.

and complexity of crew tasks could outpace crew capabilities.

Altogether, this meant that the next generation of tanks had to be designed for comfortable habitation for longer periods, with a particular emphasis on crew comfort in sealed conditions. The creation of new anthropometric standards for AFV design was to have a decisive role in this regard. This was addressed by the publication of the 1973 government standard GOST 19288-73[26], which codified the use of 90th percentile models in AFVs in the design of crew stations. It was equivalent in purpose to the MIL-STD-1472 standard[27] but smaller in scope. Its impact was small, as only a handful of AFVs entering service in the 1970's could be described as truly new developments[28].

This was followed by a comprehensive series of ergonomics design guidebooks, and most crucially, the REO-SV-80 standards[29] developed by VNII-100 in 1980, which was put into effect in January 1982. The REO-SV-80 standards were not merely specifications for crew station sizes, but a comprehensive set of standards for all man-machine interfaces taking into account psycho-physiological requirements. These developments benefitted crew workflows and layout design in new developments like the T-80U main battle tank, the BMP-3 infantry fighting vehicle, and the IMR-3 engineering vehicle. Still, even at this stage, there was no standardized design methodology among the country's tank design bureaus[30], and no major improvements were implemented in the T-72.

Soviet Tanker Anthropometry

Historically, there was a general preference for recruiting smaller men for tank crews in all tank-operating armies. The Soviet Army was no different. Even so, there was no documented policy in Soviet tank design practice to deliberately offset small interior volumes by recruiting only tankers of unusually small size and stature. Even during the Great Patriotic War, when the tanks available had generally smaller crew stations (even if the gross internal volume was larger than

---

[26] ГОСТ 19288-73 «Машины военные бронированные. Основные антропометрические показатели экипажей и десанта для определения размеров обитаемых отделений»
[27] White R. M. (1978). *United States Army Anthropometry: 1946-1977*. Massachusetts: U.S. Army Natick Research and Development Command.
[28] Исаков, П. П. (ed.) (1986). *Теория и Конструкция Танка - Том 7: Эргономическое Обеспечение Разработки Танка*. Москва: Машиностроение. p. 48.
[29] Ergonomics design guidelines for the arms and equipment of the ground forces; Министерство обороны СССР (1980). *Руководство по эргономическому обеспечению создания военной техники СВ: РЭОСВ-80*.
[30] Кудрин, И. Д., Шабалин, В. А. (1985). 'Рабочее Место Танкиста', *Вестник бронетанковой техники* 1985, сборник 3. p. 9.

postwar tanks in some cases), the height of tankers was not formally regulated in the Red Army until after the war[31].

In 1951, the medical requirements for tankers were updated with a minimum height requirement of 150 cm, which was a basic requirement for physical fitness shared by other branches of the military. This regulation applied throughout the 1950's [32], changing only in 1961 when a maximum height regulation for tankers was introduced for the first time. The limit was set at 175 cm, which was above the average for Russian men at the time.

Throughout the service of the T-72 in the Soviet Army, the official height limit of 175 cm remained unchanged, according to the 1974[33] and 1988[34] versions of the military medical regulations issued by the Soviet Ministry of Defense. Of course, like in most militaries, the height limit was not decisive. Exceptions were not uncommon, sometimes with extreme exceptions of crew members reaching or exceeding 190 cm.

Data from a 1974 survey[35] indicates that Soviet conscription-age (18-21 years old) [36] urban men had an average height of 167-168 cm at the time. For conscription-age Soviet men enlisting in 1973, the year the T-72 entered service, the average height of this cohort (18-27 years old)[37] had risen to 170-172 cm. It was only in 1980 that the average height of Soviet men reached 175 cm[38]. Strictly speaking, there are several issues in the use of urban Soviet men

---

[31] Драбкин А., Антонов В. (2014). *Фронтовой быт советских танкистов*. [Online lecture]. Available from: https://youtu.be/ylgOEA8HTGs(Accessed: 3 July 2023).

[32] *Приказ Военного Министра СССР от 6 октября 1951 г. №130: О введении в действие Инструкции по медицинскому освидетельствованию граждан, призываемых на действительную военную службу, военнослужащих и военнообязанных*. Приложение №4. Москва: Министр Обороны СССР.

[33] *Приказ Военного Министра СССР от 3 Сентября 1973 г., №185: О Введении В Действие Положения О Медицинском Освидетельствовании В Вооруженных Силах СССР*. Приложение №3: Положения о медицинском освидетельствовании в Вооруженных Силах СССР. Москва: Министр Обороны СССР. (Put into effect on 1 January 1974)

[34] *Приказ Военного Министра СССР от 9 Сентября 1987 г., №260: О Введении В Действие Положения О Медицинском Освидетельствовании В Вооруженных Силах СССР (на мирное и военное время)*. Приложение №4: Положения, введенного в действие Приказом Министра обороны СССР от 9 сентября 1987 г. №260. Москва: Министр Обороны СССР. (Put into effect on 1 January 1988)

[35] Миронов Б. (2003). *Рост и вес россиян сталинской эпохи*. Available at: https://m.polit.ru/article/2003/10/16/627075/ (Accessed: 3 July 2023).

[36] Президентская библиотека им. Ельцина, Б. Н. (2023). *Принят Закон «О Всеобщей Воинской Обязанности»*. Available at: https://www.prlib.ru/history/619624. (Accessed: 4 July 2023).

[37] *Постановление Верховного Совет СССР от 12 октября 1967 г.: О Порядке Введения В Действие Закона СССР "О Всеобщей Воинской Обязанности"*. Верховный Совет Союза Советских Социалистических Республик.
Note that the conscription age was 18-21 from 1939 to 1967. Beginning on 1 January 1968, the conscription age range was expanded to 18-27.

[38] Russian Longitudinal Monitoring Survey (1994–2004 data), in Brainerd E. (2010). *Reassessing the Standard of Living in the Soviet Union: An Analysis Using Archival and Anthropometric Data*. The Journal of Economic History, 70(01), 83. p. 105.

to represent the general Soviet population, but nevertheless, for purely incidental reasons, this group was a fairly good midpoint between the tallest (Muscovites) and shortest (rural men, mainly from collective farms) subcategories of the population.

In the USSR, the first systematic design study based on the anthropometry of tankers was carried out in 1951. The critical dimensions of 2,726 people in summer and winter uniforms were measured, and these dimensions were compared with the crew stations of 19 existing domestic armored vehicles. This study produced the first set of ergonomics design guidelines in Soviet tank building[39], some of which were reproduced in various textbooks. However, during the design of the T-64 and the "AZ" autoloader, the anthropometric data itself was not used by tank designers.

As a representation of an average tanker throughout the 1960's, the "average man" model widely used in industry was at least accurate[40], even if it was not conceptually valid. In 1967, the S. M. Kirov Military Medical Academy conducted a large-scale anthropometric study of Soviet Army tankers. When comparing the data for tankers against ground forces personnel in general, it can be seen that the mean height of tankers was slightly lower than the ground forces average (1,685 mm vs 1,697 mm) and the statistical spread in various parameters was tighter. The reason was likely to be recruitment selectivity.

For comparison, the normative anthropometric data for selected Western nations showed a major difference in height [**Table 6**]. Practice showed that designing tanks under a 5th-95th percentile framework did not necessarily guarantee that they would actually fit 95th percentile personnel. Considering the technical difficulties in actually accommodating a 95th percentile man, this standard was frequently treated as a "want" rather than a "need"[41]. Designers had to be especially creative in interpreting requirements when the question of internal height came up [**Table 5**].

---

[39] Кудрин, И. Д., Шабалин, В. А. (1985). 'Рабочее Место Танкиста', *Вестник бронетанковой техники* 1985, сборник 3. p. 8.

[40] Лубенский В. Д. (ed.) (1974). *Обитаемость Объектов Бронетанковой Техники*. Москва: ВНИИТМ. p. 116.

[41] Human Resources Research Office (1961). *Human Factors Evaluation of the Tank, Combat Full Tracked: 105mm Gun, M60*. Kentucky: Fort Knox.
Also see: BDM Corporation (1986). *Human Factors Engineering Data Base Development For Armored Combat Vehicles And Analyses Of Three Nato Tank Systems. Volume III - Human Factors Engineering Analysis of the British Chieftain Main Battle Tank*. Albuquerque: BDM Corporation.
Also see: BDM Corporation (1986). *Human Factors Engineering Data Base Development For Armored Combat Vehicles And Analyses Of Three Nato Tank Systems. Volume IV - Human Factors Engineering Analysis of the French AMX-13 Light Tank*. Albuquerque: BDM Corporation.

Table 6: Anthropometric data comparison between the USSR, FRG and USA

Demographic	USSR (1967)	FRG (1968-1974)[a,b]	USA (1966)[c]
5th percentile	1,571	1,685	1,628
50th percentile	1,697	1,790	1,742
95th percentile	1,822	1,910	1,856

a	DIN 33402-2. (1978). *Körpermaße des Menschen*; Werte.
b	Dekker, M., Molenbroek, J. (1999). 'Collecting and Deriving Data for Human Modeling Software to use in Flight Simulator Design', *SAE Technical Paper* 1999-01-1887.
c	1966 U.S. Army ground forces anthropometric survey data, used in MIL-STD- 1472A (1970) until revision D (1989)

In most cases, a shorter internal height was tolerated out of necessity. In the design of the Leopard 1, the internal height of 1,770 mm was apparently considered to be "within the lower limit" of the 95th percentile requirement[42] despite not being tall enough for a West German 50th percentile man without boots or a helmet to stand upright. In the Challenger 1, the loader's station was suitable only for a 28th percentile Royal Armoured Corps serviceman[43]. When approaching the problem of tank design holistically, it is easy to justify tempering the ergonomic requirements with pragmatism, but this was obviously not a sustainable solution. The general population grew taller (and bulkier) with each successive birth year, but at the same time, the rising lethality of the modern battlefield demanded lower silhouettes and a larger weight allowance for armor.

Table 7: Internal heights of NATO tank fighting compartments (mm)

Leopard 1[a]	Leopard 2[b]	Centurion[c]	Chieftain[d]	M48[a]	M60[a]	M1[e]
1,770	1,750	1,816	1,730	1,740	1,770	1,641

a	Spielberger, W. J. (1995). *Waffensysteme Leopard 1 und Leopard 2*. Auflage: Motorbuch Verlag. p. 59.
b	Deutsche Panzermuseum Munster (2019). Dipl-Ing. Rolf Hilmes: Wie konstruiert man einen Panzer?. [Online video]. Available from: https://youtu.be/_J1dHNtyOLI (Accessed: 21 August 2023)
c	Technical drawing, F.V. 231921 - Sheet No. 4, 'Centurion 7, 8, 9 & 10. Turret'. F.V.R.D.E.
d	Рубашкин, Д. С., Троицкий, В. В. (1983). 'Особенности Построения Рабочих Мест Экипажа', Вопросы оборонной техники, 5(111). p. 12.
e	Measured

---

[42] Hilmes R. (1988). *Kampfpanzer. Die Entwicklungen der Nachkriegszeit*. Bonn: Mittler Report Verlag GmbH. p. 97.

[43] U.K. Ministry of Defence (1988). *Armoured warfare: A Vehicles; replacement for Chieftain; new tank for the army*. DEFE 70-1890.

Small, Not Cramped

Knowing that the T-72 was made under a 50th percentile standard explains many of its nuances, but it says little about how the tank accommodated real people in that demographic. Ultimately, high combat effectiveness was entirely possible if a tank had correctly proportioned crew stations so that the crew members had enough space for their work, and were not fatigued by unreasonable sitting postures. Moreover, if smaller crew stations could be compensated by selective recruitment, and the natural variance in bodily proportions could be addressed by adjustable seats and alternative working positions, a rational layout of seats and controls was irreplaceable.

By virtue of its layout, the T-72 provided each crew member the luxury of having their own hatches. If riding from an open hatch was permissible, then it was possible for the commander and gunner to stand, or sit on the rim of the hatch opening. In fact, this was historically the main mode of riding on the T-72 during exercises and in war. Only in environments contaminated by weapons of mass destruction would all crew members be forced to remain inside, where they would be restricted to two main sitting postures; relaxed, or leaning forward to use an optical device.

In tests where T-72 crews were forced to remain sealed for three consecutive days before engaging in combat exercises, it was observed that overall performance fell by an average of 30% [44]. This phenomenon, while not surprising, is illustrative of the fact that in vehicles with limited passenger space, the inability to adopt different postures and exercise strained muscles invariably leads to fatigue over time. However, the crew stations themselves were generally quite adequate in layout and dimensions.

Although the T-72 was not originally designed to comply with the REO-SV-80 standards, a retrospective study of the crew station workspace dimensions shows that it largely met the recommendations for the "medium" size category. In the REO-SV-80 specifications, three size groups were provided: small, medium, and large. These corresponded to 5th, 50th and 95th percentile groups. Within each size group, different dimensions were specified at certain reference points.

The REO-SV-80 requirements for crew station dimensions take into account the specificity of each crew member's tasks in a Soviet postwar second-generation tank, represented by the T-64, T-72 and T-80. For the commander, this included not only torso rotation when using his cupola, but also reloading

---

[44] Тихонов М. Н., Кудрин Н. Д. (1991). 'Человеческий фактор и научно-технический прогресс в танкостроении', *Вестник бронетанковой техники* 1991, сборник 7. p. 34.

the coaxial machine gun, which was not a commander's responsibility in tanks with a human loader. Basic ergonomic considerations played their part on top of that. For instance, the gunner needed a certain amount of space to work the manual handwheels for turret traverse and gun elevation, even if those were not his primary means of controlling the turret.

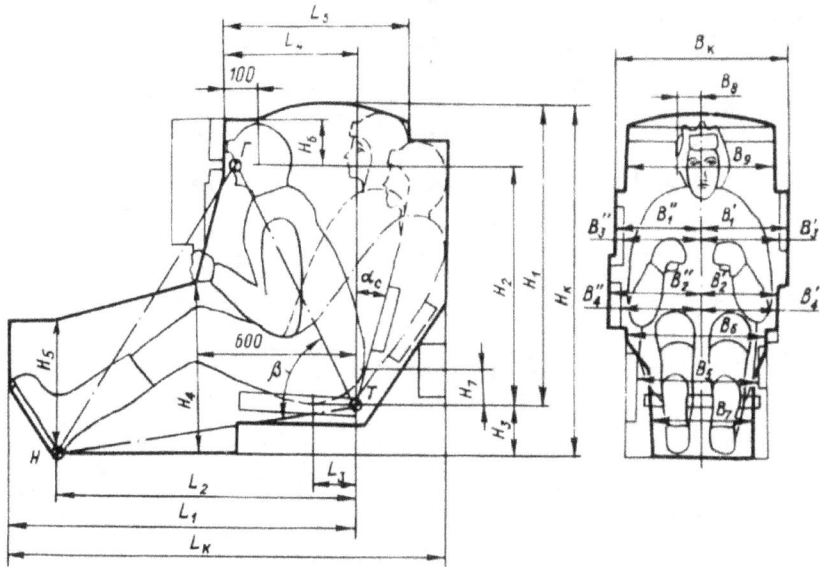

Figure 4: Crew station datum dimensions, labeled for **Table 8**. (*D. S. Rubashkin & V. V. Troitskiy, 1983*)

Overall, the T-72B[45], partially fulfilled the complete contents of the REO-SV-80 standards by 55-60% at the turret crew stations and by 75% at the driver's station. For comparison, Soviet analysts estimated that NATO tanks generally achieved 79-87% fulfillment of the REO-SV-80 standards at the fighting compartment crew stations and 79-91% at the driver's station[46]. This gap in the qualitative level of the crew accommodations was in large part because of the greater complexity of combat tasks in a T-72 due to the lower degree of automation in the fire control system [47], and the lack of an automatic transmission. Non-compliance with noise standards was also a major contributing factor. The difference in the working space available at the crew stations was, on the whole, only marginal.

---

[45] Generalized together with the T-64B and T-80B.
[46] Троицкий, В. В. (1987). 'Рабочие Места Экипажей в Отечественных и Зарубежных Танках', *Вестник Бронетанковой Техники* 1987, сборник 12. p. 10.
[47] For example, the manual nature of the commander's TKN-3 periscope, and the simplicity of the gunnery process at night, where the T-72 lagged behind tanks like the M1 Abrams and Leopard 2.

Table 8: Breakdown of REO-SV-80 crew station dimension compliance in T-72 (darkened cell: non-compliant) (all figures in mm)

Item	TC	Reference	Gunner	Reference	Driver	Reference
H_K	1,270	1,201-1,300	1,041	1,300	990	850-900
H_1	1,010	960	864	1,020	830	800
H_2	813	750	711	760	762	680
H_3	254	240	178	400	203	50
H_4	?	750	483	600	635	600
H_5	?	300	457	330	432	400
H_6	-	160	-	200	191	120
H_7	-	100	-	100	-	100
L_K	1,219	1,100	1,219	1,050	1,524	1,500
L_1	914	950	-	900	1,219	1,100
L_2	**610**	700	-	700	1,067	900
L_3	0	0-100	0	0-100	0	100
L_4	Variable	300-360	**76**	300-380	559	-30 to -80 ?
L_5	381	450	508	400	-	350
W_K	**483**	600	610	580	**660**	680
W_1	635	600	572	580	**660**	680
W_2	610	600	**483**	580	**610**	680
W_3	-	550	-	550	**610**	680
W_4	610	550	-	550	**610**	680
W_5	610	450	-	400	610	440
W_6	**356**	450	-	420	584	600
W_7	**305**	400	-	400	457	500
W_8	-	<80	-	80	-	0
W_9	-	480	-	430	508	400

Short Term Habitation

In all enclosed vehicles, the temperature differences between internal and external air temperatures were not directly proportional. In tanks, the internal temperature increases above the ambient air temperature from solar heating only after a long delay, due to the large thermal mass of the armor[48]. By the same token, internal temperatures would continue to rise even after the sun has set. In cold weather, the same remains true, but with heat flowing in the opposite direction. The main source of internal heating was the engine, separated from the crew compartment through a hollow steel bulkhead called the engine compartment bulkhead, approximately two centimeters thick.

Given that excess heating of a tank's interior in hot climates primarily arises from solar heating on the armor of the tank, and the loss of internal heat also occurs through the same armor, thermal insulation was the most direct solution for improving microclimate regulation in both temperature extremes. The anti-radiation liner, though an order of magnitude less effective than foam insulation, was still two orders of magnitude less conductive than steel, drastically reducing internal temperatures from solar heating and significantly reducing heat loss in winter[49].

Later, the anti-radiation liner was augmented by the introduction of anti-neutron cladding over the inhabited areas of the turret beginning in October 1983. Then, with the introduction of reactive armor (Kontakt-1, Kontakt-5) in the mid to late-1980's, the multilayered, spaced ERA boxes on the turret roof added more insulation from solar radiation.

The liner also eliminated the possibility of direct skin contact with the walls of the crew compartment, which, if left bare, could become dangerously hot in the summer. Tests of the M60 in the Yuma desert showed that the interior wall surface reached temperatures of up to 68.3°C[50], while desert testing of a T-80 showed that the interior wall (liner) temperature of the tank reached 38-40°C in the driver's compartment, or 36.6-39.9°C[51] in the fighting compartment. The T-72 can be assumed to have comparable results.

---

[48] Warnick, W. L., Kubala, A. L. (1978). *A Study of Selected Problems in Armor Operations*. Alexandria: Human Resources Research Organization. pp.9-10.
[49] Лубенский В. Д. (ed.) (1974). *Обитаемость Объектов Бронетанковой Техники*. Москва: ВНИИТМ. pp. 138-139, 146.
Against external heating, the anti-radiation liner has a heat transmission coefficient of 0.25 kcal/m·hr·°C whereas steel has a coefficient of 31 kcal/m·hr·°C.
[50] Suarez, I. A. (1974). *Armored Fighting Vehicle Compartment Temperatures - M60 Tank*, USAPG Report No. 203, in Frank G. R. (1982). 'Continuous Operations', *ARMOR*, (March-April). p. 21.
[51] Lower figure was at an external air temperature of 35°C before a 4-hour march, higher figure for 38°C external air temperature after.

Long Term Habitation

The T-72 essentially did not differ from any other postwar Soviet AFV in its facilities for long term habitation. The upkeep of individual tanks was carried out under a typical front line supply model, which changed little throughout the Cold War period. Battalion field kitchens were the lowest level where fresh bread could be baked, and hot foods could be prepared[52]. These field kitchens supplied hot meals and the supply unit kept a store of water[53]. Theoretically, there were to be at least two hot meals a day, distributed before dawn and after dusk together with the water ration.

The means of distributing this food and water depended on field conditions. Tank companies could make a stop at a battalion assembly point, but large gatherings invited aerial and nuclear attack, and time did not always permit such stops. The main alternative was to designate the supply truck organic to each tank company to handle the distribution of all consumables apart from fuel[54].

This supply model was not ideal, but it was the de facto standard in virtually all modern armies at the time, and it was proven to be viable – albeit not ideal – in fast-paced maneuver warfare during WW2. However, by the time the T-72 began its career in the Soviet Army, the situation had been complicated by the fact that the military and political leadership fully expected that the rear areas of large mechanized units would be among the first to be struck by tactical nuclear weapons[55]. Despite this, there appeared to be a general assumption that armored units would be somehow sustained through the same basic system.

Dry rations could be allocated to each T-72 as conserves for emergency situations, to be consumed on the explicit orders of the unit commander. The quantity of the dry rations was not specified in the ZIP[56] of the T-72, so it could be assumed that it was planned by individual unit leaders based on the local supply situation. These rations were kept in the center rear external turret stowage bin, so they were not accessible without leaving the tank and, more importantly, they could be lost through combat damage, to include nuclear blast damage. The water rations for the crew were stowed internally in two-liter aluminum flasks, one at each crew station.

---

[52] Limited to preparing soups and boiling canned products
[53] 'Гигиена питания военнослужащих', Алтайский государственный медицинский университет.
[54] Hemesley, A. E. (1976). *Soviet Tank Company Tactics.* Washington, D.C: Defence Intelligence Agency. Available at: https://apps.dtic.mil/sti/tr/pdf/ADB297807.pdf (Accessed: 30 July 2023)
Also see: Grau, L. W. (1989). *The Soviet Combined Arms Battalion Reorganization For Tactical Flexibility.* Fort Leavenworth: Soviet Army Studies Office.
[55] 'На новом этапе', in *Тыл Советских Вооруженных Сил в Великой Отечественной войне* (1977). Москва: Воениздат.
[56] "Запасные части, инструмент и принадлежности"(ЗИП); "Spare parts, tools and accessories".

Any additional provisions had to be carried in non-standard containers or obtained through improvised means - for example, in an emergency, it was possible to drain the window washer reservoirs for the driver's periscope (7 l) and the gunner's day sight (2.2 l) for water.

Dry rations were normally eaten cold, but it was possible to heat the canned component (meat, fish, porridge) through improvised means. The easiest way was to leave the cans on the cooling fan exhaust grille while the tank was parked during a short rest halt, with the engine running. The same could be done to heat water in a standard mess kit.

The escape hatch on the hull floor behind the driver could be used for refuse disposal. Unlike most escape hatches, which dropped from the floor once opened and were usually extremely troublesome to re-seal, the T-72 escape hatch opened inwards and was intended for routine use. This feature began to be added to Soviet tanks after the Battle of Lake Khasan, originally meant to allow the crew to safely dispose of shell casings during combat[57].

Another notable feature associated with the escape hatch was the small shovel stowed atop it, along with a few other items. It was noted in a TsNII-48 memorandum from 1943[58] that tankers requested changing the stowage point of the shovel from outside the tank to inside, because it was sometimes necessary to dig into the ground when exiting through the escape hatch.

All crew members typically slept in their seats. When it was possible to sleep outside, it was generally preferred to sleep on the engine compartment roof for warmth. The elimination of a human loader from the tank crew helped here – with only three people, everyone could fit on the engine compartment roof with space to spare, thanks to the extra width provided by the panniers.

---

[57] Желтов И. Г., Макаров А. Ю. (2020). *А.Я. Дик - начальник КБ, которое так и не было создано*. Available at: https://t34inform.ru/publication/p-pers-3.html (Accessed: 12 August 2023).
Also see: Пашолок, Ю. (2023). *Бронетанковый опыт на Дальнем Востоке*. Available at: https://dzen.ru/a/ZLWeXiO-xUbLHdpr (Accessed: 20 June 2024).
[58] Коломиец, М. (2009). *Т-34: Первая полная энциклопедия*. Москва: Эксмо. p. 335.

Figure 5: Escape hatch between driver's seat. It was opened by undoing the four; often a hammer was needed. It was, however, more secure than the T-64A's escape hatch with just two latches. (S. Stauber, 2024)

Self-Defense

A Soviet tanker's means of self-defense changed very little in concept since the Great Patriotic War, though his weapons evolved continually. Close-in defense was mainly reliant on hand grenades. Ten F-1 defensive hand grenades[59] were carried in two satchels, stowed next to the front right fuel-ammunition rack. Each crew member could arm themselves with a few hand grenades before evacuating, or they could be thrown from the turret hatches[60] to deter infantry.

If the crew was forced to abandon their normal duties and fight dismounted, they could do little but to cover their own retreat. A *Kalashnikov* assault rifle with a folding buttstock (AKMS, AKS-74) was strapped to the frame of the gunner's seat between his seat and his footrest - one of the only places where there was

---

[59] High-fragmentation grenades, as opposed to offensive hand grenades, distinguished by a larger fragmentation radius.
[60] Specialized hand grenade training for tankers involved learning to throw from the hatches of a tank turret mockup.

enough room for a rifle to fit, complete with 300 rounds in ten thirty-round magazines in a satchel behind the driver.

The only individual weapon was the *Makarov* PM, a compact service pistol chambered in 9x18 mm *Makarov*. It could fire eight rounds of a fairly potent steel-cored ball bullet capable of piercing a steel helmet (SSh-40) at 100 m[61], but otherwise it was of extremely dubious value as a weapon of war.

Officially, the rifle was issued to the commander[62], just as a sub-machine gun would be issued to Red Army tank commanders during the Great Patriotic War. In practice, the rifle was a communal weapon. It would be passed to whoever was in the best position to use it at any given moment, and because the rifle and its ammunition were stowed separately, and rather difficult to access, readying it for combat involved some teamwork. When evacuating under fire, one of the crew members might use the rifle to provide suppressive cover for the others, or simply to arm themselves to defend the tank from infantry at close range.

The individual firepower of the tank crew never improved beyond this meager selection. After 1979[63], the AKS-74U compact assault rifle began to see limited use as a replacement for the commander's rifle in the T-72 and in other three-man tanks and armored fighting vehicles, but it was not an improvement over a full-sized rifle in any way apart from convenience.

There were also no provisions to use any of the tank's machine guns in a dismounted role, to include the PKT coaxial machine gun. Whether this was deliberate or not is somewhat unclear. During the infancy of the Red Army's armored forces, it was expected for a tank crew to remove their tank's DT machine gun(s) and continue fighting as a dismounted machine gun team in support of the infantry[64]. The DT incorporated a retractable butt-stock, pistol grip, and bipod for this purpose[65]. It could be surmised that real combat experience demonstrated to the Soviet Army that expecting tank crews to dismount the tank's machine guns while bailing out was somewhat unrealistic, to put it mildly.

---

[61] Дворянинов, В. Н. (2015). *Боевые патроны стрелкового оружия: монография. Книга 3: Современные отечественные патроны, как создавались легенды.* Климовск: Д'Соло. pp. 177-178.
[62] Министерство Обороны СССР (1980). *Руководство по действиям экипажа при вооружении танка Т-72.* Москва: Воениздат.
[63] Popenker, M. (2023). *The history of Russian Avtomat: evolution of the Kalashnikov AK, from its early origins to the present.* E-book. pp. 74, 77.
[64] Управления Механизации И Моторизации РККА (1932). *Боевой устав механизированных войск РККА. Часть 1: Строи и боевые порядки танков.* Москва: Воениздат. p. 17
[65] Солянкин, А. Г., Павлов, М. В., Павлов, И. В., Желтов, И. Г. (2005). *Отечественные бронированные машины. XX век. 1941–1945.* Москва: «Экспринт». p. 58.

Ventilation

The crew compartment of the T-72 was ventilated by the positive pressure method. In a positive pressure system, a ventilation blower fan pumps air from the atmosphere into the crew compartment. The removal of gaseous pollutants occurs through the flow of air escaping through the imperfect seals of the tank's hatches and various moving assemblies. This is in contrast to a negative pressure system, where fresh air is introduced through the same entryways while polluted air is pumped out of the crew compartment by an exhaust fan.

The ventilation system of the T-72 was integral to its CBRN collective protection system. With all hatches closed, a filter-ventilation unit (FVU) positioned in the right rear starboard corner of the fighting compartment served as the only source of air for the crew. This device consisted of a 1 kW axial blower fan and a drum-shaped FPT-100M HEPA filter cartridge.

The blower fan was a centrifugal separator, designed to perform two functions:

1. Supply dehumidified and dust-free air (98.6% efficiency)

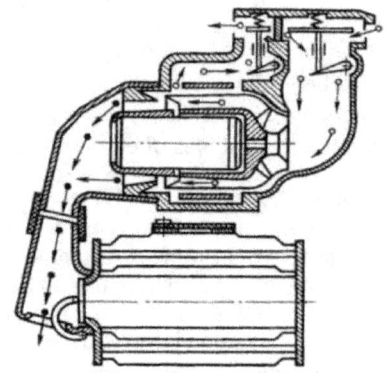

Режим вентиляции

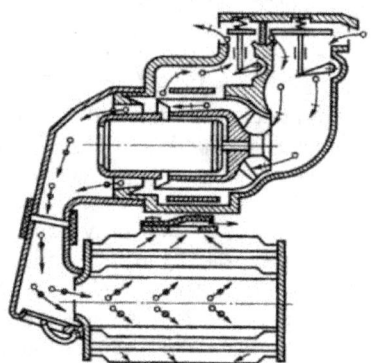

Режим фильтровентиляции

Figure 6: Air flow in FVU in unfiltered (top) and filtered (bottom) modes. (*Soviet Ministry of Defense*)

2. Generate and sustain an overpressure in the crew compartment

Due to pressure loss through the centrifugal dust removal slits in the blower casing, the pressure-flow curve of the blower was very flat. This meant that, although it could develop a respectable flow rate of 330 m^3/hr against 100 mmH2O of backpressure, this plummeted to just 100 m^3/hr against 200 mmH2O of backpressure, which was the operating condition when the blower was used together with the FPT-100M filter cartridge. This was still enough to generate an overpressure in the T-72, but ventilation efficacy suffered. The ability to build up and maintain an internal overpressure with a low flow rate was a marker of effective sealing in the hatches and turret ring. The turret ring seal was a rubber flap [**Fig. 7**, 1]. A nylon cuff kept the flap pressed lightly to the turret, enough to maintain an internal overpressure without excessive friction and wear to the rubber during turret rotation. External pressure (e.g. water, air blast) pushed into the flap, sealing the turret.

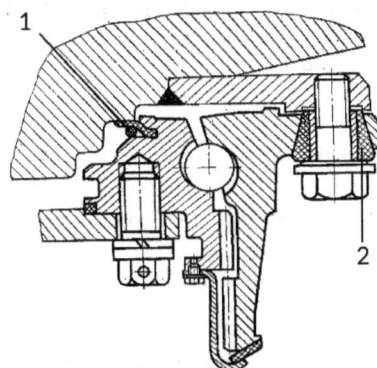

Figure 7: T-72 turret ring, showing the rubber sealing flap (1) with nylon cuff, and shock-absorbing bushing (2) isolating the turret from shock and vibrations from the hull.
(*Soviet Ministry of Defense*)

The warranty service life of the FVU blower was 500 hours [66]. However, unlike cyclonic dust filters, the fine impeller blades of the blower did not hold up well to sand and dust ingestion. In highly dusty environments abundant in coarse dust, the abrasive effect on the impeller blades shortened the FVU's service life to just 70-90 hours.

The blower had no speed selection, but the FVU itself was subordinated to a fairly complex control network. Its intake sealing mechanism and filter bypass valve were electronically controlled, and could be turned on from the commander's master console or a manual backup, but the system could also be triggered by firefighting subroutines, CBRN protection subroutines, and the armament firing circuits. In the absence of CBRN threats, the filter was bypassed by the FVU. Still, even with the blower alone, when operating a T-72 against the backdrop of a large-scale nuclear conflict, the removal of fallout by centrifugal separation could significantly diminish the long-term cumulative radiation dose of the crew.

In combat, firing the main gun or firing the coaxial machine gun would also automatically activate the FVU, if it had not been done beforehand. The continuously-developing overpressure in the crew compartment worked together with the main gun's bore evacuator to vent propellant fumes[67]. Powder fumes were removed from the open-bolt coaxial PKT machine gun by air escaping through its open barrel, also somewhat cooling it in the process.

---

[66] *Оружие и технологии России. XXI век Том 12 - Боеприпасы и средства поражения* (2006). Москва: Оружие и технологии. p. 466.
[67] Ребриков, В. Д. (1983). 'Расчет токсической дозы в танке при стрельбе из пушки без эжектора', *Вестник Бронетанковой Техники* 1983, сборник 1. pp. 21-22.

Compared to earlier tanks, the effectiveness of the ventilation system in non-combat conditions had deteriorated. Until the early postwar era, powerful negative pressure ventilation was relied upon for both ventilation and fume extraction owing to the lack of bore evacuators to prevent powder fumes from entering the tank with each shot of the main gun. Two approaches were used: placing the engine air intakes in the crew compartment (e.g. T-10, M60, Chieftain) or placing an exhaust fan in the engine compartment bulkhead (e.g. T-54), reinforced by the engine air intake and engine cooling fan holding the engine compartment itself under negative pressure. Either way, there was strong air flow through the crew compartment.

In fact, the T-72 *Ural* was provisioned for negative pressure ventilation. A small intake behind the driver's hatch, inherited from the T-64A, allowed a draft to form once the OPVT breather port was opened on the engine compartment bulkhead. The intake was deleted by 1975, but ventilating the crew compartment this way with the hatches opened remained entirely feasible.

With a positive pressure system, the FVU did not distribute air directly to each crew station[68], so each station was instead furnished with a 20-W personal fan, positioned to blow across the faces of the crew members. This gave some relief from heat stress, but these fans also had an important role in microclimate regulation. Circulating air through each crew station helped to break up zones of high humidity from perspiration, which inhibited natural evaporative cooling. During combat, the fans prevented the local accumulation of carbon monoxide.

Despite a net reduction in crew compartment air flow, the combat effectiveness of the crew markedly benefited from the switch to positive pressure ventilation because it complemented bore evacuators in expelling powder fumes, whereas holding the crew compartment at a negative pressure kept the bore evacuator from working properly[69]. The weaker air flow was generally mitigated by keeping the hatches open whenever the situation permitted, which was particularly effective for tanks like the T-72 because each crew member had a hatch of his own.

The ventilation system did not change throughout the service life of the T-72 in the Soviet Army. It served its primary purpose in that it was adequate to sustaining the tank's technical rate of fire of 8 RPM by ensuring that carbon monoxide levels did not exceed a formally established safety threshold.

---

[68] The open configuration of the T-72 fighting compartment was likely beneficial to air circulation for the entire crew.
[69] Шабалин, В. А. (1970). 'О Концентрации Пороховых Газов в Зоне Дыхания Экипажа Бронеобъектов', *Вестник Бронетанковой Техники* 1970, сборник 4. pp. 37-38.

However, like all tank ventilation systems of the period, its effectiveness as a means of microclimate regulation was marginal when the tank was sealed.

The critical factors affecting the microclimate were the sealed nature of the tank, and the air change rate. A high air change rate reduced the temperature difference between the air inside and outside the tank, and kept the buildup of humidity from sweat under control. For example, under the MIL-STD-1472C standard, outside air should be sup- plied at a minimum rate of 0.57 m^3/min per crew member. This standard was easily met by the FVU blower even in the filtered mode, but in hot climate operations (>32°C), the comfort standard was 4.2-5.7 m^3/min per person, which was beyond the capacity of the blower alone in a sealed T-72.

With that said, this was rather typical for most tanks throughout the Cold War. In both the Leopard 1 and AMX-30, a ventilation blower supplied purified air at a rate of 250 m^3/hr[70]. In the Leopard 2, it was 180 m^3/hr[71]. In the M1A1, a flow rate of 340 m^3/hr (200 CFM) was provided in the "bulk dump" mode of the NBC ventilation unit[72]. Like in the T-72, switching on the HEPA filter would halve the air flow rate[73]. Considering the total internal volumes in all of these tanks and taking into account the number of crew members, the T-72 turned out to have a favorable air change rate.

In mild weather, a tolerable internal temperature and humidity could be maintained. In the summer, however, the ventilation system could only be considered borderline at best. Moreover, in combat conditions, the ventilation capacity required for comfortable habitation were even higher. In addition to solar radiation and engine heat, which increases with intensive combat driving, the crewmen themselves were heat sources, and the gun stabilizer was a heat source when actively in use. Soviet testing in the Southern USSR showed that, for sealed static T-72 tanks, the internal dry bulb air temperature in the tank would reach 35°C with an external dry bulb air temperature of 29°C. With the gun stabilizer and FVU turned on[74], the internal temperature further rose to 39°C[75].

---

[70] Мосин, Ю. А., Ненюков, В. П., Филиппова, Н. Л., Штепанек, С. М. (1985). 'Зарубежные Фильтровентиляционные Установки', Вестник Бронетанковой Техники 1985, сборник 3. p. 54.

[71] Krapke, P. -W. (2004). *Leopard 2: Sein Werden und seine Leistung*. Norderstedt: Books on Demand GmbH. p. 91.

[72] Allen, T. A., Burrows, W. D., Terra, J. A. (1992). *Evaluation of Ventilation Inside Armored Vehicles*. Fort Detrick: U.S. Army Biomedical Research & Development Laboratory. p. 1.

[73] For the M1A1, ventilating through the M48 MCPE gas particulate filter reduced the flow rate to 100 CFM.

[74] Nearly all kinetic energy delivered by a ventilator fan ends up as heat. Refer to: Awbi, H. B. (ed.) (2007). *Ventilation Systems - Design and Performance*. London: Spon Press. p. 312.

[75] Саламахин, А. Д., Репин, А. А. (1985). 'Опытная Система Кондиционирования Воздуха Для Танка Т-72'. Вестник Бронетанковой Техники 1985, сборник 4. p. 19.

In another study, taking place in the Central USSR during summer, involving eight 10-hour marches in simulated combat conditions during daytime and nighttime, it was found that the dry bulb temperature in T-72 and T-80 tanks would be 8.7-11.3°C higher than the ambient dry bulb air temperature[76]. The fighting compartment would be slightly warmer than the driver's compartment in both tanks by 0.7-2.0°C, but under equal conditions, the temperature in the T-72 was 0.4-4.6°C higher overall than in the T-80[77], with roughly equal humidity levels. The core body temperatures of the tankers involved stayed within the norms for physical exertion (37.9°C), just shy of the critical threshold of 38.5°C for heat exhaustion.

Both studies found that the T-72 could not meet contemporary microclimate standards[78] for not exceeding an internal dry bulb air temperature of 33°C during a 6-hour operational period.

In winter conditions, crew compartment heating was provided by a radiator built into the engine preheater unit. The preheater was located in the rear starboard corner of the fighting compartment, beneath the FVU. Before starting the tank, the crew compartment would be warmed up by during the engine preheating period. With the engine running, the preheater's coolant pump was turned off and the coolant flow was pumped through the preheater by the engine. This could not be shut off, so the preheater was always hot[79], but its influence on the crew compartment temperature was decidedly marginal compared by the heat of the engine's right exhaust manifold, which was separated from the crew compartment by only the thin engine compartment bulkhead[80] - one of the less obvious ramifications of a transversely mounted engine.

To provide a proper heating effect, a 175 W forced convection fan in front of the preheater radiator could be turned on. This method of crew compartment heating was reliable, fuel efficient, and could be sustained indefinitely as long as the engine kept running. Its disadvantage was a limited heating capacity[81]. In colder temperatures, the crew typically kept their winter jackets on.

---

[76] Перегонцев, С. М., Щедрин, А. К. (1984). 'Микроклимат в танке на марше', *Вестник Бронетанковой Техники* 1985, сборник 4. p. 11.
[77] Due to the T-72 heater and the proximity of the left engine exhaust manifold to the engine compartment bulkhead. See: Шабалин, В. А., Щедрин, А. К. (1982). 'Формирование и прогнозирование микроклимата ВБТ', *Вопросы Оборонной Техники*, 20(103). p. 41.
[78] Medical and technical requirements for the habitability of military ground vehicles: *Медико-технические требования к обитаемости подвижных наземных объектов вооружения и военной техники Сухопутных войск* (МТТ СВ-81), (1983). Москва: Воениздат.
[79] To remove it, the coolant circuit had to be drained, which, when combined with its poor accessibility, made heater replacement an unpleasant task taking up 1.3 hours.
[80] And possibly several rounds of main gun ammunition.
[81] Баннов, В. В., Чернявский, В. В., Баннова, Ю. В. (2021). 'Обитаемость современных танков', *Транспортные Системы: Безопасность, Новые Технологии, Экология*. Якутск, 16 апреля. Якутск:

Chapter II
Crew Stations

The equipment in the three crew stations in a T-72 were reflective of the specialized tasks assigned to each crewman, well worth examining on their own. However, there were also several commonalities between the crew stations, the most obvious being the lighting system.

Internal lighting was provided by three PMV-71 dome lights and eight KLST-64 cabin lights[82]. While servicing the drivetrain at night, a PLT-50 wired flashlight could be plugged into an external ShR-51 power socket[83] next to the rear left convoy marker light. There were two more ShR-51 sockets inside the tank, one in the turret behind the gunner's seat and one on the hull ceiling behind the driver's seat. An FKB-2M portable flashlight was also provided.

The dome lights serve as the main and emergency lights, as they were not connected to the power supply network of the tank but were instead wired in a separate isolated circuit with the tank's batteries. One dome light was installed on the ceiling of the driver's station and two were on the ceiling of the commander's station. The KLST-64 cabin lights served as local lighting for specific control panels or areas of importance. They had dimmer lids for nighttime use, to preserve the crew's natural light vision while still having some light on the instruments.

The periscopes also deserve mention as the crew's only windows to the outside world once the hatches were closed, and for the unique design solutions found in each model to harmonize the needs of each crew member with the shape of the tank's hull and turret.

Якутский институт водного транспорта (филиал) ФГБОУ ВО СГУВТ. p. 100.
Also see: Гудков, А. И., Катин, Е. Н., Перегонцев, С. М., Фролов, А. В. (1981). 'Эргономические исследования БТТ', *Вестник Бронетанковой Техники* 1981, сборник 4. p. 12.
[82] Copy of Bosch-1-39 lamps from the 1930's.
The dome light and cabin lights used the same 10 W incandescent bulb.
[83] Derived from Bosch vibration-resistant automotive sockets commonplace in the 1930's.

Periscopes and Periscope Design

Periscopes, or "triplexes", as they were colloquially known among Soviet tankers[84], were the primary means of general vision for the crew. Their purpose was to provide the crew with a means of orienting themselves in their surrounding environment and, when necessary, to survey the battlefield for threats at close range. Although outwardly simple, tank periscopes had a number of nuanced design features to give the best possible vision at a minimal sacrifice to protection.

All Soviet tank periscopes were prismatic periscopes, reflecting an image from the objective window to the eyepiece window through the total internal reflection of light in solid glass prisms instead of mirrors[85]. Four different periscope models can be found in a T-72, each specialized for slightly different structural restrictions. The TNP-165A and TNP-160 (TNPO-160) were carried over from earlier medium tanks and even saw use in BTRs. The TNPA-65A and TNPO-168V were less suitable for general use and were subsequently less popular.

Table 9: T-72 periscope specifications[a, b]

Technical Characteristics	TNPA-65A	TNP-165A	TNPO-160	TNPO-168V
Application (No. of pcs)	All crew (5 total)	Gunner (1)	Commander (2)	Driver (1)
Type	Wide-angle hatch periscope	General purpose periscope	General purpose periscope	Wide-angle periscope
Total viewing angle (H/V)	140°/35°	74°/35°	78°/27°	138°/31°
Static FoV (H/V)	52°/6°	36°/8°	38°/5°	58°/6°
Periscope slot area (mm x mm)	108x37	122x48	140x50	250x50
Light transmission	0.6	0.55	0.5	0.45
View offset angle from horizontal	45°	20°	0°	0°

[a] Оружие и технологии России. XXI век Том 7 - Бронетанковое вооружение и техника (2003). Москва: Оружие и технологии. p. 413.
[b] Андрусов, А. В., Корнилов, В. И., Новый, Г. И., Севертока, И. И. (1988). 'Обобщение результатов оценки комплекса приборов наблюдения серийных танков и БМП', Вестник Бронетанковой Техники 1988, сборник 10. p. 11.

---

[84] Derived from the once-influential British "Triplex Safety Glass" company, which produced laminated safety glass. When laminated glass was introduced to tank vision slits in the USSR, they were called "triplexes". The term stuck for all vision blocks and periscopes.
[85] Mirrors crack easily from the shock of a powerful shell strike on the tank.

All periscopes were made from K108 photochromic glass[86] containing a cerium additive. The glass darkened under sunlight and recovered its transparency in its absence. This reduced eye strain over long hours of continuous combat, and when exposed to the intense flash of light from a nuclear detonation, the glass began to darken within 0.0002 seconds. This gave a modicum of eye protection[87], but not at all comparable to anti-flash goggles.

Optically, a periscope is directly equivalent to vision slit, and its periscopic height is equivalent to the thickness of the wall in which the vision slit is placed. In prismatic periscopes, refraction causes the light passing through the periscope to travel at an angle closer to the normal, and upon exiting back into air, regain its parallelism to its initial angle. Visually, the effect is equivalent to looking through a vision slit in a thinner wall[88]. The field of view through the periscope becomes proportionately wider as a result. Compared to mirror periscopes or direct vision slits, refraction through borosilicate crown glass (K108 refractive index: 1.518)[89] allowed a prismatic periscope to convey a ~30% larger arc of vision.

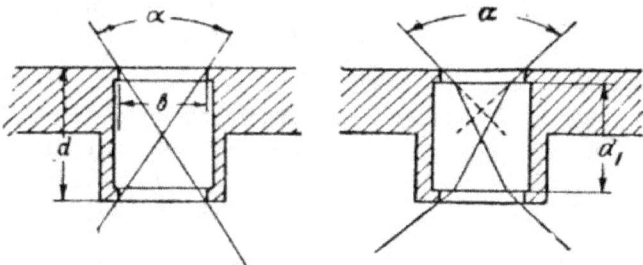

Figure 8: Viewing arc through empty vision port (left) and through vision port with glass (right). *(S. I. Vavilova & M. V. Savostyanovoy, 1948)*

The total arc of vision [**Fig. 8**] is the most common method of quantifying the degree of vision provided by a vision device. It represents the maximum size of the arc where the observer can perceive light through the device with head movement, independent of the eye's distance from the eyepiece. Another metric

---

[86] Equivalent to Schott BK7, created from K8 borosilicate crown glass. Павлов, М. В., Павлов, И. В. (2021). *Отечественные Бронированные Машины 1945-1965 гг. - Часть I: Легкие, средние и тяжелые танки*. Кемерово: ООО "Принт". p. 76.
Further reading on K108 glass is available in: Арбузов, В. И. (2008). *Основы радиационного оптического материаловедения*. Санкт-Петербург: Санкт-Петербург Государственный Университет ИТМО. pp. 145-175.
[87] Живулин, Г. А., Олихвер, А. И., Пивовар, Р. М., Снурников, А. С. (1983). 'Средства защиты органов зрения экипажей БТТ от светового излучения ядерного взрыва', *Вестник Бронетанковой Техники* 1983, сборник 5. p. 20.
[88] According to the formula $\frac{n-1}{n}$, where 'n' is the refractive index, the apparent depth is reduced to a third of the true depth.
[89] LZOS (2001). *Оптическое стекло*. Available at: http://www.lzos.ru/opglass/opgrus.htm (Defunct).

is the static field of view with a certain eye relief to the periscope. The static field of view is always smaller than the viewing angle[90].

The commander and driver relied heavily on their periscopes while performing their duties, and as such, certain creature comforts could be found on the TNPO-160 and TNPO-168V. Being intended for long periods of continuous use, both had wire frame clamps on each side of the eyepiece window to hold a darkening filter. A darkened view prevented eyestrain in bright environments, but it was most useful in the winter, where snow blindness could be an issue. The clamps could also fit a metal scratch protector, to protect the glass while doing work inside the tank. It could also function as a blackout panel to enforce light security at night, or even as a form of preemptive eye protection against nuclear flash[91].

Both periscopes were fitted with the RTS-27-4A heating system to eliminate fogging in cold weather[92]. By running a current through specially coated glass screens laid over the objective and eyepiece windows, the transparent conductive oxide coating would heat up evenly, unlike solid resistive heating elements. However, because each periscope had two of these heating screens, light transmittance was slightly diminished.

The TNP-165A and TNPA-65A were basic periscopes supplementary to other vision devices. These were not necessarily worse in terms of visibility, but they lacked heating and eyepiece covers. The TNP-165A's raised mount at least prevented snow accumulation from obscuring its window, but there were no special measures for the low-set TNPA-65A hatch periscopes.

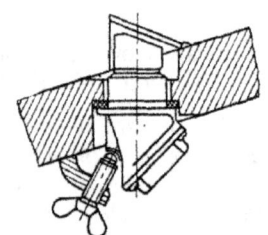

All four periscope types were characterized by relatively small overall sizes and low periscopic heights[93]. A short periscopic height is equivalent to a vision slit in a thin "wall", and thus, the field of view is correspondingly larger.

Periscopes designed to protrude above the roof of an armored enclosure had to have a periscopic height of no less than 230-240 mm to ensure sufficient headroom for the observer[94], given an average height of 178 mm between a tanker's eyes to the top of his

Figure 9: TNPA-65A hatch periscope. (*Soviet Ministry of Defense*)

---

[90] Сеннов, Н. И. (1942). *Оптика на танке*. Москва: Объединение государственных книжно-журнальных издательств. p. 7.
[91] Outside of the USSR, West German periscopes were most notable for featuring blackout flaps, which were included as part of the tank's nuclear protection scheme as flash protection.
[92] Denoted by the letter "O" in the periscope designation.
[93] The distance between any pair of mutually reflecting points on the objective and eyepiece.
[94] Сеннов, Н. И. (1942). *Оптика на танке*. Москва: Объединение государственных книжно-журнальных издательств. p. 34.

helmet. With this restriction on minimum height, obtaining a large field of view would ordinarily not be possible without enlarging the periscope, and this would increase the likelihood of combat damage to the periscope head as well as widen the areal footprint on the roof, weakening it to high explosive blast. The latter was especially important for periscopes installed in hatches, and drivers' hatches in particular. The unusually small TNPA-65A hatch periscope was created taking these hazards into account.

In postwar Soviet tank building practice, the issue of balancing vision and protection in general purpose periscopes was solved by incorporating them into domed and sloped structures, as these were considered ballistically ideal shapes for the tank body. This approach allowed standard periscopic heights to be reduced to 160-170 mm, improving vision and keeping periscope head exposure limited while staying within headroom requirements.

Where possible, all periscopes were fitted near the edge of the roof structure to minimize visual dead zones, and to maximize the useful size of the vertical arc of vision. This also increased the risk of periscope damage from bullets ricocheting off the sloped armor of the tank, so various mitigation methods were applied according to the peculiarities of each crew station.

1. Fixed forward-facing periscopes for gunners were made to be viewed by looking up at an angle; positioning the day sight at the gunner's eye level took priority. No ricochet protection was needed.
2. Fixed periscopes for the commander were made with a low periscopic height for domed cupolas.
3. Fixed periscopes for the driver were made with a low periscopic height to fit in the upper glacis. This placement also avoided interfering with the depression limit of the tank's gun. A ribbed bullet trap made from rows of welded steel bars on the upper glacis prevented ricochets.

The design of the turret and its sloping roof paid dividends to the gunner in particular, especially compared to tanks with tandem seating for the gunner and commander. In the absence of a sloping roof (e.g. Leopard 1, Leopard 2), a large periscopic height (or excess eye relief) was needed, narrowing the field of view below the level of some low-magnification sights[95]. Where periscopes were made integral to the gunner's sight (e.g. M60, Chieftain), the size of the periscope had to be limited, which again strongly impacted the field of view.

---

[95] In Leopard 2 - With an eye relief of 200 mm, the field of view was only 13° horizontally and 7° vertically. See: Krapke, P. -W. (2004). *Leopard 2: Sein Werden und seine Leistung*. Norderstedt: Books on Demand GmbH. p. 84.

If the periscope head was hit, limiting the damage to the user's eyes was the first priority. A robust clamping mechanism prevented periscopes from becoming dislodged by a strong shock, which could lead to facial injuries. The driver's TNPO-168V periscope had a shock absorbing buffer in its mount to protect the periscope if the tank's upper glacis was hit by a powerful shell.

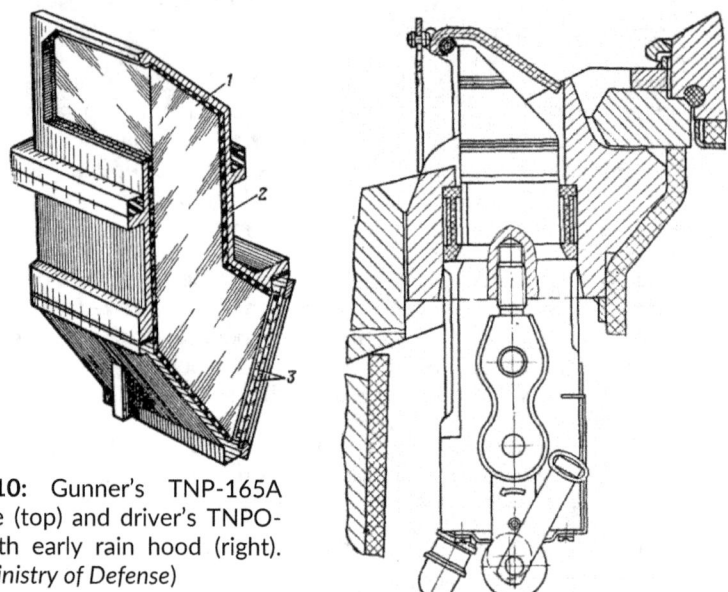

Figure 10: Gunner's TNP-165A periscope (top) and driver's TNPO-168V with early rain hood (right). (*Soviet Ministry of Defense*)

If the periscope head was struck by a sufficiently powerful bullet, the glass prism could be shattered into fragments, which posed its own set of problems. Two solutions were implemented, both carrying their own unique compromises. Tall periscopes (TNPO-160, TNPO-168V) normally had separate head and eyepiece prisms, but short periscopes (TNPA-65A, TNP-165A) were built as a single solid glass prism.

Separating two prisms prevented the propagation of cracks from the objective prism to the eyepiece prism, protecting the user's eyes[96]. Instead of an air gap between prisms, TNP-165A and TNPA-65A had a laminated ballistic glass shield over their eyepiece windows. An air gap between prisms had to be nitrogen-filled, complicating the manufacture of such periscopes, but ballistic glass seriously diminishes light transmittance.

---

[96] Ogorkiewicz, R. M. (1991). *Technology of Tanks*. Volume 1. Coulsdon: Jane's Information Group. p. 130.

The TNPA-65A, TNP-165A and TNP-160 had stress-concentrating localizer grooves cut around the periscope head to ensure a clean decapitation if the periscope head was struck by a large-caliber bullet. This gave some guarantee that a damaged periscope could be replaced in combat, if the slot itself was not severely deformed. The TNPO-160 and TNPO-168V also featured stress-concentrator grooves cut in the shape of a waffle iron on the back of their heads[97], so that a bullet strike would crack or break apart the aluminum casing instead of bulging it out.

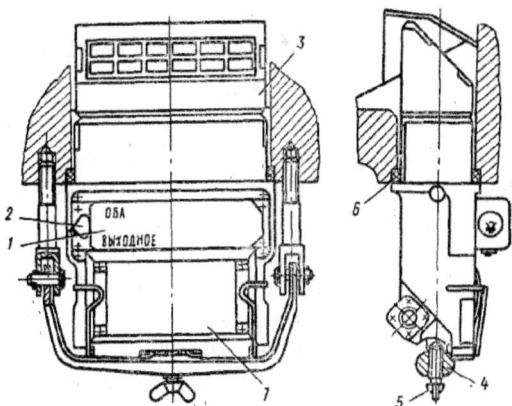

Figure 11: Commander's TNPO-160 periscopes. (*Soviet Ministry of Defense*)

The rain hood over all periscopes was made from thin aluminum sheeting so as not to deflect bullet fragments downward after penetrating through the glass periscope head. In 1979, the rain hoods on the TNP-165A and TNPO-168V were made into extended duckbill visors to reduce glare from overhead sunlight, and the commander's TKN-3 was likewise given a visor.

---

[97] First observed on the American M24 (T41) infrared driving periscope from the early 1950's.

Commander's Station

Overview

The primary role of a T-72 commander was to manage his crew, and coordinate his actions with his unit while fulfilling his obligations as a subordinate to his unit leader. At the crew level, the commander's responsibilities in combat were to make decisions on the tank's immediate objectives, monitor the battlefield, identify threats for the gunner, and direct the movement of his tank by communicating destinations or waypoints to the driver. His involvement in the duties of the other two crew members was almost entirely supervisory in nature, apart from his role in correcting the gunner's fire when engaging a target. The commander was also responsible for several secondary tasks:

1. **Coaxial machine gun**: he reloaded the PKT machine gun and cleared jams when necessary

2. **Autoloader**: he controlled the circuit breakers for the autoloader and had access to the manual and electric controls. He restocked the autoloader, set its operating mode, and operated it in the backup modes during emergencies

3. **Electrics**: he managed the tank's convoy lights and he could trigger emergency systems (shut off engine, discharge fire extinguishers)

As a member or leader of a small tactical unit, the general responsibilities of a T-72 commander were largely similar to those of a BTR or BMP commander in mounted combat[98]. The division of labor between a commander and gunner was roughly the same regardless of the type of AFV, and their roles entailed many of the same tasks. The idea of standardizing roles and responsibilities based on shared elementary tactical needs was hardly exclusive to the Soviet Army, but the degree of standardization was quite noteworthy.

Compared to his colleagues in a T-54/55 or T-62, a T-72 commander had to be trained to a higher level of technical competency to use and maintain the tank's autoloader, sighting systems, and drivetrain. In combat, however, there was virtually no difference to commanding a T-72 from commanding a T-54.

---

[98] Боевой устав Сухопутных Войск. Часть III: Взвод, отделение, танк (1982). Москва: Воениздат.

Figure 12: Commander's station. The recoil guard is not present. (*S. Stauber, 2024*)

The layout of the commander's station adhered to several basic rules for workplace convenience. From a seated view, facing forward in a neutral pose, all of the equipment was visible and within easy reach [99]. No controls or instruments were placed on the turret wall directly next to the commander's seat, maximizing the available shoulder room.

A recoil guard was attached to the commander's seat so that he was separated from the main gun regardless of the height setting of the seat. The top part of the recoil guard could be folded away to access the gun breech, and the entire recoil guard could be removed to grant access to certain parts of the autoloader. This would mainly be done to restock the carousel, but the recoil guard would also be removed when loading the gun manually.

The cupola was derived from the T-64A cupola, but its basic form was inherited from the T-54. This ubiquitous design was the product of a rather lengthy evolutionary process, with a lineage that can be traced through Morozov's design history beginning with the T-43[100]. Over time, it was refined for better visibility, more headroom, greater user convenience, and easier target-finding through several design iterations from the T-44B through the T-54 series[101].

The cupola was mounted in a barbette, 700 mm in diameter, screwed into the turret roof. It rotated on ball bearings within an intermediate ring used to support the pintle mount for the NSV-12.7 anti-aircraft machine gun. In turn, this ring rotated on the barbette on another set of ball bearings. When not in use, it was locked in the "travel" position, which placed it facing rearward with an offset of 13° to the right. Because the pintle was a simple ring clamp, the DShKM from a late-model T-55 or T-62 could be transplanted to the T-72 if the need ever arose.

The cupola could be locked to the machine gun ring in three positions: 23-00, 28-00, and 37-00. In the 28-00 position, the cupola faced directly forward when the turret fixed by its travel lock at the 32-00 position[102]. There was no specific use for the other two positions other than to keep the TKN-3 out of the way when working in the hull.

---

[99] In Western tank design practice, the radio was normally placed behind the commander in a bustle compartment.
[100] Пашолок Ю. (2019). Промежуточное звено. Available at: https://warspot.ru/14603-promezhutochnoe-zveno (Accessed: 12 August 2023).
[101] Павлов М. В., Павлов, И. В. (2021). *Отечественные Бронированные Машины 1945-1965 гг. - Часть I: Легкие, средние и тяжелые танки.* Кемерово: ООО "Принт".
[102] The turret could not be centered so that the gun cleared the driver's hatch

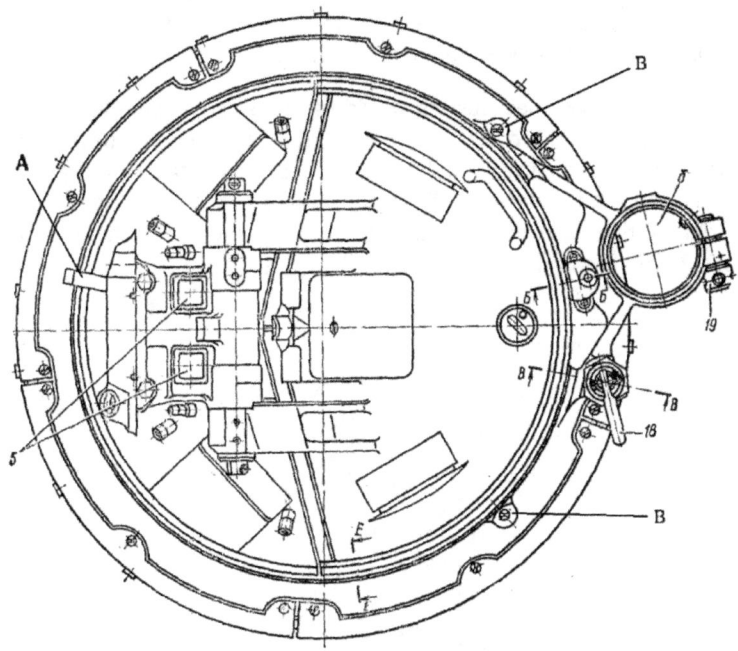

Figure 13: Top-down view of T-72 cupola with its machine gun pintle. (*Soviet Ministry of Defense*)

The front of the cupola had a fixed roof where the primary vision devices were installed. At its center was the TKN-3 combined day-night binocular periscope. Functionally, it was equivalent to giving the commander a pair of field binoculars to use from under armor. To the left and right of the TKN-3 were two TNPO-160 periscopes. Out of the 584 mm internal diameter of the cupola, only three quarters of its lengthwise space was free to accommodate the commander. The rest of the cupola was covered by the hatch, where two TNPA-65 periscopes were fitted for to broaden the view to the sides and rear[103].

By having the hatch open forward and locked in an upright position, it could function as a shield. This was a Soviet design standard during the Cold War, but such hatches were as old as the Soviet tank industry itself. The concept was born out of combat experience from the Spanish Civil War in 1936. Post-battle

---

[103] The original T-72 *Ural* (1974) cupola lacked these hatch periscopes. They were added in 1975, and in 1976, these hatch periscopes made their way to the T-64B.

analyses by the Red Army noted that once a tank was knocked out, its crew members often had to escape under fire, so additional protective measures were needed. Two solutions were conveyed by the ABTU RKKA[104] in late 1937: underbelly escape hatches[105] and "hatch-shields".

As originally envisioned, turrets would have a single large hatch that, when locked upright, could protect the crew members in the turret from small arms fire as they rode standing behind it, or protect them during an evacuation under fire. This so-called "hatch-shield" concept found its way onto nearly all Soviet armored fighting vehicles beginning in late 1937, most prominently found on early T-34 models until 1941.

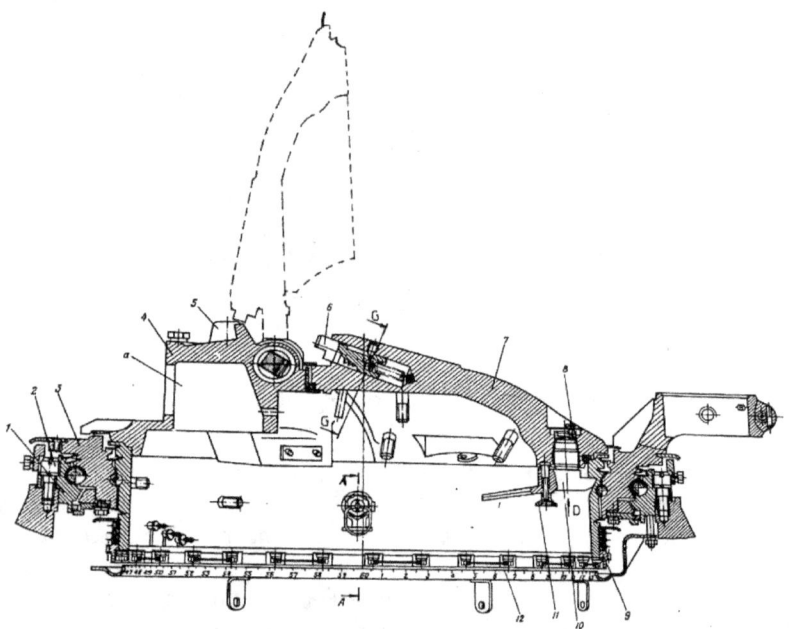

Figure 14: T-72 cupola profile cutaway, with outline of opened hatch. Note the azimuth ring fixed to the turret around the lower circumference of the cupola. (*Adapted from manual illustration, Soviet Ministry of Defense*)

Large hatches were eventually found to be inconvenient and were subsequently abandoned, but the underlying idea of using hatches to provide forward protection persisted throughout the Cold War. On the rotating cupola

---

[104] Armored Vehicle Directorate of the Red Army
[105] Желтов И. Г., Макаров А. Ю. (2020). *А.Я. Дик - начальник КБ, которое так и не было создано*. Available at: https://t34inform.ru/publication/p-pers-3.html (Accessed: 12 August 2023).

of the T-72, a hatch of this type allowed the commander to shield himself from any direction of his choice while standing up, or merely peek above the brim of the cupola with the hatch rotated to the side or rear.

The main advantage of the hatch-shield concept was that it was compatible with a low-silhouette turret when closed, and when open, it gave the commander around 400 mm of additional clearance above his cupola. When riding from an open hatch, the commander stood on his seat with his hatch in front of him. With the range of adjustment on the seat, a commander of practically any height could position himself to expose only his head over the rim of the hatch while standing. Looking over the rim of an opened hatch, a T-72 commander's eye level was elevated 2.9-3.0 meters above ground level as if he were on a pedestal.

Standing up, the commander had a good vantage point. He had a clear line of sight even with the tank concealed on reverse slopes, behind vegetation, and other local objects. If, for instance, the tank was on a grassy hill and the commander's view was obstructed by tall grass, the hatch concept allowed him to safely look well above the turret roof level, so that the tank could remain completely concealed by the hill itself while still giving the extra height needed. On a march or on combat maneuvers, the commander's view from this vantage point was practically free of dead zones all around the tank. This was especially valuable when guiding the driver through difficult obstacles.

In contrast, large cupolas with rearward-opening hatches like those found on the M60A1 (530 mm height) [106] and AMX-30 (570 mm height) [107] gave commanders a well-elevated vantage point when working from under armor and from an open hatch, but greatly raised the turret silhouette. A low-profile cupola with a side-swinging hatch like the type used on the Leopard 1 (120 mm height)[108] hardly elevated the commander's view at all.

In the late 1970's, a dust shield was introduced into the standard equipment of the T-72. It shielded the area above the open hatch, and was a minor innovation when it was first introduced on the T-72[109].

---

[106] Пеньков, М. Д., Ржевский, А. И., Шишков, Ю. А. (1976). 'Особенности Конструкции И Компоновки', *Вопросы Оборонной Техники*, 20(67). p. 7.
[107] Marzloff, J. (1971). 'AMX 30 France's Main Battle Tank', *International Defense Review* 1971, 2. p. 111.
[108] GRD: Technische Abteilung 1, Sektion 1.4 - Kampffahrzeuge (1974). *Bericht: über die technischen Versuche im Rahmen der Evaluation Panzer 68 - Kampfpanzer Leopard*. GRD report. p. 135.
Also see: Technical drawing, 2500082-000000.00.0, 'Turm mit Bewaffnung'. Wegmann & Co., Kassel.
[109] Суворов, С. (2004). 'Танки Т-72: вчера, сегодня, завтра', *Техника и Вооружение*, (July).

Figure 15: Methods of observation; mounted with tank as elevated platform (top) or dismounted, using natural terrain features for concealment (bottom). (*Soviet Ministry of Defense*)

Figure 16: The basic Do's and Don'ts of concealed reconnaissance in terrain, from Soviet tank forces manual, applicable to all tanks. These were basic rules taught to all cadets in the tank forces. (*Soviet Ministry of Defense*)

Visibility

Closed down inside the cupola, the dead zones from the T-72 cupola were still small thanks to the rounded shape of the turret in all directions, including the roof over the gunner's side of the turret [**Fig 13**]. The advantages of this design approach were especially prominent when compared to tanks with a flat turret roof like the Leopard 1, which had very large visual dead zones towards the front, rear, and to the right, and extreme dead zones when looking to the left, across the loader's part of the turret roof. Swiss testing found the dead zone to the rear to be 75-80 m in a 90° arc through the periscopes, or 100 m with the TRP panoramic sight. To the left at 7 o'clock, the dead zone was 100-145 m, and in the 8-10 o'clock sector, it was no less than 200 m[110].

On the T-72, several items on the gunner's side of the turret were local obstructions to an otherwise free view from the commander's cupola; the day sight blocked the commander's view out to 25 m, the night sight likewise blocked his view out to 35 m, and the gunner's hatch hinge and his periscope further worsen the commander's leftward view. Moreover, several other factors affect the exact size of the dead zones between each periscope, from the higher clearance of the TKN-3, small vertical viewing arc through the TNPA-65A hatch periscopes, and the forward tilt of the turret by 1°13'.

The TNPO-160 on each side of the TKN-3 was oriented 45° off center and the TNPA-65A hatch periscope on each side was oriented 120° off center. The commander looked to the sides through the hatch periscopes at an angle. The hatch periscopes were much less convenient to use than the TNPO-160 because of the location of the eyepiece windows; the commander had to look upwards and there was hardly any headroom to look downward through the hatch periscopes. These hatch periscopes were originally not present on the T-64A. They were added as part of a suite of improvements introduced on the T-72 *Ural*, and then later applied to the T-64A production line in 1975[111].

---

[110] GRD: Technische Abteilung 1, Sektion 1.4 - Kampffahrzeuge (1974). *Bericht: über die technischen Versuche im Rahmen der Evaluation Panzer 68 - Kampfpanzer Leopard.* GRD report. p. 65.
[111] Чобиток, В., Саенко, М., Тарасенко, А., Чернышев, В. (2016). *Основной танк Т-64 - 50 лет в строю.* Москва: Яуза-каталог. p. 55.

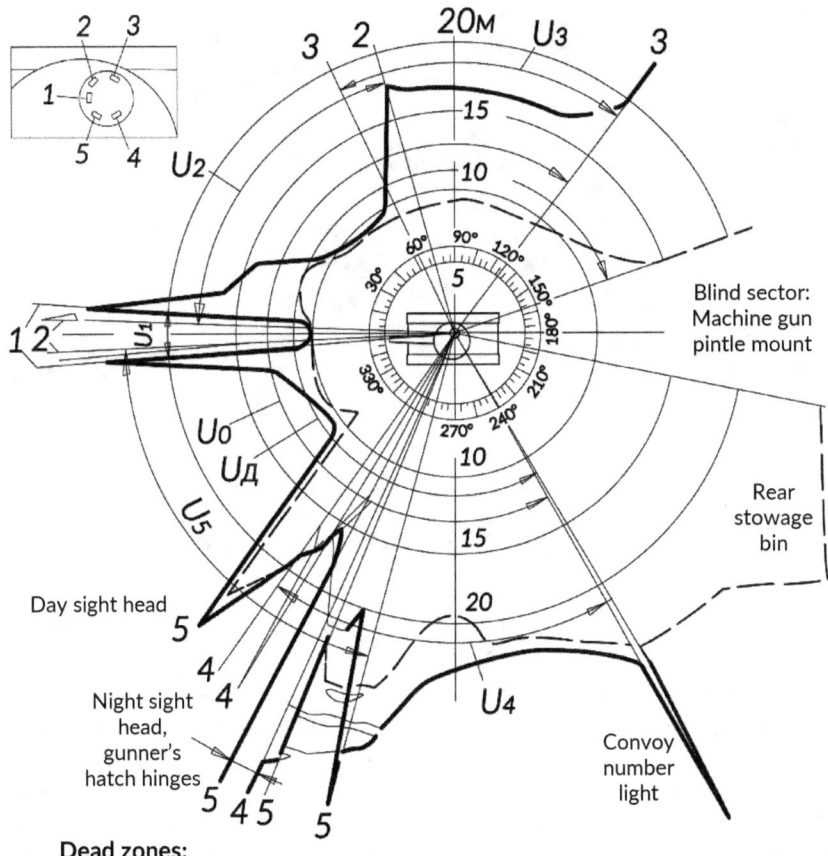

Dead zones:

——— Static view with cupola facing forward

— — Rotating cupola, dynamic view through TNPO-160 or TNPA-65A

Static view:
(sector 1-1, U_1) – TKN-3;
(sector 2-2, U_2, sector 5-5, U_5) – TNPO-160;
(sector 3-3, U_3, sector 4-4, U_4) – TNPA-65A

Fields of view:
U_1 – 10°;
U_2 and U_5 – 78°;
U_3 and U_4 – 71°

Figure 17: Viewing arcs and dead zones from T-72 commander's cupola. Obstructions that expand the dead zone are labeled. (*Drawing adapted from E. A. Germanov, N. S. Proskurova, 1981*)

Figure 18: TKN-3 and TNPO-160 from the commander's perspective. (S. Stauber, 2024)

The commander's rearward view was impeded by the pintle mount for the anti-aircraft machine gun in the "travel" position in both directions. The pintle itself blocked the view entirely, and the base of the pintle increased the size of the dead zone near the ends of the viewing arc. The most serious restriction was to the TKN-3. A tab on the cupola [**Fig. 13, A**] stopped the cupola against from turning in a full circle against two large screws on the machine gun pintle ring [**Fig. 13, B**], restricting the total turning arc to 320°. Nevertheless, the commander could still see backwards by turning the cupola to the left because of the 13° offset of the pintle. If the turret was turned to the "marching" position with an offset to the left, the commander could look straight backwards.

Compared to the T-54/55 cupola (also used in the T-62) where full-sized periscopes were embedded in the hatch and oriented at 103-104° off center, the static viewing arc was larger, though less convenient to use. The reason for replacing full-sized periscopes in the hatch with special small periscopes is still undocumented[112]. However, it is at least known that tank commanders widely preferred to keep their hatches ajar whenever possible, making hatch periscopes

---

[112] The Object 430 cupola had the same cupola shape and it had full-sized periscopes in its hatch, which rules out the possibility that full-sized periscopes may have interfered with the opening and closing of the hatch.

useless regardless of how convenient they might be to look through, and when traveling cross-country, the hatch would bounce up and down on its torsion bar. Full-sized periscopes in the hatch could come down as far as eye level, so a hatch that bounced all the way down to the closed position could very easily smash a periscope over the commander's head if he leaned too far to one side or the other.

Figure 19: TNPA-65A hatch periscope. Note the cupola azimuth ring. (*S. Stauber, 2024*)

With these five vision devices, the basic needs of a tank commander were essentially met. A good view in a large forward arc took top priority, and improving vision towards the rear by adding more periscopes generally gave diminishing returns. A Soviet study comparing Western-style cupola concepts[113] against the domestic type found that, despite the test scenario heavily favoring better all-round vision at short to medium range[114], 70% of the commander's attention was dedicated to a forward 100° arc and over 95% of his attention was contained within an arc of 200° regardless of the cupola type[115].

---

[113] A cupola with a large number of fixed periscopes for all-round vision, and a rotating cupola with the same number of periscopes.
[114] Scenario with equal likelihood of encountering enemy anti-tank weapons from all sides, at a range of 500-1,500 m.
[115] Голуб, Г. Г., Затравин, Е. И., Тютин, В. Е. (1974). 'Некоторые Статистические Характеристики Процесса Наблюдения Командира Танка', *Вестник бронетанковой техники* 1974, сборник 2. p. 27.

Comparative testing of the Soviet-style rotating cupola concept against a Leopard-type fixed cupola with a panoramic sight in West Germany by a joint French-West German research commission showed very similar outcomes[116]. Two cupola mockups were built on identical test vehicles to isolate external influences. One mockup took the cupola of the Swedish Ikv 91 light tank, analogous to a Soviet-style cupola, and the Leopard was used to create the other. Four scenarios were tested, two with the tanks at a standstill searching along a 120° sector, and two with the tanks on the move observing all around (360°). Opposing mockups searched for special targets and for each other from unknown directions, from 1-3 km away and in various positions.

From a standstill, the rotating cupola system had a slightly higher probability of finding targets and could do it slightly more quickly. On the move, both configurations were essentially comparable except that the rotating cupola performed better in the short term (1 minute). Both types showed low probabilities of target detection within the allotted three minutes. The recommendations issued by the joint French-West German commission did not touch on the number of periscopes or quality of all-round vision, only that to improve target detection capabilities, it was necessary to improve the daytime optics available to the gunner and commander, and to implement a combined daylight and thermal imaging surveillance device.

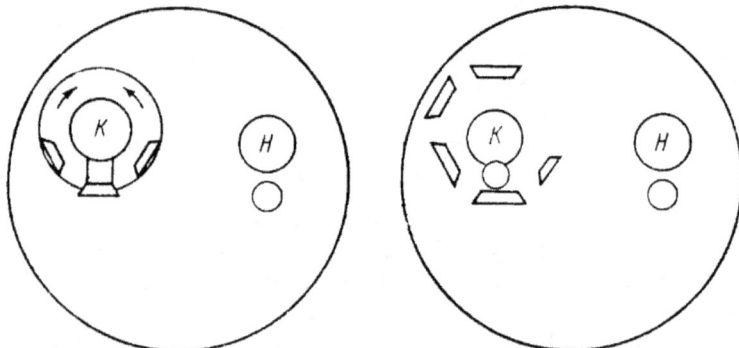

Figure 20: Mockup I (left) with one magnified optic and two periscopes in a rotating cupola. Mockup II (right) with ring of periscopes and a panoramic sight. **Notation:** *K* – commander; *H* – gunner. (*O. A. Dobisov & N. V. Kuzmina, 1984*)

---

[116] Добисов, О. А., Кузьмина, Н. В. (1984). 'Исследование приборных комплексов командиров основных танков', *Вестник Бронетанковой Техники* 1984, сборник 1. pp. 54-55.

Perhaps most interestingly, it was also recommended to have special tools to make it easier for the crew to orient themselves relative to the turret and the direction of movement. In the T-72 (and other Soviet tanks), the commander's sense of spatial awareness was intuitive in nature rather than analytical because his upper body rotated along with the cupola.

In combat, target spotting and target cuing were done exclusively through the TKN-3 periscope. The entire cupola was manually turned to scan horizontally, and the TKN-3 was manually elevated. By bracing himself on the handles of the TKN-3, the commander maintained a proper eye relief while in motion[117] and steadied his view, although it was ineffective at high speeds, especially when the view began to be affected by vibration. A counter-rotation mechanism kept the commander's view fixed on a target as the turret turned under the cupola.

The target cuing system had a simple sensor to detect if the cupola was deflected clockwise or counter-clockwise, and when the thumb button on the left TKN-3 handle was pressed, the turret turned at its maximum speed until it aligned with the cupola. As a backup, there was an azimuth ring and a pointer on the 9 o'clock position of the cupola (because there was nowhere else to put it) so that as the cupola turned, the pointer showed its azimuth angle relative to the turret. The azimuth scale was most useful for measuring large angular gaps between two landmarks.

The counter-rotation mechanism was a cardan shaft linking the cupola to a reversing gearbox at the base of the turret, where an input gear was constantly in mesh with the hull ring. Pressing the thumb button at the end of the right handle of the TKN-3 activated a solenoid that engaged a clutch, coupling the cardan shaft to the gearbox. As the turret turned, it dragged the input gear of the gearbox over the hull ring, turning the cupola via the cardan shaft at a rate exactly equal to the turret but opposite in direction.

When cuing the gunner to a target, the commander pressed and held both TKN-3 handle buttons to keep the cupola fixed on the target through the counter-rotation mechanism. The right thumb button also switched on the commander's OU-3GK spotlight as long as it was held, if the spotlight power was switched on.

This was more intuitive and much quicker than providing the commander with a turret traverse handle and a sight extension to duplicate the gunner's view, where the commander was expected to break visual contact with a target

---

[117] Вавилова, С. И., Савостьяновой, М. В. (eds.) (1948). *Оптика в Военном Деле - Сборник Статей (Том II).* Издание Третье. Москва: Издательство Академии Наук СССР. p. 51.

spotted through his cupola to take over from the gunner, and lay back onto the target. The clumsiness of this type of target cuing had been noted early on in the PT-series panoramic sight, which had been the commander's main means of observation in early T-34 models[118]. After spotting a target, the commander had to break visual contact to check an azimuth indicator on the sight body, and then attempt to reacquire the target through a secondary telescopic sight or through the same panoramic sight after resetting it to its zero position[119].

The main limitation of the T-72 target cuing system was the simplicity of its cupola azimuth sensor. It had a large intrinsic alignment error with a wide variance, reaching 17-26 mils (1.0-1.5°), so it was never possible to align the gunner's sight exactly onto the TKN-3's crosshairs. The commander would sometimes have to verbally guide the gunner on target if the target was well camouflaged. Tests showed that target handover could take 3-7 seconds for a system with a large alignment error of 22-34 mils. At the lower end (16 mils), it took only 2 seconds[120].

---

[118] Коломиец М. (2009). Т-34: Первая полная энциклопедия. Москва: Эксмо. p. 334.
[119] Желтов И. Г., Макаров А. Ю. (2017). *Общее устройство танка А-20*. Available at: https://t34inform.ru/publication/p01-7.html (Accessed: 12 August 2023)
[120] Таран, Ю. И., Фролов, Л. А. (1984). 'Новая Система Целеуказания', *Вестник Бронетанковой Техники* 1984, сборник 5. p. 16.

TKN-3

The TKN-3 was a combined day-night periscope, intended to serve as the commander's primary means of battlefield surveillance during combat. With it, a T-72 commander could search for targets, automatically cue the gunner to targets, and perform fire control tasks in both day and night. The reticle followed the standard layout for Soviet field binoculars, supplemented by an additional stadiametric rangefinding scale. The day channel had a 5x magnification and high optical clarity, giving a nominal recognition range of 3 km on tank targets[121]. In real combat conditions, the actual recognition distance was almost invariably much shorter[122]. Coated glass was used in both the day and night optics to maximize light transmittance[123].

The total width of the reticle was 80 mils, and its height was 24 mils. The scale was divided by increments of 4 mils, the same as in the gunner's day sight. With it, the commander could measure out engagement sectors, adjust the gunner's fire, and measure the range to targets with known dimensions. If the target was a tank, the stadia rangefinder scale beneath the reticle allowed a quicker

Figure 21: TKN-3 reticle with its reticle and a stadia rangefinder scale below it. (*Soviet Ministry of Defense*)

---

[121] Tank target standing openly in a grassy field, rated based on the optical resolution criteria of identifying the presence of a gun barrel and the direction in which it was pointing.

[122] A study showed that the probability of detecting a camouflaged M60A1 under ideal visibility conditions at 1.3-1.4 km was 50%, diminishing to ~0% at 3 km. See: Raisbeck, G. et al. (1981). *Design Goals for Future Camouflage Systems*. Fort Belvoir: Research and Development Command. p. II-27.

[123] The daylight lenses appear purple because the coating maximized light transmittance in the center of the visual spectrum, reflecting only some red and blue light. When looking through the optic, the image appeared yellowish-green. The night vision lenses were coated for near-infrared light, touching slightly on visible red light. The reflections were entirely blue-green light, making the lenses appear bluish-green.

and more reliable estimate to be made. The reticle was an etching on a glass cell installed in front of the eyepiece lens group, so day or night, the commander saw the same reticle. There was no reticle illumination.

The day and night modes were divided by a rotating mirror. The daylight optic was a binocular system, essentially functioning as pair of field binoculars that could be used from inside the tank. The night optic was a single-tube binocular system. It had a single large objective lens between the binoculars of the day optic, leading into a two-stage cascading image converter tube. A beamsplitter divided the image from the image converter tube into two, which were reflected into the two eyepieces. Turning the internal mirror from the day position to the night position automatically powered up the image intensifier tube[124].

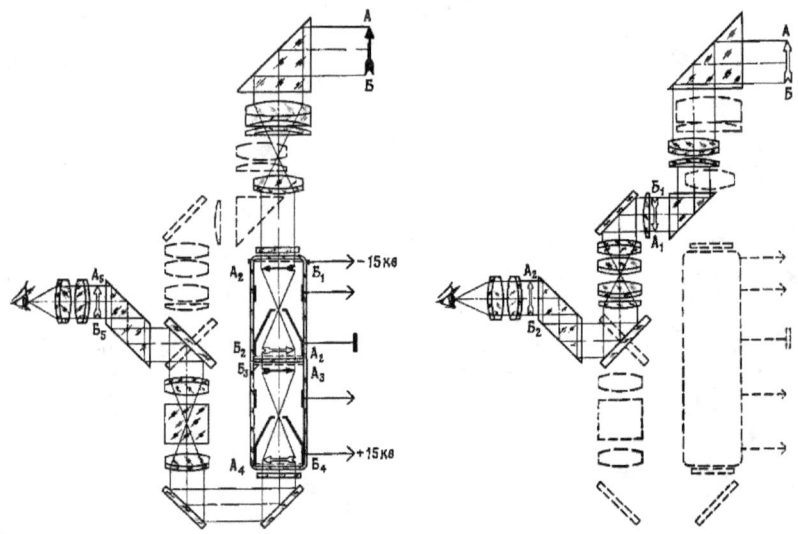

Figure 22: TKN-3 optical paths in its night (left) and day (right) channels. (*Soviet Ministry of Defense*)

Previously, the T-54/55 had been issued with separate day (TPKU-2) and night (TKN-1) periscopes. These had to be exchanged in anticipation of dawn or dusk, which was an inconvenience at best. It had not been possible to combine day and night functions into a single device because previous periscopes had been designed based on the old standard MK-4 *Gundlach* periscope mount, which was only just wide enough for a set of binocular optics.

---

[124] Министерство Обороны СССР (1972). *Танковые Приборы Ночного Видения: Техническое описание и инструкция по эксплуатации*. Москва: Воениздат. p. 30.

To accommodate the two telescope tubes (25 mm in diameter each) for the daylight binoculars, and a large objective lens group (50 mm in diameter) for the night vision optic, a new periscope trunnion was designed for the TKN-2, briefly used in the T-62 until it was replaced by the TKN-3, which inherited the same periscope trunnion. Compared to the *Gundlach* mount, the trunnion had been enlarged in radius from 33 mm to 43 mm and widened from 105 mm to 136 mm[125].

Figure 23: TKN-3 daylight lenses (sides) appear purple and the night vision lens (center) appears blue due to their anti-reflection coatings. (*M. Sylvain, 2024*)

With an objective lens twice larger in diameter than the day binoculars, the amount of light collected by the night monocular increased fourfold. But effective night vision was provided by a U-31B two-stage cascading image converter tube[126]. Structurally, a cascading tube device could be likened to two single-stage image converters laid end-to-end, although this is a slight oversimplification[127]. The U-31B had an S-1 input photocathode[128] followed by

---

[125] Measured on TKN-3.
[126] The U-31B was previously used in the TKN-2B, the predecessor to the TKN-3, found in the T-62 medium tank from 1961 to 1964.
[127] Зайдель, И. Н., Куренков, Г. И. (1970). *Электронно-оптические преобразователи*. Москва: Издательство «Советское радио». pp. 5-8. Also see: Криксунов Л. З. (1975). *Приборы ночного видения*. Київ: «Техніка». pp. 29-30.
[128] Based on its blue tint. Photocathode classification follows an international standard. See:

a cesium-antimony photomultiplier photocathode. An iris diaphragm was placed between the objective lens group and the image converter to manually control the lens exposure of the image converter tube.

The U-31B tube produced a much brighter image than a single-stage image converter[129], enough to afford it a passive night vision capability surpassing single-stage Gen 1 devices, albeit with significant noise and edge distortion (fisheye lens effect). Rather than passive night vision, however, the purpose of this tube was to enhance the viewing range with infrared illumination.

With the aid of the OU-3GK infrared spotlight, the TKN-3 offered a recognition range of 400 m on a profile silhouette of a tank-type target, reduced to 300 m on a head-on tank silhouette. This was a significant improvement over the TKN-1, which had a simple single-stage image converter paired to the same OU-3GK spotlight. The OU-3GK contained a PZh27-110 incandescent bulb with a rated power of 110 W, producing a gross luminous intensity of 2.5×10^5 Cd without its infrared filter[130].

The detection range could be higher in practice, as when optical devices of any kind are caught facing in the general direction of a spotlight, the incoming infrared light is strongly reflected toward the observer as glare. Aided by glare, the probability of detecting a target was higher than detecting the same target by its outline under both active and passive observation, and drastically quicker[131]. If an observer could be expected to have an 80% probability of detecting a tank target in 50 seconds with passive night vision, the same probability of detection could be expected in 20 seconds by searching for glare with active illumination. The critical shortcoming of this, of course, was that active illumination revealed the observer almost immediately.

Аксененко, М. Д., Бараночников, М. Л. (1987). *Приемники Оптического Излучения*. Москва: "Радио и Связь". p. 17.
[129] У-31Б - ОД0.335.442ТУ. Available at: http://www.promvpk.ru/Catalog/Product/51a269276d0fad80e402f044 (Accessed 20 August 2023).
In the early 1970's, the availability of multialkali (S-20) photocathodes with a high response in the visible spectrum allowed the S-1 input photocathode to finally be replaced. The NSPU universal small arms night sight (1971) was one of the first Soviet sights to have a multialkali cascaded image intensifier tube. See: Коронин, Ю.Н., Малинин, В.В., Попов, Г.Н. (2008). 'История создания и унификации стрелковых прицелов ночного видения на примере изделий ЦКБ «Точприбор»', *Интерэкспо Гео-Сибирь*.
[130] Павлов М. В., Павлов, И. В. (2021). *Отечественные Бронированные Машины 1945-1965 гг. - Часть I: Легкие, средние и тяжелые танки*. Кемерово: ООО "Принт". p. 83. This was four times lower than the L-2AG spotlight for the gunner's TPN-1 night sight.
[131] Гуменюк, Г. А. (1986). 'Возможность обнаружения противотанковых средств по бликам приборов', *Вестник Бронетанковой Техники* 1986, сборник 5. p. 12.

Table 10: TKN-3 technical data[a]

Magni-fication	Field of View (°)	Optical Resolution[b] (′)	Effective FoV[c] (°)	Exit Pupil Dia. (mm)	Eye Relief (mm)	Device Elevation Range[d]
5x	10	10	~4.5	5	22	+12 to -8

[a] Павлов М. В., Павлов И. В. (2021). *Отечественные Бронированные Машины 1945-1965 гг. - Часть I: Легкие, средние и тяжелые танки.* Кемерово: ООО "Принт". p. 83.

[b] Measured in arc seconds. Евсикова, Л. Г., Досужев, А. В. (1973). 'Дальность узнавания объектов БТТ', Вестник Бронетанковой Техники 1973, сборник 2. p. 34.

[c] Based on information that the rated optical resolution was applicable only to 20% of the full field of view in the center, due to edge distortion from the cascading tube. See: Спасский, Н. (ed.) (2005). *Оптико-электронные системы и лазерная техника: Энциклопедия XXI век – Оружие И Технологии России.* Москва: "Оружие и Технологии". p. 57.

[d] Pravilo Tenk M-84 i T-72: Prvi Deo (1988). Beograd: Vojna štamparija. p. 255.
The maximum elevation range permitted by the trunnions was 26°, cupola mount restricted this to 20°. This was independently verified on a functioning TKN-3 in a T-72M.

Compared to the TKN-1 in the T-54/55 series, the TKN-3 was much more useful for navigation at night under infrared illumination thanks to its extended range, but in broader terms, the true advantage of the TKN-3 was that, despite not giving a combat-effective viewing distance without illumination, it gave a better view than the naked eye, and at the very least it did not hinder the commander's vision under starlight conditions.

In any case, the viewing range was too short for the TKN-3 to be used as an independent observation device in combat, and the lack of automatic flash protection blinded the image converter with the tank's own muzzle flash. These limitations forced the commander into a more supervisory role in night combat, effectively restricting him to detecting enemy light sources, assisting the driver, and observing through the beam of the gunner's spotlight.

Broadly speaking, with the commander and gunner searching along the same frontage[132], the commander's cupola and his set of vision devices were more effective than the gunner's single periscope and his sight at ranges of up to 1,400 m. However, the value of the TKN-3 itself as a target-finding optic was significantly inferior to a high-magnification gunner's sight at practical ranges, reaching parity in target detection speed only at short range (500-750 m)[133]. Target cueing greatly increased the speed of detecting challenging targets at combat distances[134].

---

[132] Frontage of 250 meters at arbitrary range.
[133] Веселов, В. Б., Диков, С. А., Харитонов, И. С. (1982) 'Анализ Поисковых Возможностей Танков', *Вопросы Оборонной Техники*, 20(105). pp. 16-18.
[134] Hidden tank target in turret defilade from 500-2,500 m.

Radio and Intercom System

The communications system of the T-72 was standard for armored vehicles in the Soviet ground forces. There was a radio transceiver, its power supply unit, and an intercom system corresponding to the radio. The intercom system had one control box for each crew member, through which the radio, intercom and tank alarm systems were routed to the headset of the crew helmets and their laryngophones (throat microphones).

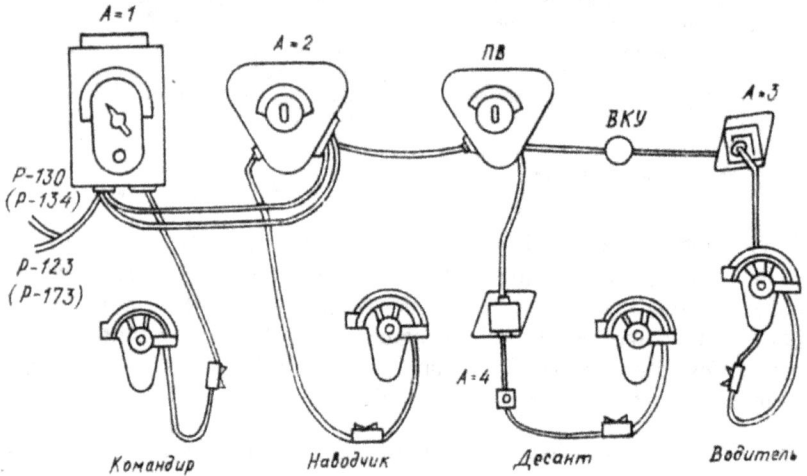

Figure 24: Intercom connections. (V. S. Starovoytov, 1990)

The intercom allowed half-duplex or broadcast communications between all three crew members. The radio was connected to the intercom system through the same switching system, so all crew members could listen at the same time, and one crew member at a time could broadcast on the radio or intercom. The commander's A-1 radio-intercom switchbox was the master switchbox through which the gunner's A-2 and driver's A-3 switchboxes were connected to the radio and intercom. The A-2 switchbox allowed the gunner to connect to the intercom, the R-123M radio, or the R-130M long-range radio [135]. The A-3 and A-4 switchboxes only connected to the intercom. Normally, only the tank commander used the radio, but the gunner could take over when needed. The standard 1.2-meter connecting cable allowed the commander and gunner to remain connected through their headset umbilical cables while standing up through their hatches.

---

[135] Устройство Переговорное Р-124. Техническое описание и инструкция по эксплуатации.

Additionally, a tank rider could plug a headset into an A-4 intercom socket located behind the commander's cupola to talk to the crew. An additional tanker helmet was included in the standard equipment of the tank for this purpose. It would be given to the squad leader and so that he could coordinate with the tank commander on field maneuvers, the timing of a dismount, and the direction of enemy fire. Once the tank riders dismounted, there was no way to communicate further without opening the commander's hatch, since the T-72 had no infantry telephone on the hull.

The radio was installed on the turret ring next to the coaxial machine gun and angled towards the commander, so to operate the radio, the commander simply reached forward. The ease of access to the radio was characteristic of Soviet armored vehicle design. In Western tanks with the radio fitted in the turret bustle, the commander had little direct interaction with the radio and instead used a separate control panel which only permitted him to switch between channels.

T-72K and T-72AK command tanks, used on the battalion and regimental levels, had an additional R-130M radio receiver to permit long range simplex communication with the regiment or division headquarters. It was installed behind the commander's seat on the turret shelf. This involved removing the ammunition racks on the autoloader carousel cover behind the seat, along with removing their propellant charges from the removal of the front right fuel tank-rack for the AB-1 gasoline generator.

The R-130M allowed the commander to continuously listen for orders while on the move and transmit if necessary, and if the battalion or regiment commander could not set up a field command post, the T-72K had a ten-meter mast antenna that could be erected to set up long-range communications, becoming a temporary self-contained command post.

The T-72 began its service in the Soviet Army with the *Magnolia* 2nd generation radio system with the R-123M radio station, a 1972 modification of the R-123 which had been the standard VHF (very high frequency) vehicle radio of the Soviet ground forces since 1961. It was paired with the R-124 intercom system. As there was only one transceiver, the radio was limited to half-duplex communication on one frequency at a time. The only difference of the R-123M from the R-123 was the removal of the voice-activated transmission switch, so users had to squeeze the chest switch box to transmit (push to talk).

In 1984, this system was replaced by the *Abzats* 3rd generation radio system, comprised of R-173 transceiver with the R-173P receiver, and the R-174 intercom system.

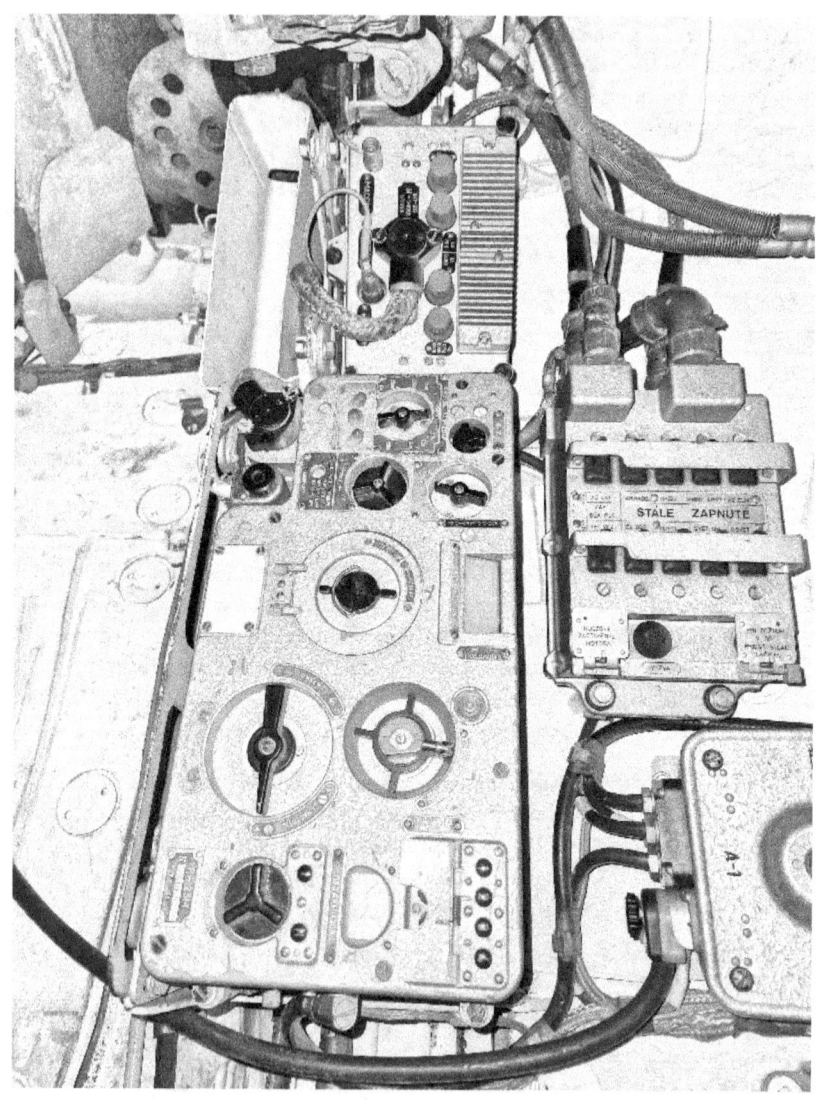

Figure 25: R-123M in T-72M. Above it is the right turret circuit breaker and the commander's A-1 intercom-radio switchbox. (*S. Stauber, 2024*)

R-123M

The R-123 was a frequency-modulated (FM) medium range radio standard for all Soviet armored fighting vehicles. It was closely comparable in capability to contemporary tank radios of the 1960's and 1970's. The previous standard AFV radio in the Soviet Army had been the R-113, which had been limited to a narrow frequency range of 20-22.375 MHz [136]. The R-123 had a wider frequency range of between 20-51.5 MHz, divided into 1,261 channels with a frequency spacing of 25 kHz. It could be tuned manually, or the commander could use its motorized automatic tuner to quickly switch between four pre-programmed channels[137]. It took several seconds for the mechanism to switch between channels depending on how far apart their frequencies were. The longest possible switching time was 15 seconds. Once a channel was set, it would stay tuned to it through an automatic frequency control circuit.

The R-123 had a transmission power of 20 W[138], higher than the R-113 (16 W). More power was needed to overcome the higher signal attenuation experienced in the higher frequencies and to improve the resistance to interference, since VHF radios operate through line-of-sight transmission and were strongly affected by forests. As such, the range of the R-123 increased only slightly compared to the lower-frequency R-113.

The higher frequency range of the R-123 allowed it to serve as a single standardized vehicle radio for all parts of a combined arms unit, and for all branches to intercommunicate with mounted and dismounted elements of other branches. The low frequencies were allocated to the tank network (20-26 MHz), which was shared with the battalion command network and overlapped with the army air defense network (21.5-28.5 MHz). The higher frequencies could link with the artillery forces (28-36.5 MHz), or the infantry (36-46.1 MHz) on their R-105M portable radios. Expanding the operating frequency range added an element of jamming protection as well because the tank network strongly overlapped with the 20-27.9 MHz band allocated to the U.S. Army armored forces.

A spark gap in the 4-meter whip antenna prevented an electromagnetic pulse from damaging the transceiver directly. A direct nuclear strike on a tank unit would still disable radio communications, however, since the antenna was found

---

[136] Given a narrow frequency range, choosing a lower range was preferred as higher frequencies were more strongly attenuated by the environment, reducing communications range.
[137] Министерство обороны СССР (1983). *Радиостанция Р-123М - Техническое описание и инструкция по эксплуатации*. Москва: Воениздат.
[138] For comparison, the West German standard SEM-25 vehicle radio had a transmission power of 15 W, and the U.S. Army's standard AN/VRC-12 vehicle radio operated at 30 W.

to be consistently damaged through its mount at a fairly modest blast overpressure of 0.1 MPa[139].

The R-123M was a large device but fairly compact for a tank radio. The transceiver and its BP-26 power supply unit were designed to occupy the standard mounting frames for the R-113 (430 x 240 x 220 mm) and its power supply unit (210 × 239 × 222 mm) so that no modifications were needed to upgrade an older tank to the R-123. The transceiver occupied a volume of 24.6 liters. For comparison, the AN/VRC-12 vehicle radio occupied a volume of 26.5 liters[140].

The communication range while driving off-road at a speed of up to 40 km/h was 20 km or more with the noise suppressor turned off, and no less than 13 km with the noise suppressor turned on. The criteria was 90% speech intelligibility at a tank speed of 40 km/h, under a background noise level of 120 dB. In Soviet testing, the radio in the M60A1 had a range of 25-30 km when rated under the same criteria[141].

R-173 and R-173P

Beginning in 1984, the R-123M was replaced by the R-173 digital radio transceiver, which originally entered service in 1979. Thanks to the use of integrated circuits in the transceiver, the volume of the device was drastically smaller than the previous radio transceivers, making it possible to integrate the power supply unit into the same casing as the transceiver, occupying the block to the right of the radio. Thus, the space previously allocated to the power supply unit was freed up, allowing an additional R-173P receiver to be fitted. The control interface was simplified by replacing dials and knobs with an electronic keypad and a digital display.

The R-173 system could be used to transmit and receive voice signals for normal communications, or transmit and receive encrypted analogue and digital data. The radio had a frequency range of 30-75.999 MHz with a frequency increment of 1 kHz, providing 46,000 operating frequencies. It could be programmed to receive and broadcast in 10 programmable preset frequencies, with a frequency switching time of 3 seconds.

The decision to further expand the frequency range was to grant more flexibility in network allocation, and the specific choice of the upper and lower

---

[139] Авдеев, В. Н., Бондаренко, В. И., Губченко, И. Н., Кузьмин, В. С. (1985). 'Танковая Взрывоустойчивая Штыревая Антенна', *Вестник бронетанковой техники* 1985, сборник 4. p. 43.
[140] Авдеев, В. Н. et al. (1976). 'Средства Связи', *Вопросы Оборонной Техники*, 20(67). p. 40.
[141] Авдеев, В. Н. et al. p. 43.

ends of this frequency range was based on matching the radio systems used in the West[142], namely the AN/VRC-12 system (30-75.95 MHz) and the British Clansman system (30-75.975 MHz). This gave the same inherent resistance to indiscriminate jamming that frequency overlapping had given in the R-123. Transmission power was raised to 30 W to again compensate for the stronger signal attenuation at higher frequencies. The R-173 also had an extremely low-power "whisper" mode[143] for more stealthy communication. Due to the higher frequency range, the R-173 came with a shorter 3-meter antenna.

The R-173 could communicate on two channels simultaneously through the single whip antenna on the tank with one channel in the 60.0-75.999 MHz range and the other in the 30.0-51.999 MHz range[144] through the use of a frequency filter. The commander could transmit or receive on the R-173 on one frequency (half-duplex) while also receiving uninterrupted voice communications on another frequency (simplex) via the R-173P.

This capability was previously only available on battalion or regimental level command tanks, which used the standard tank radio to issue orders to subordinate unit leaders while the R-130 long-range radio served as a continuous link to the headquarters of the next higher level, through which the battalion or regiment commander could receive orders. By providing a transceiver-receiver set to every tank, it became possible for Soviet tank platoons and companies to set up more elaborate tactical networks. This capability that could also be found on West German tanks equipped with the SEM-25 radio like the Leopard 1 and early Leopard 2 models, or in M60 tanks belonging to company commanders[145].

R-130M

In a T-72K, T-72AK and T-72BK command tank, there was an additional R-130M radio with its BP-260 power supply unit, a mast antenna, a TNA-3 inertial navigation device, and an AB-1 gasoline generator. The VKU-330-4 turret slip ring was replaced by the VKU-1[146].

---

[142] Калашников, Г. Г., Костылев, В. А. (1984). 'Развитие Танковой Радиосвязи в Послевоенный Период', *Вестник бронетанковой техники* 1985, сборник 4. pp. 48-51.
[143] Broadcast at 1 Watt, which is much less power than most commercial walkie talkies.
[144] Голяшов, А. В., Шамин, Б. Ф. (1983). 'Новые Танковые Средства Связи', *Вестник Бронетанковой Техники* 1983, сборник 5. pp. 37-38.
[145] Company-level M60 command tanks would have an additional R-442/VRC receiver to permit listening on one frequency (simplex) while communicating in half-duplex mode on another using the AN/VRC-12 transceiver.
[146] Ministerium für Nationale Verteidigung (1979). *Mittlerer Panzer T72. Taktisch-technische Angaben,*

The R-130M was a large device (320 x 325 x 500 mm), installed behind the commander's seat. The R-130M was an amplitude-modulated (AM) radio with single-sideband modulation. The frequency range of the R-130M was 1.50-10.99 MHZ, divided into ten sub-ranges with a total of 950 operating frequencies. At the lowest sub-range (1.50-1.99 MHz), the transmission power was 6.4-10 W, and in the higher sub-ranges, 10-40 W. The R-130M could share the tank's four-meter whip antenna with the R-123M, use a small dipole antenna, or an enormous ten-meter mast antenna. It also supported telephone communications through a two-kilometer telephone cable, which could be plugged to the A-4 receptable behind the commander's cupola.

When working with the mast antenna, the AB-1 petrol auxiliary power unit (APU) in the tank was used to supply power to the radios instead of the engine. It could also be used to recharge the batteries when needed.

Turm, Panzerbewaffnung und Panzerspezialausrüstung – Beschreibung und Nutzung. A 051/1/106. Berlin: Militärverlag der Deutschen Demokratischen Republik. pp. 250-261.

Figure 26: T-72K with deployed mast antenna. (*Soviet Ministry of Defense*)

Gunner's Station

The gunner's station took up the left half of the turret and was much more cramped than the commander's station, in large part because of the day sight. The sight completely filled the space between the gunner's seat and the front wall of the turret. The fixed recoil guard to the right of the day sight had a handlebar for the gunner to brace himself against, and the guard was marked with reference angles of 0°, 2.5° and 10° to manually position the gun. The 0° position was mainly for boresighting, the "loading angle" position (2.5°) set the gun at the correct angle to use the powered rammer to load the gun. The 10° position was to insert a 2Kh35 14.5 mm subcaliber gunnery training device.

He was responsible for controlling and maintaining all of the equipment in the tank related to its firepower, including the autoloader, stabilizer, its guns, and the sighting devices and their associated instruments. The left turret circuit breaker allowed the gunner to turn on the night sight, powered controls, electric firing circuits for the gun and coaxial machine gun, and the FVU. Beginning with the T-72A in 1979, the gunner was also responsible for aiming the smoke grenade launchers to lay a smokescreen at the commander's instructions.

In combat, the gunner's tasks were much less varied than those of the commander and driver. His basic task was simple: finding and engaging targets on the commander's instructions or under his own initiative when assigned a sector of responsibility. All of the relevant controls were amalgamated with the day sight, including the autoloader loading control panel, and the control handles to direct the turret.

Figure 27: Ammunition counter. (*Soviet Ministry of Defense*)

The autoloader loading control panel allowed the gunner to select one of the three possible ammunition types on a rotary switch and load it, or switch the autoloader to the "restocking" mode. This was the full extent of the gunner's interaction with the autoloader. Without taking his eyes off the sight, the gunner could leave his hands on the control handles and flick the "load" switch with his left thumb [147], or reach up to set a different ammunition type on a selector knob.

---

[147] The first T-72 *Ural* tanks had a less convenient Load button separate from the autoloader control panel attached to the recoil guard next to the sight.

Perhaps the most unexpectedly familiar piece of technology in the gunner's instrument cluster was the autoloader ammunition counter. It was a standard M1360 industrial milliammeter with a special backplate[148] displaying a scale for up to eleven rounds of any type in the autoloader carousel. The counter displayed the quantity of each ammunition type according to the setting on the ammunition selector knob, and it could even count the number of empty cassettes in the carousel when the "restocking" setting was selected on the knob. This was all determined by the position of the twenty-two recorder sliders in the autoloader's electromechanical memory unit, each slider matched to a cassette in the autoloader carousel and containing one resistor.

As the selector knob was turned, the counter was switched between the circuits for each ammunition type in the memory unit, each circuit passing through the sliders in series and thus acquiring a resistance proportionate to the number of rounds. The amperage arriving at the counter thus denoted the number of rounds of the selected type.

This type of electromechanical engineering characterized the T-72's automation design as a whole. Compared to the T-54/55 and T-62, the complexity of the fire control instruments rose drastically, but gunnery in a T-72 was still essentially the same. The automation of the range setting process was the single most important factor in reducing engagement times.

Indirect fire out to 10,000 m was possible on level ground, and firing out to an absolute maximum range of 12,200 m was possible by firing from a ramp. The process was entirely manual. The elevation angle was set using an aiming quadrant[149] affixed onto the gunner's side of the recoil guard, and the target azimuth offset was set using the turret azimuth clock.

The targeting process was handled by fire controllers, who communicated the elevation and azimuth values to individual tanks through radio. To aim in azimuth, a target azimuth offset was calculated relative to an aiming reference, usually a tall and easily recognizable landmark at a great distance from the firing position. Gunners laid the reticle on the aiming reference, noting the value on the turret azimuth clock, and then traversed the turret to the left or right by the offset figure. Then, the gun was superelevated by dialing in the angle in the aiming quadrant, and then elevating the gun until the quadrant was level.

---

[148] The name M1360-7 visible in **Fig. 27** denotes the backplate type. Another M1360 with a special scale (M1360-11) was used for the fuel gauge on the driver's instrument panel.
[149] A device to measure the elevation angle of the gun relative to the earth using a spirit level.

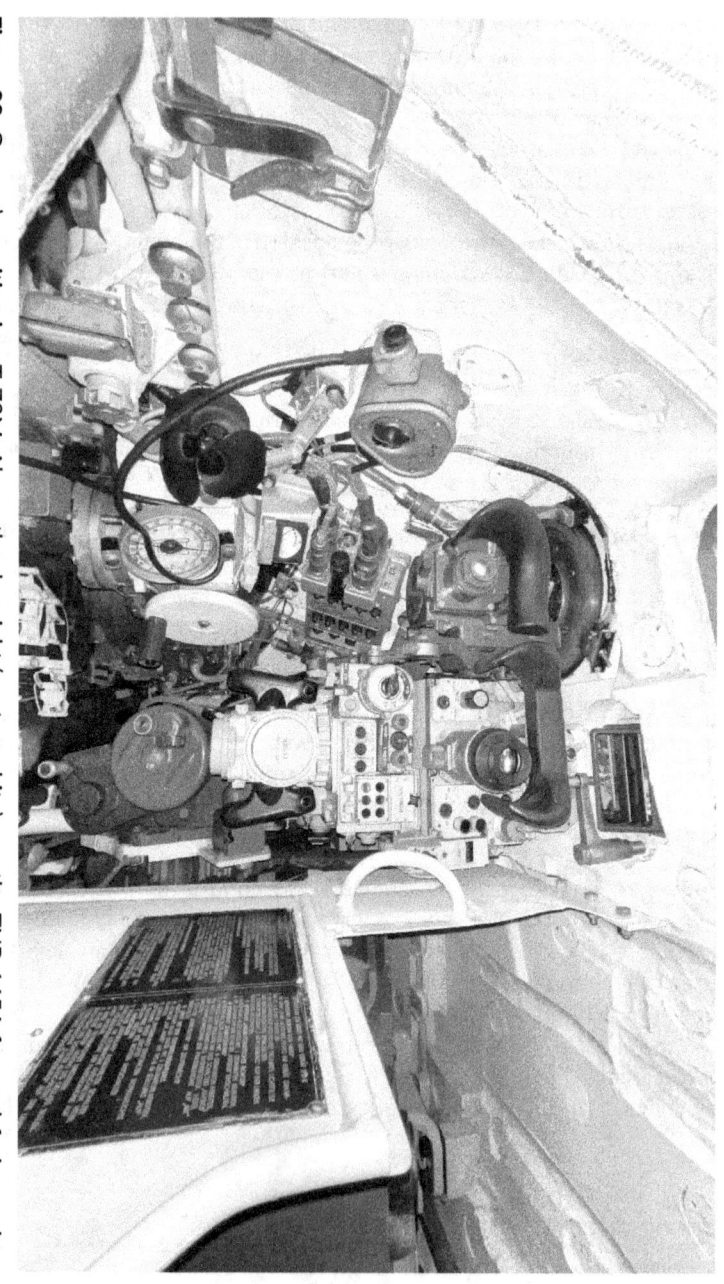

Figure 28: Gunner's position in a T-72M. Above the day sight (center, white) was the TNP-165A forward-facing periscope, and below it was the manual elevation handwheel. Below the night sight (top left) were the left turret circuit breaker, the ammunition counter, and manual traverse handwheel. The gunner's A-2 radio-intercom switchbox is the furthest to the left. (*P. Kefallinos, 2024*)

Sighting Systems

Day Sight

When the T-72 first entered service in 1973, its day sight was the TPD-2-49. It was a combined rangefinder-sight with independent vertical stabilization of the sight, and dependent stabilization of the optical rangefinder through the gun. This system allowed the gunner to measure the range to a target on the move, and fire on the move with high accuracy.

The TPD-2 was the most modern sighting system available in the USSR at the time, though the roots of its design stretch back into the 1950's. It was a refinement of the TPD-43B sight used in the T-64, which was built off of the TPD, originally made for the T-10 heavy tank (Object 267)[150]. The preliminary details of its structural design were worked out in 1955 in the form of a mock-up for the T-10's turret. The T-10 series would go on to receive the independently stabilized TPS-1 and T2S sights in service, but all work on heavy tanks in the USSR was cancelled before the TPD was ready.

The TPD evolved into the TPD-43B, and from there, the TPD-2 was born. It was the product of a long and persistent effort to improve ranging effectiveness, reliability, and user-friendliness. It contained 60 optical components and 800 mechanical components, including structural parts and fasteners, and production of the TPD-2 was around 7 times more labor-intensive than the TSh2 series telescopic sights in the T-55 and T-62.

Soon after its introduction, a laser rangefinder was added to the TPD-2-49 evolved it into the TPD-K1. The T-72 began receiving the TPD-K1 in 1978, and it became standard for the T-72A the following year. The TPD-K1 retained 78% parts commonality with the TPD-2-49, but this was, to a large extent, to its own detriment because of the number of outdated design elements carried over. Still a product of the mid-1950's at its heart, the TPD-K1 went on to serve as the standard for the T-72 until the end of the Cold War. Incremental upgrades were introduced throughout the 1980's, but none were meaningful enough to distract from the fundamental age of the sight's design.

Both the TPD-2 and TPD-K1 were virtually identical apart from their rangefinding devices, so, for all intents and purposes, they can be described as members of the same basic TPD family.

---

[150] Павлов М. В., Павлов, И. В. (2021). *Отечественные Бронированные Машины 1945-1965 гг. - Часть I: Легкие, средние и тяжелые танки*. Кемерово: ООО "Принт". p. 795

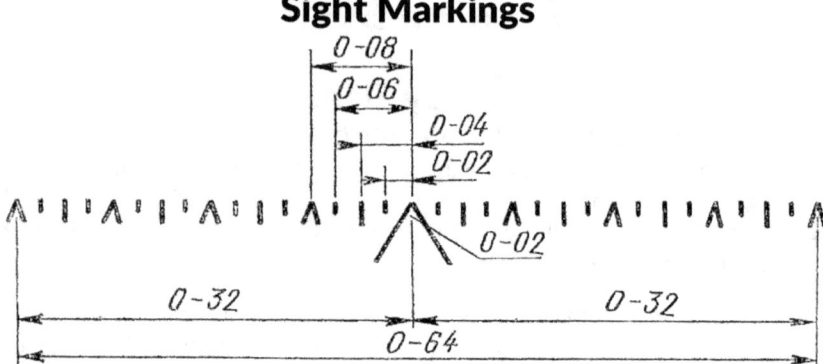

Figure 29: Standard Soviet postwar artillery sight scale design. (*Soviet Ministry of Defense*)

The term "thousandth" (тысячная) was used in the Soviet military to refer to a milliradian, while the scientific milliradian was denoted by its proper unit (mrad). Under the official military notation, 2 mils were written as 0-02[151].

The design of the reticle was standardized between tanks and artillery direct fire sights. The smallest increment on the scale was 2 milliradians (mils), each mil being equivalent to an angular width of approximately 0.057 degrees, or 3.43 minutes of arc (MOA)[152]. The smallest horizontal increment on the lateral scale was 2 mils, and the height between the tip of the center chevron to the vertical line beneath it was also 2 mils.

Starting from a working assumption on the average dimensions of a NATO tank, like a hull width of 3.4 meters, and a total height of 2.7 meters, the gunner could perform stadiametric rangefinding. This was simply to measure the range to a target by comparing its apparent dimensions to the known size of the reticle elements. An object with a height of 2 meters has an angular height of 2 mils from a distance of 1,000 m, so in combat, if a gunner lays the center chevron over a 2.7-meter target and it fits, verifying that it has an apparent height of 2 mils, the distance to the target is measured to be 1,300-1,400 m [**Fig. 30**, 7]. Stadia rangefinding with the standard reticle typically only gave a general idea of the target distance. The lateral scale could be used to measure the side profile

---

[151] Селивохин, В. М. (1962). *Танк*. Москва: Воениздат. p. 141.
[152] The standard military milliradian in the USSR was 1/6000 of a circle. The NATO mil was 1/6400 of a circle, and the true SI milliradian was 1/6283 of a circle.

of a target to greater precision, but its primary purpose was for windage adjustments and to lead moving targets.

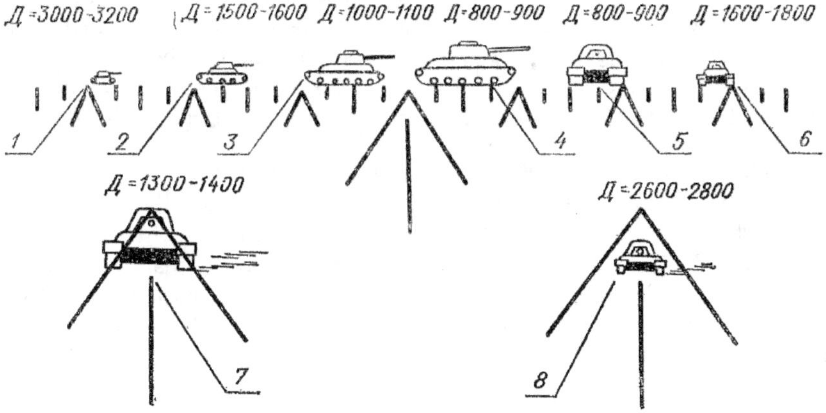

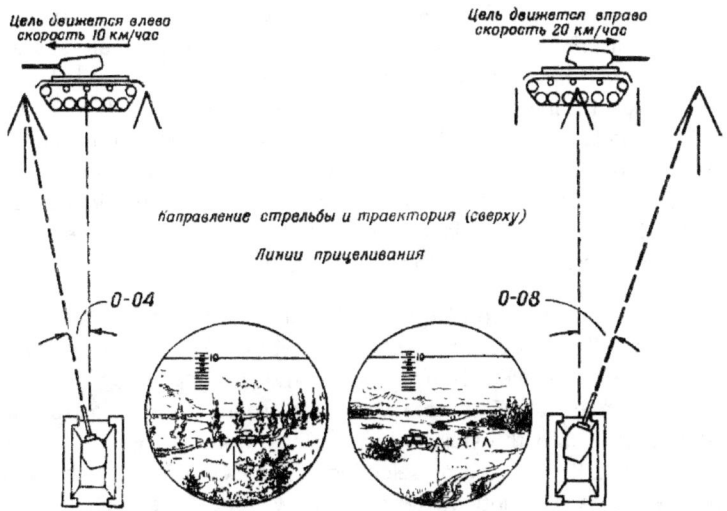

Figure 30: Stadiametric rangefinding based on assumed target dimensions. (*Soviet Ministry of Defense*)

Figure 31: Leading a moving target on lateral scale. (*Soviet Ministry of Defense*)

Without automatic leading or wind correction, these adjustments were performed manually based on simple gunnery rules, such as to lead a moving target at 1 km by 4 mils for each 10 km/h of its speed [**Fig. 31**], adjusting accordingly for different ammunition types.

The lateral scale was also used to apply an aiming offset to correct for the canting (roll) of the tank on an uneven surface. Automatic cant correction was absent in the TPD-2-49 and TPD-K1. The difficulty in manually estimating the appropriate offset mainly affected the firing accuracy with HEAT and HE-Frag ammunition. With high-velocity 125 mm subcaliber rounds, the gunner could simply aim at the edge of the target opposite the cant direction at 1 km or less, and at 1.5-2.0 km, it was enough to offset the reticle by a target half-width.

In general, the extreme velocity of 125 mm APFSDS ammunition could make up for gunnery errors of all types. The point-blank range, that is, the range where the height of the shot's trajectory does not exceed the height of the target for 125 mm APFSDS was 2,120 m. This was more than double that of the APDS from the Chieftain's 120 mm L11 gun, which was found to have a point-blank range of 982 m on a 2.7-meter target in Soviet tests[153]. The speed and flat trajectory of 125 mm APFSDS ammunition was forgiving to range estimation errors, cant error, and target lead error.

TPD-2-49

The reticle in the TPD-2-49 essentially corresponded to a standard Soviet sight reticle, differing only in that the smallest increment in its lateral scale was 1 mil instead of 2 mils. The reticle also featured a graduated range ladder to aim the PKT coaxial machine gun out to a maximum range of 1,800 m, and to aim HE-FRAG rounds out to 5,000 m, with rather coarse 200-meter increments. These marks were added to extend the direct fire range beyond the 4,000-meter limit of the ballistic computer. With HE-Frag selected, the range in the sight had to be set to 4,000 m to lower the reticle. Then, the gunner manually superelevated the gun to aim with the range ladder up to the '50' mark.

A range wheel at the top of the viewfinder [**Fig. 32**] rotated about a fixed needle to show the range to a target as the reticle dropped. The lower ring of the range wheel displayed the true range, and the upper ring was a special machine gun range scale. This scale was a workaround solution to providing the coaxial machine gun with its own range scale without adding a fourth ballistic cam to the sight. It was used by selecting HEAT/COAX in the sight, and then turning the range wheel until the target range was set in the machine gun range scale instead of the range readout scale. Because the HEAT ballistic cam was calculated for the trajectory of HEAT rounds, the machine gun range scale was drawn with non-linear increments unlike the true range scale, so that it could

---

[153] Блинов, В. П., Личковах, В. А., Николахин, В. М. (1983). 'Испытания танковой пушки', *Вопросы оборонной техники*, 5(111). p. 23.

reflect the ballistic trajectory of the coaxial machine gun instead of HEAT ammunition[154]. Alternatively, the gunner could also choose to use the range ladder, which required the range in the sight to be reset to zero. The choice of aiming method depended on whichever was more expedient at a given moment.

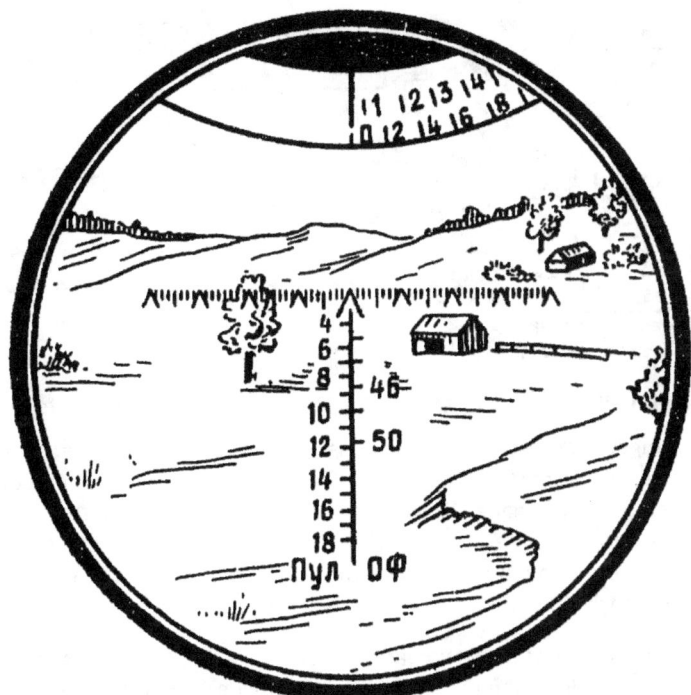

Figure 32: TPD-2 viewfinder. (*Adapted from manual, Soviet Ministry of Defense*)

The TPD-2-49 had a conventional optical rangefinder with two parallel forward-facing eyes, the only distinction being that the left eye was contained in the sight rather than in an extension toward the left of the turret like on the vast majority of tanks. The additional protection to the eyes of the rangefinder did, however, shorten the base width to 1,500 mm, which affected the ranging accuracy. Apart from cupola-mounted rangefinders, the TPD-2 rangefinder was one of the shortest of the Cold War, followed closely by the Swiss Panzer 68 with a 1,550 mm base width. On the Leopard 1, the base width was 1,720 mm and on the M60A1, it was 2,006 mm.

---

[154] The ballistic trajectory of HEAT rounds did not correspond to the machine gun whatsoever beyond 400 meters.

The viewing windows for both the main sight and the rangefinder eye were de-fogged by an air circulation system constantly blowing on the inner side of both windows with air taken from the crew compartment.

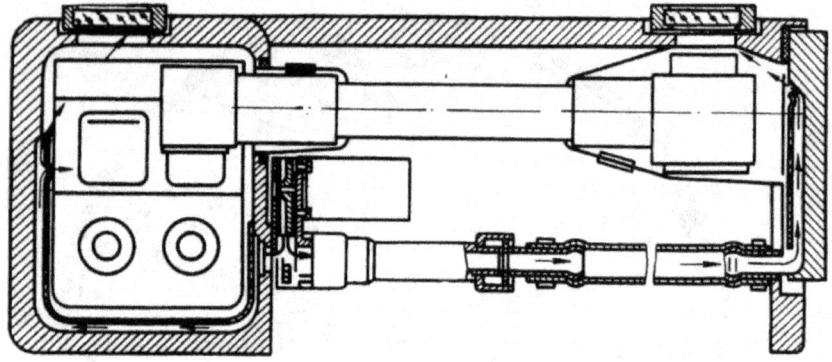

Figure 33: TPD-2-49 defogging system. (*Soviet Ministry of Defense*)

The rangefinder functioned on basic trigonometry [**Fig. 34**]. When a rangefinder has a target in its direct line of sight, a right triangle is formed, where the short side of the triangle is the base of the rangefinder and the long side is the distance to the target[155]. Knowing the base width of the rangefinder, the only other input needed to calculate the distance is one of the two unknown angles of the triangle.

In an optical system with two parallel forward-facing eyes, the angle needed to perform the calculation was conveniently provided by the fact that the image of the target in one eye appeared offset from the image in the other by parallax[156], and the larger the base width, the larger the parallax angle. Mechanically, the challenge of measuring the range to a target was therefore condensed into the much simpler task of showing the rangefinder operator the degree of parallax by superimposing or comparing the two images, and linking an image adjustment dial to a range dial. In the act of adjusting one of the two images until parallax was eliminated, the rangefinder operator measured the range to the target.

---

[155] Министерство Обороны СССР (1975). *Танк «Урал». Техническое описание и инструкция по эксплуатации (172М.ТО)*. Книга Первая. Москва: Воениздат. p. 156.
[156] Вавилова, С. И., Савостьяновой, М. В. (eds.) (1948). *Оптика в Военном Деле – Сборник Статей*. Том II. Издание Третье. Москва: Издательство Академии Наук СССР. pp. 171-175.

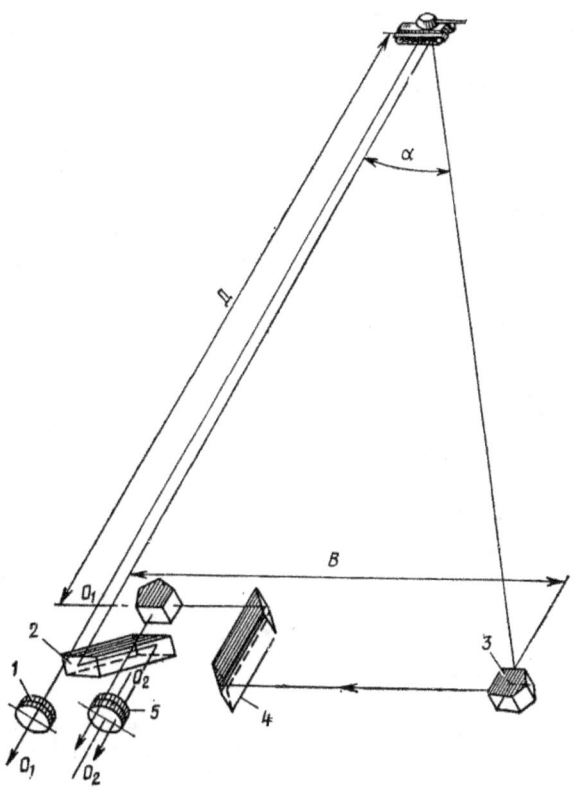

Figure 34: Basic trigonometric concept of optical rangefinding through a TPD-2-49. (*Soviet Ministry of Defense*)

In the TPD-2, the conversion of the parallax adjustment into a range measurement was accomplished with an optical compensator lens system, consisting of a convex lens behind a concave lens. When both lenses were aligned, the exit angle of the image was coaxial to the incident angle. When the convex lens was shifted laterally, the position of the image shifted at a linearly proportional rate[157]. The parallax angle was determined by the distance moved by the compensator lens to shift the displaced image until it was coincident to the fixed image. Knowing the parallax angle, the trigonometric calculation to determine the range to the target was performed mechanically by passing the

---

[157] In the naval rangefinders common in the early 20th Century, the compensating system was a pair of wedge-shaped prisms that would mutually rotate on a differential mechanism. The compensator lens was more precise than compensator prisms, but both types functioned in the same way. Relative movement between the prisms or lenses changed the thickness of glass in the light path, deflecting the rays of the image in the direction of greater thickness.

lateral sliding motion of the compensator lens to the telescope tube through the special contour of a cam [158]. Then, the rotation of the telescope tube turned the ballistic cams, so that measuring the range to a target also lowered the reticle, thus generating a ballistic solution.

As the distance to the target increased, the parallax displacement of the right image from the left image shrunk at a progressively lower rate [159], so a faraway target would appear hardly misaligned and require little to no lateral shift in the compensator lens, while the operator would intuitively expect measuring a longer range to take more turns of the range wheel. This was corrected by inverting the compensator lens sliding mechanism so that its initial position was at maximum offset, and the conversion cam pushed the lens back into alignment as the range wheel turned.

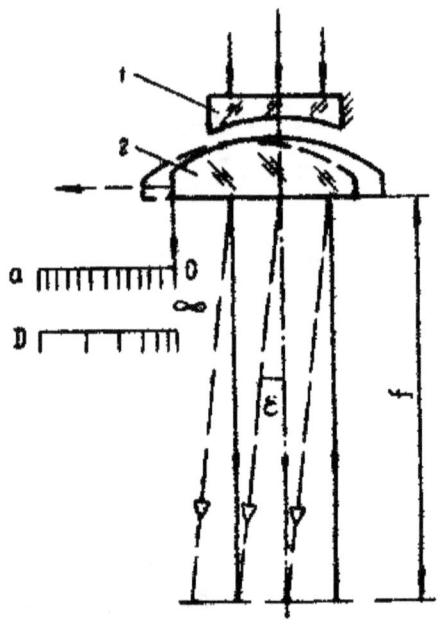

Figure 35: Light entering from above and exiting below through the TPD-2 compensator lens system, showing the change in parallax angle with lens position. (*Soviet Ministry of Defense*)

The image in the compensator lens entered the telescope system of the rangefinder eyepiece, responsible for erecting the image and giving the same 8x magnification as the main sight telescope. Inside the telescope was an image bifurcating prism. This prism split the double-image arriving from the compensator lens into separate top and bottom halves, so that instead of seeing the two images of the target superimposed in the eyepiece, one from the sight and the other from the rangefinder eye, the split image was of the target as viewed from the sight in its lower half, and the view from the rangefinder eye in its upper half. The combined image was quite narrow (2°).

---

[158] Министерство Обороны СССР (1979). *Вооружение Танка Т-72*. Москва: Военная академия бронетанковых войск. pp. 93-97.
[159] For example, when looking at extremely distant objects such as the moon, there is practically no apparent image displacement with any practlcal rangefinder base width.

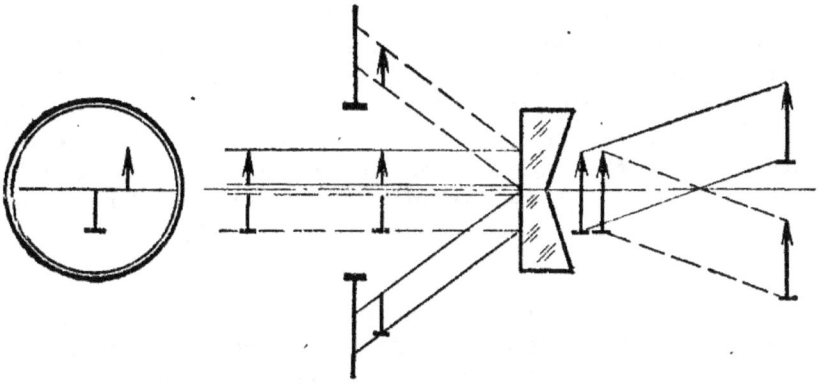

Figure 36: Image bifurcating prism. The double image enters from the right and a split image exits through the left. (*Soviet Ministry of Defense*)

By pressing the left and right thumb buttons on the control handles attached to the sight, the gunner triggered the range-setting motor to shift the compensator lens until the two halves of the target appeared to be seamlessly rejoined, thereby smoothly raising or lowering the range. If the range-setting motor malfunctioned, the gunner could manually turn the range adjustment wheel above his control handles by reaching up with his thumbs, without taking his hands off the control handles.

Figure 37: Split-image rangefinding with both eyes open, showing view with unmeasured range (left) and with measured range (right). (*Soviet Ministry of Defense*)

The split-image system was well-established in naval optical rangefinders as the most convenient way to measure the range to ships by splitting the image of their masts. On the ground, trees, telephone poles and towers were likewise easy to measure through this method, but measuring tanks and other targets posed was less straightforward, especially camouflaged tanks with an indeterminate outline. The rangefinders for the Patton series and Leopard 1 worked with the mixed-image system[160] (superimposed image) for this reason. The mixed-image was mechanically identical to a split-image system, just with the image bifurcating prism omitted.

In the TPD-2, this was solved by including an image duplication mode as a second option. The lower image of the target would be lined up against a reference mark, and the range adjusted until the target's duplicate in the upper half was also aligned to its mark. This mode was also used for targets moving across the field of view and short targets (e.g. hull-down tank).

Figure 38: Image duplication mode. (*Soviet Ministry of Defense*)

As a backup, it was possible to switch to the stereoscopic method of rangefinding. The sight had to be raised until the full size of the target was brought into view in the lower half. With both eyes open, the gunner saw the full target through both the main sight and the rangefinder eye. The range was adjusted until the reference image appeared to be at the same distance as the target[161].

---

[160] In the mixed-image coincidence system (ghost image), the sight showed the image from both rangefinder eyes, forming a double image in the eyepiece. This was less precise than split-image coincidence.
[161] Вавилова, С. И., Савостьяновой, М. В. (eds.) (1948). *Оптика в Военном Деле - Сборник Статей*. Том II. Издание Третье. Москва: Издательство Академии Наук СССР. pp. 177-181, 181-182

Regardless of the chosen method, the rangefinder had a rather long minimum range of 1,000 m. The inability to measure ranges under 1,000 m was compensated by the long point-blank ranges of 125 mm ammunition, and likewise had little impact on the coaxial machine gun. In machine gun theory, concentrations of infantry at range were most effectively engaged by aiming slightly short of the target, so that the upper half of the bullet dispersion ellipse[162] completely overlapped the target area while the lower half of the dispersion ellipse ricocheted off the ground and into the target area.

A more serious drawback was that the method of stabilizing the rangefinder eye through the gun stabilizer instead of the sight stabilizer made the rangefinder inoperable in certain circumstances. When viewing a target in a turret defilade position, the rangefinder could only depress as far as the gun (-6.21°), whereas the sight could lower by -15°, and the precision of the gun stabilizer could differ drastically from the sight stabilizer depending on the tank's speed and the condition of the terrain. The pushrod connection between the gun and the rangefinder base tube also lacked the temperature insensitivity of a parallelogram linkage. The errors induced by these factors would tend to appear as imperfect bifurcation of the target image. As a result, when moving off-road at speed, the TPD-2 rangefinder was ineffective at speeds exceeding 20 km/h[163], whereas firing on the move was effective up to 35 km/h.

It was thus necessary in certain cases for the gunner to command the driver to slow down when preparing to range a target on the move, or to first make a range measurement in advance from behind cover, and then rely on the Delta-D system while advancing towards an enemy position.

More serious still was the fact that the gunner lost the ability to perform a range measurement when the gun automatically elevated during a reloading cycle. After firing a shot and starting a loading cycle on the autoloader, the most expedient method to correct fire was by the burst-on-target gunnery method rather than repeating the range measurement process, aided by the fact that the sight calculated a ballistic solution by lowering the reticle instead of superelevating the gun.

Of the two options, lowering the reticle made for a slower engagement process because the gunner had to lay the reticle on the target twice; once to measure the range, and again before opening fire. However, lowering the reticle was the only viable method of developing a ballistic solution with this system, because superelevating the gun directly through the range control buttons

---

[162] The zone of shot dispersion, which takes the form of a vertical ellipse.
[163] Ефремов, А. С. (2010). *Уроки танкостроения*. Санкт-Петербург: «Гангут». p. 91.

would disturb the view of the target in the rangefinder eye. This mode of presenting a ballistic solution was shared by earlier models of the M48 Patton up to the M48A2. By the 1960's, the most modern fire control systems superelevated the gun while keeping the reticle on target.

The sole benefit to presenting a ballistic solution by lowering the reticle is that if the gunner observed a shell bursting short of the target, he could dial up the range setting (noting also if the burst was laterally on target) until the reticle was lowered onto the bursting point on the ground, then raise the reticle onto the target and fire again[164]. This "burst-on-target" method [**Fig. 39**] was principally effective when engaging small, covered targets that were difficult to range, and it also allowed windage corrections to be made quickly and accurately. In this case, the standard Soviet sight design practice of providing a vertically sliding reticle with a graduated windage scale facilitated precise fire correction.

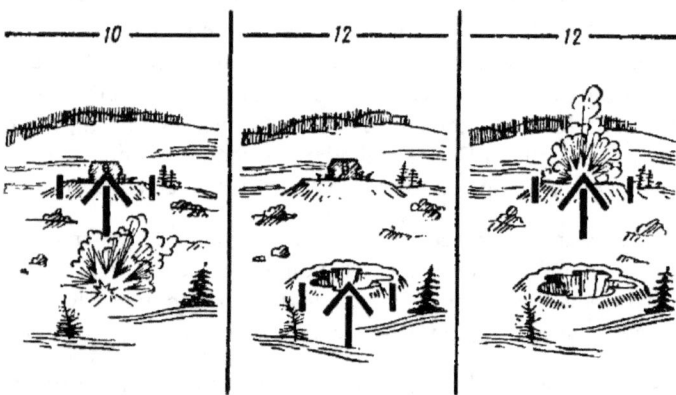

Figure 39: Basic premise of the burst-on-target method: left – observe the impact of the first shot; center – lower reticle onto point of impact; right – raise reticle back onto target, fire. (*Soviet Ministry of Defense*)

TPD-K1

The TPD-K1 viewfinder was essentially the same as the TPD-2-49 viewfinder apart from the stadiametric rangefinder in the viewfinder. In case the laser rangefinder broke down or was otherwise incapable of returning an accurate measurement for whatever reason, the stadia rangefinder allowed a 2.7-meter target to be ranged from a distance of 500-4,000 meters. Together with the range adjustment wheel to adjust the sight, the gunner could continue engaging targets without electrical power.

---

[164] This was known as the "burst-on-target" method in the U.S. Army. *Tank Gunnery* (1957). FM 17-12. Washington, D.C.: Department of The Army. pp. 87-95.

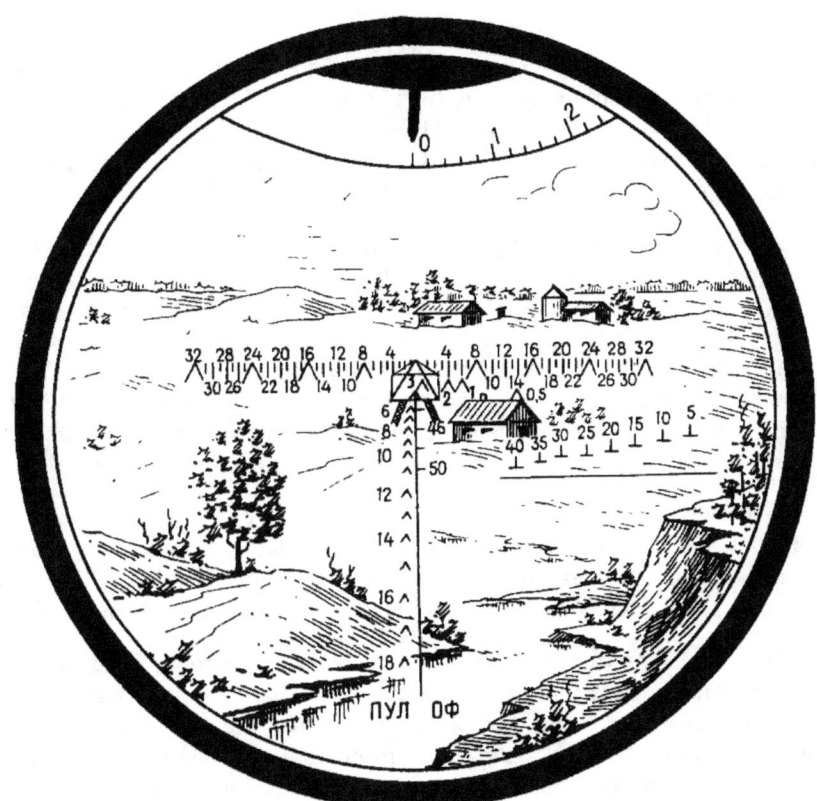

Figure 40: TPD-K1 viewfinder. (*Soviet Ministry of Defense*)

Unlike the TPD-2, the laser rangefinder in the TPD-K1 emitted through the sight periscope head. The lasing block was installed in the void left by the removal of the optical rangefinder, and the laser rangefinder computer, which was its own self-contained unit, was installed to the right of the sight. The periscope head window was widened to position the laser emitter and receiver to its right, side-by-side with the viewing telescope. The right half of the window was coated to maximize transparency to near-infrared while reflecting visible light. Inside the TPD-K1, most of the lasing block was taken up by the Nd:glass rod used as the lasing media[165]. Nd:glass rods were easy to manufacture and their thermal conductivity and thermal sensitivity characteristics were quite adequate for rangefinder lasers[166].

---

[165] A neodymium-doped phosphate glass rod (Nd:glass), enclosed within a flash tube. The flash through the Nd:glass rod produces a laser.
[166] Dakin, J. P., Brown, R. G. W. (2016). *Handbook of Optoelectronics: Concepts, Devices, and Techniques*.

Compared to the ruby laser (0.695 μm) rangefinders commonly added to tanks during modernization programs in the 1970's (T-55, T-62, M60A3, Chieftain), Nd:glass lasers were infrared lasers (1.064 μm), more resistant to atmospheric scattering and consequently managed longer ranges with less power. Like ruby lasers, Nd:glass lasers were not eye-safe[167].

To lase a target, the gunner placed an illuminated red lasing circle over it and then press the right thumb button on his control handles. The total beam divergence of the laser was 0.5 mils[168] but the majority of the energy was concentrated within a circular beam with an angular size of around 1.8', and the lasing circle projected into the sight directly represented this beam width[169]. A two-axis worm screw mechanism enabled the ranging mark to be zeroed to the center of the reticle[170]. The normal (recommended) lasing interval was 6 seconds, but rapid lasing with an interval of 3 seconds was permitted for short periods.

The laser rangefinder computer was linked to the range input motor to automatically generate a ballistic solution and display the measured range to the gunner through the range wheel in the sight. Additionally, a digital display on the computer itself showed the measured distance down to single meters[171].

Without needing the thumb buttons on the control handles to trigger the range setting motor, they were repurposed into a lasing button (right thumb) and range reset button (left thumb). The index finger triggers were unchanged. When engaging a target with the coaxial machine gun, it was no longer possible to make precision fire adjustments due to the removal of the MG scale on the range wheel. The gunner instead lased the target, reset the sight to zero, and aimed with the MG range ladder. Depending on how far the reticle must depress, which depended on the range and the ammunition type selected, the sight took 1-3 seconds to automatically generate a ballistic solution after pressing the lasing button.

The rangefinder computer could receive multiple returns (reflections) but through a range gating system, it recorded only the first return[172]. A minimum

---

Volume 1. Boca Raton: CRC Press. pp. 179-180.
[167] Dakin, J. P., Brown, R. G. W. (2006). *Handbook of Optoelectronics (Two-Volume Set)*. Boca Raton: CRC Press. pp. 1308-1309.
[168] Диков, С. А. (2011). 'Танковые дальномеры', *Мир измерений*, 2. pp. 12-22.
[169] Министерство Обороны СССР (1980). *Изделие 1А40 - Инструкция по эксплуатации*. Москва: Воениздат. p. 56.
[170] It is possible for the ranging mark to be offset from the center of the reticle due to a negligent crew, but by standard policy, the mark had to be zeroed.
[171] To troubleshoot the rangefinder and to compare against the sight range wheel.
[172] More modern systems, including the Soviet KTD-2 from the early 1980's, allow the gunner to choose between the first return or last return.

range of 500 m was built into the rangefinder computer to filter out false returns from nearby objects[173]. The gunner could set the rangefinder to filter out returns for distances shorter than 1,200 m or 1,800 m through a toggle switch on the rangefinder computer front panel.

While ranging a target at 1,900 m with the filter set for 1,800 m, the computer rejected all returns from the laser reflecting off any clutter on the terrain between the sight and the target – foliage, rocks, fences, and so on. The reflection from the target was treated as the first return, and any subsequent reflections off of objects behind the target were rejected. Target distances further than 1,800 m were more likely to incur erroneous returns, which had to be resolved with the gunner's own judgment.

Practical accuracy was improved by the introduction of a ballistic correction system to modify the ballistic solution based on external ballistic factors. Nomograms were used to graphically determine two correction half-coefficients. Summing up these half-coefficients arithmetically gave the final correction coefficient. One nomogram was used to find the coefficient for a combination of barrel wear and powder charge temperature, and another was used for a combination of barometric pressure and temperature[174].

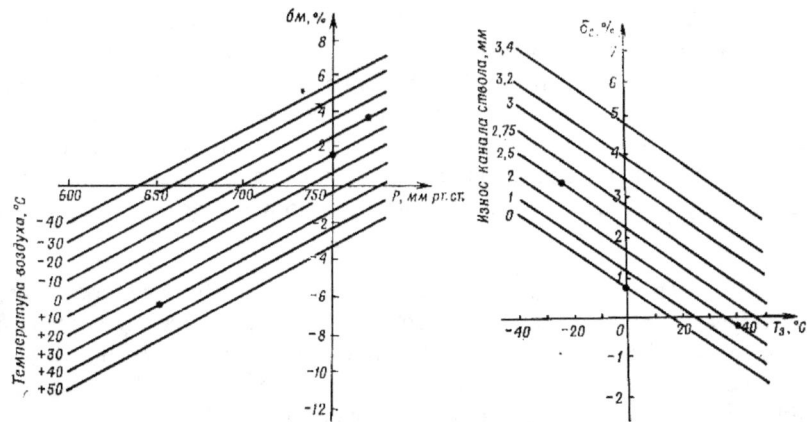

Figure 41: Air temperature and barometric pressure nomogram (left), and barrel wear and powder charge temperature nomogram (right). (*Soviet Ministry of Defense*)

---

[173] Министерство Обороны СССР (1987). *Изделие 1А40 - Техническое Описание.* Москва: Воениздат. p. 60.
[174] Obtained at the unit level and communicated to individual tanks through radio.

This correction coefficient was then dialed into a ballistic correction knob on the sight. The knob was a potentiometer that adjusted the voltage of the range signal from the laser rangefinder to the range input motor, thereby changing how much the reticle would lower for a given range input[175]. As a side effect, the range wheel no longer displayed the true range when a ballistic correction factor was applied – for that, the gunner could refer to the digital display on the computer. The ballistic correction factor was not applied if the sight was operated manually.

With the introduction of the ballistic correction system, the TPD-K1 directly matched the functions provided by the M13 and M16 electromechanical ballistic computers used in the M48 Patton and M60 series, even if the TPD-K1 was still not recognized as having a ballistic computer[176] for unknown reasons.

The TPD-K1 was later reclassified as the primary sight of the 1A40 sighting system, introduced into serial tanks in 1981[177]. In addition to the sight, there was a lateral lead generation device with an ocular display unit, and a ballistic correction device. In essence, these additions gave the TPD-K1 some of the features expected of a modern fire control system, although it was still increasingly obvious that it was too dated for the 1980's.

With the introduction of the 1K13-49 sight, the TPD-K1 was modified with a gyroscope and control connection to the 1K13. The pairing of the two sights was known as the 1A40-1. Switching the ammunition type to "U" on the TPD-K1 transferred the elevation signal of the gunner's control handles from controlling the head mirror of the TPD-K1 sight to controlling the head mirror of the 1K13 sight. It also modified the traverse signal to reduce the turret traverse speed by 2.5-3 times, to facilitate smoother target tracking at long range[178].

The primary function of the UVBU unit was to calculate the lead for a moving target based on the tracking rate, the selected ammunition type, meteorological corrections, and barrel wear. The tracking rate was determined by computing the deflection angle of the gunner's control handles in the horizontal axis together with the range data from the laser rangefinder. External factors were entered

---

[175] A side effect of this was that the range indicated on the range scale would no longer match the true range as displayed on the rangefinder's digital display.
[176] Лепешинский, И. Ю. et al. (2009). *Автоматические системы управления вооружением*. Омск: Издательство ОмГТУ. pp. 114-116.
[177] Устьянцев, С. В., Колмаков, Д. (2004). *Боевая Машины Уралвагонзавода: Танк Т-72*. Нижний Тагил: Издательский Дом "Медиа-Принт". p. 96.
[178] Министерство Обороны СССР (1987). *Изделие 1А40 - Техническое Описание*. Москва: Воениздат. pp. 42-43.

through a second correction dial, separate to the ballistic correction dial on the TPD-K1 sight.

The required lead angle was digitally displayed in a left-eye monocle in mils and the leading direction was indicated by its sign, i.e. when leading to the left by 8 mils, the figure displayed was "-8". The gunner could then aim at the target using the left 8-mil mark on the lateral scale.

The intrinsic precision of the UVBU lead indicator was 0.29 mils, but the smallest increment it could display was 0.5 mils. The tracking limit was $1°/s$[179], and the system could calculate a lead solution up to a limit of 31.5 mils in each direction, fitting just within 32-mil limit of the lateral reticle scale[180]. These limits were more than enough for engaging tracked combat vehicles of all types at practical engagement distances, but the maximum speed of a crossing target was limited at close range, and when firing at long range targets with low-velocity shells like HEAT or HE-Frag, it was limited by the size of the lateral reticle scale.

Beginning in August 1989 [181], the 1A40 was upgraded with the UVKV boresighting device, permitting the gunner to boresight the 1A40 to a 2A46M gun without leaving the tank[182]. The UVKV module was an optical device fitted just in front of the periscopic head of the 1A40 sight, designed to redirect the view to the muzzle of the gun. It consisted of a spring-loaded monocle lens-and-prism assembly, with a pull-cord for the gunner to manually lower the device over the objective window of the sight.

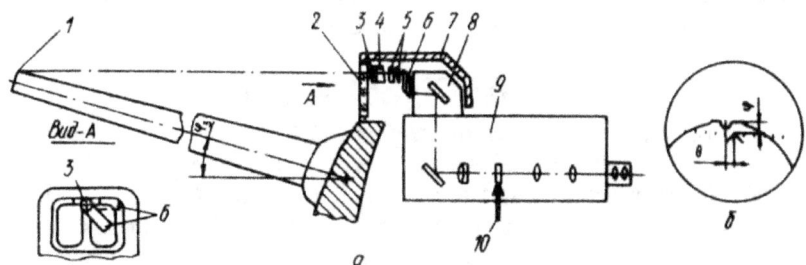

Figure 42: UVKV system layout. (*Soviet Ministry of Defense*)

---

[179] Equivalent to a crossing target moving at a speed of 32 km/h at a range of 500 meters.
[180] Министерство Обороны СССР (1987). *Изделие 1А40 - Техническое Описание*. Москва: Воениздат. p. 9.
[181] Альбом основных конструктивных изменения, проведенных на Изд. 184, 184-1, 184К, 184К-1.
[182] Потемкин, Э.К. (ed.) (1999). *ВНИИтрансмаш – страницы истории*. Санкт-Петербург: Издательство «Петровский фонд». p. 183.

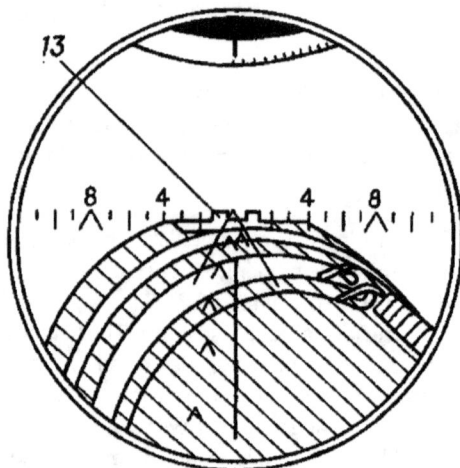

Figure 43: View of the boresighting notch through the sight, with the UVKV monocle in place. (*Soviet Ministry of Defense*)

With the device lowered into view, the gunner saw the notch at the muzzle of the gun[183] and the sight reticle aimed upon it. He could then use the vertical and horizontal adjustment screws on the sight to realign its reticle to the notch, thus boresighting the gun.

This method of field boresighting was preceded by the simple muzzle reference system (MRS) first introduced on the Chieftain, where instead of aligning the reticle directly to the muzzle reference, a reflected dot of light was the reference. Field boresighting as a feature was generally uncommon until the 1990's. For instance, the Leopard 2 was originally intended to receive an MRS, but this was deferred to the 1992-1994 because the mirror on the muzzle could not be secured robustly enough[184]. Indeed, the MRS mirror on the Chieftain was never particularly secure, needing frequent readjustment. The barrel notch of the UVKV system was an alternate take on the same idea, avoiding the general nuisances of a mirror at the end of the barrel.

Day Sight Design

The TPD-2 sight was divided into two halves. The upper half was the optical block. It contained the optical assemblies for the aiming system and the optical rangefinder, later replaced by a lasing block in the TPD-K1. The lower half was devoted to the gyroscopic sight stabilization system and its control apparatus. The gunner's control handles were made integral to the lower block, and a control panel was installed above it.

The TPD-K1 was a modification of the TPD-2. The optical rangefinder block was replaced with a laser emitter and receiver module. An external range computer was mounted to the right of the sight. Operationally, the displays and

---

[183] To give an ideal view of the notch, the focal distance of the monocle lens was 5,600mm, which was the distance between the sight and the muzzle of the gun.
[184] Lobitz, F. (2022). *Gesamptwerk Leopard 2*. Erlangen: Tankograd Publishing. p. 80.

switches on the sights were also practically the same, apart from the ones related to rangefinding.

Officially, the TPD-2 was considered a piecemeal sight, although its functions were inextricably tied to the gun controls. In fact, the sight could be more accurately described as a complete self-contained fire control system. The sight tied together a rangefinder, ballistic computer, and electronic controls in a single compact package instead of having physically separate modules for each unit.

The sight assembly was mounted to the roof of the turret by a single large bolt. The load-bearing suspension elements of the sight were protected by shock absorber bushings, and the front bracket was designed so that if the turret was pushed in by a shell strike, the bracket would rotate against the sight instead of throwing the sight rearwards into the gunner.

To fit into the contour of the turret, the sight acquired an elongated shape, but its tilt angle set the eyepiece at a downward angle, making for a more comfortable neck angle and permitting the brow pad to support some of the weight of the gunner's head, but to have the periscopic height (vertical distance from eyepiece to the center of the periscopic mirror) needed for reasonable headroom above the eyepiece, a second periscopic prism was added. Even so, the periscopic height was only 155 mm, and the height of the periscopic mirror to the bore axis of the main gun was just 306 mm. This offset was large enough that the gun barrel was not visible, which effectively expanded its effective field of view compared to the TSh2-series coaxial telescopic sights in earlier Soviet medium tanks.

The angling of the sight body made maximal utilization of the space beneath the sloping turret ceiling, and the peculiar positioning of the periscopic head relative to the sight eyepieces was determined by the slope of the turret roof. Designing sights to account for the contours of a complex turret was by no means new by the 1960's. It had already been done in the 1950's on the TPS-1 sight, mass-produced for T-10 heavy tanks and fitted to a variety of experimental medium tanks.

The periscopic head of the sight was below the roof line of the turret, so observation from a true turret defilade was not possible. This was a limitation shared by all of the T-72's Cold War contemporaries to varying degrees, but most prominent in other tanks equipped with rangefinder-sights like the Leopard 1. Ballistic protection for the periscopic head was provided by a thick

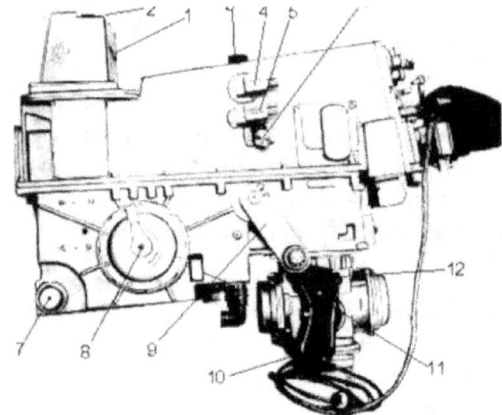

glass window[185] and a second glass window underneath the head mirror prevented fragments from ricocheting down into the sight body.

The brow pad was contoured to fit together with a Soviet tanker helmet, so the gunner did not have to push his helmet up just to use the sight, and the eyepiece had a long eye relief. This made the gunner's vision less sensitive to positional disturbances inside a moving tank[186].

Figure 44: Side view of a 1A40. (*Soviet Ministry of Defense*)

A fixed 8x magnification was a popular choice for tank sights in the 1960's, with prominent examples like the M32 in the M60A1 and the AFV No. 39 in the Chieftain. With an 8x optic, a T-72 gunner could see and identify tanks from beyond the maximum effective range of his main gun, albeit only in optimal

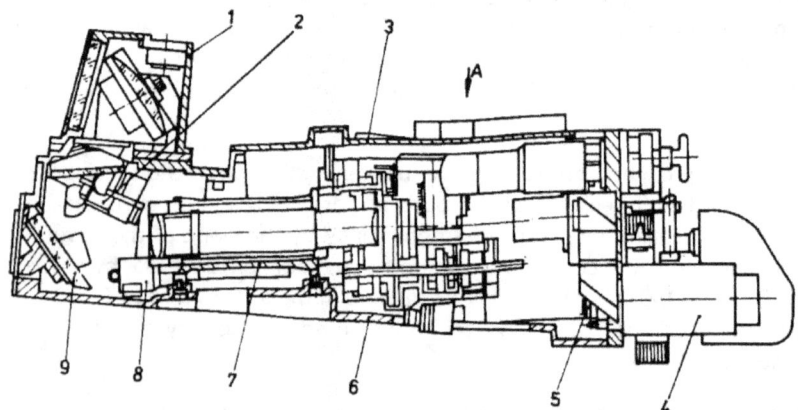

Figure 42: TPD-K1 sight cross-section. 1 – periscope head; 2 – parallelogram linkage to head; 3 – casing; 4 – eyepiece; 5 – prism; 6 – casing; 7 – telescope tube assembly; 8 – laser block; 9 – mirror. (*Soviet Ministry of Defense*)

---

[185] Ballistic glass was unsuitable because the thin glass layers shatter and delaminate from the plastic interlayers in such a way that virtually all visibility would be lost after a single bullet strike. See: Сеннов, Н. И. (1942). *Оптика на танке*. Москва: Объединение государственных книжно-журнальных издательств. p. 6.

[186] Halbert G. A. (1983). 'Elements of Tank Design', ARMOR, (November-December). p. 42.

conditions. A switch on the face plate of the sight allowed the gunner to lower a yellow-tinted filter over the telescope tube, reducing glare.

The optical assembly itself was conventional, and was very simple from an optical design standpoint. It consisted of a periscopic mirror assembly, a telescopic tube with a reticle mechanism, an eyepiece assembly, and a prism periscope to join the eyepiece assembly to the telescopic tube. The telescope tube contained the objective lens and a field lens[187], and was followed by a Zeiss Planar lens group[188], then a prismatic periscope, and ending in a Plössl eyepiece lens group[189]. Coated glass[190] played a critical role in enabling the TPD-2 sight to retain a useful level of light transmittance with such a large number of optical elements, in addition to other qualitative improvements like reducing lens flare when aiming in the general direction of the sun. All of the glass in the sight was photochromic and radiation-resistant.

Because the telescope tube was fixed, the TPD-2 and TPD-K1 had their magnification fixed at 8x. In controlled testing, the TNP-165A forward-facing unity periscope was statistically the most effective vision device for target detection, beating out a standard TSh2 sight in its 3.5x and 7.0x magnification settings in finding tank targets at ranges of up to 750 meters. It rapidly lost its usefulness at a range of 1,000 m and above, at which point a TSh2 main sight in its 7.0x magnification setting was the best performer[191]. From this point of view, it is understandable that during the creation of the TPD-2-49, a low magnification setting might not have been considered essential.

The amount of light available is proportional to the square of the scale factor of the exit pupil diameter[192], so for example, when comparing a 2 mm exit pupil to a 4 mm exit pupil, the 2 mm exit pupil is smaller by a factor of two but the amount of light available is smaller by a factor of four. The TPD-2 series had a large objective lens, providing a reasonably sized exit pupil with its 8x magnification. In the TSh2 family of tank sights[193] it replaced, the objective lens diameter was only 18.9 mm, so while its exit pupil diameter was a wholly satisfactory 5.4 mm at 3.5x magnification, switching to 7.0x magnification

---

[187] An intermediate lens to direct the light through the large gap in the optical path.
[188] A type of lens designed to greatly reduce optical aberration, or in other words, increase optical resolution. Its position behind the objective lens made it the erecting lens group in the optical system.
[189] A type of lens designed to reduce optical aberration, astigmatism, and improve contrast.
[190] Anti-reflective coatings reduce light reflections from air-glass surfaces in lenses, increasing the light transmittance of an optical system. The coated windows and lenses appear purple.
[191] Веселов, В. Б., Диков, С. А., Харитонов, И. С. (1982) 'Анализ Поисковых Возможностей Танков', Вопросы Оборонной Техники, 20(105). pp. 15-17.
[192] This is because the area of a circle is proportional to the square of its radius. Doubling the radius of a circle quadruples its area. The exit pupil diameter is determined by dividing the diameter of the objective lens by the magnification factor.
[193] Standard to the T-54/55 and T-62.

halved the exit pupil diameter to 2.7 mm[194], which had a significant effect on sighting effectiveness in low light conditions.

Input of the ammunition type was automatic via the autoloader ammunition selection dial, with a manual selector knob above the sight eyepiece as a backup. With the ballistic cam set, he can use the range adjustment wheel to perform range measurements and adjustments in an unpowered mode.

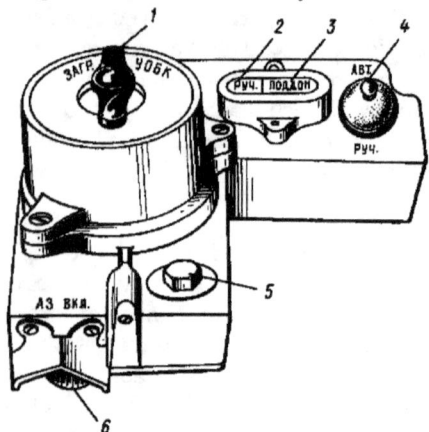

Figure 45: Gunner's autoloader control panel: 1 – ammunition type selector knob; 2 – manual mode indicator light; 3 – stub capture indicator light; 4 – autoloader activation switch (auto – on, manual – off); 5 – potentiometer tuning nut; 6 – loading switch. (*Soviet Ministry of Defense*)

The reticle group consisted of a rotating glass range disc with an etched range scale, a sliding glass reticle cell with an etched reticle, and a small lamp for reticle illumination. The range disc was spun by the range input drive, and the ballistic drive lowered the sliding glass cell, lowering the reticle in the sight image[195].

The ballistic cams were large contoured rings affixed to the telescopic tube in the sight, so that all three cams were rotated together with the tube by the range input drive. The special contour on each cam corresponded to the ballistic drop of one of the three ammunition types, creating a variable gearing ratio between the range input drive and the sliding reticle drive. For a given number of turns in the range input drive, the reticle would be lowered at a progressively higher rate as the range increased.

The sliding reticle mechanism had a roller above each of these cams, one of which was engaged at all times. As the telescope tube rotated, the reticle slide was lowered at a precalculated rate through one of these ballistic cams, thus

---

[194] Шишковского, В. М. (1978). *Огневая Подготовка, Часть Вторая: Основы Устройства Вооружения.* Москва: Воениздат. p. 192.
[195] The mechanism actually raised the sliding reticle cell, because the image was optically inverted. The image was erected along a focal point between the telescope tube and the eyepiece prism.

automatically computing the ballistic solution for any range input[196]. If the ammunition type was switched when the system already had a range set, the position of the reticle shifted as the selector knob or the selector motor turned.

Upon selecting the desired ammunition type on the autoloader ammunition selector, an electric motor in the ballistic computer engaged the corresponding ballistic cam to the reticle mechanism. When operating in manual mode, the gunner could use the mechanical selector knob at the top of the sight housing to engage one of the three cams. With a cam selected, the turning of the range wheel moved the reticle vertically by an amount corresponding to the contour of the cam[197] to an angular error[198] of 2-5'. In the TPD-2 sight, the turning of the telescope tube also turned a special cam to move the image shifting lens laterally.

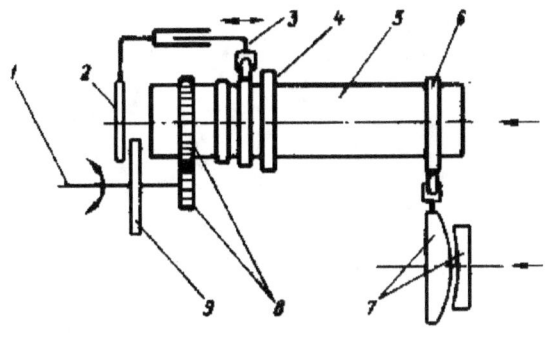

Figure 46: Scheme of TPD-2 day sight ballistic mechanism:

1 – range input drive; 2 – sliding reticle; 3 – ballistic cam selector; 4 – ballistic cams; 5 – telescope tube; 6 – image shifting cam; 7 – compensator lens; 8 – telescope rotation gear; 9 – range display wheel. (*Soviet Ministry of Defense*)

Gyroscopic control of the sight head mirror was handled in the lower half of the sight. The connection to the gun was also made here, with a parallelogram linkage connected to a bell crank on the gyroscope. The parallelogram linkage was designed to minimize the misalignment between the sight and the gun when temperature differences arose between the linkages and the tank turret[199].

Normally, thermal expansion and contraction of the turret occurred at a different rate than the sight linkages, mainly due to the fact that the linkages were inside the turret, and were thus cooler than the turret itself. With pushrod

---

[196] Министерство Обороны СССР (1987). *Изделие 1А40 - Техническое Описание*. Москва: Воениздат. pp. 81-82.
[197] Министерство Обороны СССР (1979). *Вооружение Танка Т-72*. Москва: Военная академия бронетанковых войск. p. 100.
[198] *Оружие и технологии России. XXI век Том 12 - Боеприпасы и средства поражения* (2006). Москва: Оружие и технологии. p. 375.
[199] *Танковые прицелы и системы управления огнем (СУО)*. Available at: https://38niii.ru/analitika/173-tankovye-pritsely-i-sistemy-upravleniya-ognem-suo.html (Accessed: 1 February 2024).

bell crank linkages, a small thermal mismatch resulted in the rod becoming longer or shorter than its initial length[200], creating a false rotation input to the sight. With a parallelogram linkage, the two parallel opposing rods cancel each other out, creating no net rotational input. The intrinsic mechanical misalignment was up to 0.12 mrad vertically and 0.13 mrad horizontally[201]. After a march at an average speed exceeding 20 km/h over rough terrain, the misalignment in both axes could reach 0.2 mrad[202].

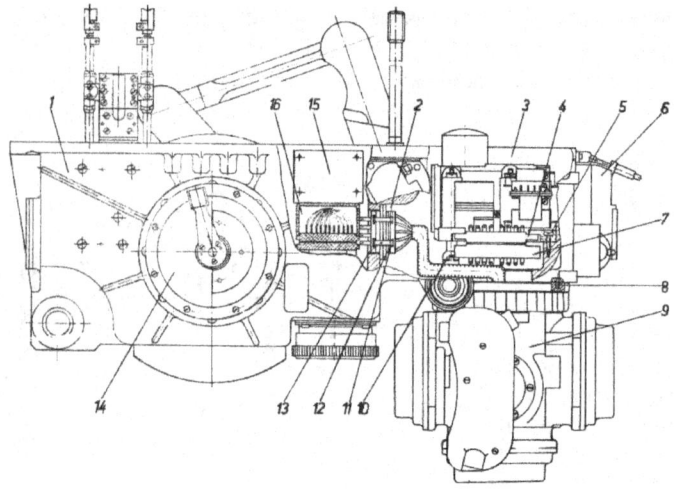

Figure 47: TPD-K1 lower block. (*Soviet Ministry of Defense*)

Gyroscopic sight stabilization in the TPD-2 operated in the vertical plane only. The gyroscope itself had three degrees of freedom: the gyroscope rotor spun along the vertical axis in its electric motor unit (gyromotor)[203], which in turn had one degree of freedom to the gimbal in the roll axis, and the gimbal was free in the pitch axis.

As the tank pitched up and down, the pitch angle of the rotor remained level at all times, and the sight head mirror acquired the same angle from its connection to the gimbal through a steel band pulley and parallelogram

---

[200] For example, under sunlight, the turret expanded and the gun grew further from the sight, but the linkages, being cooler, hardly expanded. This cranked the sight into a lower elevation angle than its initial zero.
[201] Зубарь, А. В. et al. (2018). 'Анализ Существующих Способов Выверки Нулевых Линий Визирования Прицелов Системы Управления Огнем Образцов Бронетанкового Вооружения', *Научный Вестник ВВИМО*, 3 (47). p. 58.
[202] Ibid. p. 59.
[203] Much like how the rotor in an attitude indicator (artificial horizon) instrument in aircraft spun perpendicular to the horizon.

mechanism. To elevate the sight, a low-speed electric motor controlled by the gunner's control handles rotated the rotor in the roll axis, causing it to precess in pitch until the control handles were released. As the rotor precessed, it took the gimbal with it, and by extension, elevating the sight head mirror.

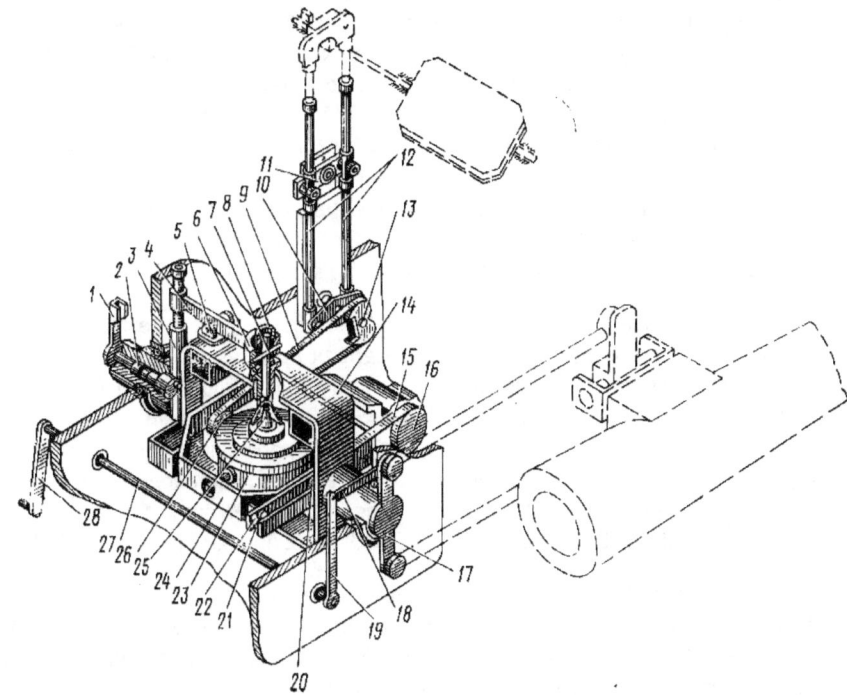

Figure 48: Gyromotor in the TPD-2 sight and its connections to the periscope head and the main gun. (*Soviet Ministry of Defense*)

A rotary transformer was mounted to the gyro frame, its rotor turned by the pitch angle of the gyro gimbal through a steel band belt. A control signal to the gun elevation drive control unit passed through the transformer, and when the stator and rotor were in alignment, the signal was neutral.

As the gyro gimbal rotated, either by control or the pitching of the tank, the rotor of the rotary transformer turned relative to the stator, changing the voltage of the control signal proportionately to the angular difference between the rotor and stator, called the misalignment angle. The signal, representing the misalignment angle, traveled to a control amplifier where it was converted into a polarized DC signal and applied to the valve control electromagnet of the hydraulic booster for the gun elevation piston. Depending on the polarity of the control signal, the gun would be raised or lowered by its piston until the rotor and stator returned into alignment.

By controlling the gun this way, the T-72 had what was known as a director-type fire control system[204]. This system improved firing accuracy on the move, and prevented accidental firing in cases where the gunner could see the target, but the gun, being more restricted in its range of elevation, had not (or could not) elevate to the same angle as the sight. When fighting from a hull-down position on a reverse slope, the inability of the gun to match the depression angle of the sight was a useful tool for ensuring the gun would not fire until the barrel cleared the slope. Seeing the target, the gunner could hold down the trigger and tell the driver to creep forward. Once the slope leveled out enough for the gun to enter the coincidence window, a shot would be fired.

In practice, signal noise and mechanical tolerances made sure that the gun could never be perfectly aligned with the sight, but the system nevertheless had a fairly high alignment precision of 0.5 mrad[205]. The sight had a fire gating system[206] to disconnect the firing circuit until a pair of contacts in the stator and rotor were aligned within this margin of 0.5 mrad, called the coincidence window. On the move, vibrations could shift the coincidence window by a further 0.1 mrad[207]. In total, the sum of the systemic errors in the fire control system from a standstill amounted to 0.3 mrad for KE and 0.37 mrad for HEAT[208].

The sight could also be independently controlled, in which case it was decoupled from the gun[209], allowing the gunner to continue controlling his view of the battlefield without moving the gun. This could be done to maintain visual contact with a target if the gun was locked for some reason.

---

[204] Ogorkiewicz, R. M. (1991). *Technology of Tanks*. Volume 1. Coulsdon: Jane's Information Group. pp. 200-203.
[205] Жирнова, Т. А. (ed.) (2004). *Устройство, эксплуатация, техническое обслуживание и ремонт стабилизатора танкового вооружения 2Э28М - Методические Указания*. Омск: Издательство ОмГТУ. p. 7.
[206] Шишковского, В. М. (1978). *Огневая Подготовка, Часть Вторая: Основы Устройства Вооружения*. Москва: Воениздат. p. 194.
[207] Зубарь, А. В. et al. (2018). 'Анализ существующих способов выверки нулевых линий визирования прицелов системы управления огнем образцов бронетанкового вооружения', *Научный Вестник ВВИМО*, 3 (47). p. 59.
[208] Баринов, Н. П., Иванов, И. К., Комаров, А. В. (1986). 'Влияние настройки механизма связи ночного прицела с пушкой на точность стрельбы', *Вестник Бронетанковой Техники 1986*, сборник 2. p. 18.
[209] Министерство Обороны СССР (1979). *Стабилизаторы Танкового Вооружения 2Э28М (2Э28М-2): Техническое Описание*. Москва: Воениздат. p. 99.

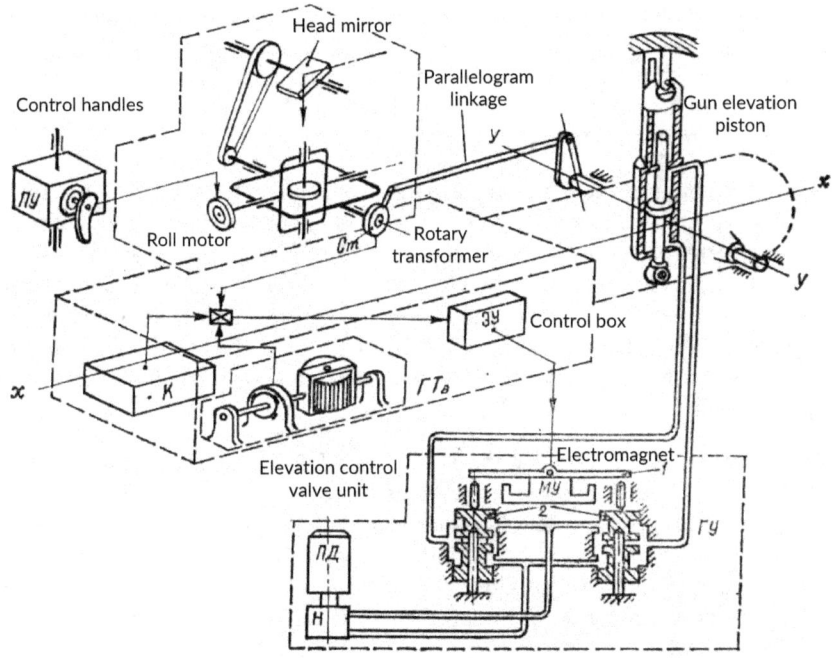

Figure 49: Components of the sight-gun stabilizer control. (*Soviet Ministry of Defense*)

The vertical and horizontal vibration of the sight was measured at under 3' (minutes of arc) on hard surfaces at up to 30 km/h[210]. This was perceptible to the gunner as a blurring of the image. Jitter in the image arose from play and elasticity in the pulley mechanism and the parallelogram linkage, with additional errors from thermal expansion with changes in the environmental temperature.

When it was not in use, the gyro rotor was "caged" – its gimbal locked to the gyro frame so that the rotor could not precess. In this circumstance, the pitch angle of the gyro rotor was coupled to the gun via the frame and the parallelogram linkage. Sight elevation was controlled solely by manually cranking the gun up and down. A two-minute startup period was needed for the gyromotor to spin up to its operating speed of 28,000 RPM, after which it could be uncaged by turning a lever on the left side of the sight.

---

[210] Жартовский, Г. С., Куртц, Д. В., Усов, О. А. (2016). *Защита оборудования и экипажа военных гусеничных машин от механоакустических и климатических воздействий*. Санкт-Петербург: Издательство «Лань». p. 135.

It was possible to reduce jitter by caging the gyroscope, keeping the sight linked mechanically to the gun. This was most useful when fighting from static positions and short halts. Modern sighting systems reduce jitter and stabilization error by reducing the number and complexity of the connections in the gyroscope-mirror coupling.

Most sighting devices accomplish this by limiting the coupling to a single pulley, and shortening the pulley by placing the gyroscope inside the sight head, which is the reason behind the trend of modern high-performance tank sights having large heads. To avoid the high vulnerability of a large head, some sights feature a fixed head mirror, stabilizing the image instead with an internal gyroscope and reflector, directing the head mirror at the target by slaving the entire sight to the main gun by a direct trunnion coupling (Leclerc) or parallelogram linkages (Type 99).

The Delta-D (ΔD) system in the TPD-2 and TPD-K1 changed the range in the sight based on the speed of the tank as it moved towards or away from the target, measured by an internal tachometer on the axle of the left idler wheel[211]. A cosine potentiometer in the turret [See Commander's Cupola] converted the tank's speed into the true closing speed by modifying the signal from the tachometer according to the orientation of the turret relative to the hull.

If, for example, the sight was pointing at the target but the tank was moving at 10 m/s (36 km/h) diagonally at a 60° angle, the system subtracted the range to the target by the actual closing speed of 5 m/s (18 km/h)[212], so the system functioned accurately regardless of whether the tank was maneuvering or driving directly towards the target. The gunner had to manually compensate for the movement of the reticle, but overall, the engagement process was shortened and the need to slow down or halt to fire while fighting on the move was sharply curtailed, which had a positive impact on survivability.

---

[211] The accuracy of the measurement depended on track slip, which varied depending on the terrain.
[212] The cosine of 60° is 0.5.

Table 11: Technical specifications of T-72 day sights

SIGHT	TPD-2-49	TPD-K1
Magnification	8x	
Field of view	9°	
Eye relief (mm)	24.2	
Exit pupil diameter (mm)	4	
RMS stabilization precision (mil)*	0.366	0.2
Elevation range (°)	-15 to +25	
Elevating speed (°/s)	0.05-3.5	
Rangefinder type	Optical coincidence with stereoscopic backup 1,500 mm optical base length	Nd:glass Laser (1.064 μm)
Ranging limits (m)	1,000-4,000	500-4,000
RMS Ranging Error		
1,000-2,000 m	3%	≤10 m
2,000-3,000 m	4%	≤10 m
3,000-4,000 m	5%	≤20 m
ΔD Calculation Error for Range 1,000-4,000 m		
Over tank travel distance of 500 m	±10% of tank travel distance	
Optical Resolution (arc seconds)		
Gyromotor off, stationary tank	7.5	
Gyromotor on, stationary tank	20	
Gyromotor on, moving tank	60	

* The root mean square (RMS) precision here refers to RMS error. It is a standard statistical measure of the magnitude of oscillating deviations from the line of sight.

Night Fighting

The night vision sighting equipment of the T-72 was standard, differing from other Soviet medium tanks only in minor structural details. At the beginning of its service, it was equipped with the TPN-1 night sight, featuring a Gen 0 image converter[213]. Gen 0 night vision devices are characterized by the need for infrared illumination to function[214]. By the time the T-72 entered service, this technology was commonplace.

The TPN-1 was an early Soviet tank night sight developed under the *Luna*-2 program during the early 1950's. It was the first mass-produced Soviet tank night sight. It was created as part of a 1957[215] army-wide revolution in night fighting capability, which saw tens of thousands of Soviet tanks standardizing on the *Luna*-2 system in the years thereafter. The T-72 was furnished with the TPN-1-49, a variant originally adapted to the T-64A. The mounting system was identical in the T-72 turret, allowing the same sight to be retained.

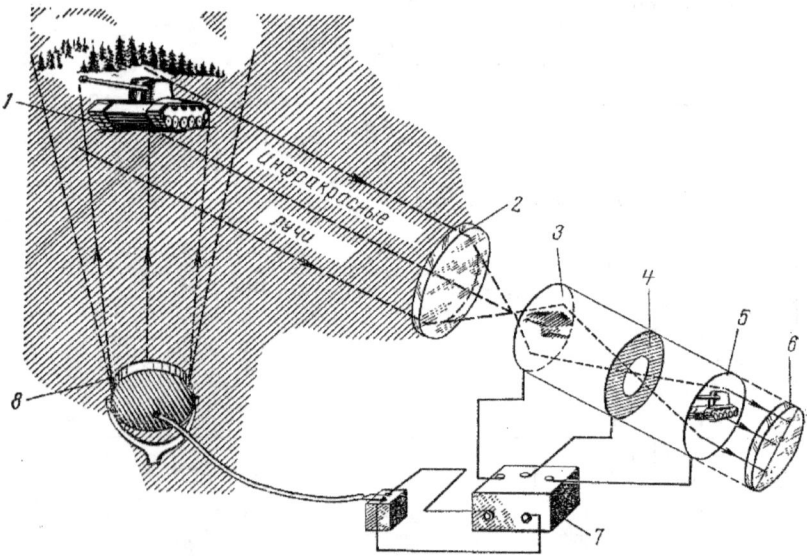

Figure 50: Principles of infrared night vision: 1 – target; 2 – objective lens; 3 – tube photocathode; 4 – anode; 5 – phosphor screen; 6 – eyepiece; 7 – power supply unit; 8 – IR spotlight. (*Soviet Ministry of Defense*)

---

[213] An image converter was called «Электронно-Оптического Преобразователя» (ЭОП).
[214] Голощапов, И. М. (1973). *Танковые приборы ночного видения*. Москва: Воениздат. pp. 62-78.
[215] Вахрушева, Ю. В., Шенбергер, А. Ю., Щеглов, Е. В. (2022). 'Танковые приборы ночного видения', *Специальная Техника И Технологии Транспорта* 2022, 14. p. 3.

The primary drawback of infrared illumination was its high visibility to observers equipped with simple metascopes[216] and infrared sights of their own. As these devices proliferated in the rank-and-file of peer threat forces[217], active infrared night vision lost much of its appeal over simple white light illumination. Target searching was also a major issue – the beam from tank spotlights was constricted to a circle approximately 1° in size by the parabolic reflector in the spotlight, so that in effect, only ≥3% of the night sight's 5.5° field of view was useful at its maximum viewing range. At shorter ranges, the light spilling over the front of the lamp illuminated more of the image, but this had the effect of illuminating the ground in front of the tank and even parts of the tank itself, making it much easier to detect by enemy observers.

Tasked with bringing a passive night vision capability to tanks, the KMZ Central Design Bureau and the *Peleng* Central Design Bureau jointly developed the TPN-3 *Kristall*-PA. Officially entering service in 1975[218], the TPN-3 entered mass production only in March 1979[219]. Officially, the T-72 began to be equipped with the TPN-3 since 1978[220]. It was an unusual hybrid sight that allowed the gunner to select between an active or passive night vision by physically switching between a single-stage image intensifier tube (Gen 1)[221] and a single-stage image converter tube as needed. In the active mode, the viewing range was extended by a new L-4A xenon arc spotlight.

The introduction of the TPN-3 coincided with the TVNE-4PA night vision driving periscope also entering service in the same year, itself also a hybrid system with two separate image converter and intensifier tubes. The fact that the TVNE-4PA did not come earlier was somewhat unusual since the viewing range requirements for a tank driving periscope were well within the capabilities of a single-stage Gen 1 tube in the early 1970's, and solving the rather miserly degree of stealth afforded by infrared headlights when tanks were driven in large columns at night had, on the whole, arguably higher priority[222].

---

[216] The simplest optical infrared device, consisting of a phosphor screen charged by either solar UV or alpha rays from a radiation source. An infrared source viewed through the screen will glow.
See: Криксунов, Л. В. (1975). *Приборы Ночного Видения*. Киев: Издательство Техніка. pp. 4-5.
[217] The PSO-1 sniper scope, fielded by the Soviet Army together with the SVD *Dragunov* rifle in 1963, had an integral metascope to enable marksmen to harass vehicles driving with IR headlights at night.
[218] *Танковые прицелы и системы управления огнем (СУО)*. Available at: https://38niii.ru/analitika/173-tankovye-pritsely-i-sistemy-upravleniya-ognem-suo.html (Accessed: 1 February 2024).
[219] Музей Рогачевского завода «Диапроектор»: у каждого экспоната – своя история (2015). Available at: https://news.21.by/society/2015/12/23/1150337.html (Accessed: 10 March 2024).
[220] Колмаков, Д. Г. (2021). *Тагильская школа. 80 лет в авангарде мирового танкостроения*. Белгород: КОНСТАНТА-принт. p. 101.
[221] An image intensifier was called «Электронно-Оптического Усилитель» (ЭОУ).
[222] The Chieftain and Leopard switched to a passive driving periscope before a passive night sight. See: Ogorkiewicz, R. M. (1991). *Technology of Tanks*. Volume 1. Coulsdon: Jane's Information Group. p. 158.

It is reasonable to speculate that there were additional factors delaying the mass production of military image intensifiers at the rate needed to outfit the enormous fleet of AFVs in the Soviet Army – keeping in mind the need to also produce PNV-57E passive night vision goggles for truck drivers and auxiliary troops. On paper, the TPN-3 was intended to be a total replacement for the TPN-1 in the T-72 family, but in practice, it was frequently substituted by the TPN-1 even during the 1980's, and the TPN-3 was absent entirely from Soviet-produced T-72M1 tanks. The TPN-3 was much more commonly found in T-64B and T-80B tanks.

Towards the middle of the 1980's, even the T-72B1 was not always spared by the TPN-3 shortage. The T-72B, on the other hand, was fitted with the 1K13-49 combined day-night sight, designed to integrate a guided missile system without changes to the day sight. The 1K13 was created as part of a unified system of laser beam-riding gun-launched ATGMs, which included the *Bastion* (T-55), *Sheksna* (T-62), *Basnya* (BMP-3) and the *Svir* for the T-72. It featured active and passive night vision modes with a hybrid system similar to the TPN-3, and it functioned as a missile sight in the daylight mode only.

Features of Night Sights

All Soviet tank night sights were mounted to the turret as a self-contained unit, with an independently sealed sight head intended for ease of replacement. An additional sight head was carried as a spare. The night sight was boresighted independently of the day sight, and the process could be carried out in day or night. In the daytime, the iris diaphragm was opened only enough for a pinhole view of the boresighting mark, protecting the tube from overexposure damage.

Although not as convenient as combined day-night sights that would later be in vogue during the 1980's, this configuration was preferable to sighting systems that required the day sight to be swapped out for a night sight, which then had to be boresighted before use. It also enabled gunners to switch between the night and day sights at will according to prevailing lighting conditions. However, the offset of the sight was uncomfortable for the gunner's neck and back[223].

The elevation angle of the gun was duplicated in the rotating mirror head of the sight with a simple open pushrod linkage and a set of sector gears on the right side of the sight casing. The push-rod was connected to the gun via a rotating pass-through rod inside the day sight.

---

[223] Троицкий, В. В. (1987). 'Рабочие Места Экипажей в Отечественных и Зарубежных Танках', Вестник Бронетанковой Техники 1987, сборник 12. pp. 10-11.

Compared to the parallelogram linkages connecting the main sight to the gun, the night sight had longer linkages and no parallelogram structure, drastically raising the sensitivity of the sight alignment to temperature differences. Large alignment errors exceeding the technical requirement of 0.5 mrad occurred with relatively small changes in temperature[224], which had an impact on long range firing accuracy.

The standard IR spotlight paired with the TPN-1 was the L-2AG incandescent bulb spotlight. Beginning in 1978, the L-4A xenon arc spotlight replaced the L-2AG on tanks equipped with the TPN-1. The TPN-1-49-23 designation denoted a TPN-1-49 paired with the L-4A spotlight. Export models like the T-72M1 which never received the TPN-3-49[225] would receive the L-4A only in 1980. With the L-4A spotlight, the recognition distance of a tank-type target with the TPN-1 was extended to 1,000-1,100 m[226].

Both spotlights had a simple fixed construction. An infrared filter made from cellulose acetate film was glued to its glass window, almost completely blocking the emission of visible light. The spotlights were almost entirely restricted to the near-infrared band in a fairly broad range (0.9-1.2 μm), but there was enough of a glow to see a working spotlight with the naked eye from close range (≤100 m). Replacing the infrared filter window with a clear glass window converted the spotlights to white light.

The L-2AG (L-4A) was a permanent fixture of the turret[227]. It was not intended to be stowed away during daytime, and unlike many other tanks with permanently affixed spotlights[228], it had no effect on the size of the tank's silhouette, which was an uncommon detail. More importantly, it also did not block the commander's forward view. The downside to this configuration is that the spotlight was below the head of the night sight, making it impossible to survey the battlefield from a turret defilade position.

A red warning lamp on the ceiling of the gunner's station lit up when the IR spotlight was turned on[229], and when it was not in use, the spotlight was capped

---

[224] Баринов, Н. П., Батян, В. И., Комаров, А. В., Медов, Б. С. (1986). 'Тепловая Нестабильность Выверки Ночного Прицела с Пушкой', *Вестник Бронетанковой Техники* 1986, сборник 3. p. 18.
[225] *Танк Т-72М1: Техническое описание и инструкция по эксплуатации – Часть I.* p. 103.
[226] Ibid. pp. 11, 107.
[227] The "A" suffix corresponded to this type of mount. The L-2G spotlight used on older tanks was designed for a different, pedestal-type mount, and was meant to be stowed away when not in use.
[228] Examples include the M48, M60, Chieftain, Centurion, AMX-30, Leopard, and more.
[229] Лепешинский, И. Ю. et al. (2009). *Автоматические системы управления вооружением.* Омск: Издательство ОмГТУ. p. 253.

with a thin-walled aluminum lid for scratch protection and to prevent the infrared filter from warping and cracking under direct sunlight[230].

Although the simplicity of these spotlights might appear to be a positive trait in terms of maintenance demands, the practical implications were somewhat less agreeable. The lack of an occluder (blackout shield) in front of the bulb to reduce light spillage towards the front of the spotlight was a significant downside in combat, because turning on the spotlight did not merely direct a focused beam downrange, it also illuminated the T-72 itself and the ground in front of it.

Most of the power supplied to the spotlight was released as heat. In continuous use, the spotlight casing and mount could reach temperatures well over 100°C[231]. Temperature control was important for both spotlights, but especially for the L-4A, especially in winter. Temperature stability and uniformity was critical for preventing lamp explosions, which was a real and persistent hazard for xenon arc lamps. Multiple spare bulbs were carried for this reason.

In terms of illuminating power, the L-4A was comparable in luminous intensity[232] to the XSW-30-U ($30 \cdot 10^6$ Cd) spotlight on the Leopard 1, but this was the weakest tank spotlight type in NATO. The IR spotlights on the M60A1 and Chieftain were several times brighter. Consequently, the detection range with the TPN-1 was only roughly on par with the identification ranges achievable by NATO night sights.

Some disadvantage persisted after the switch to the L-4A spotlight in 1979, but by this point, thermal imaging equipment had already begun to proliferate in the West. Obviously, parity in night fighting capability was never achieved. The disparity was consistently dire, but also rather inexplicable, seeing as the gap was not in image converter technology, but in spotlight design. On a tactical level, mitigation techniques would likely involve a rethinking of night combat operations, and in some cases, rejecting the use of night sights altogether in favor of converting the IR spotlights to white light.

---

[230] Шишковского, В. М. (1978). *Огневая Подготовка, Часть Вторая: Основы Устройства Вооружения.* Москва: Воениздат. p. 201.

[231] Лепешинский, И. Ю. et al. (2012). *Устройство оружия и его боевое применение. Часть вторая.* Омск: Издательство ОмГТУ. p. 81.

[232] All luminance figures refer to the spotlight output in the white light mode. Due to the infrared filter, the luminous intensity of the infrared emission was only a fraction of the white light emission.

Table 12: Technical data on T-72 night sights

NIGHT SIGHT	TPN-1	TPN-3	1K13
Generation	0	0 + 1	0 + 1
Magnification/ field of view Night	5.5x / 6°	5.5x / 6°40'	5.5x / 6°
Day	-	-	8x / 5°
Recognition distance[a] Active (m)	600-800* 1,000-1,100**	1,300	1,200
Passive 0.003-0.005 lux[b] (m)	-	500	500
Stabilization	Dependent	Dependent	Independent
Elevation arc	Matched to gun	Matched to gun	-7° to +20°
Spotlight	L-2AG	L-4A	L-4A
Light source	PZh27-200 Incandescent (27 V, 200 W)	DKsEl-250 Xenon arc (27 V, 250 W)	DKsEl-250 Xenon arc (27 V, 250 W)
Illumination power	$1 \cdot 10^6$ Cd	$30 \cdot 10^6$ Cd	$30 \cdot 10^6$ Cd
Beam width of 95% luminous intensity (H x W)	1° x 0.8°	1° x 0.8°	1° x 0.8°

a   Recognition of an object as a tank, based on the criteria of correctly determining the pointing direction of its barrel, with the tank in profile.

b   Clear, moonless night with low grass cover. Atmospheric transparency coefficient of 0.8±0.1, grassy field with a grass height of ≤0.3 m.

c   Потемкин, Э. К. (ed.) (1992). *Основы научной организации разработки*. Том 2. Москва: Издательство МГТУ им. Н. Э. Баумана. p. 208.

*   600 m on front profile of a tank, 800 m on side profile of a tank.

**   With L-4A spotlight

TPN-1-49

The *Luna*-2 system consisted of the TPN-1 sight, the L-2 spotlight, their associated power supply units, and their mechanical linkages. The TPN-1 sight was essentially identical to the old TKN-1 commander's night vision periscope in optical design, only differing in scale and hence, in its light collecting power. Both devices standardized on the same Gen 0 image converter, designated 7V[233].

The sight reticle, consisting of a center chevron with vertical dashes separated by large gaps, was etched onto the image converter's objective screen. It had no illumination. For there to be enough visual contrast to see the reticle at night, the target had to be well-lit, which was a peculiarity of several early Soviet Gen 0 devices.

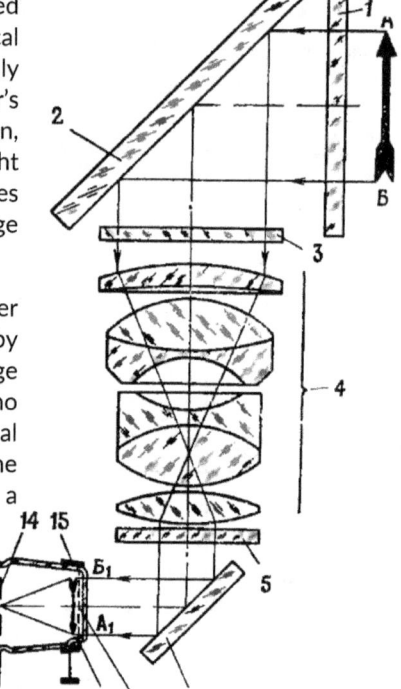

Figure 51: TPN-1-49 optical scheme. (*Soviet Ministry of Defense*)

Without infrared illumination, there was no visible image in the sight. A manually operated iris diaphragm was provided to vary the image brightness[234] and to protect the image converter tube during daytime. Together with a sunlight protection filter[235], the diaphragm was the only mechanism protecting the tube from overexposure[236]. In combat, if the gunner had to aim at a blinding light source (e.g. a tank with its own spotlight turned on), he had to manually adjust the iris diaphragm to darken the image.

---

[233] The anode terminal for the image converter tube was on the left side of the TPN-1, but on the TKN-1, it was on the right side. The tube was flipped over, so on the TKN-1, the reticle was inverted.
[234] To permit boresighting during daytime, aiming at bright light sources during night combat, etc.
[235] A protective filter inside the sight permitted only near-infrared radiation to pass into the image converter tube. This was meant to protect the sight during daytime.
[236] Muzzle flash protection was added in the (non-Soviet) TPN-1M with glare protection BS-1.

There was no means of ranging a target apart from using the graticules to perform stadiametric estimation. This was not unusual for night sights, and it was likely considered adequate for the relatively short range of the TPN-1. The viewing range of the TPN-1 was mainly restricted by the degree of illumination, but even so, a concession was made to image quality in the optical design by having a Zeiss Biotar objective lens assembly[237]. This type of lens was designed to produce a sharp image by reducing optical aberrations. With eight air-glass surfaces in the lens assembly, transmittance losses were high, which was atypical for a night vision device.

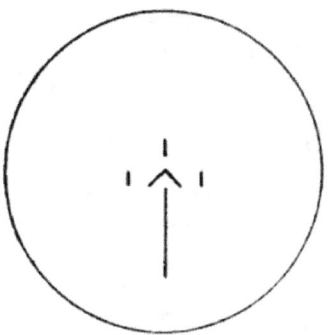

Figure 52: TPN-1 reticle. (*Soviet Ministry of Defense*)

TPN-3-49

The TPN-3 was essentially a deep modernization of the TPN-1 to obtain a passive night vision capability. Optically, it was identical to the TPN-1, and structurally, it was made to fit in the same cross-pin mount and the same sight head casing. The TPN-3 differed primarily in that it had two single-stage tubes: an image converter with an S-1 photocathode, and an image intensifier with an S-20 multialkali photocathode. By mechanically switching between these two tubes with a sliding mechanism, the gunner could choose between passive and active night vision as the situation demanded.

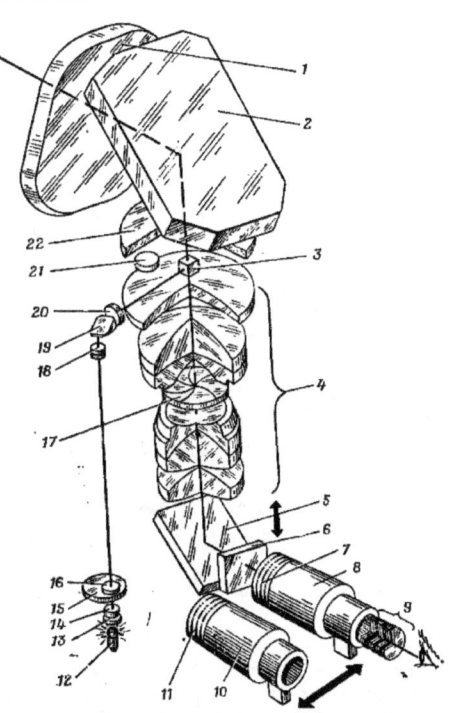

Figure 53: TPN-3 optical scheme. (*Soviet Ministry of Defense*)

---

[237] Biotar lenses for civilian cameras were also in mass production at KMZ using original German equipment from the Carl Zeiss factory.

An S-1 photocathode was better suited to active night vision mainly because S-20 photocathodes had a weak spectral response to near-infrared light, making the spotlight largely ineffective. However, passive night vision was strongly preferred. Reverting to the active mode was only justified when natural lighting conditions were too poor for a legible view.

Because the TPN-3 did not use a multi-stage cascaded tube[238], its viewing range was modest, only managing to surpass portable night vision goggles and periscopes thanks to the light collecting power of its large objective lens. It was considerably less capable than the West German PZB-200 low light television (LLTV) system[239], which could detect a tank-type target at twice the range[240].

Figure 54: TPN-3 reticles: KE (left); HEAT (center); HE (right). (*Soviet Ministry of Defense*)

The distance scales were cuttings on a disc, through which a light would shine to project an image of the scales in the gunner's view. By rotating the disc, three different sets of scales could appear in the image[241].

The TPN-3 notably featured an automatic flash protection system, triggered by a photodetector or the electric trigger for the main gun. When the protection threshold of 4 lux (twilight) was reached, or when the main gun trigger was pressed, the metal casing around the image converter/intensifier tube was magnetized, deflecting the electron path in the tube and thus preventing an image from forming on the phosphor screen.

---

[238] The newer TPN-4 *Buran-PA*, fitted to the T-80U, had a two-stage cascading multialkali tube.
[239] Köppen, U. in Kotsch, S. (2023). *Das passive Ziel- und Beobachtungsgerät PZB 200*. Available at: https://www.kotsch88.de/f_pzb200.htm (Accessed: 15 September 2023)
[240] 370 m identification range, 960 m detection range. See: Systems «Capris» (1978). *Bericht über die Untersuchung des Feind-Freund*, in Каплин, М. Е., Третьяков, В. Г., Харлашкин, С. А., Шашков, И. Н. (1984). 'Испытания Западногерманской Аппаратуры Опознавания „Свой - Чужой", *Вестник Бронетанковой Техники* 1984, сборник 4. p. 54.
[241] Беляков, С.А (2001). *Приборы ночного видения бронетанковой техники*. Омск: Издательство ОмГТУ. pp. 78-80.

1K13-49

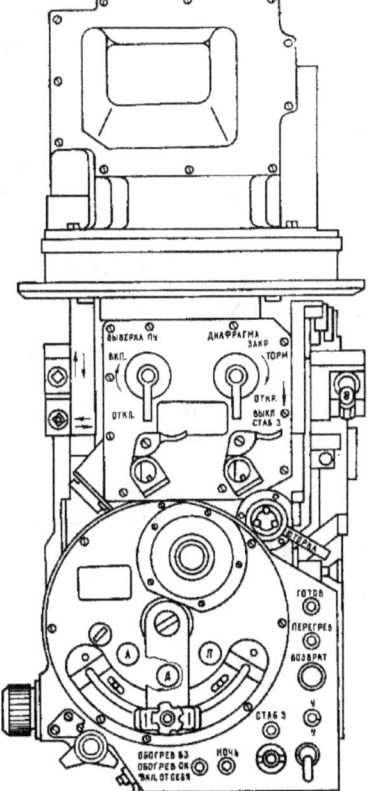

Figure 55: 1K13-49 sight. (*Soviet Ministry of Defense*)

The 1K13 combined the features of the TPN-3 with a comprehensive laser emitter system for the missile guidance system, consisting of the emitter itself, the modulator mechanism, a telescope with pancratic (stepless) optical magnification, and a refrigeration system to cool the laser emitter[242].

Through a complex set of modulators, the laser beam was split into four quadrants, each modulated to a different frequency. The modulated beam was then reflected into the optical path of the day channel. The low pulse power of all four guidance sectors and the different modulations seriously complicated the detection of the laser by the target[243].

The ATGM guidance modules were arranged in a compartment on the backside of the sight, specially shaped to fit the contour of the turret cheek[244]. This made it impossible to retain the previous cross-pin mount, so the 1K13 was instead bolted directly to the barbette for its periscope head.

The vertical stabilizer unit was housed in the sight head. Its electric motor was powered and controlled by relay from the gyroscope of the 1A40-1 system via a multi-pin cable connection. The 1K13 thus did not require its own gyroscope. Turning on the stabilizer disconnected the push-rod connection between the sight and the gun, so that the head mirror was moved only through its belt drive with the stabilizer motor.

---

[242] Ibid. p. 31.
[243] Абрамов, А. И., Гуменюк, Г. А., Евдокимов, В. И., Зборовский, А. А. (2015). 'Опыт оснащения бронетехники аппаратурой регистрации лазерного излучения', *Известия Российской Академии Ракетных и Артиллерийских Наук*, 2(87). pp. 53-54.
[244] The 1K13 models for the T-55AM and T-62M were likewise shaped for different turret contours.

As a night sight, the 1K13 provided passive and active night vision in a similar manner to the TPN-3. Switching between the two modes was done by turning a rotary selector to swing one tube in front of the eyepiece lens group for another, or, in the daylight position, to exchange it for a simple erector lens. The TPN-3-type projected reticle was carried over.

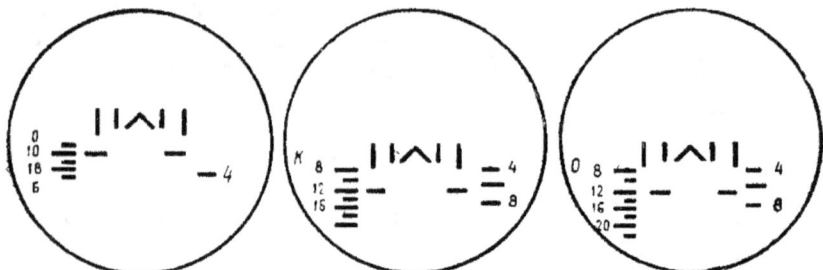

Figure 56: 1K13 night reticles: KE (left); HEAT (center); HE (right). (*Soviet Ministry of Defense*)

The day mode had its own viewfinder, with a special reticle offset to the right by 3 mils relative to the center of the sight to ensure that the ATGM flew into the guidance beam almost immediately after leaving the muzzle of the gun[245].

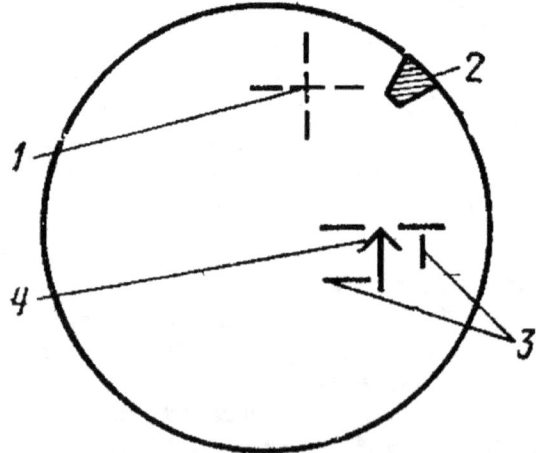

Figure 57: 1K13 day reticle. 1 – boresighting cross; 2 – ready to fire signal light; 3 – mil scales; 4 – missile aiming mark. (*Soviet Ministry of Defense*)

---

[245] Министерство обороны РФ (2001). *Танк Т-72Б: Комплекс управляемого вооружения 9К120 - Техническое описание и инструкция по эксплуатации*. Москва: Воениздат. p. 26.

Driver's Station

A T-72 driver, or driver-mechanic as tank drivers were officially called in the Soviet Army, was responsible for safely handling the tank during marches and in combat maneuvering, often while coordinating with the commander and gunner. The driver was also responsible for keeping check of the condition of the tank's drivetrain and suspension, although maintenance was always a crew effort.

Inside his station, the driver was seated directly underneath his hatch with his head in a cavity cut into the upper glacis armor. Like several Soviet tanks preceding the T-72, there was a hull oversize warning lamp on each side of the driver's periscope to warn the driver whenever the tank gun traversed over either side of the hull. The lamps ensured he was aware of which direction the gun was pointing before going through an obstacle or entering a narrow passage, allowing him to compensate or alert the gunner to move the gun.

The driver steered the tank with two steering levers. The rest of the controls were a standard set of clutch, brake and accelerator pedals, and a sequential gear selector rack to the right, just like in a car. Unlike a car, the instrument panel was to the left of the driver's seat, so reading its gauges while driving was not very easy, but under normal circumstances, the driver did not have to pay attention to the instrument panel at all. A strip of warning lamps to the right of the main periscope alerted the driver to the commander broadcasting on the intercom[246], to high coolant temperature, to the tank's ventilator (FVU) working in the overpressure mode, and to the service brake activating. This strip of lamps could be unhooked and mounted outside, above the hatch or on the driver's dust shield, so that he could remain alert to operational warnings even when driving head-out.

The two notable shortcomings of the instrument panel were that the two critical gauges – the coolant and oil temperatures – were placed at the top of the panel to be more visible when driving from a head-out position, but were still difficult to see[247], and the switches and displays were not grouped into logically separate "blocks"[248] according to purpose, though this was tangentially related to the decision to position more important displays higher up on the instrument panel for better visibility.

---

[246] In case of a malfunction in the driver's headset or insufficient volume.
[247] This was the norm for tanks, with very few exceptions. See: Заславский, Е. И., Погудин, Е. В. (1976). 'Особенности Формирования Лицевой Панели Пульта Управления Механика-Водителя', *Вестник Бронетанковой Техники* 1976, сборник 2. p. 17.
[248] Noted by Заславский and Погудин, and later, by German experts: Hilmes R. (2007). *Kampfpanzer heute und morgen: Konzepte - System - Technologien*. Auflage: Motorbuch Verlag. pp. 136-137.

Figure 58: General view of the driver's station from the driver's perspective. The strip of four warning lights on the right of the periscope are visible. (*S. Stauber, 2024*)

Figure 59: Left side of the driver's station, showing the driver's instrument panel. (*S. Stauber, 2024*)

Figure 60: Right side of driver's station, showing the control panels for the ZETs11-3 firefighting system and GO-27 CBRN protection system, partially obscured by the hatch opening mechanism. (*S. Stauber, 2024*)

The low, heavily sloped shape of the hull nose gave the driver good legroom, but little headroom. There was 1,219 mm of space between the pedals and the seat backrest, but the vertical clearance at the driver's seat measured just 928 mm tall, identical to a T-54/55 (927 mm). For a conventional upright seating posture, this was prohibitively short even according to a 50th percentile design standard. In a tank with a short hull nose like the M48 and M60, the driver's station was much taller internally at 1,054 mm (41.5")[249] but there was only 889 mm (35") of space[250] between the back of the seat and the accelerator pedal.

The height of the T-72 driver's station, modest as it was, had been a major concession to comfort, as the hull nose area was otherwise even shorter. Structurally, the internal height at the third suspension unit[251] just behind the driver's seat was only 852 mm. A tub-shaped bulge was stamped into the belly plate to lower the driver's seat by 63 mm, and a compromise to headroom was made in the driver's radiation protection by limiting the thickness of the lining on the underside of his hatch to just 22 mm[252]. With the seat in its fully lowered position, the vertical space between the seat and the underside of the hatch amounted to just 830 mm, equal to a T-54/55.

The basic requirement for the height of the T-72's driver's station was based on a rudimentary model without accounting for essential factors like allocating a certain amount of clearance for the driver to bounce off his seat during rough cross-country driving, or be thrown off it in the event of a mine blast.

For comparison, in a Leopard 2, the internal hull height at the driver's seat was 975 mm, and 85 mm of this space was taken up by the seat itself for a total usable headroom of 890 mm. If the Leopard 2 was still too short for its own 95th percentile design requirement, it at least gave more head clearance for off-road driving. Another factor was the driver's more relaxed posture, because in a Leopard 2, the driver's periscopes were installed in the rear edge of his hatch.

A T-72 driver had a classical working posture [**Fig. 61**] historically common to Soviet tanks. The large knee angle (120°) was the same as in the Leopard 2[253], but the hip angle was significantly smaller (62° vs 80°), so while a Leopard 2 driver sat in a relaxed pose at a 15° recline, a T-72 driver was, at best, sitting upright. Because the TNPO-168V periscope was installed in the upper glacis,

---

[249] Against a requirement of 40 inches. See: Halbert G. A. (1983). 'Elements of Tank Design', ARMOR, (November-December). p. 36.
[250] Klein, R., Erickson, E. (1967). 'Road Test the M48', ARMOR, (January-February). p. 45.
[251] Structural height, not including the 50 mm anti-radiation liner on the ceiling.
[252] This made the underside of the hatch approximately level with the hull ceiling. Thickness based on Object 172M hull technical drawing.
[253] Krapke, P. -W. (2004). Leopard 2: Sein Werden und seine Leistung. Norderstedt: Books on Demand GmbH. p. 143.

ahead of the hatch, its eyepiece was 480 mm ahead of the seat backrest[254], so in most cases, the driver had to lean slouch or lean forward to get a good view.

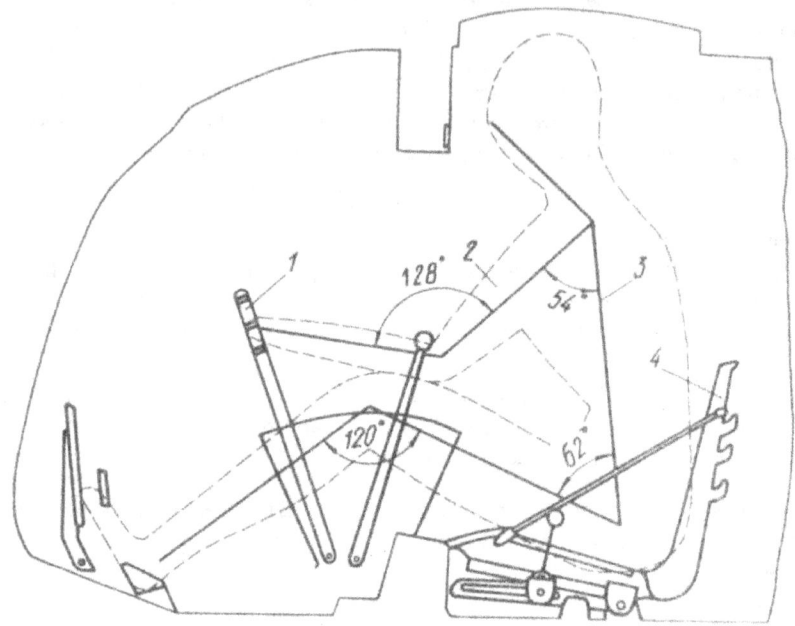

Figure 61: Joint and hip angles of a T-72 driver in a normal working posture, seat in the 90° position. (*A. P. Sofin & V. V. Smolin, 1979*)

Unsurprisingly, back pain was a chronic complaint, albeit not exclusive to the T-72. This was mitigated only to some extent by the ability to drive from an open hatch position on long marches.

The seat cushion was tilted by 15° for thigh support. The backrest could be set to three positions; 90° (perpendicular), or reclined at 105° or 120°. The seat could be adjusted forward and backward by 90 mm (in 18 mm increments) by pulling on a release hook (17), and lifted on a toothed arc by 180 mm by pulling on a release knob (13). The seat lifting mechanism was a torsion bar on a swing arm, so the driver was tilted forward to a more convenient posture when lifting the seat into a head-out driving position.

---

[254] Троицкий, В. В. (1987). 'Рабочие Места Экипажей в Отечественных и Зарубежных Танках', *Вестник Бронетанковой Техники* 1987, сборник 12. pp. 10-11.

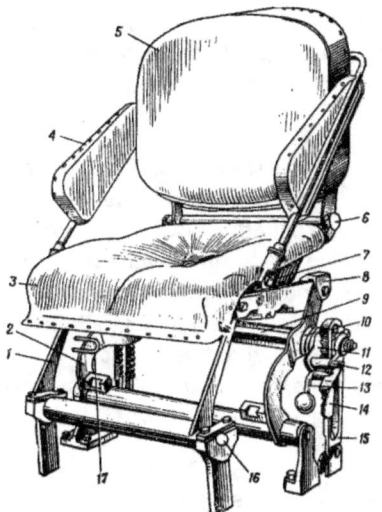

Due to his proximity to the ground, the driver was 1.5-2 times more exposed to radiation overall than the rest of the crew, particularly from secondary gamma radiation penetrating through the floor of his station. To compensate, the driver's seat had a lead plate built into the bottom, 15 mm thick and weighing 20 kg. It gave an 18-fold improvement in secondary gamma radiation attenuation [255]. For some reason, no lead plate was mentioned in any documentation or in the parts catalogue for the T-72. Instead, the weight of the driver seat's "cushion" was mysteriously listed at over 20 kg [256].

Figure 62: Floor-mounted driver's seat. (*Soviet Ministry of Defense*)

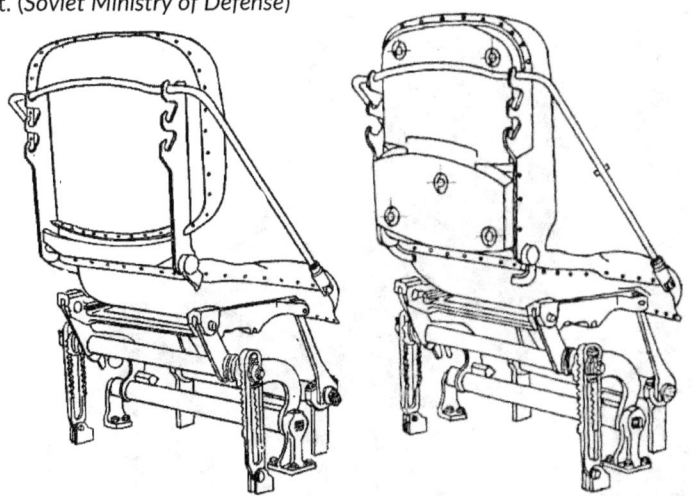

Figure 63: Late variants of floor-mounted driver's seat. Variant with removed side bolsters (left) and variant with anti-radiation panel (right). (*Soviet Ministry of Defense*)

---

[255] Решетов, А. А., Троицкий, В. В., Немцева, Г. А. (1982). 'Новое сиденье водителя танка', *Вопросы Оборонной Техники*, 20(103). p. 20.
[256] Министерство Обороны СССР (1980). *Танк Т-72 – Каталог деталей и сборочных единиц*. Москва: Воениздат. p. 237.

Radiation protection was a topic of continual improvement, so in 1978, the side bolsters [**Fig. 62**, 4] were removed in connection with the introduction of the DSP-I1 anti-radiation vest. Then, an anti-radiation panel was added to the backrest at some point in the 1980's.

In 1987[257], the mount for the seat was reworked to a ceiling-suspended frame as part of a major mine protection upgrade[258]. This modification was applied to the T-72B and to the T-72M1. The new seat was paired with a new escape hatch and additional reinforcing ribs in the hull belly next to the driver's tub-shaped bulge, designed to reduce floor deformation under blast overpressure. If the deformation of the belly plate was extensive enough to physically touch the seat, the floorplate of the suspended seat mount served as a second layer of protection. These measures protected the driver from a 10 kg TNT blast mine detonating at the inner edge of either track under the 1st road wheel.

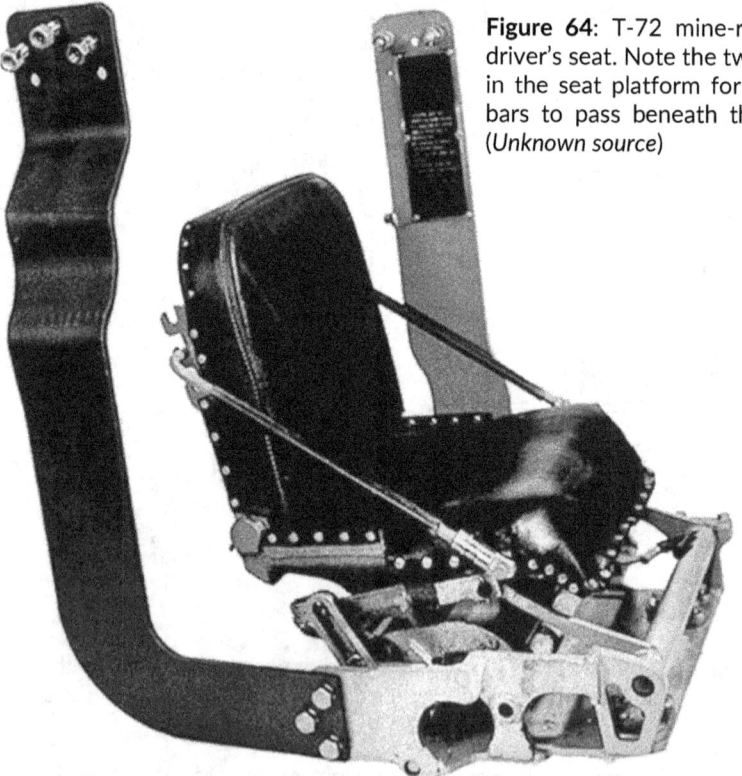

Figure 64: T-72 mine-resistant driver's seat. Note the two holes in the seat platform for torsion bars to pass beneath the seat. (*Unknown source*)

---

[257] Альбом основных конструктивных изменения, проведенных на Изд. 184, 184-1, 184К, 184К-1.
[258] Григорян, В. А. (ed.) (2007). *Защита танков*. Москва: Издательство МГТУ им. Н.Э. Баумана. p. 104.

The driver's single forward-facing TNPO-168V periscope was his main window to the outside world, supplemented by two auxiliary TNPA-65A hatch periscopes. The right hatch periscope was primarily intended as a backup for the TNPO-168V if it was damaged in combat and there was no time to replace it with a spare. Only the left hatch periscope meaningfully widened the driver's viewing arc, but nevertheless, the small size of the hatch periscopes made them inconvenient and impractical outside of low speed, close-quarters maneuvering. The lack of heating or window cleaning was also an issue on bad roads and in bad weather. The TNPO-168V had both heating and a button-activated window washer, selectable between a jet washer or a dry air jet cleaner.

The TNPO-168V alone gave broadly equivalent visibility to the two smaller driving periscopes in T-54/55 and T-62, and in fact, this equivalence is quite literal; a 1960 draft design for the Object 432 had two smaller forward-facing periscopes where the TNPO-168V would eventually be fitted.

The TNPO-168V had a slightly wider field of view (138°)[259] than the old two-periscope setup and was more convenient to look through, but it still followed the same basic requirement of affording the driver a view of the front corners (mud guards) of the hull to safely maneuver through obstacles and in tight spaces. A centering line etched in the eyepiece window helped the driver line up the tank with narrow paths. There were no requirements for side vision, so one periscope was deemed sufficient. Similar requirements existed abroad, but were often more demanding. The U.S. Army, for instance, set a requirement for vision in a 180° arc, which was almost met in the M48 and M60 series (170°)[260].

With the TNPO-168V alone, the driver's viewing arc from a normal eye relief was 116°. With the left TNPA-65A, this arc expanded to 158° [**Fig. 66**]. In theory, this was no worse than the T-80, which had an array of three periscopes to cover a static field of view of 137°, and when traveling over open ground, this was essentially true. However, comparative testing found that when crossing narrow obstacles like a path through a minefield or a path dug through an anti-tank ditch, the T-80 periscope layout accounted for a 25% higher average speed[261]. Despite an arguably worse view for general driving, it was noted that T-80 drivers could navigate more confidently through these obstacles while T-72 drivers tended to slow down more noticeably or stop before attempting to cross.

---

[259] Two periscopes with 72° field of view each, overlapping by 15°. Total field of view was ~129°.
[260] Добисов, О. А., Затравин, Е. И., Михеенков, П. И., Разумов, В. М. (1976). 'Особенности Системы Управления Огнем', *Вопросы оборонной промышленности*, 20(67). p. 29.
[261] Германов, Е. А., Заславский, Е. И., Проскурова, Н. С., Шпак, Ф. П. (1982). 'Влияние Обзорности с Места Водителя на Подвижность Танков', *Вопросы Оборонной Техники*, 20(103). pp. 15-18.

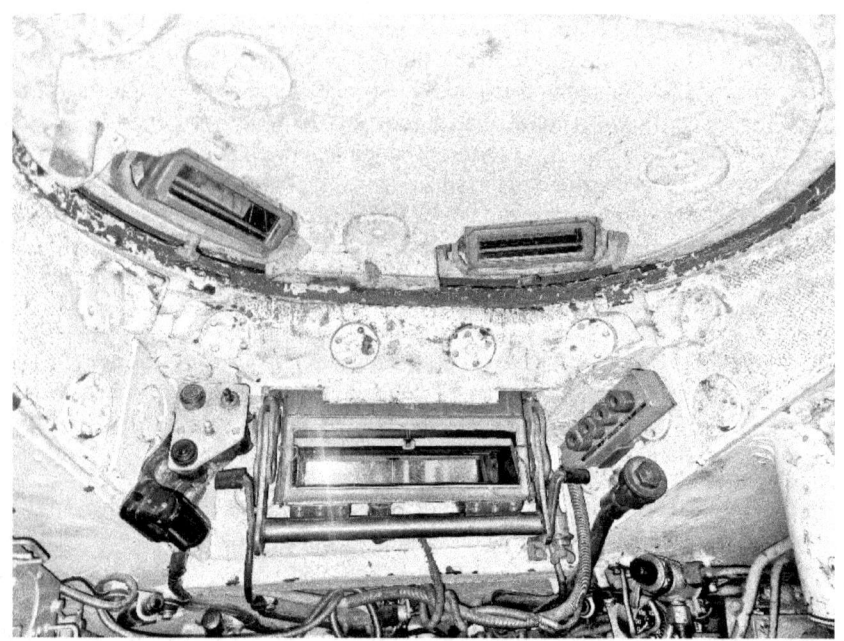

Figure 65: View of all three of the driver's periscopes. (S. Stauber, 2024)

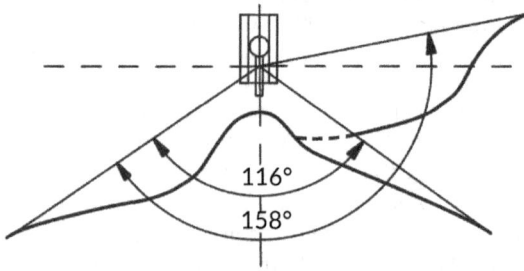

Figure 66: Viewing arcs through driver's TNPO-168V periscope alone and together with TNPA-65A hatch periscope. (E. A. Germanov et al. 1982)

Driving at Night

At night, the driver typically relied on the tank's lone FG-127 headlight on the left of the hull nose. This was an ordinary FG-126 headlight with a blackout hood and dimmer switch. In combat zones, the security risk of emitting visible light, even a thin sliver of light out of a blackout hood, could be mitigated by switching to active infrared night vision.

On the very earliest T-72s the TVN-4PA night vision periscope was provided. It could be mounted in place of the TNPO-168V main periscope, or mounted outside on a special bracket in front of the driver's hatch for head-out driving, since tank drivers were not issued night vision goggles[262]. The periscope was made for stereoscopic vision with a conventional set of binocular eyepieces. Stereoscopic vision helped drivers to judge the distance and size of incoming obstacles and terrain features, but in practice, drivers had to lean very far forward to press up against the eyepieces, which was very uncomfortable and, in the long term, dangerous. The field of view was 35°, and the driver could see up to 40-50 meters[263] away with infrared headlights.

The periscope had an internal two-position rotating shutter that could be either opened or closed. With the shutter open, the amount of light entering the converter tubes could be regulated by an external set of four-position Zeiss-type rotating diaphragms manually clipped over the objective window. These diaphragms adapted the periscope to daytime use with the largest aperture for twilight and the smallest for broad daylight. There was no iris diaphragm to regulate brightness from inside the tank.

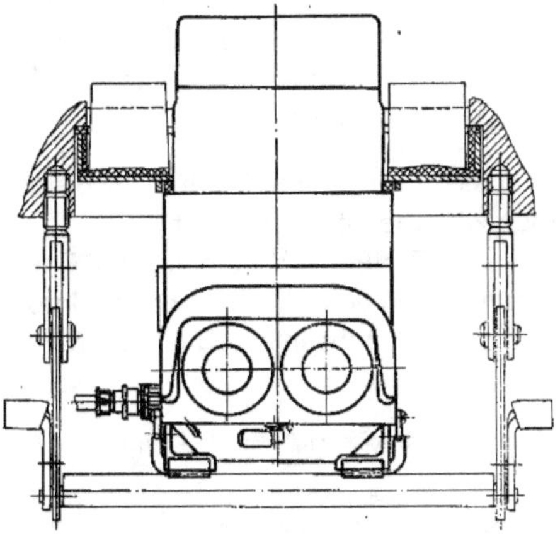

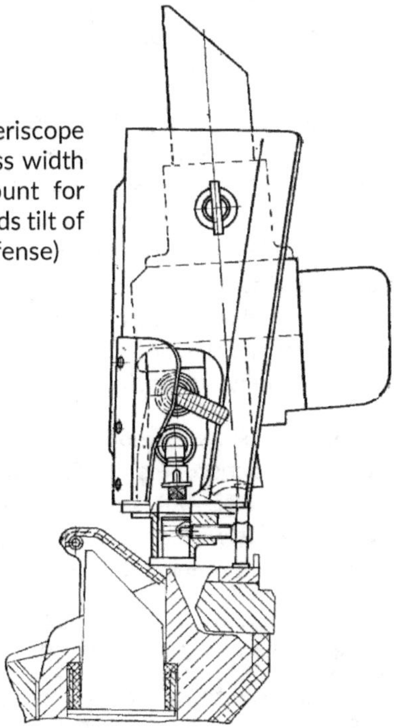

Figure 67: TVNE-4PA installed in driver's periscope slot with a plastic plug filling out the excess width (left) and TVNE-4PA on an external mount for head-out driving (right). Note the downwards tilt of the external mount. (Soviet Ministry of Defense)

---

[262] PNV-57 night vision goggles were issued to truck and car drivers, but not tank drivers.
[263] Министерство Обороны СССР (1978). Объект 172М - Пособие по Проверке Технического Состояния и Содержания. Москва: Воениздат. p. 19.

To reduce light interference from a bright horizon, a small lever on the right of the periscope could be turned to lower an internal "curtain" shutter made from black plastic fabric over the converter tube inputs, blocking out the sky like the sun visor in a car[264].

Instead of the familiar side-by-side headlight layout traditional for automobiles, the T-72 had one FG-125 (an FG-126 with an infrared filter) on the right of the hull nose, and another FG-125 on the turret, next to the gunner's night sight. This type of staggered layout was standard for Soviet AFVs, including turretless vehicles. Its purpose was to improve the uniformity of illumination in depth by angling the beams from the hull and turret headlights to coincide with an obstacle in the tank's path at 35 m and 20 m respectively[265].

Figure 68: TVNE-4B night vision driving periscope. (*Soviet Ministry of Defense*)

This arrangement also illuminated less of the tank, and when fording a stream or crossing muddy terrain, the turret headlight could be expected to remain clean and unobscured. When moving on rough terrain, off-center illumination improved road and obstacle visibility by revealing the relief of the terrain, accounting for up to a 10% increase in driving speed. However, infrared headlights were easily detectable by NATO forces[266], who were themselves densely equipped with infrared night vision devices by the 1970's. This problem was especially acute when tanks drove in large convoys.

The TVN-4PA was replaced almost immediately by the TVNE-4PA passive night vision periscope by 1975, and then with the introduction of the T-72A in 1979, the TVNE-4PA was replaced by the TVNE-4B[267] modification. These TVNE-4-series periscopes modified the basic TVN-4 binocular design into a hybrid system, keeping the V-2K image converter tube in its left monocular but

---

[264] Министерство Обороны СССР. *Прибор ночного наблюдения ТВНЕ-4Б - Инструкция по эксплуатации*.

[265] Министерство Обороны СССР (1986). *Танк Т-72А - Техническое описание и инструкция по эксплуатации: Книга Вторая (Часть первая)*. Москва: Воениздат. pp. 225-226.

[266] Department of the Army (1973). *Comparison Test of Driver's Night Vision Devices, M60 Series*. Fort Knox: Army Armor and Engineer Board. p. 11.

[267] Сафонова, Б. С., Мураховского, В. И. (eds.) (1993). *Основные Боевые Танки*. Москва: Арсенал-Пресс. p. 68.

replacing the right monocular with a V-2 image intensifier tube[268] featuring an S-25 photocathode.

Drivers saw different images in the two eyepieces, because the output of the infrared headlights closely matched the spectral sensitivity curve of an S-1 photocathode but not an S-25 photocathode, so active infrared illumination was effective for the V-2K, but not the V-2. This took some effort to get used to, and of course, stereoscopic vision was lost. The decision to retain an active system in spite of this rather serious inconvenience was to preserve a fallback option for particularly dark nights and for driving in forests, where passive night vision was ineffective. The viewing range was 60 meters with the infrared headlights, or 100 meters in the passive mode at a natural night luminance of 0.005 lux[269] (moonless night). In tests, the driving speed at night using the TVNE-4B was found to be on par with the British AV11-L7A1 night periscope and the American AN/VSS-2[270].

Batteries

Four batteries were mounted on a rack to the left of the driver, behind the front left fuel tank. Soviet lead-acid tank batteries were regulated in size, and the output was rated according to 12-volt automotive standards so that different types of batteries could be used if needed[271]. The T-72's operating voltage on battery power was thus 24 V instead of the normal 27 V from the generator. For most T-72 models, the 6STEN-140M was used, with the option of using the 12ST-70M. During the 1980's, the 12ST-85R began to proliferate. The essential difference between these types was that the 6ST-series were 12 V batteries, and the 12ST-series were 24 V batteries[272]. Both types were common, and only a trivial wiring adjustment[273] was needed to switch between them.

For combat vehicles, the critical trait of lead-acid batteries was their ability to discharge at very high currents, even if charge capacity was poor when discharging this way. If needed, the stabilizer (3.5 kW), autoloader (2 kW), and

---

[268] Беляков, С. А. (2001). *Приборы Ночного Видения Бронетанковой Техники*. Омск: Омский Государственный Технический Университет. p. 47.
[269] Министерство Обороны СССР (1975). *Танк «Урал». Техническое описание и инструкция по эксплуатации (172М.ТО)*. Книга Первая. Москва: Воениздат. p. 26.
[270] Буланкин, Н. Г., Корнилов, В. И., Михаиловский, В. Е., Погудин, Е. В. (1984). 'Сравнительные Испытания Ночных Приборов Водителя', *Вестник Бронетанковой Техники* 1984, сборник 1. p. 6.
[271] Министерство Обороны СССР (1983). *Свинцовые Стартерные Аккумуляторные Батареи - Руководство*. Москва: Воениздат.
[272] 6STEN-140M and 12ST-70M had wooden (!) cases. This was a fairly common form factor for automotive lead acid batteries in early 20th century cars like the Ford Model T. The 12ST-85R had a modern fiber-reinforced plastic case.
[273] 'Тема 3: Общее устройство системы электрооборудования. Источники электрической энергии', *Уральского федерального университета*.

the crew ventilation and protection system (1.5 kW) in the T-72 could all run solely on battery power for a short time, but this depended heavily on the condition of the batteries.

The 12ST-70M and 12ST-85R batteries were introduced to improve the usable charge capacity by parallel-wiring higher-voltage batteries, allowing each battery to discharge at a lower current. They were also more cold-resistant in both discharge power and charge capacity, which was significant for the T-72 since battery heating was absent. The temperature limit where the batteries could supply enough power to reliably start the engine[274] was -21°C with the 6STEN-140M, and -26°C with the 12ST-70M.

Table 13: T-72 battery specifications

Technical Characteristics	6STEN-140M	12ST-70M	12ST-85R
Wiring scheme	Series-Parallel	Parallel	Parallel
Battery voltage (V)	12	24	24
Maximum discharge current (A)	420	400	400
Total system capacity (Ah)	280	280	340
Individual capacity at max. discharge current (Ah)	12.6	8.0	8.0
Individual capacity at -30°C (Ah)	15.4	6.0	8.0

---

[274] Provide enough start attempts to guarantee a successful start.

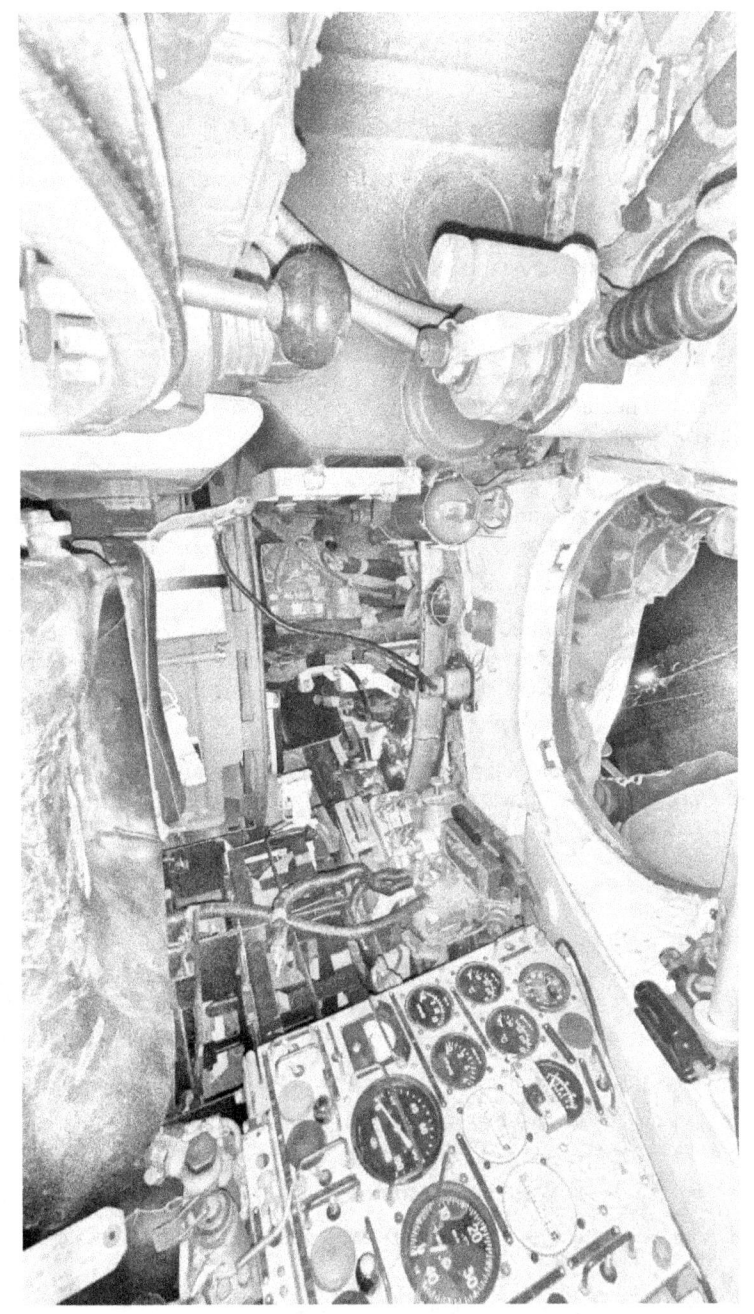

Figure 69: Rearward view of the driver's station, showing the battery rack behind the instrument panel. (*P. Kefalinos, 2024*)

Inertial Navigation

The GPK-59 gyrocompass was a navigational heading indicator designed to permit waypoint navigation without visible landmarks. The need to navigate this way could be due to low visibility weather conditions, the absence of sufficient moonlight at night, or when navigating in forests, steppes or deserts. It was also used when crossing a river by snorkeling. The drift rate of the gyrocompass in normal operation was ≤2.4 degrees in 30 minutes[275], not at all comparable to a bona fide inertial navigation system[276].

Although it added cost, the GPK-59 gyrocompass made rudimentary inertial navigation possible for all Soviet AFVs, not just the T-72. A standard military lensatic compass could not be used in or on a tank, special magnetic compasses had large heading errors, and using the stabilized turret as an inertial navigation device[277] was not always practical.

Just like aircraft attitude indicators and heading indicators, the GPK-59 had an air jet mechanism to automatically perform Schuler tuning[278] to reduce its drift rate. These air jets were produced by a turbine on the fast-spinning gyroscope rotor. When the local vertical reference axis was unchanged, there was no net torque from these air jets, but as the tank moved, the local vertical reference axis would change minutely (due to the curvature of the Earth) while the gyroscope remained unmoved in inertial space. The small asymmetry in the flow of the air jets generated a torque, leveling the gyroscope, and improving navigational accuracy – all without any additional moving parts.

On command tanks, a navigation was done through a TNA-3 or TNA-4 high-precision inertial navigation system. Inertial navigation systems saw a progressive advancement in navigational precision from the TNA-2[279] to the TNA-3, and to the TNA-4[280].

The TNA-3 and TNA-4 were inertial navigation devices based on a very high-speed vacuum gyroscope. These navigation devices had a rotating dial representing the tank to show the heading of the tank, and a rotating ring to

---

[275] Министерство Обороны СССР (1986). *Танк Т-72А - Техническое Описание И Инструкция По Эксплуатации: Книга Вторая (Часть первая)*. Москва: Воениздат. pp. 232-237.

[276] TNA-3 (1974) in T-72K - drift rate of 34 mrad over a 1-hour period; TNA-4 (1987) in T-72BK - drift rate of 80 mrad over 7-hour period.

[277] The heading is determined by compass from outside the tank, then the stabilized gun is pointed towards the objective. The gun itself then becomes a pointer needle for the driver. See: Davis, G. C. (1982). 'Low Visibility Tactical Navigation", *ARMOR*, (January-February). p. 45.

[278] Huber, C., Bogers, W. J. (1983). *The Schuler principle: a discussion of some facts and misconceptions*. EUT Report 83-E-136. Eindhoven: Technische Hogeschool Eindhoven. Available at: https://pure.tue.nl/ws/files/4325185/8407775.pdf (Accessed: 20 September 2023).

[279] TNA-2 *Setka*, fitted to T-55K tanks since 1959.

[280] Фастовский, А. Х. (1980). 'Исследование ошибок навигационных систем ВГМ', *Вестник Бронетанковой Техники* 1980, сборник 3. p. 42-43.

show the heading of the destination. The device was meant to be read together with a paper map to allow a command tank to track his position and a destination. These systems continuously tracked the coordinate position and bearing angle of the vehicle in motion[281] and continuously calculated how much distance was left to the destination and the bearing angle to the destination.

The device took a starting position as an input, which had to be determined by external methods, and then, by combining a tank speed sensor in the T-72's left idler wheel, performed digital signal processing on the tank's speed in its a gyroscope coordinator to determine the distance traveled by the tank in any direction, and then calculate how close or how far the tank was in relation to the destination. The position of the tank was plotted on a map in a handheld tablet. The starting position was set by placing the plotter on a coordinate point on a 1:50000 or 1:100000 scale paper map using a y-axis knob and an x-axis knob[282]. As the tank moved, so did the plotter.

The heading of the destination was duplicated onto a heading indicator at the driver's station. Rolling mechanical counters displayed the coordinates of the tank and the distance left to the destination. The input with the highest system error was the tank speed due to track slippage, especially on sandy or icy terrain. Pre-calculated correction coefficients were set in the system to compensate. In total, the system error reached 0.9% of the total distance.

These systems were fiddly and took well-educated officers with signals training to operate, but gave Soviet armored units a significant degree of self-sufficiency in navigating large distances. The U.S. Army lacked such a capability even into the 1980's[283].

---

[281] Кузнецов, М. И., et al. (1978). *Танковые навигационные системы*. Москва: Воениздат.
[282] Like an Etch-a-sketch tablet.
[283] Elliott, E. C. (1983). 'Soviet land navigation', *Journal of Terramechanics*, 19(4). pp. 217-223.

Chapter III

CBRN Protection

The Nuclear and Radiological Threat

In 1957-1958, the Soviet Army underwent a drastic transformation in anticipation of the nuclear threat. During this period, filter-ventilation units (FVUs) became a new standard for not just combat vehicles, but many enclosed vehicles intended to operate in rear areas, and a universal OZK[284] set was introduced for all branches of the military. When operating in areas contaminated by fallout, the main priority was to reduce by as much as possible the gamma radiation dose, which was the main external radiological hazard of early fallout[285].

For a while, the internal radiological hazard of fallout was commonly thought to be a more serious threat, as biological damage from irradiated particles in the bloodstream tends to be more acute than from external contamination. In practice, however, it is difficult for fallout to reach the bloodstream by inhalation [286], and the ingestion of contaminated food and water can be completely eliminated by eating and drinking only from sealed containers. Even if this precaution is not observed, such as in the case of the inhabitants of the Marshall Islands after the infamous Castle Bravo test[287], the short-term effects of internal irradiation from fallout were found to be minor compared to external irradiation[288].

Given that breathing filtered air through masks could only partially protect a tank crew from fallout, preventing both internal and external irradiation by sealing the tank from contamination was much more effective. For front-line vehicles, the difficulty in decontaminating crew compartments in field conditions was another strong incentive. Overpressure collective protection systems were the most effective means of accomplishing this, and consequently saw widespread application in fallout shelters and enclosed military vehicles.

In tanks and other AFVs, the stress and intense labor of field operations gave ample justification for this type of collective protection. By reducing or even

---

[284] Общевойсковой защитный костюм - interservice protective suit (CBRN suit)
[285] Glasstone, S., Dolan, P.J (eds.) (1977). *The Effects of Nuclear Weapons*. Third Edition. Washington, D.C.: Department of Energy. p. 594.
[286] Ibid. pp. 597-599.
[287] The inhabitants of the Marshall Islands were studied for acute fallout exposure after miscalculations in the March 1954 Castle Bravo thermonuclear detonation test.
[288] Glasstone, S., Dolan, P.J. pp. 602-603.

eliminating the need for crews to wear restrictive and uncomfortable PPE, a mechanized force could be expected to continue fighting after exposure to WMDs with a much lower risk of heat exhaustion. The Soviet Army was not the first to arrive at this conclusion, but interestingly enough, it was the first major military force to put it into practice on a wide scale, introducing the first comprehensive armored vehicle nuclear protection system in the T-55 medium tank. The critical feature of its protection system was the ability to independently detect a nuclear detonation and react autonomously.

The scope of the protection scheme was soon expanded to include chemical weapons, including nerve agents, leading to the development of the GO-27 system and standardized HEPA filters to provide radiation and chemical protection to all Soviet AFVs, including the T-72. The GO-27 was divided into three modules and connected to the ZETs11-3 or ZETs13-1 firefighting systems in the T-72. The GO-27 was installed in a cavity hollowed out of the right nose fuel tank next to the driver. The entire system took up just 4 liters of space[289].

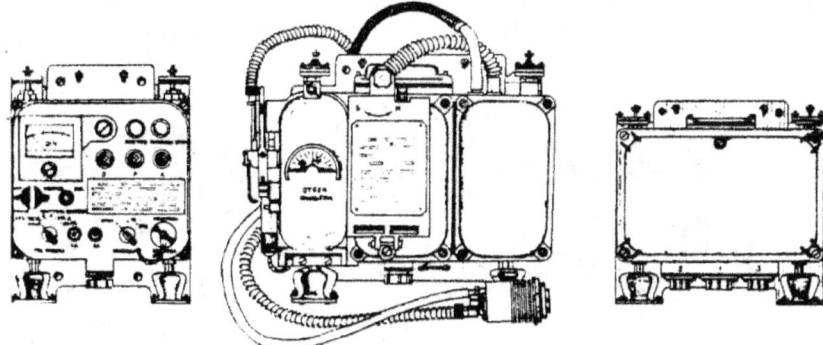

Figure 70: The constituent modules of the GO-27 system. From left to right: B-1 measuring unit, B-2 sensor unit, B-3 power supply unit. (*Soviet Ministry of Defense*)

The B-1 unit had a manual radiation measurement mode with a measurement range of 0.2 to 150 Rads/hr, but this function was normally unneeded. The GO-27 system primarily functioned in a fully automatic mode, with different reactions depending on the nature of the threat detected by the B-2 sensor unit.

---

[289] Исаков, П. П. (ed.) (1990). *Теория и Конструкция Танка - Том 10. Кн. 2: Комплесная защита*. Москва: Машиностроение. p. 132.

Type "R":   Radioactively contaminated site. When the tank was exposed to sustained gamma irradiation at a dose rate of 0.85 Rads/hr and above, the response time did not exceed 10 seconds.

Type "A":   Nuclear detonation. In the event that the tank was exposed to a gamma ray flux with a dose rate of 4 Rads/s and above, the response time did not exceed 0.1 seconds.

Type "O":   Chemical weapon attack. When chemical contaminants were detected, the response time did not exceed 40 seconds.

Once a threat was detected, the GO-27 system broadcast audio alarms into each crew headset and initiated a defense subroutine corresponding to the threat type. A Type "A" threat was a nuclear detonation. The engine was immediately stopped by cutting off its fuel, and all crew compartment air inlets were sealed. 30-50 seconds later, the FVU was turned on in the filtered mode, and the tank could resume normal operation. The reaction to type "R" threats was limited to switching the FVU to the filtered mode and turning it on, if it was not already running. Type "O" threats were handled the same way[290].

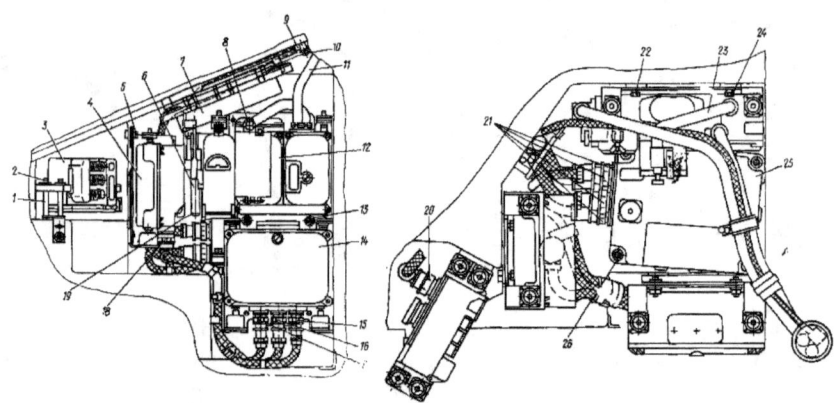

Figure 71: GO-27 installation in right nose fuel tank cutout. (*Soviet Ministry of Defense*)

The basic purpose of these defense subroutines against irradiation and chemical contamination was to reduce the danger in crossing unreconnoitered areas, and from that, erode the effectiveness of nuclear and chemical weapons as area denial assets. This capability was generally absent in the Western counterparts to Soviet tank and motor rifle units. In the U.S. Army, Bundeswehr,

---

[290] ГО-27 Техническое Описание.

and other large armies, tanks might be equipped with portable detection kits at most, relying much more heavily on CBRN protection troops to reconnoiter routes for contamination. As an example, the U.S. Army issued the M43A1 portable detector at a platoon level when a chemical attack was anticipated. It was put on top of the turret next to the commander's cupola, and it could only be used from halts[291]. The sensitivity of the M43A1 was an order of magnitude lower than the GO-27 system at a sampling threshold of 0.1-0.2 mg/m^3.

Chemical attacks were by far the most difficult to defend against. The air sampler provided detection (but not identification) of sarin, soman and V-series nerve agents in aerosol form. Its sensitivity was to the order of 10^{-5} g/m^3 (0.01 milligrams per m^3), with some variation between each of these three agents[292]. It could not detect other chemical agents. At the smallest possible detection threshold, the detection period was too long to give the crew a guarantee against lethal exposure, so in practice, the main use of the Type "O" mode was to react to high concentrations immediately following a direct chemical attack, and to provide a means of detecting residual contamination after prior alert[293].

These chemical agents were detected by an air sampling system consisting of a breather protruding from the hull roof, to the right of the driver's hatch, a filtration system, and the B-2 unit. The air sampler intake was cleaned by an air jet whenever the driver used the jet washer to clean his periscope, based on the idea that whenever the driver's periscope became visibly fouled while driving, the intake would also require cleaning.

An ionization chamber inside the B-2 unit was responsible for radiation measurement, and also chemical analysis by ion mobility spectrometry (IMS). The technology of IMS chemical detection was preferred by militaries worldwide for chemical warning systems, but its drawbacks include its susceptibility to false alarms from non-target chemical compounds (including diesel exhaust particles) and to a loss of sensitivity when the environmental conditions (temperature, pressure, humidity) deviate from a calibrated datum[294].

The GO-27 alleviated some of these issues with a two-stage filtration unit. A cyclonic filter removed coarse dust and water droplets, and a paper filter with a reel mechanism[295] captured fine particles, including water vapor and the soot

---

[291] Annex B in Department of The Army (1997). *Inquiry Into Demolition of Iraq Ammunition*. Available at: https://gulflink.health.mil/army_ig/index.html (Accessed: 12 April 2024).
[292] Лазебник, О. М., Ребриков, В. Д. (1981). 'Система Защиты Танка от Оружия Массового Поражения', *Вестник Бронетанковой Техники* 1981, сборник 4. pp. 26-29.
[293] All three listed agents are heavier than air and tend to settle, becoming absorbed by the ground.
[294] Sferopoulos, R. (2009). *A Review of Chemical Warfare Agent (CWA) Detector Technologies and Commercial-Off-The-Shelf Items*. Report DSTO-GD-0570. Victoria: DSTO. p. 20.
[295] Manually advanced at a certain time interval to maintain fresh paper over the air intake.

from smoke. Electric air intake heating prevented cold air from affecting low-temperature sensitivity. The sampled air was then pumped through the ionization chamber in the B-2 unit. Inside, a plate coated in plutonium-239 served as an alpha radiation emitter to ionize the sampled gas.

Export T-72 models intended for customers outside the Warsaw Pact were fitted with the GD-1M system[296] in the place of the GO-27. Its only sensor was the G-PP1 gamma radiation detector, with which it could initiate and control the same nuclear lockdown functions in the ZETs11-3 system provided by the GO-27, lacking only the capability to detect chemical threats. Given what was used as an alpha radiation emitter in the GO-27, one hardly needs to speculate on why an IMS chemical agent detector was omitted from export T-72 tanks, but it is important to note that these export models merely lacked the ability to independently detect chemical agents. They still retained the same FVU and the protection its HEPA filter offered.

After exiting a contaminated area, the crew could self-decontaminate small parts of the tank exterior with a decontamination sprayer bottle[297]. The small volume of decontamination fluid was enough only to decontaminate the points of the tank that the crew could not avoid touching when mounting and dismounting[298].

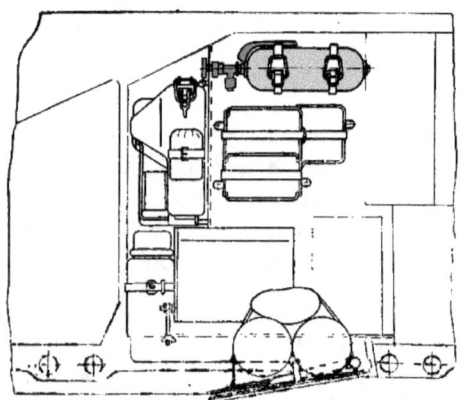

Figure 72: Stowage point for decontamination spray bottle. (*Soviet Ministry of Defense*)

Combat in contaminated areas came with its own set of problems. The T-72 autoloader ejected spent stub cases during each reloading cycle, periodically unsealing the tank, but firing the main gun had an even more dramatic effect due to its bore evacuator. Although a bore evacuator was invaluable for reducing powder fume poisoning, the negative pressure in the fighting compartment from the outflow

---

[296] *Танк Т-72М1: Техническое описание и инструкция по эксплуатации – Часть I.* p. 280.
[297] Министерство Обороны СССР (1970). *Танковый Дегазационный Комплект (комплект ТДП) - Техническое описание и инструкция по эксплуатации и паспорт №4292.*
[298] Министерство Обороны СССР (1962). *Как действовать в условиях применения ядерного, химического и бактериологического оружия - пособие солдату и матросу.* Москва: Воениздат. p. 143.

of propellant gasses worked against the hermetic seal created by an internal overpressure.

With each loading and firing cycle, the crew compartment experienced cycles of decompression and compression[299]. This effect was strongest when the bore evacuator was charged to high pressure by a subcaliber round; in tests with the T-64A, pressure drops of up to -1,050 Pa lasting for 0.8 seconds would develop, and lower but significant pressure drops would still occur with lower-pressure HEAT or HE-Frag rounds.

In fact, even when exceeding the limits of a tolerable crew compartment over-pressure (calculated from 51-204 mmH2O), mathematical modeling showed that depressurization through the gun would still create a period of rarefaction in the crew compartment lasting 0.4-0.5 seconds after a shot[300]. The higher the initial overpressure, the greater the intensity of rarefaction, and the longer it would take to return to the initial overpressure. Contamination was essentially unavoidable if the gun had to be fired in a sealed condition, which was a major tactical consideration in areas contaminated by chemical weapons. On the whole, the bore evacuator accounted for a 3-7-fold increase in the toxic contamination of the crew compartment[301], which in the case of chemical attack, could be lethal. For nuclear or radiological contamination, however, the margins for safe exposure were much more generous.

Prior to the design of the T-72, tests on the T-62's automatic case ejection system had already demonstrated that the periodic unsealing of the tank while operating in an irradiated area did not raise the radiation dose rate to the crew above acceptable limits[302]. In the T-62, the large spent cases contained a large volume of residual fumes, and ejecting the cases after a short delay would cut gas pollution in the fighting compartment by more than half[303]. In the T-72, however, ejecting stub cases did very little to improve air quality in the fighting compartment, since each stub case ejected from the gun contained just 0.5 g of carbon monoxide in residual fumes, 90% of which escape into the fighting compartment in 2-3 seconds. An ejection delay of 4 seconds was too long to give a noticeable reduction in fighting compartment pollution[304], apart from

---

[299] Ребриков, В. Д. (1983). 'Расчет токсической дозы в танке при стрельбе из пушки без эжектора', *вестник бронетанковой техники* 1983, сборник 1. p. 24.
[300] Ibid. p. 23.
[301] Compared to a D-81 with no bore evacuator.
[302] Tested by firing 30 consecutive shots and measuring the radiation dose. See: Павлов М. В., Павлов, И. В. (2021). *Отечественные Бронированные Машины 1945-1965 гг. - Часть I: Легкие, средние и тяжелые танки*. Кемерово: ООО "Принт". p. 135.
[303] '«Объект 165» и «объект 166» in Карцев, Л. Н. (2008). 'Воспоминания Главного конструктора танков', *Техника и вооружение*, (March).
[304] Нарбут, А. М. , Ребриков, В. Д., Трофимов, П. В. (1987). 'Уменьшение Загазованности

venting a marginal amount of the fumes via the crew compartment overpressure[305].

Outside of combat, the gun breech required special attention from the gunner and commander. The sliding wedge breech block did not completely close over the breech opening, so there was always a gap between where the rim of the stub case would fit between the breech and the breech block. To seal the gun, a spent cartridge casing stub had to be available in the tank at all times. For this reason, the "AZ" autoloader of the T-72 was designed to eject spent stub cases only when loading the next shot, and not immediately after a shot was fired[306].

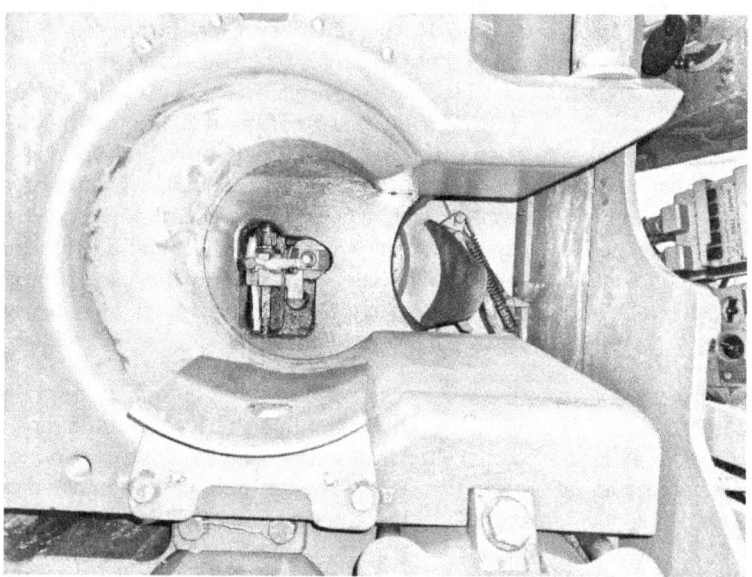

Figure 73: 2A46 with stub case to seal the breech. (*S. Stauber, 2024*)

If the gun was not reloaded, the commander removed a spent stub case from the stub catcher and placed it into the breech to seal it. When firing from the coaxial machine gun, the gun breech had to be sealed by having a round loaded or a stub case inserted, but otherwise the periodic depressurization from firing the gun was deemed acceptable. As such, wearing gas masks was not necessary while working in the tank except in the case of an FVU failure. Operationally,

---
Обитаемых Отделений БТТ при Стрельбе', *Вестник Бронетанковой Техники* 1987, сборник 1. p. 34.
[305] Briefly opening the ejection port immediately after firing a shot was more effective, but this idea was only incorporated into the T-90 and T-72BA during the 2000's.
[306] Retaining a stub case also helped counterbalance the gun for better stabilization quality.

this was the most important practical ramification of an overpressure collective protection system. Even with ventilated facepieces, respirators were the primary hindrance to the crew's sensory perception and manual work capability[307].

The absence of an overpressure collective protection system in some AFVs, predominantly American AFVs such as the M60A1 tank, was (officially) due to cost and space factors[308], though disinterest from the military leadership likely played the biggest role. After the 1973 Arab-Israeli war, the fact that all Soviet combat vehicles featured standardized automatic overpressure collective protection sparked a belated reevaluation of policy[309].

In the filtered mode, the FVU blower worked as a pre-filter for the FPT- 100M. The filter drum contained a fiber particulate filter followed by an activated carbon gas filter, arranged radially. By having outward radial flow through the filter rather than axial flow, a larger flow area was achieved without needing to increase the diameter of the drum. The filtration efficiency of the filter was sufficient to permit tanks to cross areas contaminated by chemical and biological weapons, as well as zones contaminated by fallout.

It was not advisable to use the FPT-100M filter in dusty conditions if the FVU alone gave adequate air quality, but if not, then the crew wore common R-2 face masks[310], a type of respirator equivalent to modern N95 respirator.

Under standard conditions, the nominal working capacity of the filter was 48 hours, but under highly dusty conditions (1 g/m^3), it provided 8-12 hours of use before a 25% reduction of air flow was reached (overpressure sustainment threshold). Its working capacity also depended a great deal on the type of dust encountered, which could be different from region to region. Large dust particles could not get past the FVU, but fine particles were much more difficult to separate. In Central Asian environments, FPT-100M cartridges were saturated with fine dust after 5-10 hours of operation. In milder climates, the cartridges worked for 25-30 hours[311].

---

[307] Evans, P.L. (1971). *Protection of The Crew of Armoured Fighting Vehicles*. C.D.E. Technical Paper No. 60. pp. 5-7.
[308] NBC Army Focal Point Chemical Research, Development and Engineering Center (1991). *U.S. Army Armored Systems - Nuclear, Biological and Chemical Defense Requirements and Plans to Meet Them*. p. 78 (B.4-3).
For further reading on the topic, see: Mauroni, A. J. (2000). *America's Struggle with Chemical-biological Warfare*. Westport: Greenwood Publishing Group. p. 112.
[309] Ibid. pp. 121-123.
[310] Главное Управление Боевой Подготовки Сухопутных Войск. p. 187.
[311] Горбунов, А. С., Штепанек, С. М. (1984). 'Способы Оценки Аэродинамического Сопротивления Фильтров Тонкой Очистки ФВУ', *Вестник Бронетанковой Техники* 1984, сборник 2. p. 30.

With the tank sealed and the FVU working in the filtered mode, the overpressure sustained in a T-72 was no less than 35 mmH2O (343 Pa), which was fairly robust for the purpose[312]. Failing to reach this overpressure was an indicator for filter replacement, a process that required the help of the chemical defense troops.

When replacing an expended filter cartridge, unseating the cartridge exposed the fighting compartment to contamination from the open FVU duct and the intake end of the cartridge. The cartridge could be capped off before removing it through the commander's hatch, but if the tank had been under chemical attack, the man assigned to the task had to be suited up and the area around the FVU had to be decontaminated regardless.

In some Western tanks, this was addressed by housing the filtration system externally. This could be in a bustle compartment (Chieftain)[313] or in a hull sponson compartment (Leopard 2)[314]. The removal process was quicker, safer, and less laborious in these external systems, and decontamination was easier. Apart from the need for decontamination, replacing the FVU filter cartridge in a T-72 involved dismounting the commander's seat, and if there was ammunition stowed in the slots behind the seat, those had to be removed as well.

There was no backup for the overpressure system[315] apart from individual R-2 face masks (for fallout)[316], gas masks (for chemical, biological threats), and OZK suits issued to each crew member as personal equipment. In the event that the overpressure was lost, donning PPE gave full-body protection from CBRN threats while also allowing the crew to evacuate their tank and into a contaminated environment during combat, but obviously, long-term work inside the tank was much more difficult this way. It was also possible for the crew to be wearing their OZK suits and gas masks ahead of time to march and fight from open hatches in irradiated areas[317]. The gas mask could be an ShMG (PMG mask kit), ShM-62 (GP-5 mask kit), or ShM-66 (PMG-2 mask kit)[318]. The ShMG was the most suitable type for tankers, because its unique faceplate design with forward-facing eyepieces was much more convenient for optical sights.

---

[312] This was a typical overpressure level. In the Leopard 1, it was 30 mmH2O. See: Hilmes, R. (2019). *Kampfpanzer Leopard 1: Entwicklung - Serie - Komponenten*. Stuttgart: Motorbuch Verlag. In the Leopard 2, it was 40 mmH2O. See: *Kampfpanzer Leopard 2. Teil 1. Beschreibung*. TDv 2350/033-10.
[313] Ministry of Defence (1966). Tank, Combat, 120-mm Gun, Chieftain, All Marks: User Handbook 1966.
[314] Hilmes, R. p. 142.
[315] In NATO tanks, hybrid systems were more common. In addition to an overpressure system, individual ventilated facepieces were provided.
[316] Министерство Обороны СССР (1970). *Респиратор Р-2. Техническое описание и инструкция по эксплуатации*.
[317] Главное Управление Боевой Подготовки Сухопутных Войск (1989). *Учебник Сержанта Танковых Войск*. Москва: Воениздат. p. 187.
[318] Ibid. p. 190.

Nuclear, Radiological Threat

Despite the effort placed in collective protection, fallout was not the primary radiological threat to tank crews in a nuclear war, and simply providing filtered air did not define a nuclear-hardened tank. For Soviet combined arms ground units, tactical nuclear weapons (TNW) like 203 mm artillery-fired tactical nuclear shells and tactical ballistic missiles with a nuclear warhead (e.g. Honest John, Lance, Pluton) were the primary threats. Tactical nuclear weapons (TNW) were characterized by low explosive yields and an increased reliance on penetrating radiation to produce casualties compared to strategic nuclear weapons (SNW). Weapons of this class were intended to be used in the airburst mode[319] under both NATO and Warsaw Pact doctrine[320]. Due to limited explosive yields and a doctrinal preference for fallout-free low airbursts, fallout was, for the most part, not a militarily significant aspect of TNW usage.

Regardless of the explosive yield, TNWs produced a large amount of ionizing radiation, the most dangerous type being penetrating radiation (gamma, neutron). Transient irradiation from the nuclear burst lasted for 10-15 seconds, but the total radiation dose could accumulate further from irradiated soil. For virtually all TNWs, the radius of effective transient radiation damage (acute radiation poisoning) on armored targets was considerably larger than the blast damage radius. For warheads with a yield lower than 1 kt (e.g. the W79 artillery round), radiation overtook blast as the primary means of damage even against personnel in the open[321]. At the more extreme end, enhanced radiation (neutron) warheads produced casualties almost entirely via neutron bombardment[322], doing relatively little damage by blast or thermal energy.

For a conventional 1 kt fission warhead in a low airburst, the distance limit for blast survival for a conventional tank with steel armor was considered to be around 100-150 m. At the same time, against tanks of this type (M60A1, T-

---

[319] Чугасов, А. А. (1969). *Ядерное Оружие*. Москва: Воениздат. p. 29.
The same working assumption was made in: 'Biophysical and Biological Effects of Ionizing Radiation' in Department of Defense (1996). *NATO Handbook on The Medical Aspects of NBC Defensive Operations - AMedP-6(B)*. Washington, D.C: Department of the Army.
[320] Doughty, R. A. (1979). *The Evolution of US Army Tactical Doctrine, 1946-76*. Fort Leavenworth: U.S. Army Command and General Staff College, Combat Studies In- stitute. Available at: https://www.armyupress.army.mil/Portals/7/combat-studies- institute/csi-books/doughty.pdf (Accessed: 27 September 2023)
Also see: Nichols, T., Stuart, D., McCausland, J. D. (eds.) (2012). *Tactical Nuclear Weapons and NATO*. Carlisle: U.S. Army War College, Strategic Studies Institute.
[321] Реукова, Т. Ф. (ed.) (1980). *Учебник Сержанта Мотострелковых Подразделений*. Москва: Воениздат. p. 45.
[322] Главное Управление Боевой Подготовки Сухопутных Войск (1989). *Учебник Сержанта Танковых Войск*. Москва: Воениздат. p. 182.

55)^{323}, for which a transmission factor of 0.5 against neutrons could be used as a baseline ^{324}, immediate permanent incapacitation ^{325} of the crew was guaranteed within a radius of 500 meters ^{326}, with latent acute radiation poisoning^{327} out to 760 m, and radiation illness of varying degrees at greater distances. Considering that penetrating radiation from a 1 kt burst is effective on personnel in the open out to a radius of 860 m, killing through its shock wave only out to a radius of 290 m^{328}, steel armor did little but delay certain death.

The basic military purpose of nuclear hardening on tanks was to minimize the loss of combat effectiveness in tank units after a nuclear strike. Primary nuclear protection through radiation shielding gave the survival rate needed at the tactical level, and secondary nuclear protection through filtered ventilation gave the long-term resilience needed to achieve operational and strategic goals. In the strategic balance of power, a mechanized force with stronger nuclear hardening could thus gain the upper hand after an exchange of fire. Secondary protection without primary protection would be tantamount to equipping tanks with the means to continue fighting during a nuclear holocaust, but not survive the nuclear exchange that caused it. On those grounds, the emphasis on radiation shielding in Soviet tank development was purely logical. The very same conclusions informed armor development in the American "Main Battle Tank" proposals from the late 1950's to the early 1960's^{329}, and in the West German KPz-70 from the U.S-FRG MBT-70 program^{330}.

Surviving most of the transient effects of a nuclear attack came naturally to tanks. The main structural considerations for nuclear protection, i.e. blast sealing and dissipation, thermal hardening and resistance to overturning, strongly overlapped with preexisting requirements for conventional threats.

A nuclear blast was characterized by sustained pressure on all tank surfaces facing the detonation. Turrets could be torn off their hulls by the blast pressure,

---

[323] Davidson, C. N. (1979). Nuclear Notes Number 8: Armored Vehicle Shielding Against Radiation. Virgina: U.S. Army Nuclear and Chemical Agency. pp. 5-6.

[324] Hudson, C. I., Haas, P. H. (1976). 'New Technologies: The Prospects' in Holst, J. J., Nerlich, U. (eds.) (1976). *Beyond Nuclear Deterrence: New Aims, New Arms*. New York: Crane, Russak & Company. p. 138.

[325] Tank crews caught in this radius would be incapacitated within five minutes, and remain so until death within a day or two.

[326] Miettinen, J. K. (1977). 'Mininukes and Neutron Bombs: Modernization of Nato's Tactical Nuclear Weapons. Introduction of Enhanced Radiation Warheads', Instant Research on Peace and Violence, 7(2). p. 52.

[327] Incapacitated within two hours, death within weeks

[328] Чугасов, А. А. (1969). Ядерное Оружие. Москва: Воениздат. p. 135.

[329] Hunnicutt, R. P (1990). 'The Ultimate Main Battle Tank Design' in *Abrams: A History of the American Main Battle Tank*. Novato: Presidio Press. pp. 93-116.

[330] Spielberger, W. J. (1995). Waffensysteme Leopard 1 und Leopard 2. Auflage: Motorbuch Verlag. p. 397.

so streamlined or rounded turrets with a smaller cross-sectional area were favorable, and such shapes were generally preferred for ballistic shaping to begin with. To prevent the tank from being thrown or overturned, a low center of gravity was desirable, and a low center of gravity was already a design priority for better cross-country stability and obstacle-crossing ability. All surfaces with a direct line of sight to the detonation would be bombarded with enough thermal radiation[331] to set external rubber, wooden, or plastic components alight, but at the same time, napalm protection was a more difficult requirement to meet, and sealing against incendiary liquids was more difficult than directional heating.

For instance, external covers on gun masks (plastic, rubber) could not be the only means of sealing a gun embrasure. In a T-72, if the plastic gun mask cover was torn off or burnt away due to a nuclear detonation[332], or even due to ordinary battle damage, an internal cover behind the gun mask allowed the tank to still generate and maintain a crew compartment overpressure. This internal cover was made from the same material as the external cover, and it was sealed along its periphery with waterproof putty at all times[333]. Double-sealing the gun embrasure was standard practice internationally for tanks with overpressure NBC protection.

Even the threat of electrical damage from an electromagnetic pulse (EMP) was negligible. Exceptionally thin metal sheeting can effectively shield internal circuits and wiring from induced voltages at large incident field strengths[334], which is also why the electrical spark-activated primers and fuzes for 125 mm ammunition don't go off under an EMP.

These design solutions preserved the tank from nearly all of the destructive influences of a nuclear detonation, save for radiological damage. Secondary nuclear protection was at least fairly straightforward to secure. It was enough to install an FVU, and gamma rays from induced radioactivity in the terrain were already strongly attenuated (by approximately 25 times) even by the modest thicknesses of steel on tank bellies[335]. Penetrating radiation was a different matter entirely.

---

[331] Чугасов, А. А. p. 116.
[332] A 1 kT airburst ignites rubber roadwheels at 400 m, see: Талу, К. А. (1963). *Конструкция и Расчет Танков*. Москва: Военная академия бронетанковых войск. p. 97.
[333] Министерство Обороны СССР (1969). *Техническое Описание Танка Т-64*. Москва: Воениздат. p. 556.
[334] Department of Defence (1987). *Military Handbook - Grounding, Bonding, And Shielding For Electronic Equipments And Facilities*, MIL-HDBK-419A. Volume 1. p. 10-9.
[335] Davidson, C. N. (1979). Nuclear Notes Number 8: Armored Vehicle Shielding Against Radiation. Virgina: U.S. Army Nuclear and Chemical Agency. p. 1.

Table 14: Distance limit in km for various types of radiation casualty in a tank (0.5 transmission factor), from ground bursts (G) and air bursts (A)

Burst power, kt	Casualty Type					
	Lethal		Heavy		Medium/Light	
	G	A	G	A	G	A
1	0.25	0.28	0.63	0.69	0.7	0.76
2	0.3	0.32	0.71	0.77	0.78	0.83
3	0.32	0.34	0.77	0.81	0.83	0.87
5	0.35	0.37	0.81	0.87	0.89	0.93
10	0.4	0.4	0.9	0.95	0.98	1.0
20	0.48	0.5	1.0	1.1	1.1	1.15
30	0.53	0.55	1.05	1.15	1.2	1.2
50	0.6	0.6	1.15	1.25	1.3	1.3

Чугасов, А. А. (1969). Ядерное Оружие. Москва: Воениздат. p. 138.

In the Soviet Army, the threshold for no reduction in the combat effectiveness of a soldier was considered to be 50 rad (53 roentgen) in a single event (with an exposure interval no shorter than four days)[336]. An upper threshold of 100-200 rad was survivable with temporary radiation sickness.

Beginning in the early 1960's, the requirements for protection against penetrating radiation for new tanks stipulated all-round protection with a safe radius of 1,000 m under a dosage criterion of 28 rads (30 roentgen)[337] from a 30 kt bomb[338]. Considering that the total penetrating radiation dose from a 30 kt fission detonation at 1,000 m was 6,000 roentgen (5,580 rads), a radiation attenuation factor of 200 (!) was needed[339], which was plainly impractical.

Improving the neutron transmission factor of an all-steel tank fourfold from 0.5 to 0.125 could already reduce the radiation casualty area of a typical 1 kt TNW strike by 30-55%. With a 10 kt burst, the reduction reached 20-40%[340]. An improvement of this magnitude was achieved by adding an anti-radiation liner of limited thickness to the T-55 tank, creating the T-55A. The T-72 achieved a higher degree of protection through purposeful design and a more efficient lining material and liner distribution. On the T-72B, the total penetrating radiation attenuation factor was 12.3 (transmission factor 0.0813),

---

[336] Главное Управление Боевой Подготовки Сухопутных Войск (1989). *Учебник Сержанта Танковых Войск*. Москва: Воениздат. p. 181.
[337] 1 roentgen measured in the air is equal to an absorbed dose of approximately 0.95 rad in tissue.
[338] Павлов М. В., Павлов, И. В. (2023). 'Отечественные Бронированные Машины 1945-1965 гг. - История ракетных танков', *Техника и вооружение* 2023, (September). p. 29.
[339] Рисунок III-26 in Чугасов, А. А. p. 81.
The distance required for the radiation to naturally decline to 30 roentgen was 2,150 m.
[340] Davidson, C. N. (1979). *Nuclear Notes Number 8: Armored Vehicle Shielding Against Radiation*. Virgina: U.S. Army Nuclear and Chemical Agency. p. 11.

averaged across all crew stations. Against secondary gamma radiation, the attenuation factor reached 35[341].

This level of attenuation was quite costly from a technical standpoint. It added weight and diminished the space available in the crew stations, when tank development projects already routinely blew past weight and size limits. Considering that there was no way to ensure that the tank would face a nuclear detonation head-on, all-round radiation shielding was an obvious requirement, and strong shielding was also needed on the roof due to neutron and gamma ray scattering in the atmospheric, due to which a significant dose of radiation arrived from odd directions[342].

Anti-Radiation Shielding

The greatest efforts were focused in neutron shielding. The relative biological effectiveness (RBE) of neutrons produced by nuclear weapons is not greater to that of gamma radiation in acute biological effects[343], but neutrons were more penetrating than gamma rays, and were thus a more serious threat to armored vehicle crews. Neutron shielding was especially important for preserving a tank's electronics. Fast neutrons are also responsible for most of the permanent nuclear radiation damage to electronics[344].

Most of the electrical network of the tank functioned was wired to a negative ground system, where the body of the tank itself functioned as the negative path to complete a circuit between a device and the generator. For the turret, the electrical path was the turret ring and the body of the traverse mechanism[345]. There was therefore only a single wire connection to most of the electrical units. Supplementing this was an isolated network with fully wired positive and negative terminals, used for the bilge pump, the internal dome lights, and the internal power sockets. This was to ensure that emergency power would always be available to critical devices during snorkeling operations.

---

[341] Гусев, С. А. (1991). 'Боевая машина поддержки танков', *Вестник бронетанковой техники* 1991, сборник 7. Available at: http://btvt.info/5library/vbtt_1991_bmpt.htm (Accessed: 20 January 2024).
[342] Чугасов, А. А. (1969). Ядерное Оружие. Москва: Воениздат. pp. 87-88.
[343] Glasstone, S., Dolan, P.J (eds.) (1977). *The Effects of Nuclear Weapons*. Third Edition. Washington, D.C.: Department of Energy. p. 577.
The relative biological effectiveness of neutrons in causing long-term biological damage (cancers) is drastically greater (10-80), but due to the long time spans involved, this was of little combat relevance.
[344] Hnatek, E. R. (1969). *Nuclear Radiation Hardening for Electronic Components*. California: Lockheed Missiles & Space Company. pp. 10-23.
[345] In some tanks, it was forbidden to rest the base end of a cartridge on the floor because of the risk of a current passing through the electric primer.

Using the tank body as a negative ground was not considered ideal for EMP hardening[346], but for a tank it was unproblematic simply due to dissipation in a large metal mass. Integral radio interference filters were present in sensitive electrical equipment, such as the TPN1 night sight. All electric wires were sheathed in metal braiding[347] for radio interference shielding, and the hull and turret each had a frequency filter to suppress high-frequency currents, preventing interference with the radio.

Table 15: Half-value layer thicknesses of various armor materials (cm)

Material	Gamma	Neutrons
Polyethylene	21.8	2.7
Glass-reinforced plastic	12	4
Lead	2	12
Steel	3.5	11.5

Чугасов, А. А. p. 84.

Dense materials, having a large density of atoms, were effective for gamma shielding, whereas materials rich in nuclei, such as hydrogen and carbon, gave good neutron attenuation [348]. Diesel fuel, a hydrocarbon, and hydrogenous (hydrogen-rich) polymer liners were thus applied strategically to maximize protection. Even the driver's lower body was shielded by the water reservoir for the periscope jet washer between the lower glacis of the hull and the driver's pedals. The use of fuel as a means of radiation shielding was one of the main technical justifications for locating the driver centrally with two fuel tanks on his flanks. High radiation attenuation for the driver was achievable with a fuel barrier thickness of just 120 mm[349] between him and the 80 mm steel side hull armor. Including the external fuel tanks, fuel accounted for a reduction in the all-round neutron dose at all crew stations by an average of 1.37 times[350].

Glass textolite in the hull upper glacis composite armor served as neutron shielding[351] via the boron content (as boron oxide) in its borosilicate glass fibers.

---

[346] Dolan, P. J. (ed.). (1972). Capabilities of Nuclear Weapons, Part 2, Damage Criteria, Effects Manual 1. Washington, D.C: Defense Nuclear Agency. p. 9-177.
[347] Multi-pin connectors were adapted from the aviation industry, allegedly originating from Bendix Aviation Corp connectors reverse engineered from the B-29.
See: Клапауций. *Разъемы ШР*. Available at: http://www.155la3.ru/passiv.htm (Accessed: 13 September 2023).
Also see: Frei, J. R., Uline, W. A. (1951). Electrical connector having resilient inserts. United States Patent and Trademark Office No. US2563712A. Available at: https://patents.google.com/patent/US2563712 (Accessed: 13 September 2023).
[348] Чугасов, А. А. (1969). *Ядерное Оружие*. Москва: Воениздат. p. 82.
[349] Питанов, В. В., Чурилин, А. В. (1981). 'Использование топливной системы для противорадиационной защиты ВГМ', *Вестник Бронетанковой Техники* 1981, сборник 4. p. 40.
[350] Воробейчик, Г. М., Кондратьев, В. И., Потемкин, Б. В. (1972). 'О влиянии внутренних баков с топливом на уровень противорадиационной защиты экипажей танков и БМП', *Вестник Бронетанковой Техники* 1972, сборник 6. p. 14.
[351] Старовойтов, В. С. (ed.) (1990). *Военные гусеничные машины*. Том 1: Устройство, Книга 1.

The composite armor design gave better shielding than the T-55A upper glacis, which was a 100 mm steel plate backed by a 30 mm POV-20 liner.

Steel on its own was already fairly effective against high-energy gamma radiation, having an effective half-value thickness[352] of 35 mm [**Table 17**], but did not provide effective neutron shielding. A layer of steel followed by a hydrogenous material was a textbook example of effective shielding for fast neutrons. The iron in steel armor provided a medium for the inelastic scattering of fast neutrons, and as these neutrons passed into a hydrogen-rich layer, they were thermalized through elastic scattering and then absorbed[353]. Thicker steel armor improved the effectiveness of a hydrogenous layer behind it[354].

In the process of neutron absorption, low-energy secondary gamma radiation was emitted through induced radioactivity. Low-energy gamma rays posed less of a biological hazard than high-energy gamma rays, and were easier to attenuate[355] through the internal equipment of the tank, the crew seats, and protective clothing. The liner was also a barrier against this radiation[356].

The design, thickness, and distribution of anti-radiation material was created through mathematical modeling, verified through tests with a VVR-L-02 portable reactor and verified by full-scale nuclear tests on real tanks[357]. The thickness and composition of the liner was optimized according to the thickness of the turret and hull armor. On thinner parts like the turret rear and ceiling, 45-50 mm of POV-45/50SV was applied to compensate for the relatively low gamma ray attenuation. Thickly armored zones like the turret front and the driver's glacis cutout were lined with just 10-20 mm of POV-45V. The cavity in the left fuel tank for the driver's instrument panel and the cavity in the right fuel

---

Москва: Издательство МГТУ им. Н. Э. Баумана. p. 327.
[352] Also known as half-attenuation thickness; the thickness of material where radiation dose is halved.
[353] Glasstone, S., Dolan, P.J. pp. 346-348.
[354] Ямпольский, П. А. (1961). *Нейтроны атомного взрыва*. Москва: Госатомиздат. p. 113.
[355] Glasstone, S., Dolan, P.J (eds.) (1962). *The Effects of Nuclear Weapons*. Second Edition. Washington, D.C.: Department of Energy. p. 394.
[356] McAlister, D. R. (2016). *Neutron Shielding Materials*. Illinois: PG Research Foundation. Available at: https://www.eichrom.com/wp-content/uploads/2018/02/neutron-attenuation-white-paper-by-d-m-rev-2-1.pdf (Accessed: 10 September 2023).
Up to 30% of the total cumulative radiation dose in a tank can be from induced radioactivity.
[357] Ирдынчеев, Л. А., Фрид, Е. С. (1985). 'Расчет доз гамма-излучения наведенной радиоактивности в танках при их облучении нейтронами ядерных взрывов'. *Вестник Бронетанковой Техники* 1985, сборник 6. p. 26.
Also see: Балашов, И. В., Малофеев, А. М., Чистяков, М. В., Хазов, Н. Н. (2013). 'Противорадиационная Защита: Вчера, Сегодня, Завтра. Иллюстрации предоставлены ОАО «НИИ Стали»', *Техника и Вооружение*, (March).
Also see: Алексеев, М. et al. (2012). *НИИ Стали 1942-2012*. Москва: Издательство СканРус. p. 33.

tank for the CBRN protection system and firefighting system were compensated by anti-radiation liners, which were otherwise unneeded.

The commander's recoil guard and the backs of the seats in the turret had 30 mm panels of the same anti-radiation material. Later, the driver's seat backrest received a 30 mm panel as well. Local shielding via panels on the crew seats, previously absent from the T-55A, was the most weight-efficient method of neutron shielding[358]. Compared to a bare steel hull and turret shell, the lining weakened the neutron dose to the crew by 1.5 times[359].

Keeping in mind the advantages of local shielding, DSP-I1 anti-radiation vests [360] were developed in the late 1970's and procured (in unknown quantities), to be issued only in case of a nuclear war. The vital organs were covered by hard rubber inserts containing lead dust – quite literally placing an anti-radiation liner directly on each crew member. They were heavy (7-8 kg) and were intended to be worn only when a nuclear attack was anticipated.

From October 1983, anti-radiation cladding[361] was added to the driver's hatch area and to the turret, furthering improving radiation attenuation. The design of the cladding was based on neutron bombardment tests carried out from 1974 to 1983 with underground nuclear detonations[362].

Radiation protection even extended to the tracks. Unlike fallout, induced radiation could not be removed by decontamination scrubs. A gadolinium rare earth metal additive was added to the chemical composition of tank's tracks to reduce secondary radiation by 2.5 times[363], offsetting the propensity of Hadfield steel to intense induced radioactivity compared to other steel alloys. Irradiated tracks were a major contributing factor in the radiation hazard to personnel working around tanks exposed to nuclear attack[364].

---

[358] Рейтблат, В. Л., Сержантов, Е. П., Студниц, М. А., Фрид, Е. С. (1975). 'Противорадиационная защита экипажей танков', *Вестник Бронетанковой Техники* 1975, сборник 1. p. 38.
[359] Старовойтов, В. С. (ed.) (1990). *Военные гусеничные машины*. Том 1: Устройство, Книга 1. Москва: Издательство МГТУ им. Н. Э. Баумана. p. 374.
[360] Developed by NII Stali. See: Балашов, И. В., Малофеев, А. М., Чистяков, М. В., Хазов, Н. Н. (2013). 'Противорадиационная Защита: Вчера, Сегодня, Завтра', *Техника и Вооружение*, (March). p. 11.
[361] Colloquially called "подбой / надбой", but in technical documentation, they were simply referred to as "подкладка / накладка" (lining or cladding).
[362] Due to the 1963 Partial Test Ban Treaty (PTBT), to which the USSR was a signatory.
[363] Ирдынчеев, Л. А., Кудин, В. Т., Рейтблат, В. Л., Шерстюк, А. А. (1978). 'Снижение наведенной радиоактивности высокомарганцовистой стали', *Вестник бронетанковой техники* 1978, сборник 1. pp. 23-24.
[364] Ирдынчеев, Л. А., Рейтблат, В. Л., Фрид, Е. С. (1977). 'Наведенная радиоактивность как фактор радиационного поражения личного состава танковых войск', *Вестник Бронетанковой Техники* 1977, сборник 2. pp. 26-31.

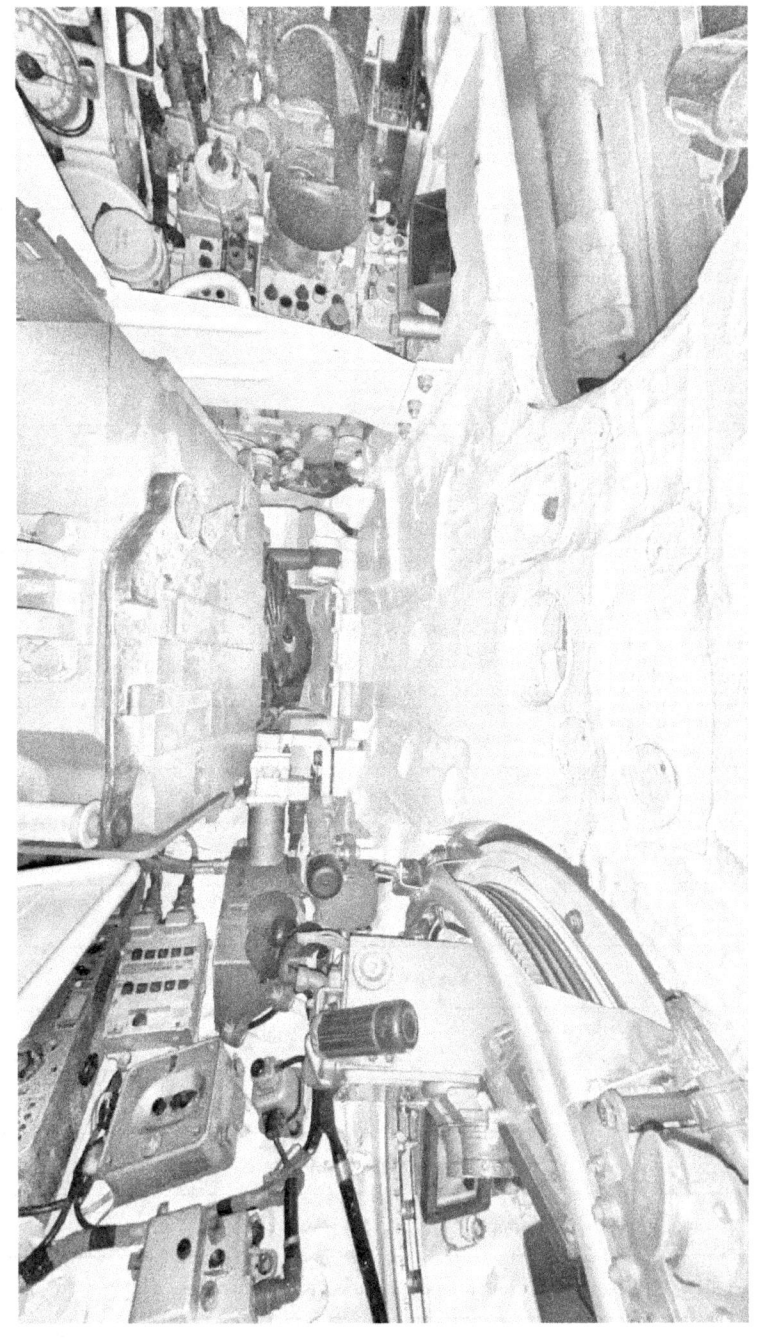

Figure 74: General view of turret interior. (*P. Kefallinos, 2024*)

Table 16: Anti-radiation lining location and thicknesses at various points

	Item	Description and measured thickness
		Hull
a	Driver's cutout front, top, sides	13 mm
b	Driver's hatch underside	16 mm
c	Hull ceiling	54 mm
d	Right hull nose side	54 mm
e	Right hull side	54 mm basic thickness • Curved around the autoloader carousel to 38 mm minimum along a ~400 mm section • Local cutout for the FPT-100M filter, ~20 mm minimum
f	Left hull side	54 mm basic thickness • Curved around the autoloader carousel to 38 mm minimum • Local cutout for ammunition, ~30 mm minimum
g	Carousel cover	20 mm atop carousel cover, additional local 13 mm layer underneath carousel cover
h	External panels on panniers	30 mm
		Turret
a	Periscope cladding	50 mm
b	Commander's hatch	38-40 mm
c	Right wall	12 mm at very top edge, thicker towards bottom
d	Case ejection hatch	50 mm
e	Ceiling at turret peak	43 mm
f	Ceiling near hatch	35 mm on right hatch perimeter, 43 mm at front perimeter (*)
g	Gunner's hatch	48 mm
h	Left wall	22 mm, measured at A-2 intercom panel

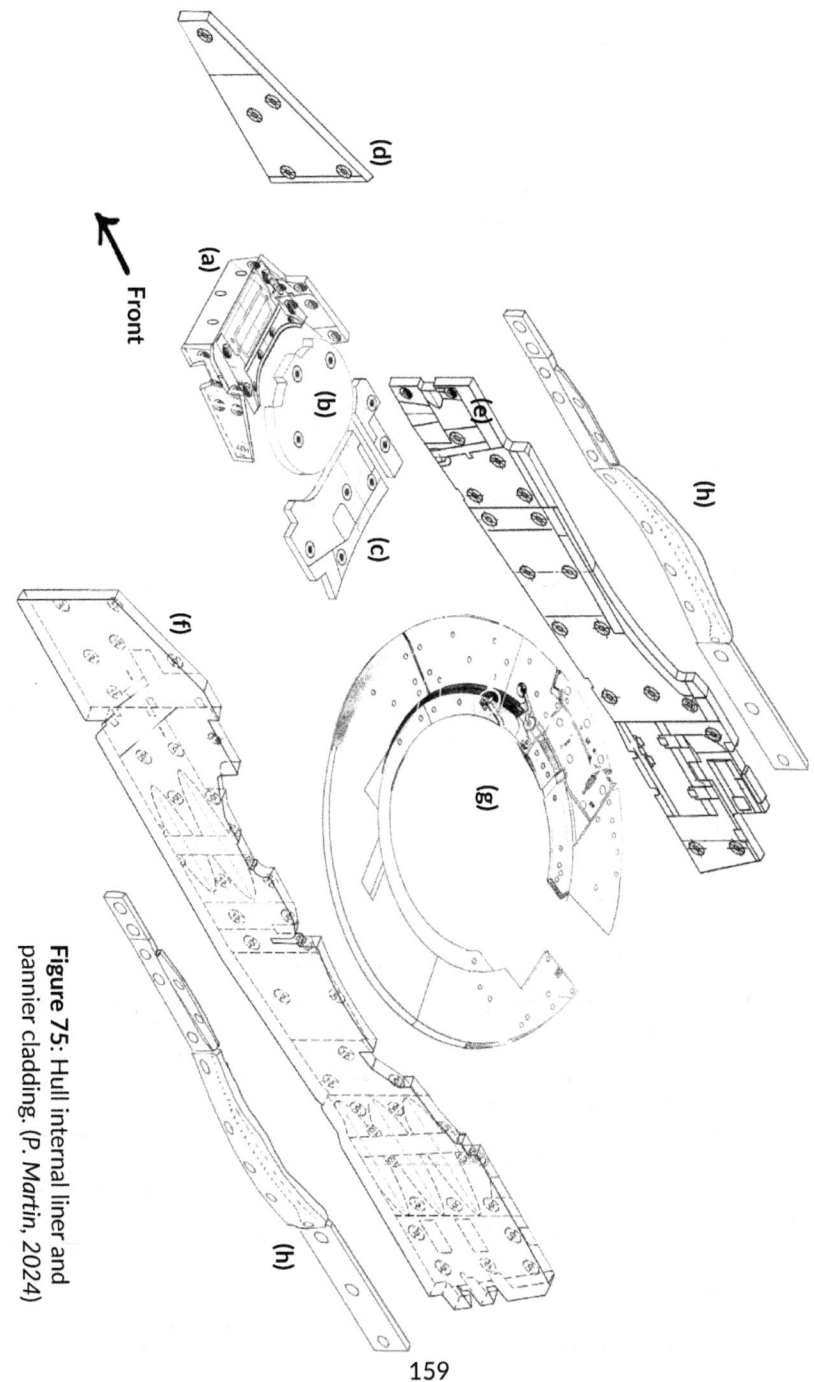

Figure 75: Hull internal liner and pannier cladding. (P. Martin, 2024)

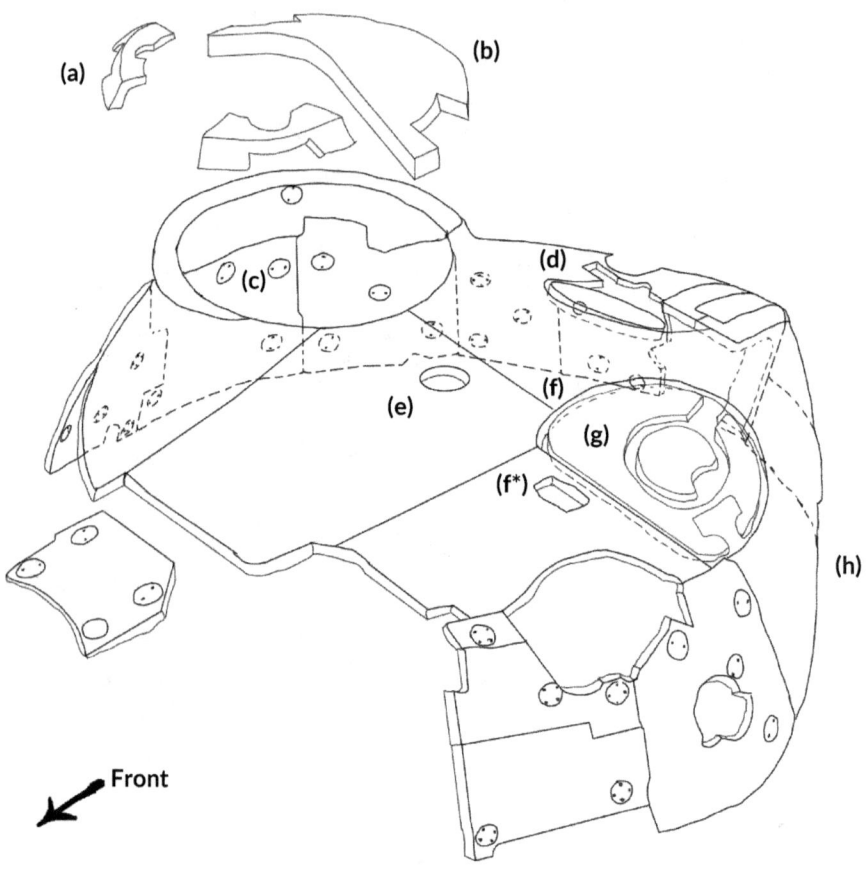

Figure 76: Turret internal lining. (*P. Martin, 2024*)

Table 17: Anti-radiation cladding location and thicknesses at various points

	Item	Description and measured thickness
a	Cupola rim	22 mm
b	Roof	22 mm
c	Left wall	22 mm
d	Sight panels	22 mm
e	Hatch and roof	30 mm, likely applicable to the entire roof panel

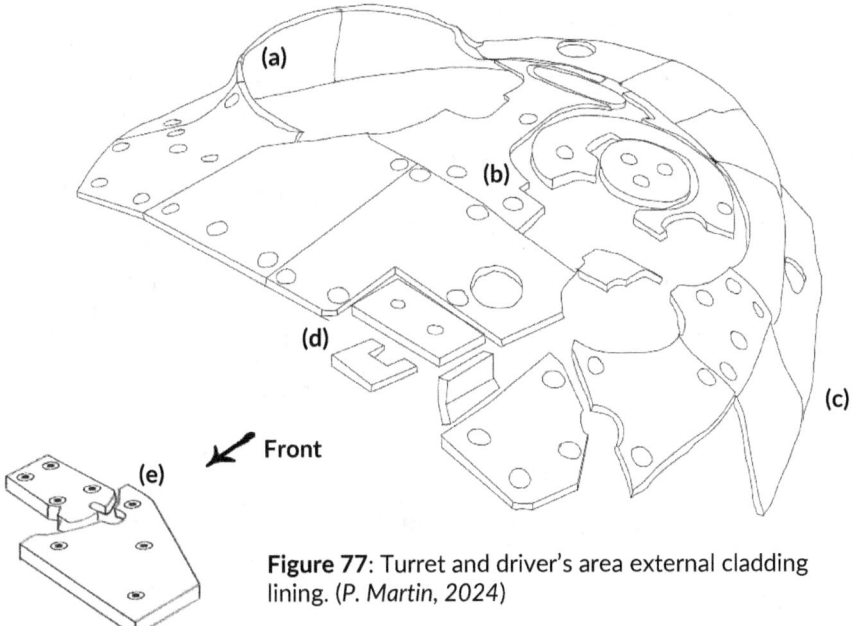

Figure 77: Turret and driver's area external cladding lining. (*P. Martin, 2024*)

Since anti-radiation materials had to be installed directly on armor, the choice of materials had to comply with mechanical and chemical requirements. The POV-45 material was a polymeric composite of butyl rubber (polyisobutylene, PIB) with polyethylene, the polyethylene serving to augment the strength and heat resistance of the compound. High molecular weight PIB grade P-200 was used[365] together with low-density PE[366], both containing a large amount of hydrogen for neutron attenuation. The POV-45/50SV material was the same compound but containing a large mass of evenly distributed lead powder, doubling its effectiveness as gamma radiation shielding. Both compounds contained boron nitride as a neutron moderator.

The neutron attenuation of the material was enhanced by inner and outer layers of 21F/17F heavy duty fabric, known literally as "Belting" («Бельтинг») fabric[367]. It was a chemically treated coarse fabric made from twisted cotton

---

[365] Воробьева, Г. Я. (1975). *Коррозионная стойкость материалов в агрессивных средах химических производств-Химия*. Москва: Издательство «Химия». p. 151.
Also see: НПО Еврохим. *Полиизобутилен*. Available at: http://ehim.spb.ru/ru/ximicheskoe-syre/ximicheskie-produkty/poliizobutilen (Accessed: 10 December 2023).
[366] Кузнецов, Е. В., Прохорова, И. П., Файзуллина, Д. А. (1976). *Производства Полимеров и Пластических Масс на их Основе*. Издание Второе. Москва: Издательство «Химия». p. 5.
[367] Referring to the fabric commonly used on conveyer belts through the early 20th Century.

coated with flame retardant additives. The outer layer was additionally impregnated with 3% boron. The inner layer between the armor and the POV-45 gave better adhesion to the armor's surface, and the boron content in the outer layer was calculated to efficiently improve neutron attenuation power[368].

The panels for both the lining and cladding were glued onto the tank's armor and firmly fastened by special screws with a 2-hole or 4-hole flat head, which looked (and functioned) like large washers[369].

Apart from radiation protection, the constituent materials of the liner had a number of desirable features, including gas-impermeability, high corrosion resistance [370] low chemical reactivity, low moisture absorption [371], and good thermal stability. This made the lining and cladding practical permanent fixtures of the tank. They did not absorb and retain water or CBRN contaminants, and decontamination sprays could be freely used on them.

Table 18: POV-45 composition by weight[a]

Material	Polyisobutylene	Polyethylene	Boron nitride	Lead
POV-45V	48.5%	40.0%	11.5%	-
POV-45/50SV	27.5%	22.5%	7.4%	42.6%

[a] Кеслер, И. Я. et al. (1974). 'Влияние Физико-Механических Свойств Материалов На Эксплуатационные Характеристики Деталей Противорадиационной Защиты', *Вестник Бронетанковой Техники* 1974, сборник 2. pp. 43-44.

With a hardness equivalent to a solid rubber boot heel[372], the POV-45 material was not particularly effective as safety padding, and it certainly did not qualify as a ballistic material. However, even relatively weak materials such as solid polyethylene[373] could be effective at absorbing fragment energy, and even stopping light fragments. In fact, these anti-radiation panels contributed to the protection of the crew as a spall liner[374], and Swedish tests of T-72M1 tanks in

---

[368] Григорян, В. А. (ed.) (2007). *Защита танков*. Москва: Издательство МГТУ им. Н.Э. Баумана. p. 196. Research on neutron moderating efficiency was presented in: Пугачев, Б. Л., Сысоев, В. Н. (1982). 'Влияние концентрации бора на противорадиационную защиту танка', *Вестник Бронетанковой Техники* 1982, сборник 1. p. 28.

[369] Wnlike in the T-55A, where the liner was studded with large round screw heads.

[370] Гармонова, И. В. (ed.) (1976). *Синтетический Каучук*. Москва: Издательство «Химия». p. 340.

[371] Кузнецов, Е. В., Прохорова, И. П., Файзуллина, Д. А. (1976). *Производства Полимеров и Пластических Масс на их Основе*. Издание Второе. Москва: Издательство Химия. p. 10.

[372] Shore A hardness 80 ± 5.

[373] On its own, low-density PE was stronger, but with the higher elongation of the POV-45 compound, the total strain energy capacity saw a net increase.

[374] Баннов, В. В., Чернявский, В. В., Баннова, Ю. В. (2021). 'Обитаемость Современных Танков', *Транспортные Системы: Безопасность, Новые Технологии, Экология*. Якутск, 16 апреля. Якутск: ФГ-БОУ ВО СГУВТ. pp. 101-102.

1992 even found them to be quite effective against light anti-tank grenades[375]. In addition to stopping fragments, a liner of low acoustic impedance in contact with the armor surface suppresses spallation. In general, similar or better results could be achieved with any thick, dense and elastic material bonded to the back surface of an armor plate[376]. Such liners were also effective at reducing crew injury from the shockwave transmitted into the crew compartment through the armor by a high explosive shell.

Studies on crew survivability with non-penetrating damage on a T-54 showed that high-explosive and shaped charge shells detonating on the armor would almost always severely injure crew members through blunt force trauma if they were physically in contact the armor at the moment of detonation[377]. Even without physical contact, the shockwave transmitted through the armor and into the air would still consistently inflict moderate to mild trauma injuries at a distance of 200-1,000 mm. A foam liner made from PVC-E plastic, 20 mm thick, was found to be an effective means of damping the shock energy transmitted into the crew compartment, but did not enter service[378]. The anti-radiation liner fulfilled this purpose instead.

The dual-purpose nature of the liner as radiation shielding and blast-fragment shielding was, at least in part, serendipity rather than design intent, and quite interestingly, this path was closely paralleled at the Ballistics Research Laboratory (BRL) in the U.S. While testing polyethylene as a potential dual-purpose composite armor filler for radiation and shaped charge protection, BRL researchers also investigated how incorporating polyethylene as a liner affected post-penetration fragmentation. It was found that even a thin, one-inch molded polyethylene panel secured firmly to the back surface of a steel armor plate would significantly reduce the number of fragments ejected from the armor under shaped charge attack, and that increasing the thickness of the liner improved its effectiveness over a wider range of exit hole diameters[379].

While the liner in the T-72 does not qualify as a true, purpose-built spall liner, it nevertheless fulfilled the same function in cases where the armor was penetrated to a low degree of overmatch.

---

[375] Svantesson, C. G. (2009). 'När fienden kom till Sverige', in Lindström, R. O., Svantesson, C. G. Svenskt Pansar: 90 år av svensk stridsfordonsutveckling. Available at: http://www.ointres.se/ryska_strf_till_sverige.htm (Accessed: 27 August 2023).
[376] Department of Defense (1995). Design of Combat Vehicles for Fire Survivability, MIL-HDBK-684. p. 3-33.
[377] Rabbits were used as test subjects.
[378] Павлов М. В., Павлов, И. В. (2021). Отечественные Бронированные Машины 1945-1965 гг. - Часть I: Легкие, средние и тяжелые танки. Кемерово: ООО "Принт". pp. 125-126.
[379] Schmidt, J. G. (1976). Ballisticians in War and Peace: A History of the United States Army Ballistic Research Laboratories. Volume 2, 1957-1976. Aberdeen: Aberdeen Proving Ground. p. 142.

Interestingly enough, the anti-radiation liner was widely publicized as a combined anti-radiation and spall liner soon after its existence in the T-55A became known to Western analysts during the 1970's. The liner in the T-72 was subsequently identified by the same term in major publications[380].

Spall liners could scarcely be encountered in serial tanks during the late 1950's when POV-20 and POV-20/50 was first introduced, and were still vanishingly rare by the end of the Cold War. In some foreign tanks, an anti-fragmentation curtain could sometimes be found behind gun embrasures as a replacement for steel bullet splatter shields, such as the nylon curtain behind the gun rotor in the M60A1[381] and Chieftain, but the same protection was invariably absent from the rest of the crew compartment, and this remained unchanged for all Western tanks throughout the rest of the Cold War.

---

[380] The most influential was: Jane's (1989). 'Inside the T-72', *Jane's Soviet Intelligence Review*, 1(11). p. 525. Other publications include:
'Warschauer Pakt: Die Modernisierung des Kampfpanzerbestandes'; *Oesterreichische mil. Zeitschrift* 1980. pp. 263-264.
Backofen Jr., J. E. (1982). 'Armor Technology, Part 1: Armor, (May-June). p. 39.
[381] *Operator's Manual: Operator Controls and PMCS. Tank, Combat, Full Tracked, 105-MM Gun, M60A1 (RISE). Tank, Combat, Full Tracked, 105-MM Gun, M60A1 (RISE Passive)* (1980). TM 9-2350-257-10-1. Washington, D.C.: Department of the Army. p. 2-128.

Chapter IV
Protection

Concealment and Signature Management

For tankers who were required to maintain constant battle readiness, almost all their time had to be spent inside their tanks, and there would be practically no time for uninterrupted rest. If, however, there was an opportunity to spend time outside the tank during halts and in unit assembly areas, a tank dugout could be prepared, and attached to it would be a shelter for the crew[382]. Such shelters had to be dug out manually, but the dozer blade could be used to entrench the tank. According to Soviet Army norms, a tank crew was expected to dig a full tank dugout with a shelter in medium density soil with shovels in 12 hours. With the help of bulldozer equipment like a dozer blade, the same task would take just 30-35 minutes, 15 minutes of which was spent digging in the tank itself.

Without the aid of cover or concealment to hide it from view, a T-72 still had the advantage of its diminutive size. Concealment through a low silhouette in open terrain is provided by cumulative effect, like for example, an imperceptibly small slope in the terrain that, over a large distance, covers a significant height. In flat country, the value of a low silhouette is only obvious when local concealment is available; large bushes, tall grass, depressed paths (dried streams), and so on.

Signature reduction is principally dependent on controlling the absorption and emittance of electromagnetic radiation so as to minimize the contrast between the tank and its environment. Visual camouflage was the most important part of the T-72's signature reduction like any other tank, but next to it, infrared signature reduction was given the most attention for the fact that the detection of armored units through infrared photography and active night vision was widely available and had proven military significance. Near-infrared signature reduction had been a noted consideration for the Red Army as early as the 1930's when false-color photography became a widespread method of battlefield reconnaissance.

Reconnaissance by false-color photography continued in the postwar era with special infrared films or "camouflage detection" films[383] like Kodak Aerochrome

---

[382] Чугасов, А. А. (1969). *Ядерное Оружие*. Москва: Воениздат. р. 282.
[383] Raisbeck, G. et al. (1981). *Design Goals for Future Camouflage Systems*. Fort Belvoir: Research and Development Command. p. II-9.

and Ektachrome IR[384]. Rudimentary green paints did little to hide ground vehicles in vegetation when viewed through these films simply because the reflectance of these paints in the near-infrared (NIR) spectrum did not at all resemble the scenery. Where trees, grass and other chlorophyll-rich plant matter appeared in a rich, deep pink, painted vehicles were captured only in visible light and thus appeared in a bluish-gray monochrome.

During WWII, all Soviet military vehicles featured a simple red iron oxide primer (undercoat) with a top coat of 4BO oil enamel[385]. In the postwar years, this combination was replaced by more durable synthetic polymer enamels with better resistance to solvents like kerosene. The first was NPF-10, introduced in 1951, later replaced by KhV-518 in 1975. Both came in a similar shade of dull green, and a khaki-colored option was available for KhV-518. A brown-colored[386] FL-03K phenolic undercoat[387] was paired with both of these polymer enamels to ensure proper adhesion to metal surfaces.

NPF-10 was an alkyd (pentaphthalic) resin enamel with a conventional pigment formulation based on yellow and blue pigments to create green. KhV-518 was a chlorinated polyvinyl chloride (PVC) resin enamel obtaining the same shade of green by mixing orange, yellow and blue pigments. The NIR reflectance of NPF-10 was very low (<10%) and its reflectance curve was practically flat. Green KhV-518 had a much higher NIR reflectance, somewhere between that of black soil and sandy soil, and well below that of green vegetation[388].

T-72s were almost always painted in a standard green KhV-518 base color at the factory before delivery to the Soviet Army. Two-tone or three-tone camouflage patterns could then be applied by individual units according to their local environment using various water-based paints, oil paints, or dry mineral paints – all with their own unique infrared characteristics[389], making the deforming camouflage as effective against infrared night vision as it was visually.

This difference would turn out to have important ramifications under active infrared night vision observation. Approaching the reflectance of green vegeta-

---

[384] Both sensitive up to 0.9 μm. Ektachrome Aero was used by the US military during WWII.
[385] Enamel in this case refers to a durable paint resistant to weather.
[386] ГОСТ 9109-81. «Грунтовки ФЛ-О3К И ФЛ-О3Ж. Технические Условия». Tanks kept in open-air storage might not rust, but could begin to appear brownish after the top coat weathers away.
[387] Гуревич, Б. Г., Чепулис, Л. Л. (1985). 'Шероховатое лакокрасочное покрытие танка', Вестник бронетанковой техники 1985, сборник 1. p. 52.
[388] Аксененко, М. Д. et al. (1967). 'Некоторые рекомендации по улучшению маскировки объектов бронетанковой техники', Вестник Бронетанковой Техники 1967, сборник 4. pp. 1-5.
[389] Министерство Обороны СССР (1985). Руководство по инженерным средствам и приемам маскировки сухопутных войск. Часть I – Средства и Приемы Маскировки Войск. Москва: Воениздат. pp. 17, 222.

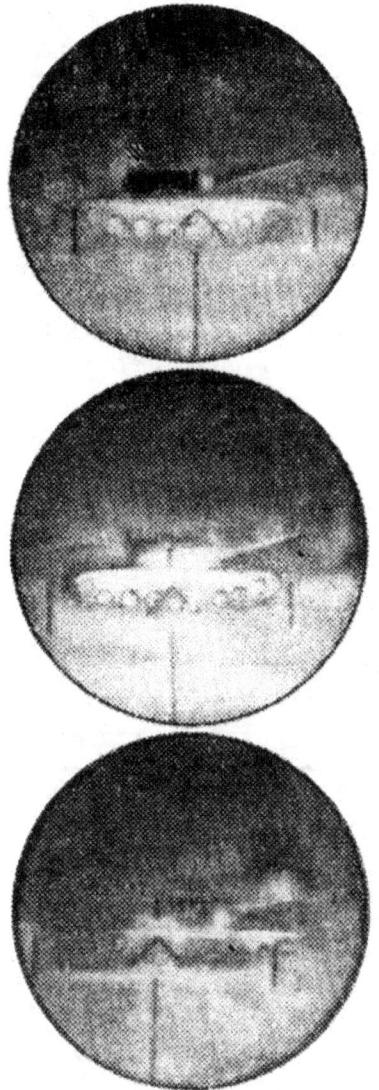

tion in the NIR band might seem at first glance to be desirable for blending into a verdant woodland background, but in actual tests, a tank caught in the beam of an IR spotlight was more difficult to identify and detect when painted in low-reflectivity NPF-10 than in KhV-518[390]. Even at close range where a tank painted in NPF-10 might be expected to be revealed by its high negative contrast against a brightly-lit foreground and a backdrop of bushes [**Fig. 78**, top], the turret was relatively difficult to detect because it was silhouetted against the dark night sky, and the hull and suspension would still approximate the brightness of grass because of accumulated dust.

KhV-518 was marginally better against passive night vision [391] because night luminance was highly diffuse, and a higher NIR reflectance helped to blend in with greenery. However, this was irrelevant for early passive devices (S-20 photocathode) because of their minimal IR response[392]. Regardless of the base enamel, the best results were obtained with a low NIR-reflectance enamel was used in a deforming pattern with a high NIR-reflectance enamel.

Figure 78: Tanks painted in NPF-10 (top), KhV-518 (middle) and in a NIR-deforming pattern with both enamels (bottom) observed through a TPN-1 night sight with IR spotlight, in a summer woodland environment, viewed at a distance of 450 meters.

---

[390] Аксененко, М. Д. (1965). 'О маскировке объектов бронетанковой техники в инфракрасной области спектра', *Вестник Бронетанковой Техники* 1965, сборник 2. p. 44.
[391] Антипов, А. П. et al. (1981). 'Двухцветная окраска БТТ', *Вопросы оборонной техники*, 20(98). p. 40.
[392] See: Höhn, D. H., Büchtemann, W. (1973). 'Spectral Radiance in the S20-Range and Luminance of the Clear and Overcast Night Sky', *Applied Optics* 12(1). p. 53.

Interestingly enough, the standard military paints in a number of NATO member armies since the 1950's were formulated solely for camouflage against false-color photography with filters rather than infrared night vision. The requirement was for the spectral reflectance curve of the enamel to match the rapid rise in IR reflectance of chlorophyll [**Fig. 79**, 3] from 0.7 μm to 0.8 μm, but without exceeding a reflectance of 40%[393]. The U.S. Army had a similar requirement against false-color photography, but detectability was secondary to the main goal of reducing cabin temperatures, which was achieved by standardizing on a solar heat-reflecting enamels[394] in the mid-1960's. The standard monotone olive drab color was specified for a very high reflectance of 60% at 0.8 μm[395]. In fact, the pigments used in KhV-518[396] were identical to that used in solar heat-reflecting enamel[397], which is largely why its NIR reflectance was so much higher than NPF-10.

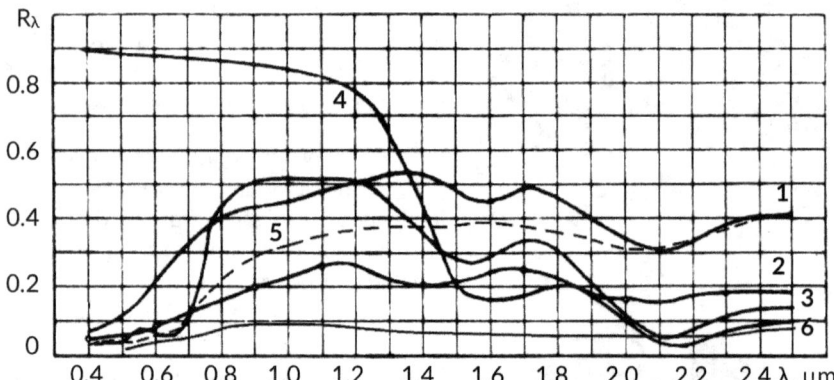

Figure 79: Spectral reflectivity curves of various surfaces. 1 – sandy soil; 2 – dark soil; 3 – green vegetation (grass, leaves); 4 – snow; 5 – green KhV-518; 6 – NPF-10. (*Adapted from M. D. Aksenenko, 1967*)

In the mid-1980's, the Soviet Army half-heartedly followed an international trend of transitioning to factory-painted multi-tone camouflages, along with the introduction of a new KhS-5146 enamel with AK-070 primer. KhS-5146 was a

---

[393] Central Weapons Laboratory (1970). *Dealing with Green Camouflage Color for Vehicles*. Report.
[394] MIL-E-46096 solar heat reflecting olive drab top coat with MIL-C-46127 solar heat reflecting primer.
[395] Krewinghaus, A. B. (1969). 'Infrared Reflectance of Paints', *Applied Optics* 8(4). p. 808.
The same requirement was later carried over to the CARC enamel: MIL-DTL-53039 and MIL-DTL-64159 CARC top coat with MIL-DTL-53022 or MILDTL-53022 epoxy primer.
[396] Lead chromate (yellow) with indigo (blue) and lead molybdate (red), mixed into green. Talc used as pigment extender. The paint had an extremely high lead content.
[397] Pickett, C. E. (1969). 'Solar Heat Reflecting Coatings', *International Automotive Engineering Congress*, Detroit. January 13-17. pp. 3, 6.

two-component [398] polyvinyl chloride acetate (PVCA) enamel with greatly improved abrasion resistance and a high tolerance to special chemical decontamination solvents used in degassing scrub-downs[399]. In this regard, KhS-5146 was directly comparable to the U.S. Army's CARC enamel[400], but instead of reflecting solar heat, KhS-5146 was formulated for a minimal NIR reflectance to suppress laser guidance devices, particularly SALH (Semi Active Laser Homing) weapons working in the 1.064 μm wavelength like the Hellfire. Its reflectance at 1.0 μm was a mere 8%[401]. A coat of KhS-5146 was estimated to decrease the target capture range of a SALH weapon by almost a third[402].

Thermal[403] signature reduction was also taken into consideration, albeit without having a decisive influence on the T-72's design in any way. While some attention was devoted to long-wave infrared (LWIR, 7.5-14 μm) signature reduction in research, real action was only taken to reduce the tank's medium-wave infrared (MWIR, 3-5 μm) signature, most probably because it was simply a realistic goal at the time.

By the 1960's, MWIR detectors had proven practical for constructing passive homing heads for guided weapons. Heat-seeking air-to-air missiles like the American AIM-9 Sidewinder are the most notable examples of such weapons, but less known are the early experiments on heat-seeking anti-tank missiles like the Soviet *Glaz* ATGM, also based on an MWIR homing head. The limited detection range of MWIR detectors on ground vehicle exhausts kept these experiments grounded, but it had been demonstrated that such weapons could realistically target tanks and other armored vehicles out to a distance of 1.5 km by their exhaust emissions. To the American and Soviet industry experts, the technology was perceived as a likely, or even imminent threat.

Infrared signature suppression revolves around the reduction of contrast by reducing infrared radiance to background levels. Radiance is determined by temperature and the emissivity of the surface or particle, and the most important aspect of MWIR signature suppression is exhaust temperature

---

[398] Mixed with a hardener before painting.
[399] Драк, П. И., Касьян, В. А., Калашникова, Н. М., Калиночкина, Е. В. (1985). 'Новое защитное покрытие для ВГМ на основе двухцветной эмали ХС-5146', *Вестник бронетанковой техники* 1985, сборник 5. pp. 41-42.
[400] Chemical Agent Resistant Coating paint.
[401] Антипов, А. П., Бажанов, Г. В., Калашникова, Н. М., Максимов, Ю. В. (1985). 'Маскировка БТТ от лазерных средств обнаружения и наведения', *Вестник Бронетанковой Техники* 1985, сборник 5. pp. 41-42.
[402] Драк, П. И., Касьян, В. А., Калашникова, Н. М., Калиночкина, Е. В. (1981). 'Новое защитное покрытие для ВГМ на основе двухцветной эмали ХС-5146', *Вопросы оборонной техники*, 20(98). pp. 11-13.
[403] Thermal radiation is often used in reference to infrared waves longer than the NIR band.

management. The T-72 configuration was decidedly inferior to exhaust entrainment systems like the T-64A, Leopard 1 and M60A1, where the engine exhaust flow was naturally intermixed, or entrained with the relatively cool air from the cooling system exhaust. The radiance of the T-72 exhaust at 2.7 μm and 4.2-4.3 μm [404] was seven times higher than for tanks with exhaust entrainment, almost entirely because the temperature of the exhaust flow emerging from the T-72's exhaust spout was 770°C, barely lower than it was at the exhaust manifolds of the engine. In contrast, the M60A1, which was the first production tank to have been built with MWIR signature suppression as a core part of its design, saw an exhaust temperature of just 400°C [405].

The only thermal suppression measures on the T-72 were on its engine compartment roof. Though the roof resembled that of previous Soviet medium tanks, the grilles over the radiator intake were designed in such a way that there was no direct line of sight to the radiator from an overhead view. This was a major departure from the louver-type intake grilles of the T-55 and T-62, which had wide gaps between each slat when they were left in the open position for an unrestricted airflow[406]. Spaced metal heat shields over the engine and the exhaust functioned as cool barriers hiding the two most important hot spots on the engine compartment roof[407]. This solution was developed through joint research with VNII-100 during the late 1960's[408].

The exhaust pipe was double-walled and covered from above by an oil tank, holding a reserve supply of engine oil[409]. The exhaust spout was concealed beneath a multi-plate heat shield[410]. Both ends of the shield were open for air to flow across its plates as the tank moved, keeping the topmost plate cool.

The same overhead multi-plate heat shield concept was later applied to the T-55 and T-62 during the T-55AM and T-62M modernizations, but to conceal the

---

[404] The absorption band of water and carbon dioxide respectively, two of the main combustion gasses in the exhaust flow.
[405] Исаков, П. П. (ed.) (1990). *Теория и Конструкция Танка - Том 10. Кн. 2: Комплесная защита*. Москва: Машиностроение. pp. 51-52.
[406] Мельников, Р. И. (1967). 'Аэродинамические характеристики броневых решеток входов воздуха в воздушный тракт танковых систем охлаждения', *Вестник Бронетанковой Техники* 1967, сборник 4. p. 6.
[407] Потемкин, Э. К. (ed.) (1999). *ВНИИтрансмаш - страницы истории*. Санкт-Петербург: Издательство «Петровский фонд». p. 195.
Also see: Исаков, П. П. (ed.) (1990). *Теория и Конструкция Танка - Том 10. Кн. 2: Комплесная защита*. Москва: Машиностроение. pp. 48-49.
[408] Later on, the T-80 series would receive mineral fiber insulation over the same zones.
[409] Because the engine had a very high oil consumption rate.
[410] These shields are explicitly stated to be intended for infrared signature reduction. See: Старовойтов, В. С. (ed.) (1990). *Военные гусеничные машины*. Том 1, Книга 2. Москва: Издательство МГТУ им. Н. Э. Баумана. p. 96.

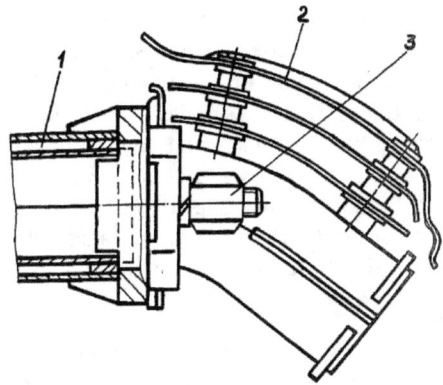

Figure 80: T-72 exhaust: 1 – double-walled exhaust pipe; 2 – heat shield; 3 – mounting screw. (*Soviet Ministry of Defense*)

exhaust pipe and not the spout. The downward angling of the T-72 exhaust spout concealed its opening from an aerial view, and was a simple and popular of doing so, as observed on the exhaust systems of the M60A1, Abrams, Leopard 1 and Leopard 2, and Chieftain, to name a few of the best-known examples. However, directing the exhaust flow downwards would later prove to be counterproductive against modern thermal cameras since the faint heat trace on the ground was visible from an aerial view[411].

The engine was shielded from above by an OPVT (deep wading kit) cover panel, a storage cover for the radiator sealing lids used for deep wading. The two lids would be stacked on the engine deck, and then the cover panel was closed over it, creating a three-layered spaced barrier like on the exhaust.

These measures also had a marginal effect on lowering the apparent temperature of the T-72 from an aerial view, but had little effect on ground-level observers. No special measures were taken to reduce the radiance of the tank's frontal aspect by lowering its temperature and nothing was done to control the emissivity of its surfaces. The LWIR emissivity of KhV-518 was ordinary, which is to say that it was very high (~0.96)[412], and no low-emissivity coatings entered service for the T-72 before the end of the Cold War.

Thermal detectors, even early types, had a much longer range than MWIR detectors. A thermal detector with a minimum resolvable temperature (MRT) of 0.2-0.3°C can detect a tank at 3 km, and thermal imaging devices like forward-looking infrared (FLIR) reconnaissance pods on aircraft have an MRT of 0.1-0.2°C[413], extending their detection ranges even further. The design measures in the T-72's engine compartment roof had a positive effect against aerial reconnaissance, and ground reconnaissance was addressed by various measures like the addition of side skirts on the T-72A to obscure the suspension.

---

[411] Gonda, T., et al. (2003). 'An exploration of vehicle-terrain interaction in IR synthetic scenes', *The International Society for Optical Engineering*.

[412] From the relationship between reflectivity and emissivity in Kirchhoff's law of thermal radiation. Based on reflectivity at 10.6 μm described in: Невзоров, В. А., Непогодин, И. А. (1986). 'Контраст танка Т-72 при лазерной локации', *Вестник Бронетанковой Техники* 1986, сборник 1. pp. 26-27.

[413] Исаков, П. П. (ed.) (1990). p. 36.

Later, the addition of explosive reactive armor (ERA) brought a certain masking effect by essentially covering the tank in thin steel boxes mounted with an air gap. Unless the ERA boxes were allowed to heat up under prolonged direct sunlight, the apparent temperature of the T-72 was significantly lowered, especially for moving tanks. Even without the side skirts and any additional armor, the average thermal contrast[414] of a bare T-72 was overall lower than the T-54/55 and T-62. Soviet testing[415] found the average thermal contrast, measured by the average change in temperature (ΔT), to be 2°C across its frontal arc. U.S. Army signals intelligence corroborated this figure with a reported ΔT of 2°C while the T-54/55 and T-62 were attributed with a ΔT of 5°C[416].

The visibility of the T-72 to radar was somewhat controlled by the streamlined shape of the pannier fuel tanks and stowage boxes, but no other measures were taken. The odd, unordered shape of the fuel tanks and spare parts boxes on the hull panniers had been a major source of increased radar returns on the T-55[417], and shapes with right angles were to be avoided, especially flat vertical faces perpendicular to the ground because of the tendency to reflect radar waves off the ground[418].

Incidentally, the placement of tools in thin-walled pannier bins was not ideal for mine resistance. An anti-tank mine consistently destroyed the first few pannier bins and scattered their contents, which restricted these bins to non-essential tools. The panniers themselves had to be straightened out or replaced after a sufficiently powerful mine blast[419], but this was much less of a problem than if the turret was struck by a powerful high-explosive shell, deforming the pannier into the suspension. This is unlikely to prevent the tank from driving, but it was the type of damage to require immediate rectification. The exhaust pipe was also a problem. It was welded to the hull side, so if it was bent or pierced, replacing it was not possible without cutting it off and welding a new pipe in its place[420].

---

[414] Based on the ΔT metric, determined by the average of the apparent target temperature to the average apparent temperature of the background within a given field of view. This is the most rudimentary metric of thermal contrast and tends to overrate targets that have a low ΔT from extreme hot and cold spots.

[415] Гуменюк, Г. А. (1986). 'Влияние внешней среды на показатели обнаружения танка по тепловому контрасту', *Вестник Бронетанковой Техники* 1986, сборник 5. pp. 13-15.

[416] Rosa, S. P., Lindsley, T. (1989). 'Tank Thermal Signatures: The Other Variable In the Gunnery Equation', *ARMOR* (September-October). p. 32.

[417] Неверовский, Н. А., Овчаров, В. Г., Рещиков, И. Ф., Терехин, И. И. (1966). 'Снижение радиолокационного отражения танков за счет выбора оптимальных наружных форм', *Вестник бронетанковой техники* 1966, сборник 1. p. 24.

[418] This was a notable disadvantage of large metal side skirts or so-called "bazooka plates", such as those found on the British Centurion and Chieftain tanks.

[419] Глухов, Г. К., Кочегаров, П. П. (1980). 'Особенности Ремонта Подорванных на Минах Танков', *Вестник бронетанковой техники* 1980, сборник, 6. pp. 36-37.

[420] Соболев, Е. Г. (1981). 'Восстанавливаемость танков при боевых повреждениях', *Вестник*

Tucha-2 Smoke Grenade System

Later T-72 models were equipped the *Tucha-2* 81 mm "offensive" smoke grenade system, intended to enable individual tanks or units to maneuver behind smoke concealment. Each launcher tube was a small mortar loaded with a 3D6 smoke grenade. The T-72A (1979) had the 902A *Tucha-2* system with twelve launcher tubes. The T-72AV (1985) and all variants of the T-72B (1984) were fitted with the 902B *Tucha-2* system, the main difference being that there were only eight tubes. The 902A and 902B systems were otherwise identical, sharing the same control panel [**Fig. 81**].

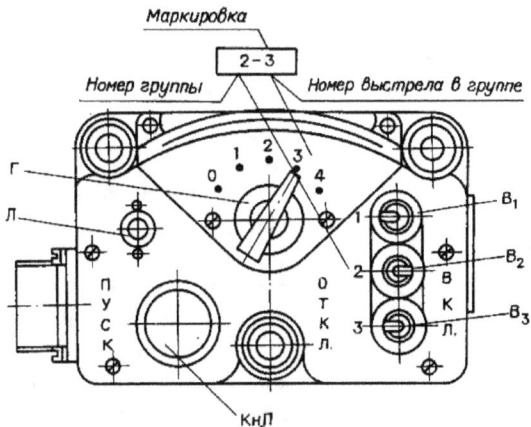

Figure 81: *Tucha-2* control panel. B₁, B₂, B₃ – salvo select switch; Г – salvo size (no. of shots) selector knob; Image shows panel with salvo 2 selected, salvo size 3 shots (2-3). (*Soviet Ministry of Defense*)

The 3D6 smoke grenades produced HC smoke, a basic type of military obscurant smoke. It only provided effective obscuration in the visual spectrum (0.4-0.76 m)[421]. The smoke compound was based on emitting zinc chloride, which captures moisture in the air to form a white smoke. The zinc chloride was produced by burning hexachlorobenzene with zinc oxide and aluminum[422]. Each 3D6 produced a smokescreen with an average width of 27 meters and average height of 8 meters within 10-20 seconds of launch[423].

The 902A system arranged its twelve tubes in two banks on the left and right turret cheeks, fanning out into a very narrow forward arc [**Fig. 82**]. To achieve the 250-300 m range, the tubes were tilted at a 45° elevation angle. The simplest

---

бронетанковой техники 1981, сборник, 3. p. 13.
[421] НИИ прикладной химии (2016). *Тучи, которые защищают*. Available at: http://www.niiph.com/ru/novosti/stati/24-tuchi-kotorye-zashchishchayut (Accessed 2 February 2024).
[422] Hexachlorobenzene produced a better-moderated smoke than older and more common hexachloroethane smoke compounds, still found in U.S. Army HC smoke devices.
[423] This includes a grenade flight time (and pyrotechnic fuze burnout time) of 7-12 seconds and smoke build-up time of 6-10 seconds.

way to set up the two banks of launchers for the intended salvo size would be to have six tubes on each turret cheek, but the left bank gained the seventh tube because of the IR spotlight to the right of the gun. That tube was ever-so-slightly tilted to the right to compensate. On the 902B system, all of the launcher tubes were clustered on the left of the turret but they were still aimed into the same narrow forward arc.

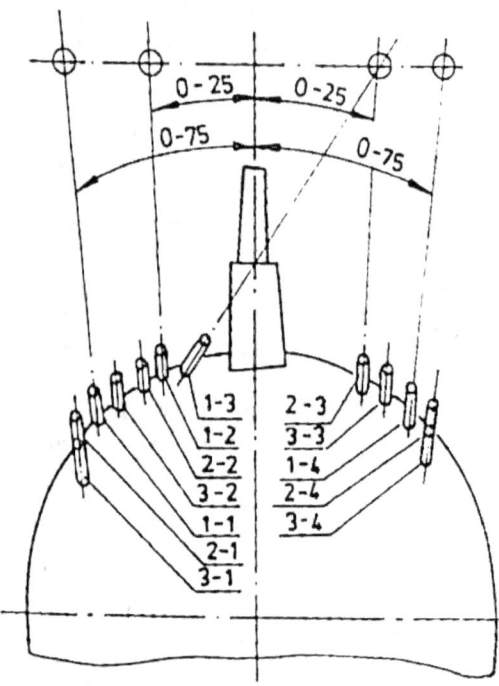

Figure 82: 902A tube layout on T-72A and T-72M1, marked for 3 salvoes of 4 shots each. (*Soviet Ministry of Defense*)

The simplest way to use the *Tucha-2* system was set the control panel to launch a salvo of four smoke grenades, and the number of salvoes available depended on whether the tank was equipped with the 902A system or the 902B system. With each salvo, a dense smokescreen would be formed 250-300 meters away along a width of around 80 meters. The long throw distance provided maneuvering space behind the smokescreen, and the grenades burned for 1-1.5 minutes, which gave the persistence needed for a tank to move by bounding between smoke clouds under most weather conditions.

Larger smokescreens could be made by firing multiple salvoes by rotating the turret, and even manually aiming individual grenade shots across a wider frontage. A large smokescreen up to 220 meters wide could be produced this way with twelve tubes, and eight tubes could manage 150 meters[424].

---

[424] Гриненко, С. В., Кравченко, Ю. М., Трещевский, Н. П. (1984). 'пути совершенствования танковых систем постановки аэрозольных завес'. *Вестник Бронетанковой Техники 1984, сборник 2*. pp. 33.

This "offensive" method of smoke projection fundamentally differed from the individual "defensive" type used on NATO tanks since the 1960's. The British-American type, for instance, threw a large salvo of six smaller 66 mm grenades into a wide 90° arc at 30 meters to conceal an individual tank, and by using red phosphorous as the smoke agent, near-infrared sensors and laser rangefinders could be blocked effectively. The *Tucha-2* could be compared more closely to the early smoke projector systems fitted to British tanks during WWII and other gun-type smoke bomb launchers, only with a much longer range[425].

Firefighting Systems

The firefighting equipment of the T-72 consisted of an automatic firefighting system and one or more portable fire extinguishers. An automatic system reduced the likelihood of a total loss from fuel fires as well as ammunition fires resulting from fuel fires, which was noted as the main source of tank losses in major battles during the Great Patriotic War[426]. The ZETs11-3 automatic firefighting system came as a basic feature for all T-72 models since its debut into service until January 1990[427], when it was replaced by the ZETs13-1 *Iney* explosion suppression system.

These systems could detect a fire, run a simple defense subroutine and discharge a fire extinguisher bottle into the affected compartment without crew intervention. The system was required to provide two attempts for at least one of the compartments. If any of the crew members detected a fire before the system could react, they could manually discharge an extinguisher bottle with the press of a button. The driver was in charge of the main control panel, and the gunner and commander had emergency buttons on their circuit breakers.

For safety reasons, the firefighting system was wired to the tank's batteries in a two-wire (closed loop) circuit. The system was turned on when the tank's master power button was pressed and there was no way to turn it off, but the engine stopping subroutine and the fire extinguisher bottles could be blocked by switching the system to the deep wading mode (OPVT). Otherwise, both the ZETs11-3 and ZETs13-1 systems were always running in the background[428].

One of the main difficulties with combating fires in a tank with an automated system is that with a limited number of discharge nozzles to cover the most fire-prone parts of the tank, it was still impossible to extend equal coverage to the

---

[425] 4-inch smoke generator projector – 125 yards. 2-inch bomb thrower – 150 yards.
[426] Горбунов, А. С., Мелихов, Т. Н., Тамбовцев, Ф. Д. (1980). 'К истории создания автоматической системы ППО', *Вестник Бронетанковой Техники* 1980, сборник 3. p. 54-55.
[427] Альбом основных конструктивных изменения, проведенных на Изд. 184, 184-1 и 184К.
[428] They ran on just 35 W and would function as long as the tank had some electrical power.

entire interior space of the tank. This essentially limited automated systems to the total flooding method, which had been standard since the first Soviet tank automatic fire extinguisher system was introduced on the T-54 in 1947. By discharging enough extinguishing agent to keep the gaseous concentration at a certain level, the system was guaranteed to be effective on small fires in the nooks and crannies of the hull, or even inside punctured fuel tanks and punctured electrical high-voltage units.

Of course, this came with the rather unfortunate side effect of depriving the crew of oxygen, which was only made worse by the intrinsic toxicity of high-performance fire extinguishing agents. Historically, Soviet tank fire extinguishing systems first relied on basic carbon dioxide, but since the mid-1950's this was replaced with a special mixture called Composition 3.5, named for the fact that it was 3.5 times more effective than carbon dioxide. Composition 3.5 was formulated for reduced toxicity by using ethyl bromide (Halon 2001) in a 30% mix with 70% carbon dioxide instead of the more conventional methyl bromide (Halon 1001), occasionally used in early aircraft and industrial fire extinguishing systems[429].

In 1976, Halon 2402 (R-114B2) replaced Composition 3.5[430]. Compared to other standard fire extinguishing agents like the U.S. Army's Halon 1301 and the USAF's Halon 1211, Halon 2402 was more effective at putting out fires but it was also much more toxic[431]. The ratio of the concentration required to extinguish a fire over the concentration which induces symptoms of narcosis, known as the F-factor, was 14. The F-factor for Halon 1301 was just 0.8[432]. The USSR accounted for nearly all global production and deployment of Halon 2402, mostly in aircraft, building and ship fire extinguishing systems[433], so it was not unusual to see its adoption in ground vehicles, but as an extinguishing agent for inhabited spaces, it was a rather questionable choice. There were, nevertheless, advantages in the ease of handling Halon 2402 because, like Composition 3.5, it was a liquid at room temperature and the servicing infrastructure in the Soviet Army to fill and maintain liquid fire extinguishers was already well established[434].

---

[429] The lethal concentration of Halon 2001 is 148,000 ppm, but for Halon 1001, it was 5,900 ppm. Department of Defense (1995). *Design of Combat Vehicles for Fire Survivability*, MIL-HDBK-684-1. p. 7-20.

[430] Потемкин, Э.К. (ed.) (1999). *ВНИИтрансмаш – страницы истории*. Санкт-Петербург: Издательство «Петровский фонд». p. 193.

[431] Humphrey, B.J, Smith, B.R, Skaggs, S.R. (1990). *Toxicity of Halon 2402*. ESL-TR-88-59. pp. 2-3.

[432] Gann, R. G. et al. (1990). *Preliminary Screening Procedures and Criteria for Replacements for Halons 1211 and 1301*. NIST Technical Note 1278. Gaithersburg: National Institute of Standards and Technology. p. 181.

[433] United Nations Environment Programme (2006). *Montreal Protocol On Substances that Deplete the Ozone Layer*. Report. pp. 13-17.

[434] Ребриков, В. Д., Ширман, Б. А., Сподак, В. В. (1988). 'Особенности зарядки баллонов системы

In recognition of these toxicity issues, the ZETs13-1 *Iney* system developed in the 1980's adopted a hybrid system replacing Halon 2402 in the crew compartment with Halon 1301 (R-13B1) while retaining Halon 2402 for the engine compartment. Unlike Composition 3.5 and Halon 2402, Halon 1301 boiled when it was transferred from a storage tank into another vessel at high pressure, so a new infrastructure of filling equipment accompanied the changeover. This likely contributed to the late rollout of the *Iney* system.

In all cases, the fire extinguishing agent was heavier than air. This mitigated some of the danger to the crew, but nevertheless, if a fire extinguisher bottle discharged in the crew compartment, it was important for the crew members to hold their breath while bailing out to avoid asphyxiation and frostbite in the lungs. If bailing out was not possible, they had to don their gas masks. To evacuate the fire extinguishing gasses from the crew compartment, the ventilator was turned on and, if possible, the floor drainage ports or the escape hatch were opened.

A portable fire extinguisher was also included to fight any small fires overlooked by the automatic system or to handle fires outside the tank. In most T-72 models, including export and license-produced models, one OU-2 carbon dioxide fire extinguisher was stowed on the front perimeter guard of the autoloader carousel. It was most accessible to the driver, since the fire risk was highest in the hull. At some point, the manual firefighting equipment was revised to two OKh-2 fire extinguishers containing 2.0-2.1 kg of liquid Halon 2402 nitrogen-pressurized to 4.41 MPa (45 kg/cm^2). One of the halon extinguishers was carried internally, replacing the CO2 fire extinguisher, and the other was stowed in the center rear external turret bin to put out external fires, to include even napalm fires.

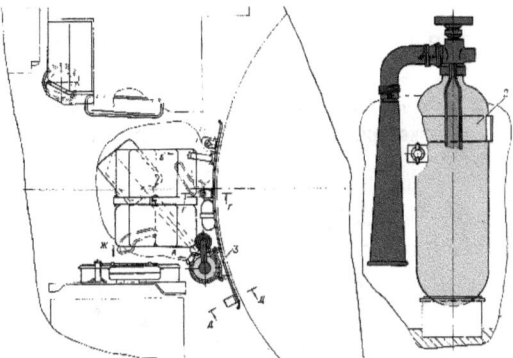

Figure 83: OU-2 fire extinguisher stowage point in T-72 models before the 1980's. (*Soviet Ministry of Defense*)

противопожарного оборудования хладоном 13В1', Вестник Бронетанковой Техники 1988, сборник 10. pp. 26-27.

ZETs11-3

The ZETs11-3 was a variant of the ZETs11 series adapted to the specific internal layout of the T-72. There were three identical extinguisher bottles, one in the crew compartment and two in the engine compartment, filled with 1.2-1.3 kg of Halon 2402 and nitrogen-pressurized to 6.86 MPa (70 kg/cm^2). All three bottles were interconnected and could discharge into either compartment by selectively setting off the squibs in a special distributor valve. The extinguishing agent would then be discharged through nozzles arranged close to the hull floor to cover the most likely sites of a pool fire, since all of the T-72's fuel tanks and most of the hydraulic fluid reservoirs were in the hull.

The control logic for the firefighting system was a finite state machine built entirely on electromechanical relays and capacitors inside a B11-5 control unit, with the GO-27 CBRN protection system piggybacking off some of its subroutines. Fire detection was provided by four TD-1 thermopile heat detectors in the crew compartment and five TD-1s in the engine compartment.

Beginning with the T-72A, the crew compartment was upgraded with five more TD-1 sensors and a new B11-5-2S1 control unit. The five new sensors were arranged around the autoloader carousel and wired to a stronger amplifier for increased sensitivity[435], and the new control unit was more reliable.

Thermopiles generate a voltage proportionate to the temperature difference across its junctions, which made the TD-1 effective even in a hot environment like the T-72's engine compartment, aided by the ZETs11-3 system having the sensitivity to react to a temperature difference as small as 50°C. After starting the tank's engine, for instance, the

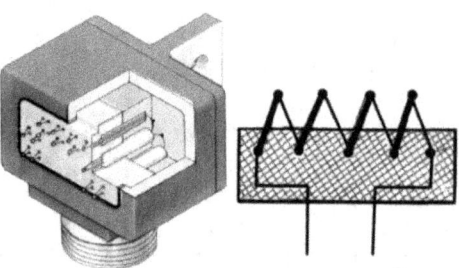

Figure 84: Thermocouples in the thermopile of a TD-1 sensor. (*Soviet Ministry of Defense*)

protruding "hot" junctions of the fifteen chromel-copel thermocouples in a TD-1 were heated slowly enough that the "cold" junctions embedded in plastic could warm up, so the temperature difference was never high enough to register as a fire. When exposed to rapid heating by a nearby flame, the thermopile output

---

[435] Борисюк, М. Д., Жадина, О. В., Харланова, В. П. (1976). 'Анализ качества и совершенствование танковых систем противопожарного оборудования', *Вестник Бронетанковой Техники* 1976, сборник 3. pp. 14-16.

signal could cross an amplification threshold, and from there, the amplified signal would trigger a relay and set the fire extinguishing subroutine into motion.

1. The tank's engine is stopped.
2. The ventilation intake is closed, and if it was turned on, the filter-ventilation unit (FVU) is stopped.
3. The discharge squib for extinguisher bottle No. 1 is triggered, releasing the fire extinguishing agent into the affected compartment.
4. The fire alarm is set off.
5. Expended bottle is marked as empty.

From the start of a fire until extinguisher discharge, the reaction time of the ZETs11-3 system was no more than 10 seconds[436]. If the fire was successfully extinguished, then after 30-50 seconds, a ventilation subroutine turned on the FVU in the overpressure mode to ventilate the crew compartment, and the fire alert ended[437]. If the TD-1 sensors continued signaling a fire, the cycle was repeated until all three bottles were expended, and if a second fire appeared in a different compartment while the system was in the middle of its subroutines for the first fire, the system discharged into the second compartment immediately.

ZETs13-1 *Iney*

The ZETs13 *Iney* system began development in 1982, entered mass production in 1988[438] and was introduced into the T-72 production line in 1990. *Iney* was both an extinguishing system and an explosion suppression system, capable of detecting an incipient fuel mist fire and suppressing it before it could progress into a fireball explosion.

Ten OD1 optical infrared sensors in the crew compartment monitored the free spaces around each of the internal fuel tanks where a fuel mist cloud could form from ballistic penetration. The response time of the OD1 sensor from the eruption of a fire until its detection did not exceed 2 milliseconds, and one of the two dedicated quick-discharge extinguisher bottles would be discharged by 90% within 100 milliseconds, disrupting the combustion of the would-be fireball through special high-flow rate nozzles. Each bottle contained 1.9-2.0 kg of liquid Halon 1301, nitrogen-pressurized to 7.35 MPa (75 kg/cm^2).

---

[436] Бондарь, А. И. et al. (2015). 'К вопросу совершенствования систем противопожарной защиты отечественных боевых машин', *Механіка та машинобудування* 2015, 1. p. 4.
[437] Макаренко. А., Кузнецов, Ю. (1999). *Электроспецоборудование танка Т-72*. Омск: Военная кафедра Омский Государственный Технический Университет. pp. 189-197.
[438] Потемкин, Э.К. (ed.) (1999). *ВНИИтрансмаш – страницы истории*. Санкт-Петербург: Издательство «Петровский фонд». p. 340.

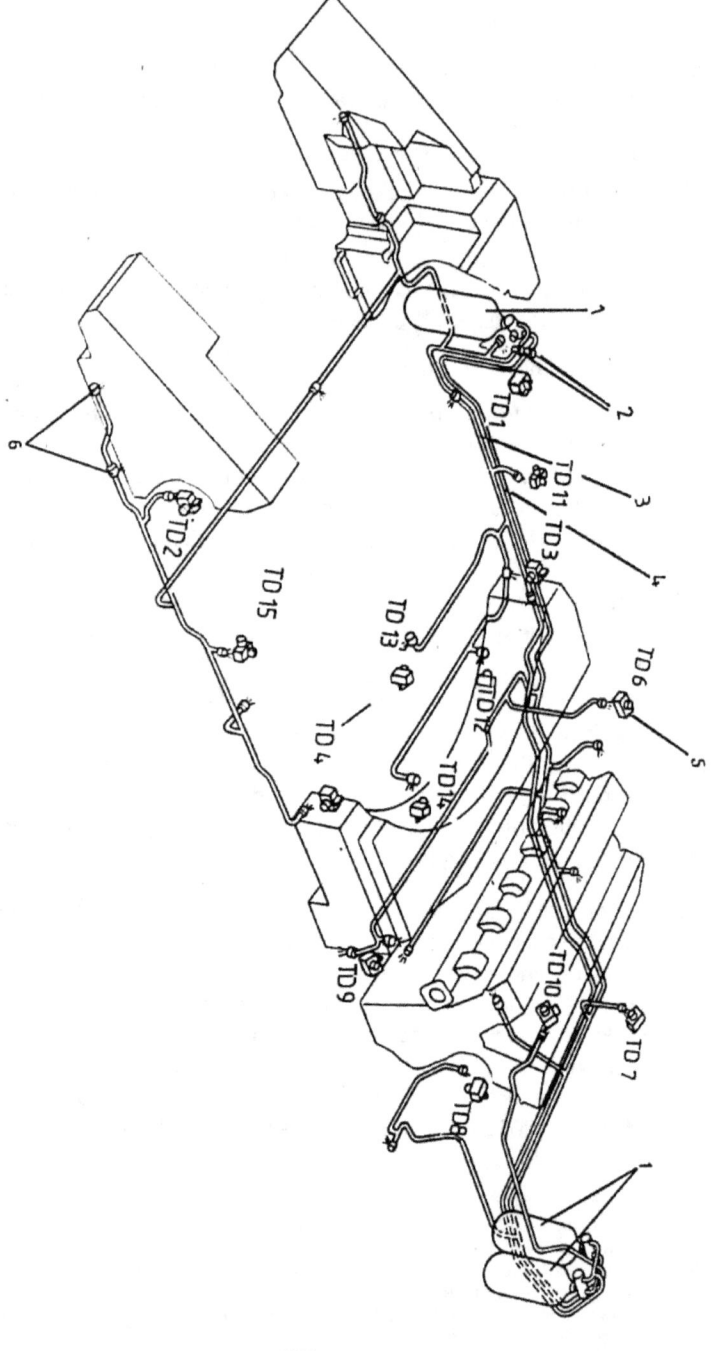

Figure 85: Thermal sensor layout in T-72A with ZETs11-3 automatic firefighting system. Each TD-1 sensor is numbered, up to TD 15. The early ZETs11-3 crew compartment sensors correspond to TD 1, TD 15, TD 4, and TD 12. (Soviet Ministry of Defense)

Sensor coverage in the crew compartment was extensive, but this was mainly a reflection of how the T-72 stored its entire fuel supply in the crew compartment. In the M1 and M1A1 Abrams, there were just four optical sensors[439], and the same was true of the Leopard 2A4 as well[440]. The reaction time was comparable to the explosion suppression system in the Leopard 2A4 (80-150 ms)[441] and well within the minimum total reaction time threshold of 250 milliseconds to avoid first degree burns on crew members[442].

The engine compartment was limited to an extinguishing system. Five TD-1 sensors were laid out like in the ZETs11-3 system, and the two bottles for the engine compartment each held 0.58-0.62 kg of liquid Halon 2402 nitrogen-pressurized to 6.86 MPa (70 kg/cm^2). Since the free volume in the engine compartment was much smaller than the crew compartment and the extinguisher bottles were not shared between compartments, smaller bottles were quite adequate.

OD1 sensors had been tested for the T-72's engine compartment, but were found to be unsuitable because the receiver windows were fouled by soot, dust, and fuel and oil particles during tests[443]. Nevertheless, the nonviability of optical sensors in engine compartments was not unique to the T-72 or other tanks with open engine compartments. The tight packaging of the drivetrain components makes it difficult for a fireball to develop and there are few unobstructed lines of sight, so continuous heat detectors are often preferable, despite their much slower reaction times. Tanks like the Chieftain and Leopard 2A4 were also fitted with a type of continuous heat detector – the Graviner Firewire™ linear thermistor cable sensor, originally made for aircraft engine bay fire suppression systems.

Besides explosion suppression, *Iney* also introduced a small improvement to the alarm system. Once a fire was detected, the driver was alerted by all of the lights next to his TNPO-168V periscope turning on at once, and a visual alarm unit warned the commander if the fire was in the crew compartment or the engine compartment, on top of the basic audio alarm through the intercom.

---

[439] Department of Defense (1995). *Design of Combat Vehicles for Fire Survivability*, MIL-HDBK-684-1. p. 7-59.
[440] Krapke, P. -W. (2004). *Leopard 2: Sein Werden und seine Leistung*. Norderstedt: Books on Demand GmbH. p. 87.
[441] Kowalski, K. (2018). 'Automatyczne systemy przeciwpożarowe w wojskowych pojazdach bojowych', *Zeszyty Naukowe SGSP 2018*, 65 (2). p. 33. Also see: Krapke, P. -W. p. 91.
[442] Booz, Allen & Hamilton Inc. (2003). *Final Technical Report Fires Experienced and Halon 1301 Fire Suppression Systems in Current Weapon Systems*. p. 17.
[443] Бондарь, А. И. et al. p. 11.

Armor

Structural Overview

The hull was welded from various grades of low alloy chromium-nickel-molybdenum (CrNiMo) rolled homogenous armor (RHA) steel plates, heat-treated to medium hardness. The plates were produced by hot rolling to the desired thickness, which homogenized the microstructure of the steel.

The thickest plates were made from 42 SM steel, and the thinnest were made from 43 PSM steel. Plates of intermediate thickness were made from 49 S steel. To optimize nickel expenditure according to the required hardenability of the alloy, one of these three grades was selected for each detail on the tank based on the required plate thickness, and the alloy composition was tuned as needed.

Most crucially, the hardenability of the alloy was much less critical for thinner plates, so the nickel content could be pared down quite severely for the roof and belly plates on the tank without any loss of toughness. The practice of using low-nickel alloys for thinner plates was also followed in some countries outside of the USSR. The hull belly and roof plates were made from 43 PSM, the hull rear and turret roof were made from 49 S, and the rest was made from 42 SM.

Table 19: Alloy composition of T-72 rolled steel plate grades

Thickness range (mm)	C	Mn	Si	Cr	Ni	Mo
42 SM						
60-90	0.28-0.34	0.30-0.50	0.18-0.35	1.80-2.30	1.40-1.90	0.28-0.38
49 S						
35-75	0.28-0.34	0.30-0.55	0.18-0.35	1.40-1.90	1.00-1.50	0.25-0.35
43 PSM						
Up to 30	0.25-0.31	0.30-0.55	0.18-0.35	1.80-2.30	0.50	0.25-0.35

All structural plates in the T-72's body were joined by automated welding with manual welding for the interior details due to space restrictions[444]. Soft and tough austenitic welds[445] allowed the welds to withstand direct hits on the

---

[444] Боткина, Г. Я., Кочергин, А. К., Щелкунов, Г. М., Подрезов, В. Г. (1986). 'Применение автоматической сварки броневой стали танка Т-72', *Вестник Бронетанковой Техники* 1986, сборник 5. p. 41.
[445] 10Kh20N7CT austenitic weld

joints between armor plates without cracking. Compared to the T-55A, the share of labor spent on welding increased by 1.5 times[446].

Medium hardness steel was optimal in terms of weldability, ballistic resistance, weld survivability, and long-term fatigue resistance. High hardness armor (HHA) steel showed higher resistance to penetration from subcaliber penetrators and shaped charges, but cracks during welding and cold embrittlement at the required -40°C operating temperature was a persistent problem for many HHA grades.

The medium hardness armor steels used in the T-72 changed little since their creation in the early 1950's. Soviet analysts found all three grades to be essentially equivalent to foreign (unspecified) medium hardness steel compositions[447].

Hull stiffness in all three axes was an important consideration for surviving a nuclear blast or mine blast without systemic structural damage. In principle, tank hulls can handle marginal elastic warping from shocks transmitted through the suspension during off-road driving without any special measures, but under blast loading, the plastic warping could be extreme enough disrupt the alignment of the autoloader and drivetrain parts, leaving the tank toothless and immobilized despite outwardly appearing fine[448]. Long mechanical control rods that extended the entire length of the hull were particularly problematic if they were placed atop the hull floor, which generally experienced the strongest deformation.

On the T-72, hull rigidity was intrinsically high from the tub-shaped belly and the support to the hull roof provided by the upper glacis armor. Additional reinforcement was provided by stamping a complex pattern of stiffening ribs and bumps into the belly plates throughout the entire hull. The reinforcement to the engine compartment floor was especially prominent. Between the rounded bulges for the BKP transmission units, the deep ribbing under the engine and the oil primer pump access port at the very center of the compartment, there were nearly no flat surfaces at all. The bulge stamped into

---

[446] T-72 adaptacja licencji: Prace wdrożeniowe 1978-1982' (2016). *Szybkobieżne Pojazdy Gasienicowe*, 3(41). Gliwice: Ośrodek Badawczo-Rozwojowy Urządzeń Mechanicznych „OBRUM" sp. z o.o. p. 4.
[447] Высоковский, С. Н. et al. (1983). 'Сравнение требований зарубежного и отечественных стандартов на поставку листовой противоснарядной брони', *Вопросы оборонной техники*, 2(108). pp. 52.
[448] Абрамов, Б. А., Доронин, В. П., Лазебник, О. М., Прокуряков, В. Б. (1972). 'Некоторые Пути Повышения Прочности и Жесткости Днищ Танков', *Вестник Бронетанковой Техники* 1972, сборник 1. pp. 15-18.

the floor for the driver's seat served as local reinforcement reducing the deformation floor height underneath the driver.

The engine access panel and the radiator panel were both bolted down to the hull so that the engine compartment roof was structurally contiguous, at some cost to convenience from the sheer number of bolts that had to be unscrewed for a drivetrain inspection. Small inspection hatches were present on the engine access panels of earlier T-54/55 and T-62 models, but these were removed since 1967 as part of a new standard that the T-72 shared. Internally, the engine compartment bulkhead was welded on all four sides and largely closed apart from two small access panels, also firmly bolted, but for hermetic sealing reasons as well as strength. The engine compartment was additionally stiffened with two cross beams, one under the hull roof behind the turret ring, and another directly between the engine access panel and the radiator.

High-productivity casting methods were used to form armored covers with complex shapes. Investment casting was used to form smaller complex parts, and shell-mold casting was used to form medium sized structural parts, like the final drive casings, the drive sprockets, the shock absorber casings, crew hatches, and more. Shell-mold casting was almost exclusively used in the automotive industry for the precision casting of engine and transmission components, allowing complex shapes to be produced at a lower material and labor cost than by machining billets. The continual expansion of the automated casting facilities at *Uralvagonzavod* made it possible to continually reduce the price of the T-72 throughout its production[449].

---

[449] Устьянцев. С. В., Колмаков. Д. Г. (2013). *T-72/T-90. Опыт создания отечественных основных боевых танков*. Нижний Тагил: ОАО «НПК «Уралвагонзавод» имени Ф. Э. Дзержинского». pp. 177-180.
Also see: Иванов, Н. С., Панов, В. Т., Швед, А. Д. (1982). 'Возможности экономии металла в производстве ВГМ', *Вестник Бронетанковой Техники* 1982, сборник 4. pp. 49-50.

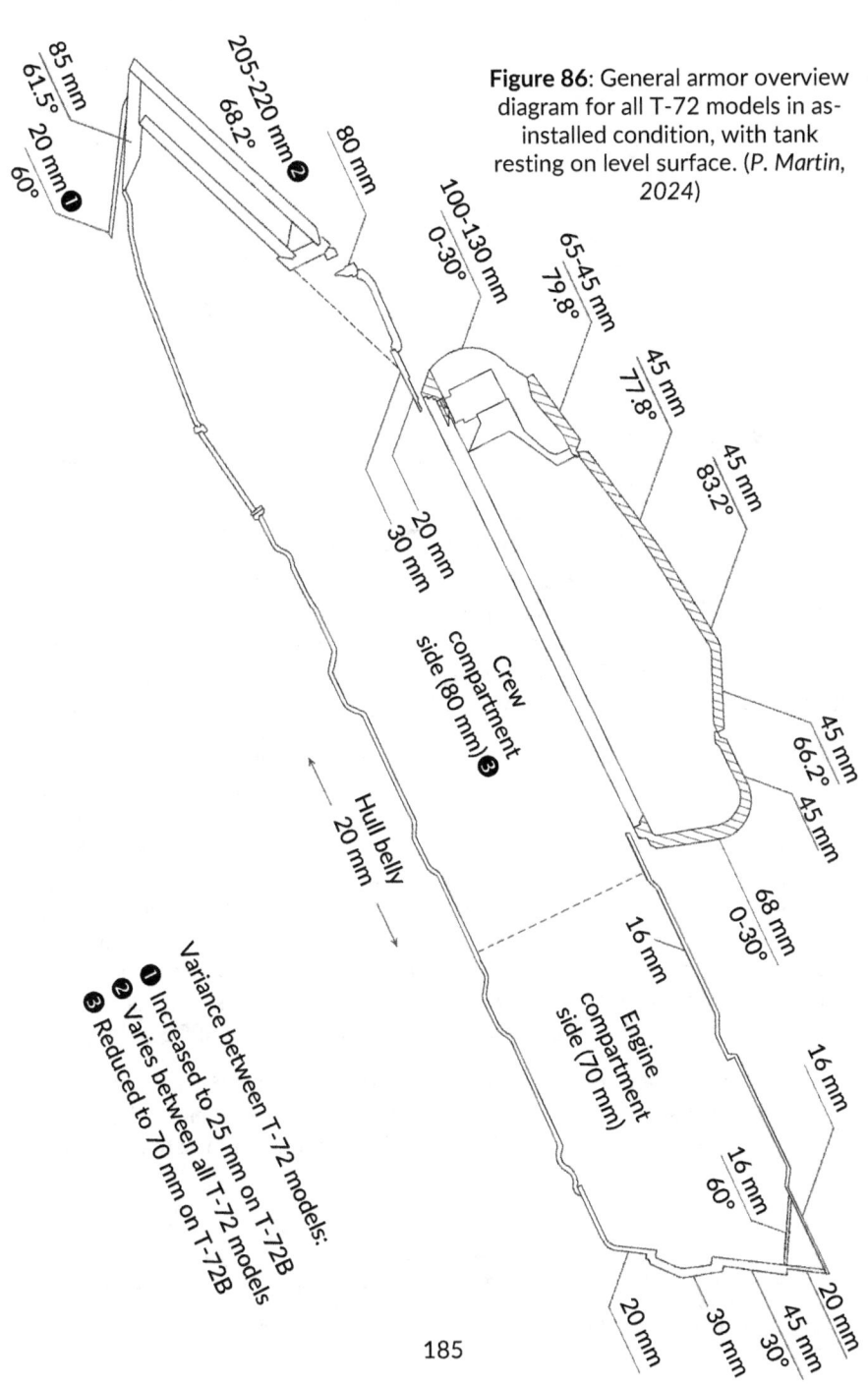

Figure 86: General armor overview diagram for all T-72 models in as-installed condition, with tank resting on level surface. (*P. Martin, 2024*)

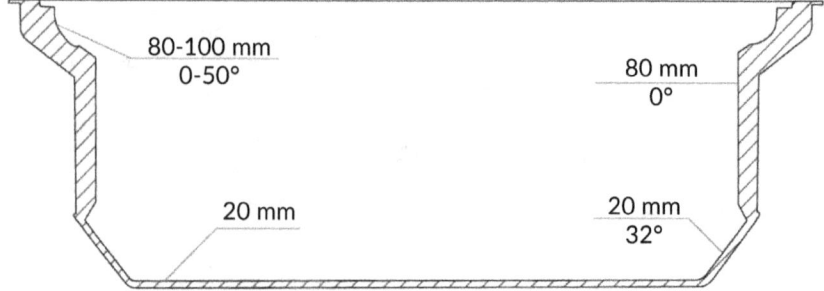

Figure 87: Crew compartment hull side armor profile for T-72 models prior to the T-72B, where the plate thickness was reduced to 70 mm. (P. Martin, 2024)

Composite Armor Design

The impetus for the development of composite tank armor in the USSR was the existential threat posed by shaped charge (HEAT) technology in the 1950's. The development of solid armor-piercing ammunition for high-powered guns never stopped, but a HEAT shell could penetrate a thickness equal to 3-4 times its own diameter in armor steel, practically independent of its impact velocity.

With such amazing qualities, shaped charge technology seemed to be the ultimate rebuttal to tank armor as a concept. The extreme efficiency of shaped charges was instrumental to the birth of anti-tank guided missiles, and in NATO, towed anti-tank guns, a legacy workhorse of WWII, were replaced entirely by infantry-portable recoilless guns firing HEAT grenades in frontline units by the 1950's[450]. At the same time, shaped charge weapons did not at all diminish the continuing importance of solid armor-piercing rounds, as both types evolved side by side for tank guns in most militaries. In the U.S. Army Armored Force, for instance, the standard 90 mm gun of the M48 Patton was superseded by the 105 mm gun of the M60, delivering both HEAT and a potent APDS round.

In response, a government decree was issued on May 30, 1960 on the creation of special armor solutions to protect a tank from the 105 mm ammunition of the M60[451]. The task was assigned to the country's central research institute on tank

---

[450] In the U.S. Army, the 105 mm M27 recoilless rifle entered service in 1952 to replace the 57 mm M1 and 3-inch M5 guns under the Battalion Anti-Tank (BAT) program, followed by the 106 mm M40 recoilless rifle in 1955.
[451] Алексеев, М. et al. (2012). *НИИ Стали 1942-2012*. Москва: Издательство СканРус. p. 36.

protection, NII-24 (NII *Stali*) in cooperation with VNII-100. This was the official birth of composite tank armor in the USSR.

In October 1960, the state of existing tank armor in the USSR was summed up none too happily by the Scientific and Technical Council of the GKOT[452]:

> Имеющиеся на вооружении средние и тяжелые танки являются незащищенными от подавляющего большинства современных кумулятивных средств поражения под курсовыми углами, заданными по защите от бронебойных снарядов.
>
> Усиление защиты танков от кумулятивных средств поражения путем утолщения стальной брони требовало более чем двукратного увеличения толщин и весов основных бронедеталей, что практически осуществить не представлялось возможным.

Translated, the report stated:

> The medium and heavy tanks in service are unprotected from the majority of modern shaped charge weapons at the angles specified to protect against armor-piercing projectiles.
>
> Strengthening the protection of tanks from shaped charge weapons by thickening the steel armor required the thickness and weight of the main armor details to be doubled, which practically did not seem possible to implement.

The armor for the T-64 was required to provide complete protection from 105 mm APDS and 105 mm HEAT. The armor had to be equivalent to 330 mm RHA against APDS and stop 105 mm APDS at a distance of 500 m. It also had to be equivalent to 450 mm RHA against shaped charges and stop 105 mm HEAT[453].

In the absence of actual 105 mm ammunition for armor verification, surrogates were identified. 100 mm APDS, fired from a BS-3 field gun or D10-T tank gun was adopted as an analogue for 105 mm APDS. Smoothbore 100 mm HEAT shells fired from the Soviet T-12 anti-tank gun served as an analogue for 105

---

[452] GKOT – State Committee of the Council of Ministers of the USSR on Defense Technology. See: Устьянцев, С., Колмаков, Д. (2007). *Боевые Машины Уралвагонзавода: Танки 1960-Х*. Нижний Тагил: «Медиа-Принт». p. 54.
[453] Tarasenko, A. (2016). *Armor protection of the tanks of the second postwar generation T-64 (T-64A), Chieftain Mk5P and M60*. Available at: https://btvt.info/3attackdefensemobility/432armor_eng.htm (Accessed: 2 March 2024).

mm HEAT. Supplementary verification was performed with 85 mm HEAT shells and 100 mm blunt-nosed, full-caliber AP shells, fired from a BS-3 or D10-T. These shells were much more valuable for checking the structural integrity of the turret under extreme shock loads because of the much larger mass, and thus momentum carried by these AP shells. Welded armor structures that survive attacks from high armor penetration power can collapse under bombardment from AP shells.

Work on studying various filler materials for composite tank armor had already begun in 1958, so a practical solution was rapidly prepared and implemented on the T-64, and from there, the same design solutions migrated onto the T-72. The T-72 would begin its service with virtually the same armor as a T-64A produced in the late 1960's, with the exception of the turret, which featured a simplified monolithic cast armor instead of a multi-layered steel construction.

The armor fulfilled the initial requirement for protection against 105 mm weapons to satisfaction. However, towards the mid-1960's, the requirement was stretched to include 115 mm HEAT shells with a reference penetration value of 450 mm RHA, i.e. the 3BK4 shell, as fired from the T-62's 115 mm gun. At the same time, the KE protection requirement was also stretched to include protection against 115 mm APFSDS fired from the *Molot* gun from a distance of 1,500 m.

This basic level of protection was formidable enough that in the U.S, the realization that the T-64 and T-72 had composite armor kickstarted the development of the Improved TOW modification of the TOW anti-tank guided missile[454] and the follow-on TOW 2 missile to deal with the much-feared T-80[455], which was assumed to have even better armor than the T-64 and T-72.

The commitment to explosive reactive armors as the most effective and most promising direction for tank armor development was reflected institutionally at NII Stali. The reactive armor department at NII Stali became the institute's top department, and the head of the department became deputy director in 1985, and, in 1991, was appointed general director[456].

---

[454] Warford, J. M. (1983). 'The T95: A Gamble in High-risk Technology'. *ARMOR*. (September-October) 1983. p. 41.
[455] Furlong, R. D. M. (1980). 'Delay in Improved TOWS for Europe?', *International Defense Review*, 9 (September). p. 1343.
[456] Калиниченко, В.И. (2017). *Защита для брони*. Москва: ИПО «У Никитских ворот». p. 63.

Side Screens

On the T-72 Ural, each side of the hull was protected by a set of four flip-out flaps, called "spring-out screens" (взводные экраны) or "anti-shaped charge screens" (противокумулятивный экраны). These flaps were more popularly known as "gills" in some Warsaw Pact members like Czechoslovakia[457]. The purpose of these flaps was to detonate shaped charge warheads at a great distance from the sides of the tank, allowing the shaped charge jet overstretch and dissipate before reaching the sides of the hull.

The concept of these "gill" armor flaps could be traced to a 1964 project investigating spaced nets and swing-out screens called ZET-1. The objective of the project was to explore promising types of add-on protection to improve the survivability of existing tanks lacking a built-in defense against HEAT shells[458]. The T-72 inherited its "gills" from the T-64A and used the same flaps, receiving the same changes as updates were made to the T-64A[459].

The flaps were sheet aluminum panels (3 mm) with riveted rubber flaps (6 mm)[460], mounted on a spring-loaded frame. The flaps had two positions: traveling and combat. In the traveling position, the flaps were hooked to the hull sides, unfastening the hooks deployed them to the combat position. The spring-loaded frame prevented damage in a collision with a tree, a fence, walls, and so on, but colliding with a shorter obstacle could still twist the frame off its bolts.

In the combat position, the flaps were nearly perpendicular to the hull, just swept back slightly at a 15-20° angle[461]. Each flap created an air gap of 1.8-3.5 meters to the hull side plate within a protected arc of ±22°, depending on the exact point of impact. At larger angles, gaps in the coverage appeared. The air gap was large enough that the "gills" could protect the sides from 115 mm HEAT[462].

---

[457] The term "žábry" or "gills" was notably used to refer to the flaps in the Czechoslovakian Army technical magazine ATOM: Velek, M. (1989). 'T-72', ATOM 1989, (April), 16(112).
Other mentions in other magazines and post-USSR internet discussions give the impression that the name originated from either Poland or Czechoslovakia.
[458] Солянкин, А. Г, Желтов, И. Г., Кудряшов, К. Н. (2010). Отечественные бронированные машины. XX век: Том 3. Отечественные бронированные машины. 1946-1965 гг. Москва: ООО «Издательство "Цейхгауз"». pp. 88-94.
[459] Чобиток, В. В. (2022). Бортовые противокумулятивные щитки. Available at: http://armor.kiev.ua/Tanks/Modern/T64/shields/ (Accessed 1 June 2024).
[460] Measured.
[461] Павлов М. В., Павлов, И. В. (2021). Отечественные Бронированные Машины 1945-1965 гг. - Часть I: Легкие, средние и тяжелые танки. Кемерово: ООО "Принт". p. 538.
[462] Солянкин, А. Г., Желтов, И. Г., Кудряшов, К. Н. (2010). Отечественные бронированные машины. XX век - Том 3: Отечественные бронированные машины. 1946-1965 гг. Москва: Издательство «Цейхгауз». p. 90.

Conventional side skirts were first installed on the T-72 on an experimental basis in 1975, later becoming standard for the T-72A in 1979. These side skirts were 10 mm thick, made from a reinforced laminate with six layers of plastic fabric alternating with six layers of fine steel wire mesh, pressed into rubber and surfaced with rubber. The mounting points for the "gills" remained, even surviving through the T-72B series, but the flaps themselves were never seen fitted during peacetime.

The rubber side skirts offered a modicum of shaped charge protection, but their original purpose was to help suppress the dust cloud thrown up by the tank's suspension on dry terrain. The need for these skirts was identified very early on for the Object 219 (T-80) due to the high dust ingestion rate of its gas turbine engine in dry environments. Similar skirts migrated onto the T-72 and T-64 from there. The first appearance of rubber skirts in the T-72 family was in 1975 on the Object 176, a product improvement project for the T-72 that later became the basis of the T-72A.

Unlike the "gill" flaps, these skirts were found to be quite practical in everyday use, especially compared to conventional metal skirts. Solid metal "bazooka plate" skirts like on the Centurion and Chieftain collected mud in the suspension and could be bent quite easily, but a rubber skirt was flexible enough to avoid both of these problems, and at the same time possessed the rigidity to set off HEAT grenades on impact. The rubber and plastic composition also gave a certain degree of neutron shielding[463].

Even without the additional space given by "gill" flaps, the T-72's side skirts were still spaced 745 mm away from the hull sides, which at a 22° side angle was enough to protect against most 105 mm HEAT charges, including the tank-fired 105 mm M456 HEAT shell[464].

The only threat that the side skirts was not intended to stop was 115 mm HEAT[465]. Officially, the side armor of a T-72A (T-72M1) was rated at 500 mm RHA to shaped charges within a ±22° frontal arc[466], inclusive of the side skirts.

The skirts were ineffective when shot perpendicular to the hull sides. In Polish tests, 10 mm rubber side screens were shown to degrade the penetration of PG-

---

[463] NII Stali.
[464] Based on T-55 tests with 100 mm HEAT as an analogue to 105 mm HEAT, and the performance of the 105 mm M456 HEAT shell during Centurion side skirt tests. The low coherence of the jet produced by the M456 shell can also be seen in: Whitley, D. O. S. (1977). *HEAT vs HESH Paper*. Memorandum.
[465] According to descriptions, 115 mm HEAT with a copper liner, i.e. 3BK4M.
[466] NII Stali poster, T-72M1.

7VM grenades to 155 mm and of 100 mm 3BK-5M shells to 220 mm[467], not enough to protect the 70-80 mm hull sides.

The survivability of rubber skirts to the blast-fragmentation effect of a tank-fired HEAT shell was quite poor, but nevertheless better than light metal skirts. During live fire tests on a Centurion, one or both of the 6 mm steel skirt plates were blown off with each 105 mm HEAT strike, exposing the entire hull side or leaving it only half-covered[468].

---

[467] Wišniewski, A. (1996). 'Ocena Efektywności Ochronnej Opancerzenia Dodatkowego Zmodernizowanego Czolgu T-55A', *Problemy Techniki Uzbrojenia i Radiolokacji*, 58. pp. 68-69.
[468] War Office (1966). *Assessment of lethality of US 105 mm HEAT shell against space-plated Centurion tank*. Technical report WO 194/472. Chertsey: FVRDE.

Figure 88: T-72 Ural with early anti-shaped charge screens. (Soviet Ministry of Defense)

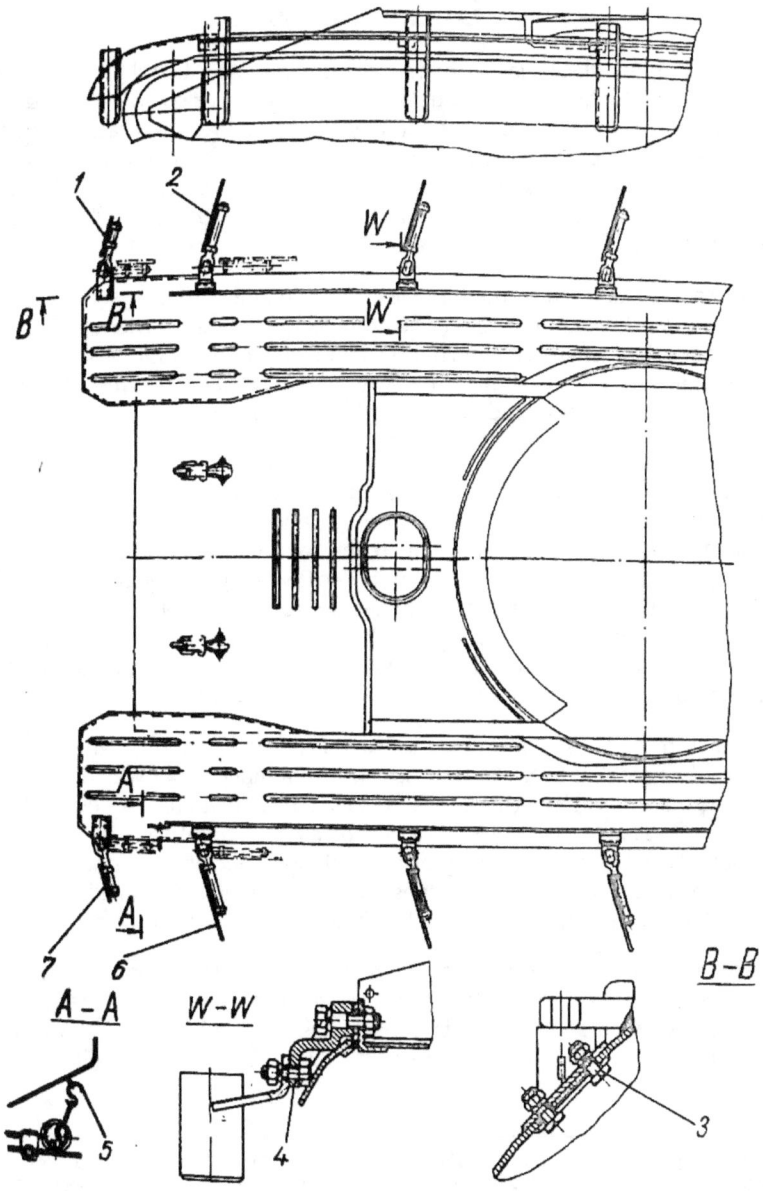

Figure 89: T-72 Anti-shaped charge screens. (*Soviet Ministry of Defense*)

Upper Glacis

The T-72 upper glacis was a sloped structure with an integral front and back plate, with an interlayer placed in between. The interlayer could be a solid filler or an array of thinner plates. The interlayer was secured by brackets on the sides. The entire hull nose was assembled as a single self-contained item, the upper and lower glacis plates welded together and the driver's cutout fitted to the upper glacis. The nose assembly was then welding to the tank hull belly, side, and roof plates.

This was, for all intents and purposes, the same method of assembly as the hull nose armor plates on a conventional hull. There was no way of accessing the interlayer to exchange it even during a total hull rebuild, or to replace it after combat damage. A perforation in the armor could be filled in with a steel plug, and the only option to upgrading an existing tank for new threats was to attach additional armor on top.

The main criteria for effectively defeating a shaped charge jet with multilayer armor with a passive filler is to destroy or expend the tip of the jet within the first layer. The tip penetrates armor in the hydrodynamic mode, where the critical property is the ratio of armor density to jet density due to the overwhelming pressure exerted by the jet tip from its extreme velocity. Once the tip is expended, the jet interacts with subsequent armor elements through its midbody, traveling at a much lower velocity. Material properties, strength in particular, become more relevant to resisting the jet as a result.

If the front plate is followed by a low-density filler with a higher specific strength than steel, the jet can be stopped with a much lower weight of armor than otherwise possible with a barrier of solid steel. The initial models of the T-72 were created according to this concept by incorporating a filler of glass textolite, a type of glass-reinforced plastic with high specific strength. It was also feasible to place a layer of high hardness steel behind the front plate[469], but only a modest increase in mass efficiency is achieved with this approach[470].

Despite the inclusion of low-density filler materials, the armor itself was by no means light. In fact, the T-72 had some of the heaviest armor in the world at the time it entered service. The structural mass of the T-72A hull nose (with towing hooks, but no other features) was 3,517 kg. The upper glacis alone

---

[469] Брызгов, В. Н., Симаков, И. К. (1985). 'Вклад тыльного слоя высокотвердой стали в противокумулятивную стойкость брони', *Вестник Бронетанковой Техники* 1985, сборник 2. pp. 26-28.

[470] In the early T-64A turret, high hardness armor (HHA) steel inserts gave the turret an 8-12% increase in resistance than the same weight of homogeneous steel armor. See: Григорян, В. А. et al. (2006). *Частные вопросы конечной баллистики*. Москва: Издательство МГТУ имени Н. Э. Баумана. p. 255.

weighed 3,018 kg[471]. The armor itself was equivalent to 330 mm of solid steel in line-of-sight thickness, or a 120 mm steel plate sloped at 68°.

To obtain similar or better protection at a lower weight while at the same time meeting updated protection requirements, new armor configurations and novel solutions were explored. In total, five iterations of the T-72 upper glacis were created and two types of explosive reactive armor (ERA) were developed, creating a total of nine possible armor combinations between them.

Table 20: T-72 hull armor iterations

Model	T-72	T-72A	T-72B		
In-service	1973	1979	1984		
Period	1973-1975	1975-1984	1984	1985	1988
Armor type	Three-layer composite	Improved three-layer composite	Five-layer spaced steel	Six-layer spaced steel	Three-layer spaced steel
Add-on armor	Otrazhenie-2 (1983)		-		
Hull side	"Gill" panels	Full-length reinforced rubber side skirt			
ERA	None	Kontakt-1 (1984)	-	Kontakt-1	Kontakt-5
Protection Level, HEAT (RHA equivalent, mm)					
Basic	450	500	N/D	N/D	N/D
w/ Add-on armor	500	550	N/A		
w/ ERA	-	-	850-900*		
Protected arc	±22°	±35° (w/ ERA)			±22°

A steeply sloped plate erodes and fractures a penetrator by its high resistance and high obliquity, or angle to the penetrator trajectory. The destructive effect on the penetrator occurs in the impact and breakout phases. In the impact phase, the penetrator faces asymmetric resistance from the plate because the lower edge of the penetrator is the first to strike the plate surface. The nose of the penetrator meets an axial resistance force, but also a lateral resistance force proportional to the impact angle. The lateral load on the penetrator deflects the nose of the penetrator away from the plate. After traveling through the plate, the penetrator again meets asymmetric resistance in the breakout phase

---

[471] Hull nose technical drawing. Нос корпуса 172.01.002с6-6с6.

because there is a smaller thickness of material below the penetrator than there is directly ahead of it and above it.

The effectiveness of an armor plate in breaking up the penetrator depends heavily on both thickness and obliquity. The T-72 upper glacis armor was designed to break up a penetrator with its front plate and absorb the residual elements in the subsequent layers. Advancements in APFSDS technology forced the exploration of alternative fillers in the early 1980's and the front plate began to be reinforced to more effectively break up sophisticated penetrators, culminating with the introduction of Kontakt-5 explosive reactive armor.

The high slope angle of the upper glacis also contributed to shaped charge protection. Due to the splattering of the shaped charge jet on highly sloped surfaces, especially narrow jets typically created by light, shoulder-fired anti-tank grenades, there was a tendency for the penetration power at high obliquity (>60°) to be significantly lower than the normal values[472]. Ideally, if the plate is struck by a grenade without a graze-sensitive fuze, the grenade fails to detonate and no damage is done at all.

This essentially did not apply to contemporary ATGMs, often capable of functioning at impact angles of 75° or more, but the actual impact angle rarely corresponded directly to the structural obliquity of the target, but rather, to the obliquity of the target plus the angle of attack of the missile – a nuance that is not reflected in penetration figures determined from static tests. The angle of attack varied according to the aerodynamic characteristics of various ATGM models, older missiles in particular often having a small coefficient of lift and low velocity, needing a large angle of attack to maintain level flight. A positive angle of attack translates to a reciprocal decrease in the penetrating power of the warhead, and to further complicate matters, the exact angle of attack often varied significantly throughout a typical flight profile.

ATGMs with a two-stage motor rely on a booster rocket motor to accelerate quickly to a high velocity before a sustainer motor kicks in to either maintain a stable cruise velocity or slowly accelerate the missile. A 2nd generation ATGM like the MILAN would begin the stable phase of its flight profile at 4°, and under the acceleration of its sustainer motor, reach its peak velocity at its 12th second of flight (1,850 m flight distance) where it required an angle of attack of only around 2.3° [473]. Its penetrating power on a 68° upper glacis could thus diminish by 11% at long range or by as much as 21% at short range.

---

[472] Брызгов, В. Н. (1985). 'Исследование навесной динамической защиты танков израильской армии', *Вестник Бронетанковой Техники* 1985, сборник 3. pp. 51-53.
[473] The drop-off after the 12th second was due to motor burnout from fuel exhaustion. The thrust

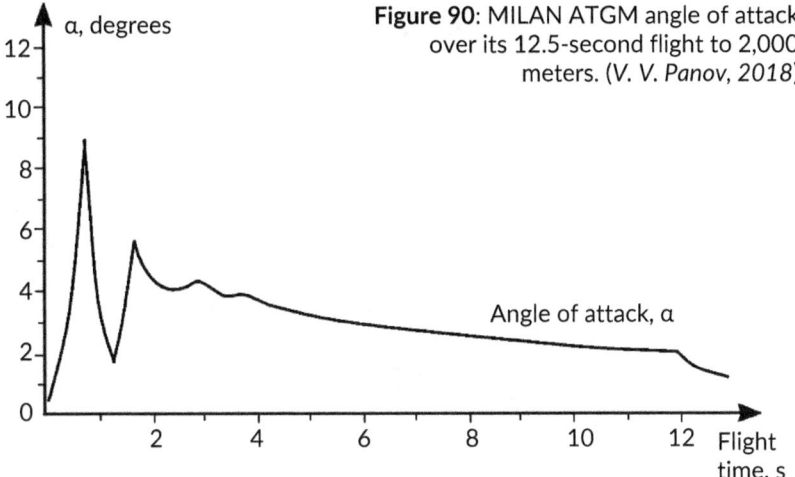

Figure 90: MILAN ATGM angle of attack over its 12.5-second flight to 2,000 meters. (*V. V. Panov, 2018*)

Early 1st generation systems like the SS.10 and ENTAC[474] were much slower, flying at an average angle of attack of 6-7°, and even a relatively speedy 1st generation missile like the Vickers Vigilant flew at 5°[475]. Despite static penetration figures nearing 600 mm RHA, the effect of the high angle of attack may impede these missiles from piercing the T-72's basic upper glacis armor, equal in protection to just 450 mm RHA against shaped charges.

ATGMs with a single-stage motor have a gliding flight profile (e.g. TOW), where a powerful motor rapidly accelerates the missile to a high velocity and then cuts off to let the missile glide for the remainder of its flight. Missiles of this type fly with a low angle of attack initially, but the angle of attack rises quickly as the flight distance grows, especially if velocity was lost through maneuvers on the flight path to the target.

Of course, bullets could also ricochet off the upper glacis, and into the driver's periscope. This was mitigated by four anti-ricochet ribs on the upper glacis, later reduced to three ribs as the armor thickness increased (1976) so as not to interfere with the driver's view, and then reduced again to just two ribs when the *Otrazhenie*-2 plate was added (1983).

---

vectoring steering system lost control authority, and the missile entered a ballistic trajectory.
Панов, В. В. et al. (2018). *Высокоточное оружие зарубежных стран. Том 1. Противотанковые и многоцелевые ракетные комплексы*. Second Edition. Тула: АО «Конструкторское бюро приборостроения им. академика А. Г. Шипунова». p. 165.
[474] Stauff, J. E. (2008). *COMHART - Tome 10: Armements Antichars Missiles Guidés Et Non Guidés*. Paris: Comité pour l'histoire de l'armement terrestre. p. 98.
The ENTAC had a downwards-tilted warhead to compensate for its high angle of attack, but this solution did not garner much popularity worldwide.
[475] Forbat, J. (2012). *The 'Secret' World of Vickers Guided Weapons*. Stroud: The History Press. p. 245.

Three-layer STB armor (T-72)

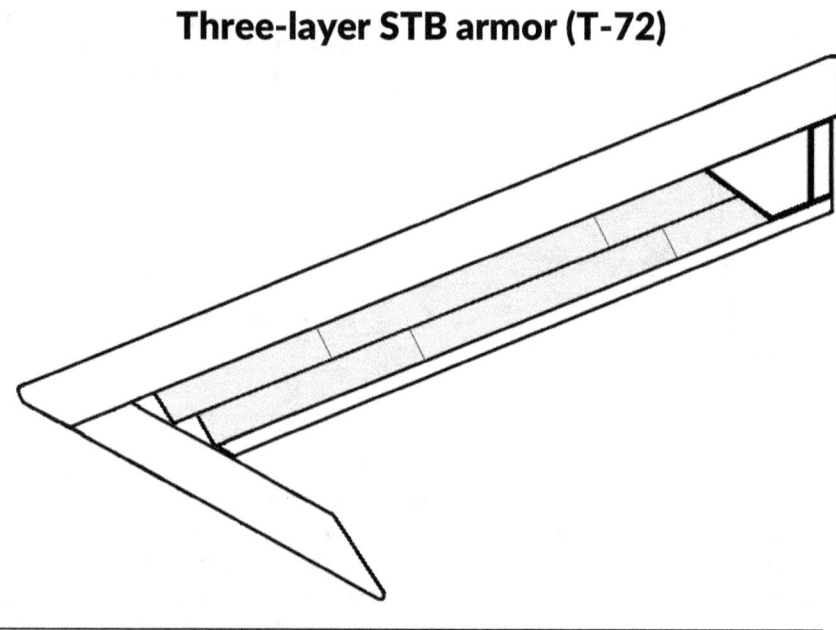

Steel	Glass Textolite	Steel
80	105	20

All figures in mm

Figure 91: Three-layer array with STB interlayer. (R. A. Then, 2024)

In principle, the armor functioned as a classical "disruptor-absorber" system for both shaped charges and KE penetrators. The steel front plate erodes the penetrator and breaks it up through its high obliquity. The residual penetrator is then arrested in the STB layer. The armor was evaluated to be capable of stopping all 105 mm APDS, and could stop 120 mm APDS at a distance of 1,000 m[476]. It also stopped 115 mm steel APFSDS at a limit velocity of 1,400 m/s[477], equal to a firing distance of 1,500 m.

The 80mm front plate alone was a challenging target for early 105mm APDS, leaving the penetrator with almost no energy left to deal with the subsequent layers. West German testing of captured T-55 tanks with 105mm DM13 rounds (L28A1 produced under licence) allowed a graph of the safety limits of the side

---

[476] Устьянцев, С., Колмаков, Д. (2007). *Боевые Машины Уралвагонзавода: Танки 1960-Х*. Нижний Тагил: Издательский Дом «Медиа-Принт». pp. 158-159.

[477] Устьянцев, С. В., Колмаков, Д. (2004). *Боевая Машниы Уралвагонзавода: Танк Т-72*. Нижний Тагил: Издательский Дом "Медиа-Принт". p. 91.

armour of the tank, 80mm of RHA, to be produced for ranges of 800 meters and 200 meters. The steel plates were made from 42 SM steel, hardened to 285-311 HB[478].

Known as armor glass texolite (стеклотекстолит, броня), abbreviated to STB. Glass textolite is a laminate of woven glass fiber textile sheets in a matrix. Glass textolite is not the same as fiberglass, which is manufactured from loose, unaligned strands of continuous or chopped glass fibers suspended in resin. Fiberglass is largely isotropic – that is, its mechanical properties are the same regardless of the directionality of the forces acting on it. Fiberglass made from oriented fibers are anisotropic, in that they possess high strength in one direction but are much weaker in all other directions. In the USSR, this type of fiberglass was known as SVAM (anisotropic fiberglass).

Glass textolite was produced from glass fiber textile in a plain weave, laminated by hot-pressing in a phenol resin matrix. As an oriented fiber material, it was anisotropic, but not to the degree of SVAM. The textile was woven out of borosilicate glass[479] fibers spun into rovings[480]. It was functionally identical to conventional E-glass[481], historically the earliest type of glass fiber used in glass-reinforced plastics and relatively weak in ballistic protection. Arranged in a sandwich structure between the steel front and back plates of the armor array, the degree of radiation attenuation was higher than a simpler two-layer structure[482]. The fiber volume fraction was in the range of 60-65% based on the known density[483] of 1.85 g/cm^3. Compared to an epoxy matrix, phenol resin degraded at higher temperatures and retained its ductility at lower temperatures, also being more ductile overall at some cost to strength.

---

[478] Алексеев, О. И., Терехин, И. И. (1976). 'О Некоторых Закономерностях, Определяющих Защитные Свойства Трехслойных Преград при Обстреле Сплошными Оперенными Бронебойно-Подкалиберными Снарядами', *Вопросы оборонной техники*, 20(63). p. 20.
[479] Modern ballistic textolite grades use S-glass with both higher tensile strength and elongation.
[480] Bundles of fibers.
[481] "Electrical" glass fiber, originally created for electrical insulation fiberglasses. In the USSR, electrical-grade glass textolite was STEF (СТЭФ – Стеклотекстолит, Эпоксифенольное связующее), which differed from STB in having an epoxy-phenol matrix.
[482] Пугачев, Б. Л., Фрид, Е. С., Шашкин, В. И. (1986). 'Особенности выбора материалов и толщин слоев в многослойных элементах противорадиационной защиты танков', *Вестник Бронетанковой Техники* 1986, Сборник 4. pp. 24-27.
For instance, the MBT-70/KPz-70 was planned to have an anti-radiation filler between the front and back steel plates of its armor, demonstrating an example of an optimal radiation shield.
[483] Алексеев, О. И., Терехин, И. И. p. 20.

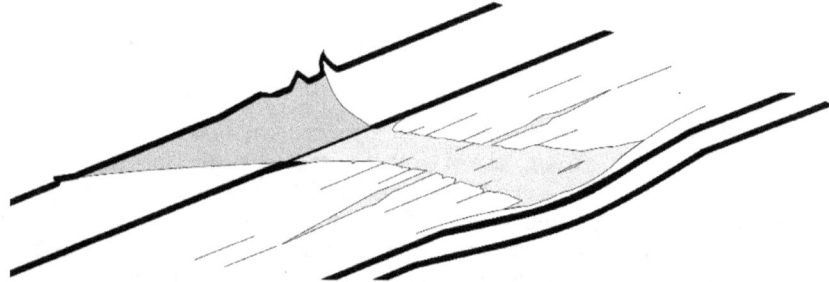

Figure 92: Visualization of damage to glass textolite and thin back plate after stopping an APDS projectile. (*P. Martin, 2024*)

Energy absorption in glass textolite is primarily based on expending the penetrator energy in fiber crushing and delamination, and to a much lesser extent by matrix crushing. The primary factor is crushing strain, which is a parameter where glass fibers greatly outperform high-modulus fibers like Kevlar in rigid composite panels[484]. In the initial stage of penetration, the STB interlayer would be penetrated by crushing/shearing of the fibers, and towards the end, it delaminates and experiences local deformation.

Laboratory tests with long rod penetrators showed that, by weight, STB exhibited a higher resistance to penetration than medium-hardness armor steel[485] when impact velocities exceeded 1,000 m/s[486], up to 40% higher at 1,800-1,900 m/s. The dependence on penetrator velocity is because of the intense strain-hardening experienced by fiber-reinforced composites, especially in oriented fiber composites[487].

However, the specific configuration of the 80-105-20 armor was perhaps the least optimal way of incorporating glass textolite. At a high obliquity, the anisotropy of glass textolite[488] severely reduced its strength (at some gain to elongation), and the thick steel front plate sapped a penetrator of much of its velocity after breakout. The residual penetrator entering the STB interlayer was practically guaranteed to be traveling at a velocity below 1,000 m/s, greatly diminishing the efficiency of the STB layer. A more modern long rod penetrator can overcome the front plate without losing much velocity, but such a round would overcome the interlayer quite easily.

---

[484] Fleck, N. A. (1997). 'Compressive Failure of Fiber Composites', *Advances in Applied Mechanics*, Volume 33.
[485] BSST armor steel, generally similar to tank armor grades.
[486] Григорян, В. А. (ed.) p. 293.
[487] This was one of several justifications for the choice of glass textolite over GFRPs with non-oriented fibers.
[488] Exhibition of different strength and elongation along different directions due to fiber orientation.

A somewhat more favorable configuration, as determined by Soviet researchers[489] and verified by their Chinese counterparts[490] through full-scale tests, was to invert the 80-105-20 array so that the penetrator would be met by the STB layer as soon as possible. However, the gain in resistance was fundamentally limited[491].

Moreover, in the basic 80-105-20 configuration, some energy was dissipated by the deformation of the STB layer and the thin back plate, which, despite being medium hardness armor steel, was thin enough to deform a great deal without cracking. The interaction is comparable to a composite helmet deforming from a bullet strike, and just like a helmet, it was undesirable for the armor to bulge inward directly into the driver's workspace and into internal equipment.

This was addressed by building additional clearance between the upper glacis and the internal equipment directly behind it. No equipment of significant weight was mounted directly onto the back face of the upper glacis in the driver's station[492], since the size of the gap needed for shock isolation was prohibitively large[493]. The armor directly in front of the driver's periscope was solid steel, both to reinforce the area and to eliminate the possibility of the armor pushing into the periscope, throwing it into the driver's head. The driver's space was additionally protected by eleven steel studs anchoring the 20 mm back plate of the array to the 80 mm front plate through the interlayer.

To improve the performance of three-layered STB armor against APFSDS rounds, replacing the thin back plate with a stiffer barrier was critical. At the very least, the back plate was to be no less than 35-40 mm thick[494]. Research in this direction started in the early 1970's and was implemented at some point prior to the T-72A.

---

[489] Григорян, В. А. (ed.). p. 295.
[490] 陆祥璇 (1989). '"两炮"火力系统总体论证', in 高膛压火炮系统论文集. 机电部兵器科学研究院. p. 27. Lu Xiangxuan (1989). 'Overall demonstration of the "two guns" firepower system'.
[491] The limit of penetration with an experimental composite cored penetrator on an 80-104-20 armor mock-up was 1,520 m/s, and for a 20-104-80 mockup, it was 1,612 m/s.
[492] Потемкин, Э.К. (ed.) (1999). *ВНИИтрансмаш – страницы истории*. Санкт-Петербург: Издательство «Петровский фонд». p. 197.
[493] Гапон, В. В., Гусев, О. П., Нанава, И., Тетельбаум, Р. Д. (1987) 'Исследование Динамики Амортизированного Внутреннего Оборудования Танка', *Вестник бронетанковой техники* 1987, сборник 8. pp 20-22.
[494] Алексеев, О. И., Терехин, И. И. p. 24.

Improved three-layer STB armor (T-72A)

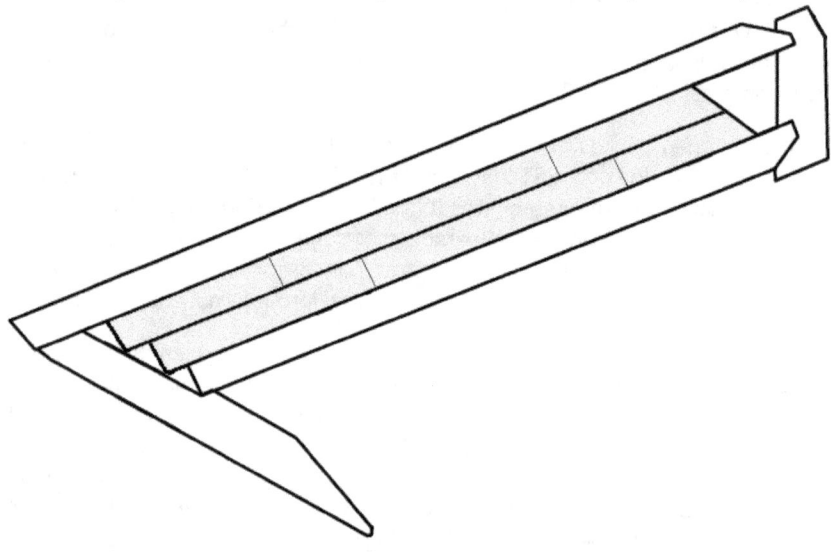

Steel	Glass Textolite	Steel
60	105	50

All figures in mm

Figure 93: Improved three-layer armor with STB interlayer. (R. A. Then, 2024)

A revised array design with a 60 mm front plate and 50 mm back plate was put into production in 1976. The most meaningful change brought about by this modification was the likely improvement in protection from 120 mm APDS. The armor was tested against 115 mm APFSDS, but it would not meet its match until the early 1980's when the M111 *Hetz* round appeared in the arsenal of Israeli forces during the Lebanon war. At the time, Soviet analysts believed M111 to be a license-produced clone of the M735 round, the first in-service American 105 mm APFSDS round[495]. News of the potency of the M111 round was thus quite understandably taken with great alarm.

At a range of 2,000 m, M111 *Hetz* achieved partial perforation according to the criteria of back face deformation or cracking[496] – that is, the armor was not pierced all the way through, but reached its protection limit. As a reference, the

---

[495] Account given by Тумасов, В. Д. in Баранов, И. Н. (2010). *Главный конструктор В.Н. Венедикто: Жизнь, отданная танкам.* Нижний Тагил: ООО «Рекламно-издательская группа «ДиАл». p. 167.
[496] кондиционного поражение

60-105-50 armor was equivalent in weight to a 120 mm steel plate sloped at 68°, and in Chinese tests with M111 during the 1980's, the penetration limit for a 150 mm RHA plate set at 60° was 3,300 m, and the limit for a 150 mm RHA plate set at 65° was 1,380 m[497]. This verifies that the 60-105-50 armor had a mass efficiency coefficient close to 1.0, which is to say that its protection was only equivalent to its own weight of steel.

A more efficient filler against KE threats was needed to keep up with improving penetrator technology, but at the same time, the margin of defeat was small enough that a plate of add-on armor would solve the vulnerability of the armor to the latest available 105 mm APFSDS.

Under this premise, the *Otrazhaemost*[498] (Reflectivity) R&D program was launched on the 4th of November 1982. It consisted of two parallel projects: *Otrazhenie-1* and *Otrazhenie-2*[499]. *Otrazhenie-1* was a comprehensive effort to stop the M111 round with a new filler design that would not change the overall mass of the armor, its thickness, or its structural layers. Two spaced armor configurations were created under this program. *Otrazhenie-2* sought to upgrade existing tanks to the same standard with the smallest possible weight gain, which was achieved by welding a 16 mm high hardness steel plate onto the existing armor. With the plate, the limit of back face damage returned to 500 m against M111, as per the original requirement for 105 mm APDS.

---

[497] 梁禾, 芮兰德 (1989). '"86式100滑高膛压反坦克炮研制的回顾'', in 高膛压火炮系统论文集. 机电部兵器科学研究院. p. 49. Liang He, Ruilan De (1989). 'Review of the development of the Type 86 100 smoothbore high-pressure anti-tank gun'.
[498] Отражаемость. See: Колмаков, Д. Г. (2021). *Тагильская школа. 80 лет в авангарде мирового танкостроения*. Белгород: КОНСТАНТА-принт. p. 108.
[499] Account given by Марасев, М. И., Терехин, И. И. in Баранов, И. Н. (2010). *Главный конструктор В.Н. Венедикто: Жизнь, отданная танкам*. Нижний Тагил: ООО «Рекламно-издательская группа «ДиАл». pp. 190-191.

Spaced Armor (T-72B)

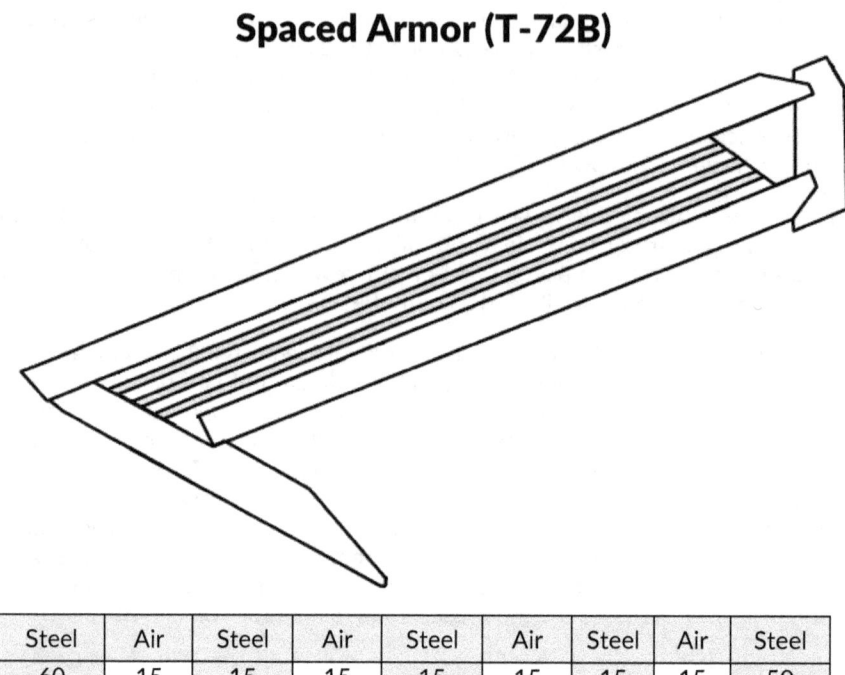

Steel	Air	Steel	Air	Steel	Air	Steel	Air	Steel
60	15	15	15	15	15	15	15	50

All figures in mm

Figure 94: Early *Otrazhenie*-1 armor array with five spaced layers. (R. A. Then, 2024)

The mass production of T-72 hulls with an early spaced armor design created under the *Otrazhenie*-1 program began in early 1983 and gradually accelerated. A pack of three high hardness steel plates spaced equally by 20 mm air gaps[500] replaced glass textolite as the filler material. The plates were secured by side brackets, similar in appearance to the side brackets for previous armor arrays. The choice of relatively thin spaced steel plates[501] instead of a low-density filler was likely a compromise between protection from shaped charges and from high aspect ratio long rod penetrators.

The steel grade for the spaced armor pack is unknown. Out of the available grades of high hardness armor steel, one of the most likely candidates is SK-2Sh electroslag remelted steel (ESR), a contemporary high hardness steel achieving minimal loss of toughness through high micro-cleanliness from the ESR process.

---

[500] Based on the exposed armor of a T-72 destroyed in Grozny, Chechnya.
[501] Thin relative to typical long rod penetrator diameters of 20-30 mm for 105-120 mm guns.

SK-2Sh provided 1.3-1.38 times increase in resistance compared to the cast steel used in the turret, and 1.1-1.15 times higher resistance than medium hardness RHA like 42 SM[502].

As an absorber layer for long rod penetrators, an array of thin spaced plates is less efficient than a filler of steel plates alternating with a low-density material. Lacking back face support from a filler, spaced thin steel plates are perforated with minimal energy[503]. Simple two-layer spaced arrays consisting of stacked thick front and stacked back plates and variations thereof are much more mass-efficient against long rod penetrators[504]. Nevertheless, multiple strikes on spaced plates deform and fracture the tip of a long rod penetrator, degrading it much more effectively for the back plate to absorb than a solid glass textolite filler. Spaced plates are used in other heavy armor systems for the same reason[505].

Moving into the T-72B proper, a more comprehensive armor design spawned from the same *Otrazhenie*-1 program was introduced. The glacis was thickened to 220 mm for a wider interlayer cavity, giving space for new spaced plate array, much heavier and much more densely packed than the early type. The array consisted of two 10 mm HHA plates followed by two 20 mm HHA plates, all individually spaced apart by 10 mm.

It is likely that thin 10 mm plates were chosen to further improve shaped charge protection. The thickness of a spaced shield should be equal to 2-3 jet diameters, thick enough to disrupt the jet tip, but thin enough to prevent efficient steady state penetration, and the air gap between plates should also be 2-3 jet tip diameters. This was estimated to provide sufficient space for the disintegrating jet particles to disperse before they could begin holing the next plate. For large caliber HEAT shells and ATGMs, where jet tip diameters were typically around 3-4 mm, the choice of 10 mm spaced plates seem appropriate.

---

[502] Чепурной, А. Д. (2000). 'Разработка и промышленное освоение производства сварнокатаной башни боевого танка'. *ОАО "ГСКТИ"*. p. 70.
[503] Григорян, В. А., Ермаков, В. И., Мачихин, С. А., Терехин, И. И. (1987). 'Исследование стойкости комбинированной брони к воздействию бронебойных подкалиберных снарядов', *Вестник бронетанковой техники* 1987, сборник 8. p. 19.
[504] Friesecke (1995). ‚Neuartige Werkstoffe und Materialien für den ballistischen Schutz', *Wehrtechnisches Symposium Thema: "Aktuelle Schutztechnologien für Gepanzerte und Ungepanzerte Fahrzeuge"* : 15.11. - 17.11.1995. Conference proceedings. Mannheim: Bundesakad. für Wehrverwaltung und Wehrtechnik.
[505] Hazell, P. J. (2015). *Armour - Materials, Theory, and Design*. Boca Raton: CRC Press. p. 166. Also see: Rosenberg, Z, Dekel, E. (2016). *Terminal Ballistics. Second Edition*. Heidelberg: Springer Berlin. p. 244.

Against long rod penetrators, a spaced armor layout with thinner plates followed by thicker plates is also more efficient[506].

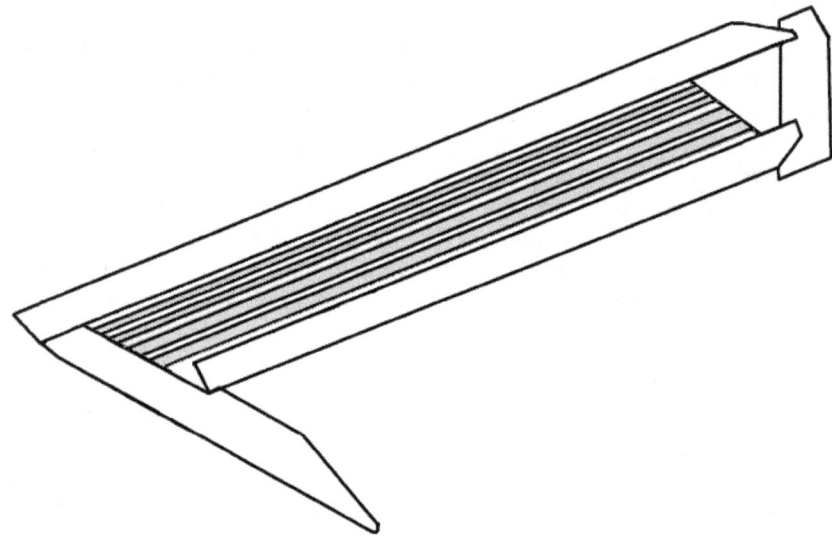

Steel	Air	Steel	Air	Steel	Air	Steel	Air	Steel	Air	Steel
60	10	10	10	10	10	20	10	20	10	50

All figures in mm

Figure 95: Late *Otrazhenie*-1 armor array with six spaced layers. (R. A. Then, 2024)

During trials in 1985, armor solutions developed under the *Otrazhenie*-1 project successfully resisted 3BM32 *Vant*[507] at a velocity exceeding its normal muzzle velocity[508]. The *Vant* round was used as a surrogate to represent a contemporary NATO 120 mm depleted uranium penetrator[509]. To represent the capabilities of such a penetrator, *Vant* was fired with an overcharge. It is unclear if the 1985 configuration was among the ones to have proven invulnerable to *Vant*, but with the sheer mass of steel present in the array, it is highly plausible.

---

[506] Friesecke (1995).
[507] Vant was developed as a service round, but like other Soviet depleted uranium munitions, its production only fed long term storage depots. Soviet policy was to avoid using depleted uranium munitions except in the event of WW3.
[508] Account given by Хейфиц, Г. А. in Баранов, И. Н. (2010). *Главный конструктор В.Н. Венедикто:*. *Жизнь, отданная танкам*. Нижний Тагил: ООО «Рекламно-издательская группа «ДиАл». p. 217.
[509] Ефремов, А. С. (2010). *Уроки танкостроения*. Санкт-Петербург: «Гангут». p. 157.

Spaced plates perform well as part of the disruptor layer by destroying the penetrator tip with each impact and breakout[510], and the effectiveness of the plates is dependent on their hardness. A small air gap between plates gives space for the penetrator tip to be deflected off each successive plate.

Under high strain rates[511], a tungsten alloy penetrator work hardens intensely, becoming stronger (harder) but losing ductility[512]. Sintered tungsten-nickel-iron (WNiFe) alloys like the Soviet VNZh-90 alloy used in various 125 mm APFSDS ammunition show a tremendous increase in dynamic yield strength above 1.9 GPa at high strain rate, but relative elongation plummets to 2%[513].

In steady-state eroding penetration – that is, penetrating a target without a change in velocity at the tip-target interface, the drastic embrittlement of the penetrator tip was not debilitating. However, where steady-state penetration does not occur, i.e. in the impact and breakout phases, armor solutions imparting lateral loads could break up penetrators with great efficiency. In a spaced oblique armor array such as this, the significant ductility advantage of depleted uranium (DU) alloys over tungsten heavy alloys gave DU penetrators an advantage[514] given that the other relevant variables (like rod diameter) are controlled, even if the two types have the same penetration in an RHA target.

Conversely, it is obvious that with more modern, tougher heavy metal alloys, the performance of complex armor is degraded. The twin approach of increasing the aspect ratio of long rod penetrators and improving dynamic mechanical characteristics has been the main focus of development efforts for APFSDS ammunition since the 1970's.

Of course, the front and back plates of the array should ideally be high hardness steel as well, but this was not possible in these two spaced armor arrays for structural reasons. In the T-72B turret and in other modern tank armor designs, high-hardness steel or other hard and dense barriers were placed in front of a medium hardness steel back plate as a compromise solution. The most

---

[510] Anderson, C. E., Littlefield, D. L. (1994). *Pretest predictions of long-rod interactions with armor technology targets*. Technical report, SwRI Project No. 07-5117. San Antonio: Southwest Research Institute. p. 61.
Incidentally, the test penetrator for the study in this report was made from X27X, a 91% tungsten alloy very similar to VNZh-90 and other tungsten alloys in service ammunition during the 1980's.
[511] High rate of deformation under stress
[512] Kunze, H. -D., Meyer, L. W., Staskewitsch, E. (1983). ‚Dynamic strength and ductility of a tungsten-alloy for KE-penetrators in swaged and unswaged condition under various loading', *Proceedings of 7th International Symposium on Ballistics 1983*.
[513] Балакин, С. М., Белков, П. А., Данилов, П. Н., Ломов, С. В. (1987). 'Механические свойства сплава внж-90 при повышенной скорости деформации', *Вестник Бронетанковой Техники* 1987, сборник 8. р. 60.
[514] Andrew, S. P., Caligiuri, R. D, Eiselstein, L. E. (1992). 'A review of penetration mechanisms and dynamic properties of tungsten and depleted uranium penetrators' in *Computational Modeling of Dynamic Failure Mechanisms in Armor/Anti-Armor Materials*. Menlo Park: Failure Analysis Associates, Inc. pp. 143-144.

weight and volume-efficient means of imparting such loads was with heavy reactive armor. In 1987, Kontakt-5 integral reactive armor was introduced to the T-72B and with it, a new, optimized upper glacis structure that no longer compromised KE protection for shaped charge protection.

Integral ERA (Kontakt-5)

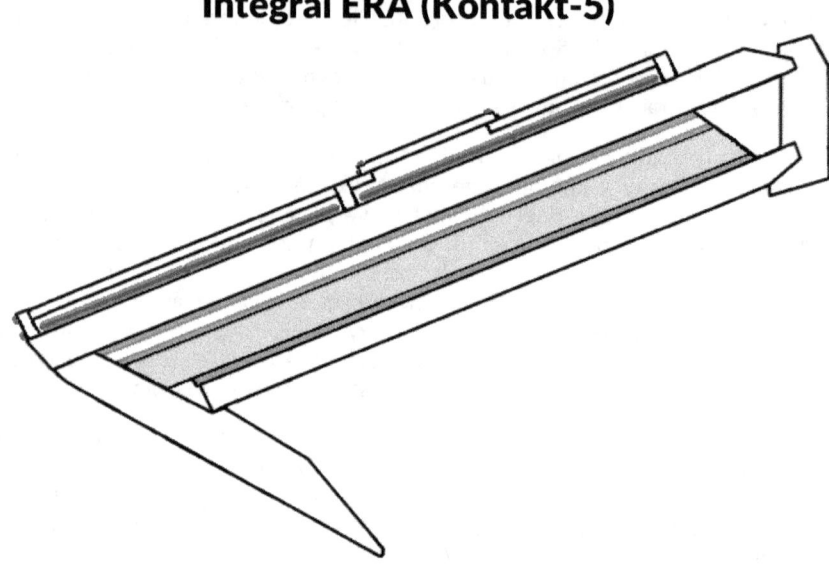

Kontakt-5	Steel	BP	Air	BP	Steel	Plastic	Steel
55-60	60	8	19	8	60	10	50

All figures in mm
Kontakt-5 – 15 mm cover plate, 15-20 mm air gap, 26 mm (two 4S22 explosive elements)
BP – rubber layer with steel bulging plate
Plastic – anti-radiation layer, made from unknown plastic

Figure 96: Optimized armor array with Kontakt-5. (R. A. Then, 2024)

The composition of the upper glacis was optimized for stopping potent long rod penetrators by harmonizing its layout with a layer of Kontakt-5 reactive armor. Thin spaced plates were abandoned in favor of an efficient two-layer stacked spaced array. Instead, a single 60 mm high hardness steel plate[515] was stacked on top of the medium hardness back plate, sandwiching between them a thin layer of hydrogen-dense plastic for neutron shielding. A pair of reflecting

---

[515] In this case, the plate material is assumed.

plate NERA elements were incorporated onto the surfaces of the front plate and filler plate to shore up shaped charge resistance.

A classic disruptor-absorber system remained, where now the reactive armor and the 60 mm front plate together constituted the disruptor layer, and the stacked plates behind it collectively made up the absorber layer. The internal 60 mm plate was very likely a BTK-1 increased-hardness steel, with a design hardness of 352-401 HB and ductility characteristics at the low end of 42 SM steel[516].

High hardness steel improved the array's resistance to both KE threats and shaped charges[517]. Soviet testing demonstrated that as the hardness of armor steel increased from 110 HB to 650 HB, the penetration depth diminished by approximately 40%, matched by British testing[518]. Other testing showed that a change in strength (hardness) from 0.7 GPa (212 HB) to 2.05 GPa (555 HB) decreased penetration by 32%[519].

Prior disruption of the jet by the ERA layer and the front plate increased the influence of material strength in resisting the shaped charge jet, which made high-hardness back layers desirable. Even a simple two-layered structure with layer of high hardness steel behind a medium hardness front plate could leverage the increased sensitivity of the jet to material strength[520]. In the early T-64A turret, high hardness armor (HHA) steel plate inserts behind medium-soft cast steel gave the turret an 8-12% increase in resistance to shaped charges for the same weight of homogeneous steel armor[521].

Kontakt-5 was added onto the upper glacis as semi-integral armor[522], requiring more installation work than add-on solutions like Kontakt-1 but having considerably more accessibility than integral ERA, where the explosive elements are embedded inside the armor itself (e.g. Merkava 4). The Kontakt-5 panels on the turret front, turret roof, and the sides of the hull were add-on panels.

With very little allowance to further thicken the upper glacis, the Kontakt-5 panels were designed to throw a head-on flyer plate instead of working

---

[516] Власова, И. И. et al. (1984). 'Свойства листов брони БТК-1 повышенной твердости', *Вестник Бронетанковой Техники* 1984, сборник 5. pp. 41-42.
[517] Rosenberg, Z, Dekel, E. (2016). *Terminal Ballistics*. Second Edition. Heidelberg: Springer Berlin. pp. 240-244.
[518] Backofen Jr., J. E. (1983). 'Armor Technology (Part III)', *ARMOR*. (March-April). p. 19.
[519] Rosenberg, Z, Dekel, E. p. 244.
[520] Брызгов, В. Н., Симаков, И. К. (1985). 'Вклад тыльного слоя высокотвердой стали в противокумулятивную стойкость брони', *Вестник Бронетанковой Техники* 1985, сборник 2. pp. 26-28.
[521] Григорян, В. А. et al. p. 255.
[522] Григорян, В. А. et al. p. 437.

bidirectionally. This was much less efficient, but reimagining the upper glacis as two-layer spaced armor made the most of the weight and space budget.

A long 15 mm cover plate made from brittle high hardness steel over the upper glacis served as a heavy flyer plate, designed primarily to defeat KE threats. Shaped charges were defeated by the light 2 mm flyer plates thrown by the 4S22 explosive elements loaded into the Kontakt-5 panels.

As a long rod penetrator or shaped charge jet struck the cover plate, small particles were ejected from the back surface of the plate at a very high velocity through spallation, penetrating the first layer of 4S22 elements and initiating detonation in both layers. Because the explosive charge began reacting before the penetrator physically reached the elements, this method of ERA initiation, known as remote sensing, was faster than direct initiation. The 4S22 explosive element was designed with increased sensitivity to improve detonation reliability under KE attack and to reduce the detonation delay.

Reducing the detonation delay curtailed the length of the jet tip that could escape the light flyer plates before they acquired significant velocity[523], and was thus an effective means of enhancing the disruptive effect on shaped charges. This mechanism did not, however, improve the performance of Kontakt-5 compared to Kontakt-1, but rather compensated for the relatively poor effect of head-on flyer plates compared to in-pursuit flyer plates[524]. By the point the first light flyer plate struck the 15 mm cover plate, its final velocity was 2,000 m/s, and the impact was of such violence that a ~4 mm layer of spall would usually be ejected forwards off of the cover plate. This became a third light flyer plate against the shaped charge jet. The period of action against shaped charges ended when the second light flyer plate, 4 mm thick, struck the cover plate and heavy flyer plate, and the period of action against KE penetrators began.

The interaction between the heavy flyer plate and the long rod penetrator is dominated by momentum transfer between deformable bodies[525], but the

---

[523] Held, M., Schwartz, W. (1994). 'The Importance of Jet Tip Velocity for the Performance of Shaped Charges against Explosive Reactive Armour', Propellants, Explosives, Pyrotechnics, 19(1). pp. 15-18.
The basic premise of this mechanism is adequately outlined, for further reinforcement, see: Held, M. (2005). 'Shaped Charge Optimization against ERA Targets', Propellants, Explosives, Pyrotechnics, 30(3). pp. 216-223.
[524] Kontakt-5 on the T-80U upper glacis arranged the first layer of 4S22 elements obliquely like in Kontakt-1. This introduced one in-pursuit flyer plate, but worsened detonation reliability due to the loss of the air gap towards the upper part of the panel.
[525] Hazell, P. J. (2015). Armour - Materials, Theory, and Design. Boca Raton: CRC Press. p. 277. Also see: Held, M. (2005). 'Defeating Mechanisms of Armours for Main Battle Tanks', Journal of China Ordnance Society, 1(1). p. 6.
The instantaneous pressure at the point of collision is generally insufficient for erosion to occur between the two bodies.

interaction was not purely between the flyer plate and the penetrator. Due to the long delay before the flyer plate began to accelerate (35-40 μs), a KE penetrator travelling at any relevant ordnance velocity (>1,500 m/s) will have traveled through the entire depth of the Kontakt-5 panel, reaching the front plate before the heavy flyer plate could meaningfully act on the penetrator body. The subsequent interaction with the flyer plate occurred while the penetrator already began to pierce the front plate, where asymmetric resistance acted to deflect the tip. Only then did the motion of the flyer plate transfer momentum to the penetrator and began to deform and rotate it. The long length of the flyer plate (500 mm) increased the interaction period.

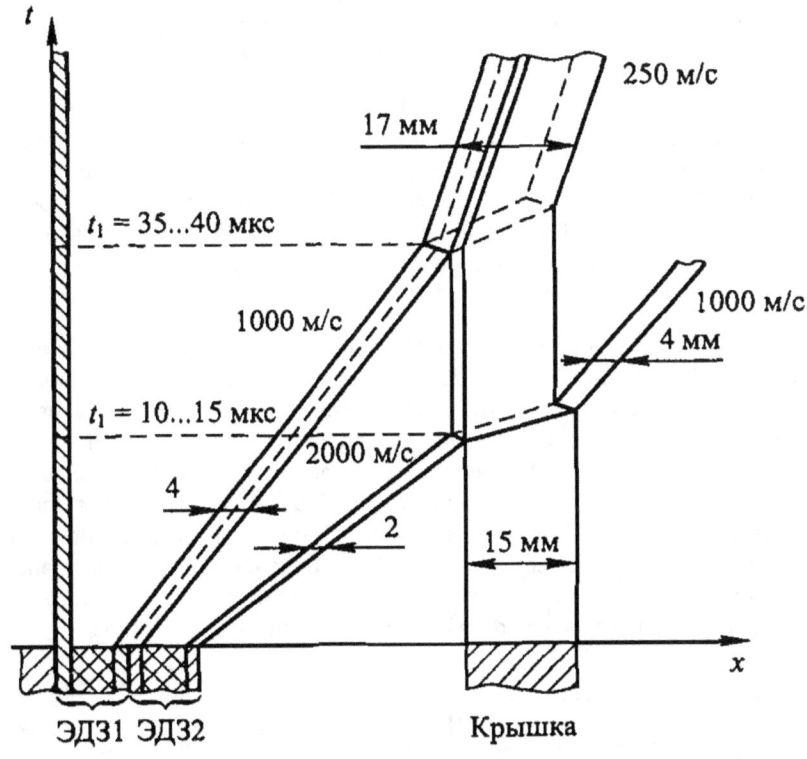

Figure 97: Time-displacement graph of Kontakt-5 integral to T-72B since 1987 upper glacis. The final heavy flyer plate was an aggregate with 4S22 cover plates. It was 17 mm thick and traveled at a reference velocity of 250 m/s. (V. A. Grigoryan et al., 2006)

Naturally, the subsequent armor-penetrator interaction was very complex. The Kontakt-5 panel and the 60 mm upper glacis front plate were positioned to function collectively as a disruptor layer to deform and damage the rod, which is subsequently absorbed by the middle and back plates. The momentum transfer from the flyer plate pitches the penetrator upwards, but the rod body resists a change in attitude because of the enormous amount of inertia it brings through its velocity. While the rod begins to rotate during penetration from the impetus of the flyer plate, the more immediate result of the collision with the flyer plate is the deformation of the rod body. Consequently, the strength of both the flyer plate and the penetrator plays a critical role in the condition of the penetrator at the end of the interaction. Momentum transfer was increased by having a thick plate at a large obliquity to the penetrator trajectory[526], and a high flyer plate hardness was more effective in deforming a penetrator.

Naturally, the toughness of a long rod penetrator is a critical factor in surviving heavy ERA, and it is obvious that different penetrators will behave differently against Kontakt-5. During its development, Kontakt-5 was tested out at the *Donguz* firing range against all available APFSDS ammunition, seeing success against even the latest domestic depleted uranium (DU) rounds[527] like *Vant*, which featured a thick-bodied monobloc rod that, in principle, would perform well against heavy ERA.

The low speed of the heavy flyer plate and long delay before its movement was effective on tandem warhead ATGMs. In tandem warheads, the primary or rear charge was typically given a fuzing delay of 150-300 μs[528] to allow the flyer plates of the ERA to completely clear the path of its jet after the ERA is detonated in advance by the warhead's precursor charge. This worked, but only against ERA with light, short and fast flyer plates[529]. Alternative tandem warhead weapons with a non-detonating precursor[530], also designed against light ERA, may not be effective on Kontakt-5 due to its thick cover plate and its system of initiation by remote sensing. Kontakt-5 could thus be fairly considered to be effective against all modern penetrator technologies of its time.

---

[526] Паластров, П. С., Мелешко, И. А., Платов, А. И., Рототаев, Д. А. (1991). 'Исследование устройств динамической защиты от бронебойных подкалиберных снарядов', *Вестник бронетанковой техники*, 1. Available at: https://btvt.info/5library/vbtt_1991_vdz.htm (Accessed: 2 June 2024).
[527] Алексеев, М. et al. (2012). *НИИ Стали 1942-2012*. Москва: Издательство СканРус. p. 97.
[528] Партала, С. В. et al. (2004). *Конструкция средств поражения, боеприпасов, взрывателей и систем управления средствами поражения: Конструкция и функционирование ПТУР*. Пенза: Издательство ПАИИ.
[529] Растопшин, М. (2014). *Причины несовершенства отечественных ПТУР*. Available at: https://nvo.ng.ru/armament/2014-08-29/1_ptur.html (Accessed 2 June 2024).
[530] A shaped charge made from plastic, designed to blow a wide hole in ERA without detonating it.

Turret

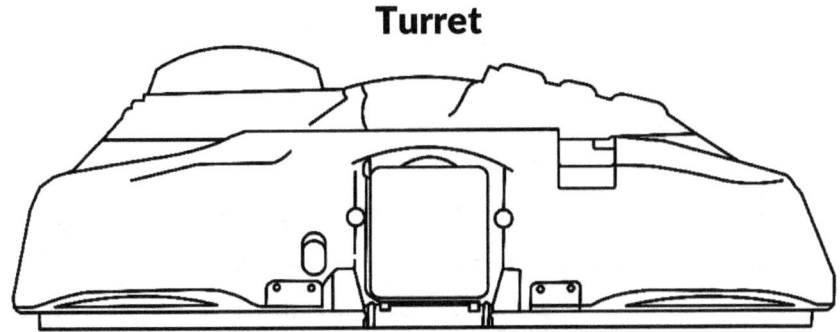

Figure 98: T-72M1 (T-72A) turret from the direct front. (P. Martin, 2024)

The T-72 turret had a streamlined, rounded shape characteristic of Soviet tank turrets and turret concepts developed in the late 1950's and early 1960's. Starting from a circular plan, the turret cheeks were bulged out by the thick armor, leaving a teardrop shape. The roof was steeply sloped and bowed to form a shape that, while not exactly a dome, might be recognized as one. The resultant silhouette was distinctly smaller than Western tank turrets, particularly in the upper half where the domed shape made the largest difference in size.

Though the base of the T-72 turret was quite wide because of its bulbous cheeks and wide turret ring, its low silhouette greatly minimized the exposed frontal area in a hull-down position. The projected frontal area of the T-72A "sand bar" turret was 1.48 m^2, comparing very favorably to the 2.5 m^2 of frontal area presented by the M60A1 turret[531], the 2.4 m^2 of the M1A1 Abrams turret and even the Leopard 1's much lower turret, with 2.0 m^2 of frontal area[532]. The low projected area of the turret was maintained across a wide frontal arc due to its "teardrop" shape.

The frontal projection of all T-72 turrets was very well protected, usually on par with, or surpassing the heaviest armor on its direct Western counterparts. Looking at its armor weight, this should be no surprise. Based on the T-72B hull structural weight of 15.3 tonnes and turret weight of 8.1 tonnes[533], the turret weight was around 55% of the hull weight. This was the same as the M1A1

---

[531] Without cupola. See: Романов, Н. И. (ed.) (1973). *Теория стрельбы из танков*. Москва: Военная академия бронетанковых войск. p. 127.
With cupola, the M60A1 turret had a projected area of 2.9 m^2.
[532] Measured from known dimensions.
[533] Вульфельдт, Э. И., Ганчо, Ю. Г., Жуков, В. Ф., Касьянов, В. Д. (1988). 'Объемно-массовый анализ защиты серийных танков', *Вестник Бронетанковой Техники* 1988, сборник 10. p. 22.

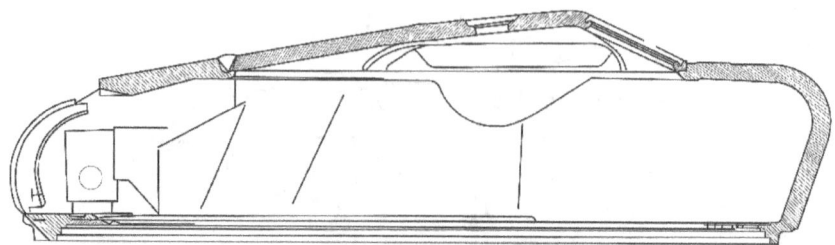

Figure 99: T-72A turret side profile. (*Adapted, Soviet Ministry of Defense*)

Abrams and very similar in general to modern Western tank models, despite the radical difference in turret size and construction. Where the T-72 turret lacked was in the thickness of its sides, primarily due to its reliance on its "teardrop" shape.

The choice of a cast turrets was a matter of following established industry standards. In the postwar era, cast armor was customarily treated to a low-medium hardness, which, when arranged at a moderate to high obliquity was optimal against full-caliber AP rounds. High obliquity armor was also effective against common types of APDS ammunition.

The design of the T-72 turret was a continuation of these conventions, modified by thickening the cheeks to achieve the desired protection. A non-metallic interlayer was incorporated into the cheek armor by using a cast-in approach. The T-72 used so-called "sand bar" cores, while the T-64A used balls formed out of high-purity alumina, known as "ultraporcelain" balls. This approach was abandoned only by the early 1980's because a cast-in approach was not viable for integrating inserts of non-energetic reactive armor (NERA).

Like the hull, the turret was required to resist 105 mm APDS and HEAT, but in an arc of ±35°. This meant that, even at a side angle of 35°, the entire turret had to meet the protection requirement. There was no requirement for shaped charge protection at a side angle exceeding 35°. The extreme differentiation in armor thickness between the front, sides and rear of the turret was the natural outcome of these requirements.

Protection from direct hits on the gun embrasure was not emphasized and the commander's cupola was also exempted from the protection rating, being considered an intrinsic weakened zone. Protection on the sloping turret roof was evaluated separately from the turret, and was considered resistant to 105 mm APDS and HEAT[534]. The lack of significant modernization potential for the

---

[534] Комяженко, А. Г., Тимохин, В. И., Тренина, Н. К. (1974). 'Влияние ослабленных зон на поражение броневой защиты', *Вестник бронетанковой техники* 1974, сборник 6.

roof armor would later become an issue when 105 mm APFSDS began to proliferate in the early 1980's, along with graze-sensitive fuzing on HEAT shells. This limitation was partially addressed by the introduction of ERA, which performed extremely well at the high obliquity provided by the turret roof.

The protection rating for the turret was thus biased towards the cheek areas, which, considering that the cheeks covered only around 40% of the turret area within the ±35° arc, somewhat misrepresented the protection level averaged across the entire turret[535]. This was an issue of considerable gravity in designing a turret for shaped charge protection. During the 1970's, Western clean-sheet tank designs with high shaped charge protection universally favored welded turrets with a high degree of composite armor coverage. Gun mantlets with composite or light spaced armor were sometimes used to ensure at least a modicum of shaped charge resistance, even if it was not possible to ensure completely uniform protection across the turret.

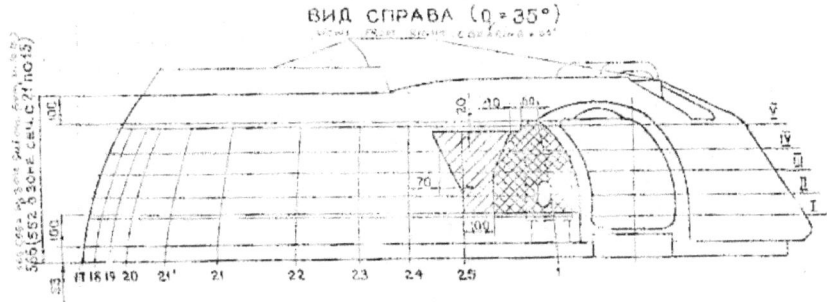

Figure 100: T-72 turret from a side angle of 35°. Checkered region – zone vulnerable to 100 mm blunt-nosed AP shell at 615 m/s; Shaded region – zone vulnerable to 100 mm APDS at 1,400 m/s. Vulnerability criteria: back face bulge with hairline cracks. (*Soviet Ministry of Defense*)

Still, the rounded shape of T-72 turrets retained practical design elements like the absence of reverse bevels or inwardly rounded surfaces. This eliminated the possibility of deflecting bullets or shells into the hull roof or into the turret ring, and ensured that under high-explosive shell attack, the blast and fragment flow was not funneled into the space between the hull and turret. This significantly reduced the damage from direct-fired HE-Frag shells[536]. Tank-fired HE-Frag

---

[535] This was belatedly recognized during the mid-1970's, leading to the creation of a turret with higher cheeks and a more steeply sloped roof for the Object 219 (T-80) series. This turret design went into service only in the late 1980's on the T-80U.
[536] An 85 mm HE-Frag shell striking a T-54 mod. 1947/48 turret at the reverse bevel bent the hull roof near the driver's hatch by up to 30 mm, and could jam the turret by deformed the turret ring ball bearing

shells in particular impart a great deal of additional energy into their fragments through a high shell velocity[537].

That said, the decision to continue producing cast turrets was, without a doubt, strongly influenced by industry inertia. As the demand for protection grew, the thickness, complexity and size of the cast armor pieces grew, but such castings were invariably accompanied by difficulties in controlling wall thicknesses and crack formation. If the product exceeded the specified weight, excess metal had to be removed manually with grinders, with periodic weight verification. Cracks had to be filled in with welds and then smoothened over by grinding. This was laborious and time-consuming, diminishing somewhat the advantage of casting over welded structures.

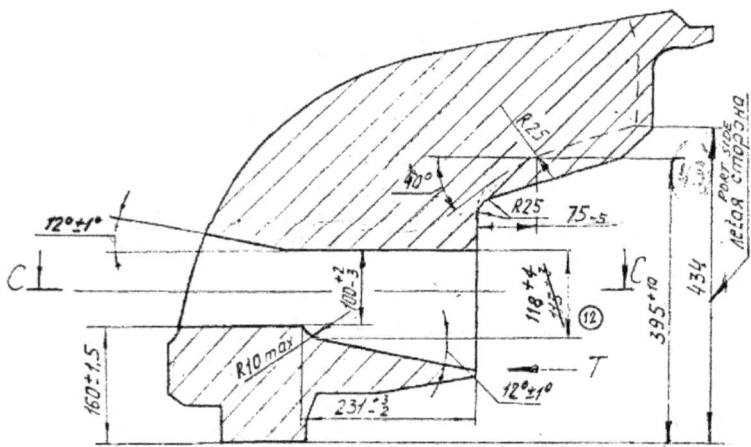

Figure 101: Coaxial machine gun port in T-72M1 turret. (*Soviet Ministry of Defense*)

From 1973 until 1976, the T-72 Ural was produced with a monolithic turret. The lack of composite armor was compensated by the extreme thickness of solid steel armor. This turret was an adaptation of the standard T-64A turret for the late 1960's, containing a filler of stacked high-hardness steel plate inserts. The dimensions of the two turrets were identical apart from a minor change in

---

races. See: Павлов М. В., Павлов, И. В. (2021). *Отечественные Бронированные Машины 1945-1965 гг. - Часть I: Легкие, средние и тяжелые танки*. Кемерово: ООО "Принт". pp. 361-362.

[537] Гаюн, В. В. et al. (1983). 'Действие бронебойно-фугасного снаряда по броне', *Вопросы оборонной техники*, 5(111).

geometry to the turret sides and a noticeable extension to the "tail" of the turret, which was an adaptation for the T-72's autoloader[538].

Beginning in 1977, T-72 tanks began to be delivered to the Soviet Army with "sand bar" turrets. The monolithic turret continued to be produced only for export purposes after 1977[539]. The sand rod turret was considered to be more advanced than the T-64 pattern turret with a filler of "ultraporcelain" balls[540], but more in terms of its cost-benefit balance rather than from a purely technical standpoint. The production of these sand cores was adaptable to existing foundry equipment and did not require expertise in ballistics-grade ceramics, while still managing to provide the same level of protection.

Passive and reactive armor and their hybrid approaches became possible thanks to a new understanding of how penetrator energy could be used against itself by transferring energy into a material with a low modulus of elasticity[541]. Pierced by a shaped charge jet, low-modulus materials expand violently from the shock energy delivered by the jet tip. By controlling the direction of expansion, a jet penetrating through such materials supplied the energy to throw a disruptive plate or even force a flow of material to destroy its own body. This concept was categorized under the umbrella term of non-energetic reactive armor, or NERA. The breadth of possibilities with NERA led to a divergence of design approaches in the late 1970's.

NII Stali explored NERA through several unique armor concepts, using a low-modulus material as both a filler and within cells, interlinked only by light thermoplastic polyurethane (TPU) as the material of choice. The T-55AM and T-62M received "metal-polymer" armor, where armor blocks containing a TPU filler with embedded steel sheets worked against shaped charge attack by using the TPU as both a passive filler and as a medium of energy transfer into the sheets, throwing them obliquely against the jet body. "Cellular armor" was implemented in the turret of the T-80U. A filler of TPU was confined in special pockets so that, once the jet entered the armor, it displaced the filler against the hemispherical end of the pocket, where it had nowhere else to go but to flow around and back into the body of the jet, severely disrupting its shape.

The T-72B received "reflecting plate" armor, the third direction of research and ultimately the most successful. By sandwiching a thin layer of rubber between a thick high-hardness steel front plate and a thin, soft steel back plate,

---

[538] Based on a comparison of real turrets modelled in 3D by photogrammetry.
[539] Устьянцев, С. В., Колмаков, Д. (2004). *Боевая Машины Уралвагонзавода: Танк Т-72*. Нижний Тагил: Издательский Дом "Медиа-Принт". p. 69.
[540] Account given by Тумасов, В. Д. in Баранов, И. Н. (2010). *Главный конструктор В.Н. Венедикто: Жизнь, отданная танкам*. Нижний Тагил: ООО «Рекламно-издательская группа «ДиАл». p. 167.
[541] A material that can be deformed with a low load

a jet passing through the "reflecting plate" at an angle delivered a large amount of shock energy into the rubber layer. This forced the rubber to expand violently and in turn, propel the thin steel back plate outward, bulging it out across the jet body. Outside the USSR, this was a familiar form of NERA known as bulging plate armor, popularized by the work of Manfred Held in the 1970's at IBD, West Germany. It is also closely related to the widely publicized "Chobham" armor developed since the 1960's and finally put into service in the 1980's on the M1 Abrams and Challenger 1.

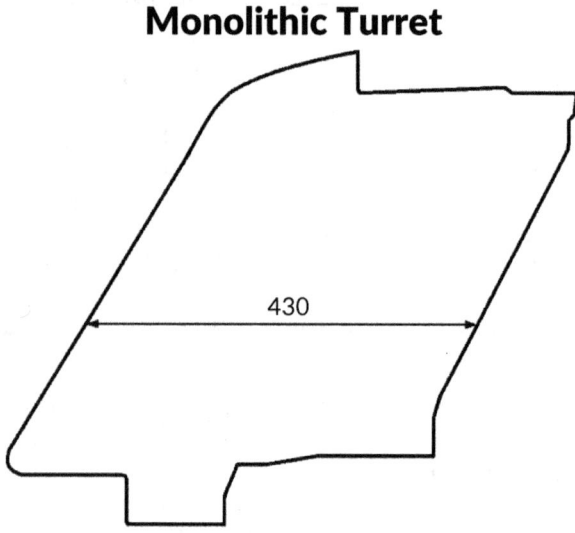

Figure 102: Section 3-3 of T-72 Ural turret, equal to a side angle of 40° to the turret. (*Adapted from factory drawing, Soviet Ministry of Defense*)

The turret was made from SBL-2 steel, a high-chromium, high-nickel vanadium steel alloy. This grade was used for all cast T-72 turrets, while T-64A turrets switched to MBL-1 in the late 1970's[542]. Compared to the steel in the M60A1, which had an all-cast hull and turret, the SBL-2 grade had a higher nickel content, which is commensurate with its much larger intended thicknesses. Most crucially, SBL-2 was produced to a higher level of micro-cleanliness based

---

[542] (СБЛ-2 – Сибирская броневая, литая), and MBL-1 (МБЛ-1 – Мариупольская броневая, литая) was developed at Mariupol. See: Солянкин, А. Г, Желтов, И. Г., Кудряшов, К. Н. (2010). *Отечественные бронированные машины. XX век - Том 3: Отечественные бронированные машины. 1946-1965 гг.* Москва: Издательство «Цейхгауз». p. 86.

on the low sulphur (0.016% S vs 0.022-0.032% S) and phosphorous content (0.016% P vs 0.020-0.032% P) achieved in Soviet production.

Table 21: Alloy composition of T-72 (SBL-2) turret and of its close equivalents

Alloying Elements	C	Mn	Si	Cr	Ni	Mo	V
T-72 (SBL-2)	0.29%	0.84%	0.041%	1.49%	1.66%	0.24%	0.15%
M60A1	0.28-0.30%	1.06-1.15%	0.32%	0.9-1.02%	1.25%	0.49-0.57%	-
Centurion (IT90E)	0.30%	1.45%	0.42%	0.50%	0.88%	0.39%	-

Alloy elements:
C – carbon; Mn – manganese; Si – silicon; Cr – chrome; Ni – nickel; Mo – molybdenum; V - vanadium

The strategic expense of the SBL-2 steel associated with its high chromium and nickel content was directly tied to the difficulties in uniformly hardening very thick structures. In this case, the large thickness of the turret cheek armor complicated the quenching of its center, which reduced the proportion of the steel austenite microstructure transformed into martensite. Quenching is also accompanied by rapid shrinking of the workpiece, which for a complex cast turret with a massive difference in thickness between the front and rear walls results in a great deal of stress accumulating along nearly every surface. Suppressing cracks and crack growth from these stresses required strict quality control and smoothly curving transitions between the cheek armor and the much thinner sides and roof parts. The roof plate itself was much too thin (45 mm), so the cast was simplified by welding the roof plate separately. It was one of a few details on the T-72 to have been produced by stamping with a 10,000-tonne press[543].

To produce high-quality cast armor of such an enormous thickness, the turret underwent a complex multi-stage heat treatment process lasting several days after casting. In total, the turret was heated three times above the temperature of total austenitization. Preliminary heat-treatment involved homogenization, normalization, and tempering.

The steel was first homogenized (diffusion annealed) by heating it to 1,120-1,150°C and holding it for 18-20 hours. It was then air-cooled. This helped to reduce crack formation in the subsequent steps. The steel was then normalized (normalization annealed) by heating it to 1,000°C and holding it for 12-14 hours

---

[543] Устьянцев, С. В., Колмаков, Д. (2004). *Боевая Машины Уралвагонзавода: Танк Т-72*. Нижний Тагил: "Медиа-Принт". p. 99.

to obtain a complete phase change to austenite, and to ensure the transformation of the coarse-grained structure to a fine-grained structure, associated with increased toughness. The steel was then air cooled. Finally, the steel was tempered at 680-700°C for 14-16 hours and air-cooled. All three of these steps reduced the strength of the steel but at a rate controlled to dramatically increase its impact toughness, which improved its ballistic resistance[544]. The tempering temperature was also high enough to serve a double purpose as a stress relief stage, again to reduce cracking.

Final heat-treatment started with heating the turret to 960-980°C, holding it for 12-14 hours to ensure through-heating, followed by water-quenching. The turret was then tempered a second time at 680-700°C for 14-18 hours[545]. The resultant armor was tempered martensite with some bainite in the center, where quenching was less effective[546]. Tempered martensite was very typical for steel armor in general, and it was the ideal form for a low carbon alloy steel like SBL-2, but the bainitic center, a consequence of the extreme thickness, was undesirable[547].

The armor was tempered at a high temperature for toughness, with a correspondingly low hardness of 229-241 HB[548], but with a more brittle center where the bainite concentration was highest[549]. The armor was nevertheless still slightly harder than M48 and early-production M60A1 turrets (202-212 HB)[550] and similar to late-production M60A1 turrets (229-255 HB)[551].

---

[544] Unlike tensile and yield strength, which are statically measured properties, impact toughness is a test at a high strain rate, and has a stronger correlation to ballistic resistance.
[545] Маслова, Ю. Н., Оголюк, В. И. (1976). 'Отработка оптимального режима термообработки башен танка Т-64А, изготовленных из стали СБЛ-2', *Вопросы Оборонной Техники*, 20(63). pp. 43-48.
[546] Nevertheless, decomposition of austenite into ferrite and cementite was not observed in serial tank turrets. See: Ежов, А. А., Левин, Л. С., Маслова, Ю. Н., Чикаленко, Г. А. (1981). 'Сравнительные исследования броневых сталей МБЛ-1 и СБЛ-2', *Вестник бронетанковой техники* 1981, сборник 1. pp. 42-43.
[547] The bainite observed in SBL-2 castings for T-64A turrets was upper bainite, which was the least desirable type in a low carbon steel. See: Niccols, E. H. (1976). Literature Review: Impact toughness of bainite vs. martensite. Technical report. Watervliet: Benet Weapons Laboratory. pp. 3-6.
[548] Галанова, Н. М. et al. (1983). 'исследование трещин под прибылями танковых башен', *Вестник бронетанковой техники* 1983, сборник 5. p. 44.
Ultimate tensile strength: 80-86 kg/mm^2 (784-843 MPa);
Impact strength: 1.29-1.39 MJ/m^3;
Elongation: 12-16%;
[549] In test samples with an equivalent thickness, elongation fell from 11-17% to 5-10%, Poisson's ratio fell from 25-48% to 16-32%. See: Ежов, А. А., et al. p. 43.
[550] Королёв, Г. Е., Наумик, Н. М., Трикоз, Е. И. (1979). 'Броневая Защита Американского Танка М48А3', *Вестник бронетанковой техники* 1979, сборник 3. p. 52.
[551] Дрибинский, А. М., Мисюк, А. Ф., Олизаревич, Л. В. (1976). 'Броневая Защита', *Вопросы Оборонной Техники*, 20(67). p. 34.

As tough as it was, soft armor at a moderate slope was not ideal for any particular type of threat. It was most suitable for contemporary APDS with a tungsten alloy core, but only in the sense that it was the least inefficient. Softer armor was also less effective against shaped charges, albeit to a lesser extent.

However, by virtue of its extreme thickness, even the most powerful 120 mm APDS rounds struggled against this armor, not to mention 105 mm APDS and early APFSDS. The main shortcoming of the monolithic steel turret was its low protection value against shaped charge weapons. It did not meet the stretched requirement for 115 mm HEAT protection. It did, however, still partially meet the basic requirement for 105 mm HEAT.

These turrets were produced until 1977[552], when they were replaced by a turret with so-called "sand bar" armor cores. A sintered casting core was incorporated into the armor as a non-metallic layer, resulting in a cheap, light, but effective form of composite armor.

Sand Bar Turret

The composite turret appearing on T-72 tanks since 1977 was cast from SBL-2 steel with integral cores in the turret cheeks, referred to as "sand bar" (песчаным стержнем) cores. American observers noticed the visibly thicker turret and gave it the nickname "Dolly Parton"[553]. The turret itself was sometimes described as having a "sand bar filler" (с залитым песчаным стержнем) in technical documents and scientific articles. Despite this, the core embedded in each cheek was not actually a filler of loose sand, but rather a hard sintered block, essentially an investment casting mold.

The T-72A was outfitted with this turret[554] since the beginning of its service in 1979, and continued to be manufactured with this turret for five more years until 1984. In the USSR, the production of "sand bar" turrets only continued after 1984 for exported T-72M1 tanks. This turret was also produced under license outside the USSR. The turrets of these tanks were practically identical to that of Soviet Army T-72 models.

The "sand bar" cores were a type of sand casting mold characterized by the use of sodium silicate as a polymeric binder for quartz refractory material, made from a mixture of foundry sand and quartz dust. The recipe called for 75-78% foundry sand, 17.0-18.5% quartz dust, 10.5-12.0% sodium silicate, and a binder of 2.8-4.5% molding clay and 1.8-2.0% graphite to create the shape of the wet core before it was baked. The material was shaped into lefthand and righthand

---

[552] Устьянцев, С. В., Колмаков, Д. (2004). *Боевая Машины Уралвагонзавода: Танк Т-72*. Нижний Тагил: Издательский Дом "Медиа-Принт". p. 69.
[553] The bulging cheeks were apparently reminiscent of popular singer Dolly Parton's signature bust.
[554] Turret drawing index 172.10.073SB.

blocks with the help of the clay binder and then fired at a temperature of around 1,200°C in a furnace[555], producing an aggregate with a bulk density of 2.5 g/cm^3 and with a maximum thickness of 100 mm. The sodium silicate binder is vitrified[556] into a glassy binder for the quartz during baking but the quartz itself was not vitrified, and thus the "sand bar" core is an aggregate rather than a ceramic or a glass. Its critical feature was its ability to hold a fairly precise shape during the subsequent casting of the turret; its mechanical properties were otherwise unremarkable.

These finished "sand bar" cores were placed in an inverted turret casting mold, supported on three steel pedestals, and the turret was cast over them. After the turret cooled, it was removed from the mold, inverted, and the pedestals, protruding above the turret cheeks, were cut flush to the turret roof and ground down; three circular bumps may still be seen above each cheek on a finished turret. The turret, with the cores embedded in each cheek, then proceeded to a heat treatment process very similar to the heat treatment applied to the monolithic turret, but with modified holding times to account for the effect of the "sand bar" cores on the steel depth.

The cores had a large thermal inertia, which complicated the cooling of the surrounding steel. Cross-sections found that the low cooling rate led to the retention of ferrite and perlite in the metal interface with the cores, and cracks formed to a depth of 3 mm. Crack growth was naturally stopped by the rapid change to martensite, so the cracking was superficial and considered acceptable. Apart from the mild weakening of the steel immediately surrounding the "sand bar" cores from these metallurgical limitations, the turret was stronger than the monolithic type. The steel on each side of the core was individually thinner, which allowed the hardness to be increased slightly to 241-277 BHN (3.65-3.90 mm)[557]. This likely had a positive effect on its resistance to all types of penetrators.

In spite of only making up a fifth of the total armor thickness along the turret cheek, the addition of the "sand bar" was enough to raise the protection level to meet the requirement for 115 mm HEAT protection in an arc of ±35°, set nearly a decade earlier. The T-64A turret with ultraporcelain balls only partially complied with this requirement. It was rated for 75% protection from 115 mm HEAT in an arc of ±35°. The "sand bar" turret was rated to a thickness equivalent to 500 mm RHA against shaped charges. Given that the total line-of-sight (LOS) thickness of the armor at a 35° side angle reached 530 mm and of that thickness,

---

[555] Unknown holding time. Process reported by Wolski, J. from Bumar-Łabędy casting facility.
[556] Vitrified sand is a glassy material.
[557] Гладышев, С. А., (1982). 'Характеристики стали СБЛ-2 при изготовлении башни', *Вестник бронетанковой техники* 1982, сборник 1. p. 53.

415 mm was solid steel, the "sand bar" core was an effective low-density filler, even if the protection level of the turret was still fairly modest. The turret was completely immune to 125 mm APFSDS (3BM15), and tests of ex-NVA T-72M1 tanks in unified Germany during the 1990's showed that it was categorically immune to 105 mm APFSDS.

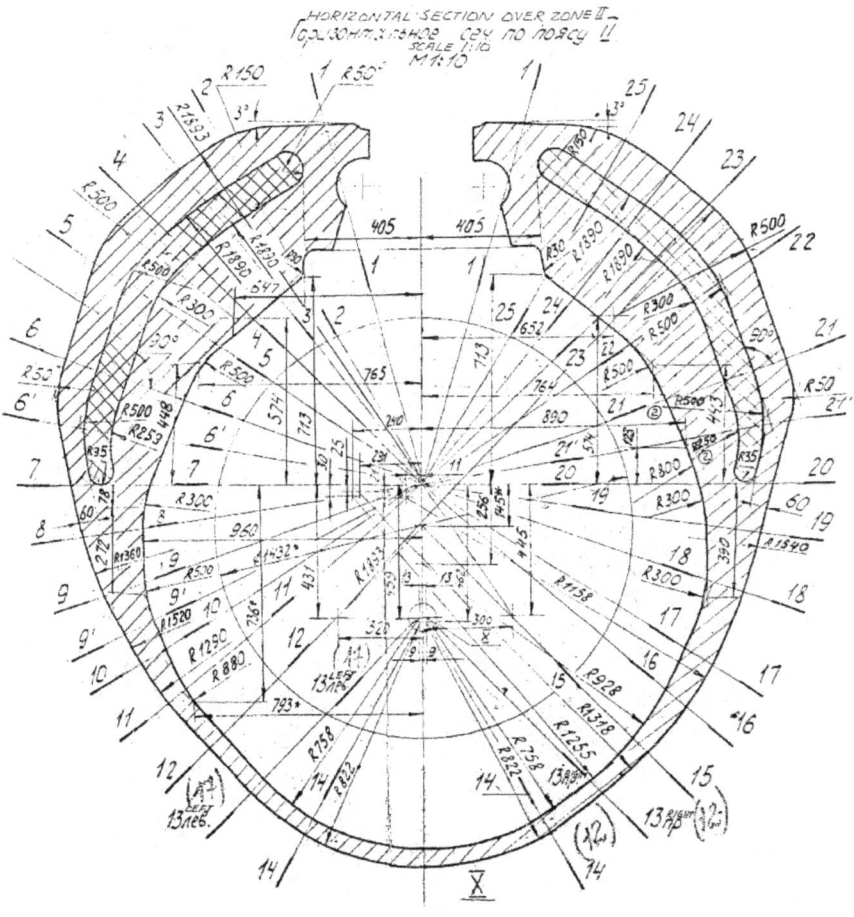

Figure 103: Plan view of T-72M1 turret at Belt II. Sections of the left half of the turret wall from 1-1 to 14-14 are shown in the next two pages. Section 2-2 can be used to represent the armor of the entire turret projection at a side angle of 30°. The 35° side angle used as a protection reference is between Section 2-2 and Section 3-3. (*Soviet Ministry of Defense*)

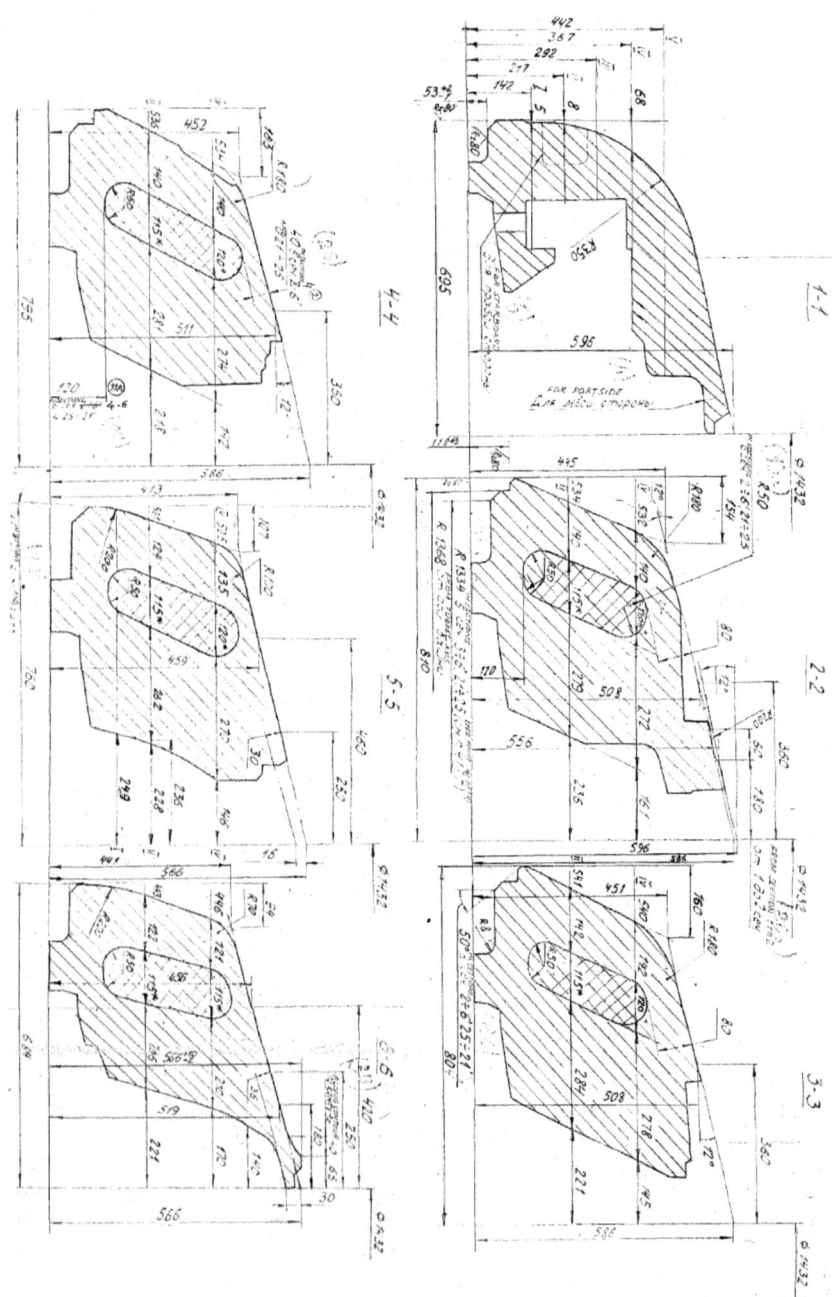

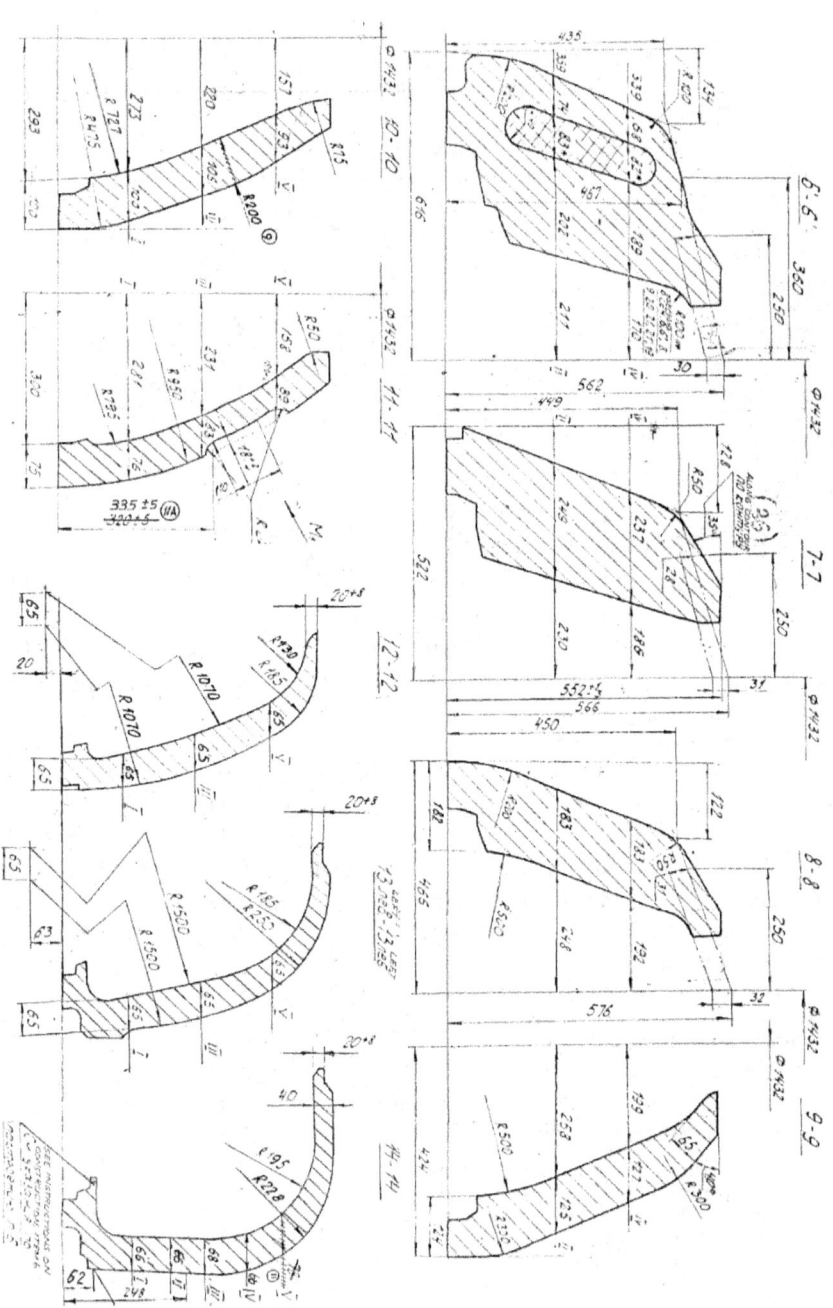

Reflective Plate turret

The "reflective plate" armor was a type of bulging plate armor, which had a literal origin from the German term, "Beulblechpanzerung". T-72s with "reflecting plate" armor turrets began to be produced and delivered to the Soviet Army in early 1983, but did not fully replace the "sand bar" turret until January 1, 1984.

Figure 104: "Super Dolly Parton" turret, as seen by American observers in 1986. The tank model was assigned the designation of M1986/1. (CIA, 1986)

The T-72B's reflective plate armor was contained in an improved turret with a thickened gun mask and bulging turret cheeks, humorously referred to as the "Super Dolly Parton" turret by American observers after the new turret made its appearance in the November 1986 Red October parade[558]. Few details are known about the turret structure, but it can be assumed that the thinner walls did not face the hardening difficulties of the previous types. The armor consisted of an array of 21 plates inserted into a cavity. Each panel was a sandwich of a 6 mm rubber layer between a 21 mm high hardness steel front plate and a 3 mm soft steel rear bulging plate[559]. This array was backed by a 45 mm high-hardness steel plate. Like passive armor, the effectiveness of bulging plate armor against shaped charges was inversely proportional to jet velocity[560]. HEAT warheads with greater penetration power were also less affected by bulging plates.

In principle, the bulging plates themselves had a minimal influence on any KE penetrator relevant to the T-72B. The low thickness of the sheet and the choice of the metal grade was principally determined by the bending action of the bulging plate into a shaped charge jet.

---

[558] CIA (1987). *T-72 M1986/1 Identification Features*. Available at: https://www.cia.gov/readingroom/document/cia-rdp87t00758r000102480001-3 (Accessed 20 June 2024)
[559] Warford, J. M. (2002). 'The Soviet T-72B Main Battle Tank: The First Look at Soviet Special Armor', *Journal of Military Ordnance*, 12(3). pp 5-7.
[560] Held, M. (2005). 'Shaped Charge Optimisation against Bulging Targets', *Propellants, Explosives, Pyrotechnics*, 30(5). pp. 363–368.

As the jet reaches the back surface of the heavy front plate, the deformation of the plate sets the rubber and bulging plate layers in motion. When the jet perforates the front plate and enters the rubber layer, the rubber expands violently, propelling the bulging plate to a high velocity obliquely against the body of the jet[561].

The lateral momentum causes the jet to break up, and the longer it is in motion, the less coherent the jet becomes as its particles move further from the main body. However, large segments largely remain unaffected because of the intermittent nature of the bulging plate interaction. In the T-72B turret there was little room for the jet to travel and break up within the limited dimensions of the turret array, which was a typical limitation in practical armor designs. Multiple panels were placed in the path of the jet to compensate.

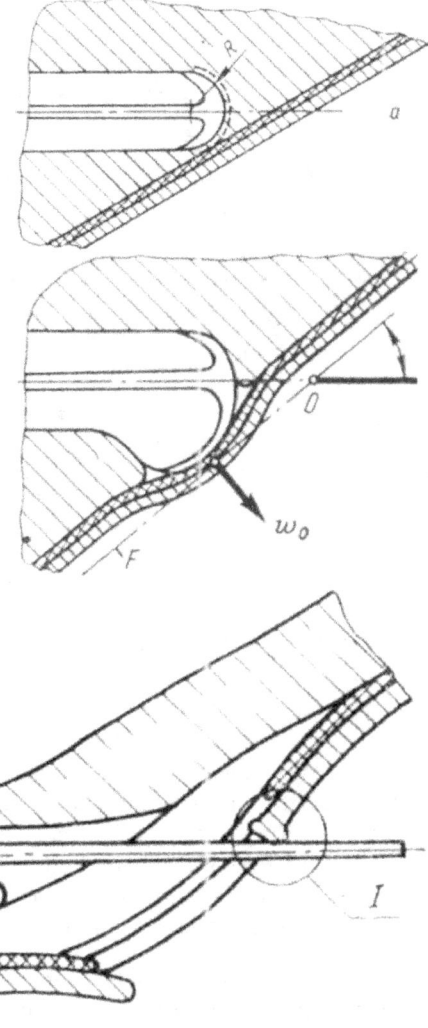

Figure 105: Steps of bulging plate action on a shaped charge jet during the penetration of a reflecting plate panel. (*L. N. Anikina & I. I. Terekhin, 1987*)

---

[561] Аникина, Л. Н., Терехин, И. И. (1987). 'Взаимодействие кумулятивной струи с трехслойной броневой преградой', *Вестник Бронетанковой Техники* 1987, сборник 2. p. 34.

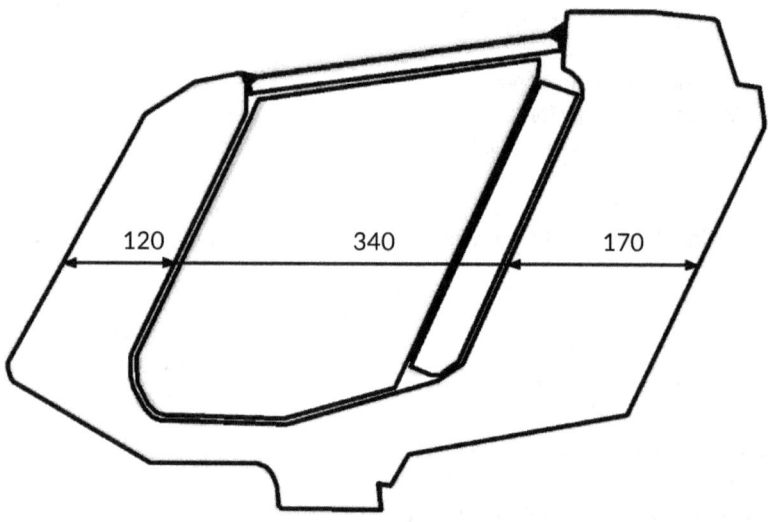

Cast steel	Reflecting plate	Back plate	Cast steel
108	21-6-3	45	155
25°	55°	25°	25°

Figure 106: T-72B turret cross section. (R. A. Then, 2024)

The reflecting plate array in the T-72B turret was angled to place four panels in the path of a penetrator when attacked from the direct front, but from this perspective the panels were presented at a modest obliquity of 35°. From a side angle of 35°, the array presented only two panels, but at a much more effective angle of 70°. The configuration of the "reflecting plate" armor in terms of the ratio of layer thicknesses was optimized for a shaped charge of medium caliber[562]. On its own, without Kontakt-1, the T-72B turret is claimed to have resisted the Soviet *Konkurs* ATGM in tests[563].

In principle, the bulging plates were effective on shaped charges but had a negligible influence on long rod penetrators. Against these and other KE threats, the array behaved solely as robust spaced armor, similar to the T-72B upper glacis array.

In 1988, the composition of the filler was changed to thicker reflecting plates with a 50 mm front plate and double in-pursuit bulging plates. Like the 1987 hull

---

[562] Григорян, В. А. (2006). *Частные вопросы конечной баллистики*. Москва: Издательство МГТУ им. Н. Э. Баумана. pp. 281-283.
[563] Claimed by NII *Stali* webmaster in NII *Stali* website guestbook (forum). Defunct.

array, this design followed the basic principle of meeting improved penetrator rod metallurgy with thicker plates, which were more effective at breaking up strong rods.

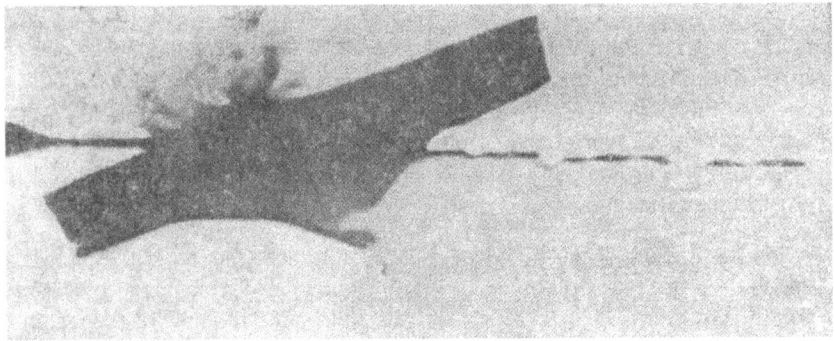

Figure 107: Disruption of a shaped charge jet by the bulging plate of a reflecting armor plate, corresponding to the layout used in the T-72B turret. (L. N. Anikina & I. I. Terekhin, 1987)

Kontakt-1 ERA

Work on the integration of the reactive armor with the T-72 was completed in the summer of 1982 and testing of experimental tanks with this new reactive armor kit were carried out in November 1982. Since 1984, the large-scale fitting of Kontakt-1 on T-72 tanks began. New production tanks would have the ERA mounts installed during final assembly at the factory, and existing tanks would be retrofitted during a scheduled overhaul at repair facilities across the USSR. The installation of Kontakt-1 blocks did not differ between tanks that had the 16 mm add-on armor plate on the upper glacis and those that lacked it.

Due to the replacement of the Object 172M-1 with the Object 184 on the UVZ production line in 1984, all T-72AV tanks were upgraded from existing tanks rather than new-builds. After the delivery of Kontakt-1 kits to repair facilities, the upgrading of tanks began. The first T-72AV tanks entered service in 1985 after their scheduled overhauls in late 1984, and after 1985, most T-72A tanks had received ERA during scheduled repairs. By the end of the year, a supply of Kontakt-1 kits had been established to tank repair facilities across the USSR to upgrade existing tanks of all models, and the first T-72AV tanks entered service in 1985 after their scheduled overhauls in late 1984.

The effective thickness of the T-72A and T-72B armor against ATGMs and grenades was increased to 850-900mm RHA in a 70° frontal arc on both the hull and turret. The protection against tank-fired HEAT shells was only equivalent to 730-750mm RHA due to the high impact velocity and sturdy projectile design allowing shells to partially perforate Kontakt-1 blocks after impact.

The design of the layout took into account the probable distribution of hits from shaped charge weapons of all types. Shoulder-fired anti-tank weapons like the 84 mm Carl Gustaf recoilless rifle were relatively weak (400 mm penetration), but were most often targeted at the sides at favorable angles. Powerful ATGMs like the TOW-2 (900 mm penetration) were long-ranged and most often struck the frontal armor.

Considering the likely range of angles of attack, the typical shot-to-shot variation in penetration power, and the incomplete ERA coverage of the tank (70% frontal, 30-40% side), Kontakt-1 was calculated to improve the weighted average probability of resisting penetration by 1.8 times [564]. The internal arrangement of explosive elements in a V-shape strongly reduced the sensitivity of the disruptive effect of Kontakt-1 blocks to their inclination angle [565].

Kontakt-1 came in full-sized rectangular blocks and triangular blocks. The triangular block was shaped to fit in areas where the full-sized block would otherwise block the crew's field of vision; in front of the driver's periscope, and on the turret roof, and other places. A full set of Kontakt-1 for the T-72AV and T-72B consisted of 227 blocks. There were 48 blocks mounted on the side skirts on each side of the hull, 70 blocks on the frontal arc and the roof of the turret, and 61 blocks on the upper and lower glacis of the hull. The total weight of the armor kit including the additional fittings and mounting frames was around 1,500 kg.

On the T-72AV, the blocks on the turret cheeks were mounted onto special metal frames inclined to set the blocks at the optimal angle of 68°. On the T-72B, the Kontakt-1 blocks on the turret cheeks were fitted directly to the armor surface, limiting the maximum vertical obliquity to 20-30°, but improving the ERA coverage around the turret ring and the gun embrasure.

Kontakt-1 blocks on the turret and hull structure were screwed onto steel studs welded to the armor surface. The blocks on the side skirts were bolted onto steel strips, to give the flexible skirt a measure of lengthwise rigidity.

---

[564] Комяженко, А. Г. et al. (1984). 'Методический подход к выбору характеристик динамической бронезащиты танка', *Вопросы оборонной техники*, 20(116).

[565] Xiangdong Li, Yanshi Yang, Shengtao Lv (2014). 'A numerical study on the disturbance of explosive reactive armors to jet penetration', *Defence Technology*, 10(1). pp. 66-75.

T-72M1 models exported with a Kontakt-1 kit like later models of the Indian T-72M1 *Ajeya* and the T-72S had a lightened layout with just 165 blocks, achieved mostly by omitting the blocks mounted to the side skirts. The weight of this modified set was 1,200 kg.

A disadvantage of the light metal mounting frames used to affix the Kontakt-1 blocks on the turret cheeks is that the detonation of the block on one half of the frame is enough to destroy the frame itself, thus removing the other block in the process. This was the cost of ensuring that the block was installed at the optimum 68-degree angle.

The ease of installing and replacing the blocks meant that the entire modification could be carried out as part of regular scheduled maintenance and blocks lost to battle damage can be easily replaced. The 4S20 explosive elements that made up the working parts of the ERA were safe in handling and insensitive to bullets, napalm, and artillery fragments. They could, however, be reliably initiated by the direct impact of a shaped charge jet[566].

Each element weighed 1.35 kg and was made up of a 6 mm explosive layer sandwiched between two 2 mm soft steel sheets. The charge was PVV-5A[567], a low-sensitivity plastic explosive composed of 85% hexogen and 15% phlegmatizer[568]. PVV-5A was originally an ordnance explosive most notably found in the MON-50 directional mine, an analogue of the M18 Claymore directional mine. Using soft steel for the casing ensured that the front and rear faces detached as intact, contiguous flyer plates instead of fragmenting into shards.

The disadvantage to using light metal boxes for the blocks was that powerful tank-fired shells or artillery shells could clear off large portions of the tank through blast and fragmentation effect. It was found in tests that, on average, the area covered by Kontakt-1 that was left exposed after the impact of a single 125 mm 3BK14M shell was 70-85% on the upper and lower glacis of the hull, 20-30% on each side of the turret, and 50-55% on the sides of the hull[569].

Another downside to individual blocks was that they drastically increased the total surface area of the tank exposed to contaminants, and the underside of the boxes, which were spaced by a short distance above the tank hull and turret,

---

[566] Because of this, it is also evident that despite its very low sensitivity, PVV-5A was not analogous to modern insensitive explosives.
[567] PVV (ПВВ – Пластичное Взрывчатое Вещество) simply means "plastic explosive".
[568] Гребенюк, А.М., Одинцов, Л.Г., Васильев, В.А., Шеломенцев, С.В (2016). *Производство взрывных работ при проведении аварийно)спасательных и других неотложных работ в различных чрезвычайных ситуациях*. Москва: ФЦ ВНИИ ГОЧС. p. 48.
[569] Костин, Ю. Н. et al. (2014). 'Аналіз Живучості Динамічної Захисту Вітчизняних Танків', *Механіка та машинобудування*, 1. p. 93.

Figure 108: T-72B1 in Soviet military photoshoot. (Soviet Ministry of Defense)

was inaccessible to scrubbing. This introduced complications in the standard shower-and-scrub decontamination process[570] used by the chemical troops[571], which was an excellent reason for keeping the boxes dismounted during peacetime exercises. How decontamination units would cope with these difficulties during war, however, remained an open question.

Kontakt-5 ERA

The explosive element in Kontakt-5 was the 4S22, containing a PVV-12M plastic explosive compound. Initial studies found that the existing PVV-5A compound was too insensitive for reliable detonation from APFSDS strikes, which was solved by tuning the explosive to phlegmatizer mix by 5%[572] to create PVV-12M compound (90% RDX and 10% plastic binder)[573]. Its properties were directly analogous to the C-4 plastic explosive[574]. The dimensions of 4S22 were identical to 4S20.

The basic handling safety features of the 4S22 element were the same as 4S20. It was rated for a 1.5-meter drop onto concrete, and did not detonate under small arms fire, including 7.62 mm and 12.7 mm B-32 (AP-I) bullets, autocannon fire, tested with 30 mm AP-T, and artillery fragments at 10 meters, although the artillery shell in question was not specified. It also did not detonate after being doused in napalm.

Due to the low shock sensitivity of the explosive compound, relying on shock initiation [575] was impractical, which made it necessary to pivot to spall initiation[576], also known as a type of remote sensing.

When a projectile or shaped charge jet passes through a barrier placed over an explosive charge, a highly energetic burst of spall is generated at the back surface of the plate and travels towards the explosive charge before the penetrator itself exits the barrier. The spall particles, though light, are very small, and thereby exert a high instantaneous pressure in the explosive. This initiates detonation. Remote sensing through spall initiation improved detonation

---

[570] Wisniewski, A. Pirszel, J. (2021). 'Protection of armoured vehicles against chemical, biological and radiological contamination', *Defence Technology* 17(2).
[571] Radiological, Chemical and Biological Defence units, organic to combined arms units.
[572] Алексеев, М. et al. (2012). *НИИ Стали 1942-2012*. Москва: Издательство СканРус. p. 96.
[573] Кобылкин, И. Ф. (2016). 'Распространение Детонации в Тонких Слоях Взрывчатого Вещества с Инертными Перегородками' *Физика горения и взрыва*, 52(1). p. 116.
[574] Кобылкин, И. Ф., Петюков, А. В. (2015). *Проявление эффекта ударно-волновой десенсибилизации при возбуждении детонации в тонких слоях взрывчатого вещества высокоскоростными ударниками*. Москва: МГТУ им. Н.Э. Баумана. p. 4.
[575] Detonation via the shock transmitted through a cover plate.
[576] In the post-Soviet 4S23 explosive element for *Relikt*, practical shock initiation was achieved by mixing shock-concentrating plastic microspheres inside the explosive compound.

sensitivity and reduced the detonation delay time. The layout of all Kontakt-5 panels on a T-72 except the turret roof blocks was made to initiate through remote sensing. The turret roof blocks were solely for shaped charge protection, with a particular focus on anti-tank bomblets.

If the 4S22 elements failed to detonate through remote sensing, the subsequent penetration of the elements by the penetrator was a secondary means of initiating the ERA. For long rod penetrators, the confinement of the explosive charge and the reflection of shock energy from the back plate of the ERA panel improves the likelihood of successfully detonating the charge.

Tests with domestic APFSDS penetrators with a tungsten alloy cap showed that the likelihood of initiating Kontakt-5 was reduced to 50% or less at a critical impact velocity of 1,500 m/s and below[577]. For 120 mm and 125 mm guns (muzzle velocity 1,650-1,700 m/s), this velocity threshold translated to a firing distance of approximately 1,500-2,000 m. However, for 105 mm guns, which operated on a muzzle velocity of around 1,500 m/s, Kontakt-5 could be expected to function less than half the time at combat distances.

---

[577] Маркачев, Е. В., Рототаев, Д. А., Чублров, В. Д. (1991). 'Возбуждение детонации ВВ в составе динамической защиты при воздействии бронебойного подкалиберного снаряда', *Вестник Бронетанковой Техники* 1991, сборник 1. Available at: http://btvt.info/5library/vbtt_1991_01_dz_detonazia.htm (Accessed 15 August 2023).

Chapter V

Armament

Control & Aiming

The full range of elevation of the D-81 gun was 20°, physically limited by the size of the gear sector built into the gun. This range motion was divided into +15° of gun elevation and -5° of gun depression. Because of the structural tilt of the turret, the gun could actually elevate by +13.78° (13°47') and depress by -6.21° (6°13'). The gun depression limit was within the lower end of the -5° to -8° range considered desirable for a tank gun, but the elevation limit fell quite far from the desired +16 to +18° range[578].

The gun was aimed by powered drives, backed up by a set of manual handwheel controls. Normally, these manual controls were usually used for routine tasks like bringing the turret to the correct orientation to engage the travel lock, but they were also used to aim the gun for indirect fire.

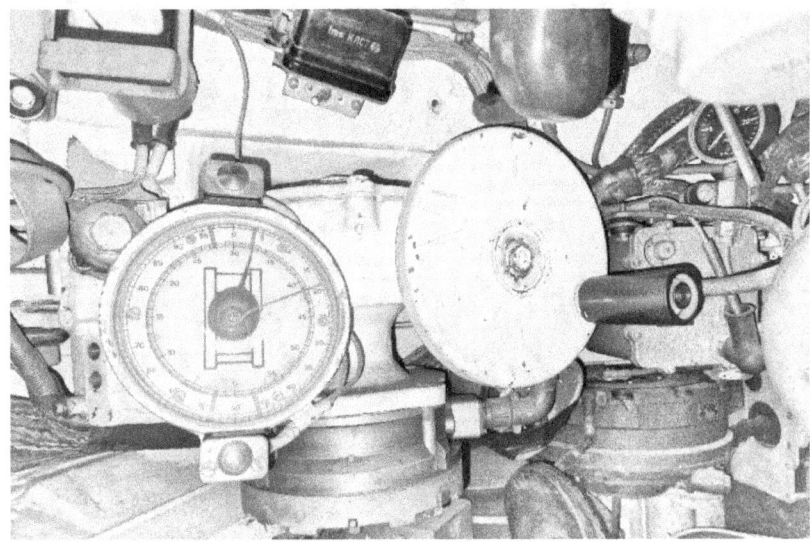

Figure 109: Traverse handwheel with turret azimuth clock to its left. (S. Stauber, 2024)

---

[578] Старовойтов, В. С. (ed.) (1990). *Военные гусеничные машины*. Том 1: Устройство, Книга 1. Москва: Издательство МГТУ им. Н. Э. Баумана. p. 52.

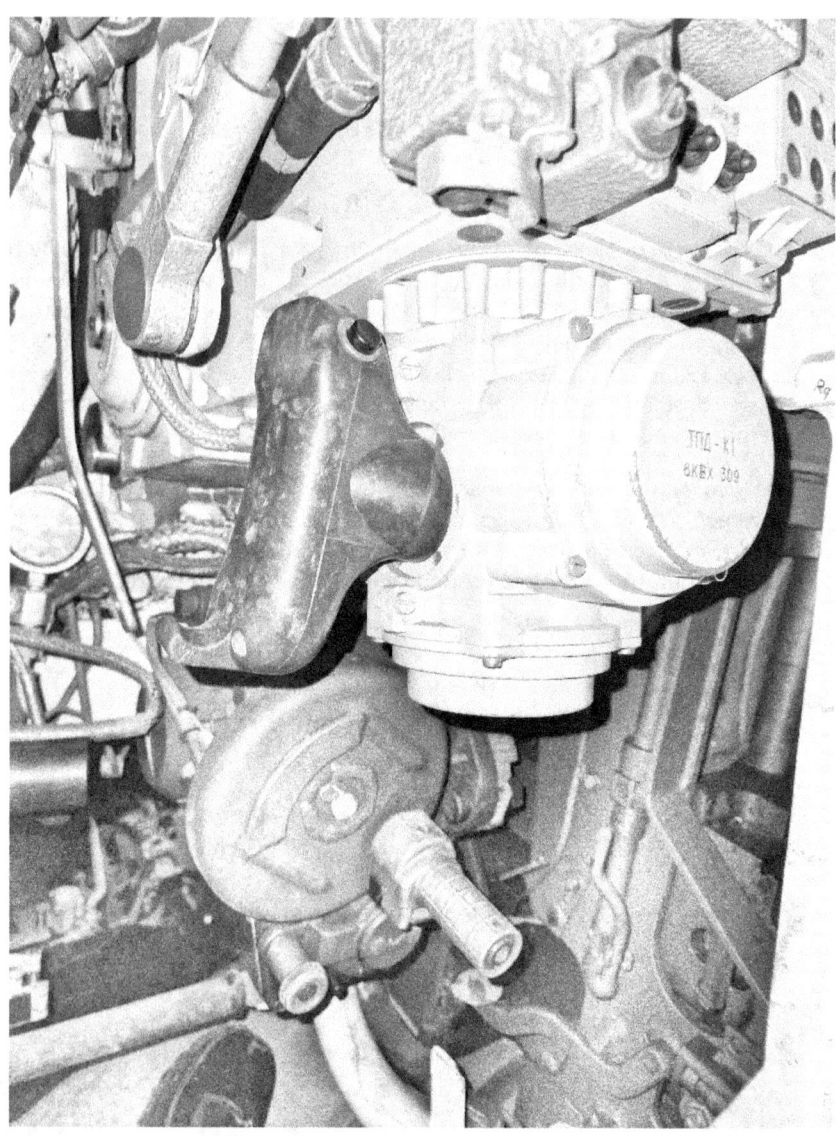

Figure 110: Gunner's control handles in the foreground, with the elevation handwheel below and ahead of it. There was an electric trigger on the elevation handle left. (*S. Stauber, 2024*)

If the gunner was actually forced to fall back on the manual controls in combat, the weight and imbalance of the turret made turret rotation a slow and laborious process. Tests with the T-64A[579] found that on level ground, the maximum turret traverse speed was 0.6-0.8°/s with a force of 1.6 kg on the handwheel. At a roll angle of 15°, the force required on the handwheel shot up to 20 kg and the maximum traverse speed was halved to 0.3-0.4°/s. The amount of effort needed on the handwheel became prohibitive at larger roll angles. If at all possible, the T-72 was always operated with powered controls, though this was typical of modern tanks anyway.

The T-72 was initially equipped with the 2E28M *Siren* stabilizer, followed by the 2E42 *Zhasmin* in 1984[580]. In both systems, the automatic mode was the primary operating mode. Twisting the gunner's control handles side-to-side turned the turret and twisting them up-and-down elevated the sight head on the TPD-2-49 or TPD-K1 sight, and the gun would follow. This mode was applicable to all firing conditions - stationary, in motion, and on short halts.

A semi-automatic mode was also available as an emergency mode in the event of stabilizer failure. In this mode the gun was not stabilized and the gun elevation drive was disconnected, so the gun had to be elevated manually. Only the powered traverse drive was retained.

The traverse system had track and slew speed modes. The tracking mode was controlled exclusively through the gunner's control handles to perform a fine lay on a target and to track moving targets. It gave smooth, stepless speed control from 0.07°/s to 6°/s, marked by a soft stop in the control handles from stiff springs. Twisting the control handles past the soft stop engaged the slewing mode. The TKN-3 traversed the turret in the slew mode only, as did the driver's emergency override button (counter-clockwise only). The driver's emergency turret override worked as long as the tank's electrical mains were turned on, even if the stabilizer was switched off[581], as long as the circuit breaker was closed. The slew speed depended on the stabilizer model.

The performance of the gun elevation drive changed little between different stabilizer models. In any case, the balance of the D-81 gun had a major impact on stabilization quality. A well-balanced gun tends to stay at rest through its own inertia even as the tank pitches up and down under it. Gross deviations in

---

[579] Мазуренко, А. И., Морозов, Е. А. (1985). 'Один Из Путей Повышения Надежности Комплекса Танкового Вооружения', *Вестник Бронетанковой Техники* 1985, сборник 6. p. 1.
The T-72 turret was identical in terms of imbalance, though the traverse mechanism was not.
[580] Устьянцев, С. В., Колмаков, Д. (2004). *Боевая Машины Уралвагонзавода: Танк Т-72*. Нижний Тагил: Издательский Дом "Медиа-Принт". p. 96.
Also see: Барятинский, М. Б. (2008). *Т-72. Уральская броня против НАТО*. Москва: «Яуза».
[581] Министерство Обороны СССР (1979). *Стабилизаторы Танкового Вооружения 2Э28М (2Э28М-2). Техническое Описание*. Москва: Воениздат. p. 30.

barrel attitude and minute deviations imparted through vibration are both minimized accordingly. This is sometimes – and inaccurately – called inertial self-stabilization[582]. An unbalanced gun places large loads on the elevation drive, which must repeatedly raise or lower the gun by working against trunnion friction and the oil pressure induced in the elevation piston[583].

Normally, the stabilizer was turned off entirely when combat was not expected, and the time needed to get the stabilizer to operational condition was two minutes. The stabilizer operating time was considered unlimited during combat, but outside of combat the stabilizer operating time was administratively limited to four hours. Powered traverse or stabilized operation on battery power alone was possible but not sustainable due to the very high power consumption of the system.

---

[582] Шаповалов, А. Б., Солунин, В. Л., Костюков, В. В. (2017). *Системы управления, наведения и приводы: История создания и развития*. Москва: Издательство МГТУ им. Н. Э. Баумана. p. 235.
[583] Pressure that forces the gun to elevate or depress as the hull pitches up or down, generated by the hydraulic coupling between the elevation piston and hull

Table 22: Technical data for T-72 gun stabilizers

Technical Data	2E28M		2E42-2	
	Gun (Vertical)	Turret (Horizontal)	Gun (Vertical)	Turret (Horizontal)
Nominal precision (mrad)	0.8	2.0	0.4	0.6
Min. speed (°/s)	0.05	0.07	0.03	0.05
Max. speed (track) (°/s)	3.5	6.0	3.5	6.0
Max. speed (slew) (°/s)	-	18 (22)[a]	-	16 (24)
Stabilizing rigidity (kg-m/mrad)	65	300	70	300
Damping characteristics*	1-4 mrad; 50 mrad	3-5 mrad; 75 mrad**	1-4 mrad; 50 mrad	1-3 mrad; 75 mrad
Drift rate (mrad/s)	16	16	16	16
Time to readiness (min)	2.0		2.0	
Time of continuous work (hr)	4.0		4.0	
Total system mass (kg)	320		240	

[a] Глазунов, С. Д., Макаров Б. Ф., Мельников, В. И. (1999). *Привод наведения и стабилизации танкового вооружения*. Rospatent No. RU2138758. Available at: https://patents.google.com/patent/RU2138758C1/ru (Accessed: 14 September 2023).

* Both stabilizers were underdamped. Data specifies the number of oscillations before the gun returned to its indicated axis, and the maximum gun deflection angle during the first oscillation.

** After 1.5-2.0 hours, damping strength degraded to 6 oscillations; 100 mrad.

2E28M *Siren*

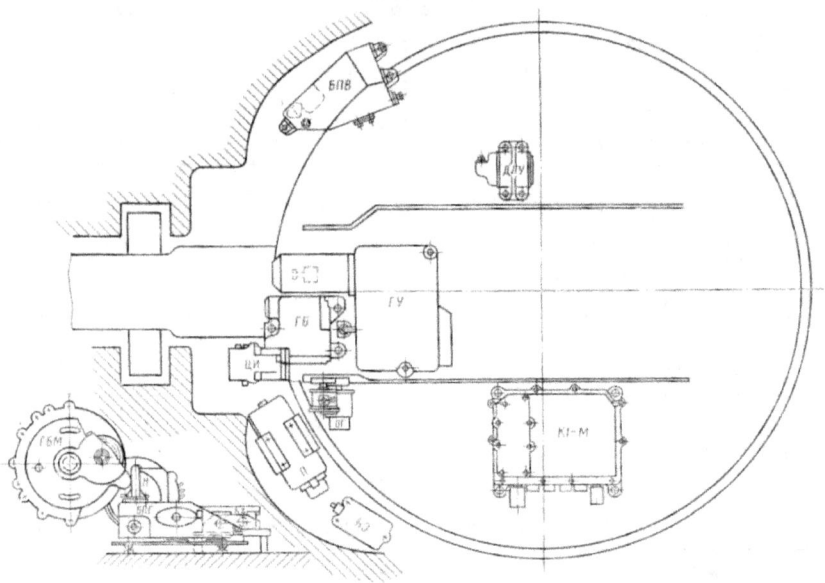

Figure 111: 2E28M-2 stabilizer layout, showing its hydraulic traverse motor in the hull, distributor box under the gunner's seat, and vertical stabilizer units under the 125 mm gun. (*Soviet Ministry of Defense*)

The 2E28M *Siren* represented the second generation of Soviet tank stabilizers[584]. The turret was traversed by a hydraulic motor and the gun was elevated by a double-ended hydraulic piston.

Like all other Soviet tank stabilizers, 2E28M was worked out jointly between TsNIIAG[585] and TsNII-73. The design of the 2E28M was new and distinct from previous Soviet tank gun stabilizers, but it still adhered to a common industry practice of sourcing components from other sectors, most conspicuously seen in its GMA-4P motorized gyroscope (gyromotor). Originally an aviation device, it was created in the late 1950's for the AGI-1 artificial horizon indicator, one of the first of its type in the USSR, notably found in fighter aircraft like the MiG-17 and Su-7.

This practice dated back to the first tank stabilizers in the country, to include the T-54A (GMA/30 gyromotor[586]) and T-54B/T-55 (GM-4 gyromotor). Even

---
[584] Шаповалов, А. Б., Солунин, В. Л., Костюков, В. В. p. 234.
[585] Central Research Institute of Automatics and Hydraulics.
[586] Ibid. p. 239.

the electric turret traverse drive motor in these stabilizers was generic, chosen out of a catalogue of industrial motors to find the best match for the performance requirements. For the 2E28M, however, this was not possible because the T-64 and T-72 turret was severely unbalanced. The moment of inertia was not prohibitive on level ground[587], but if the tank sat on a side slope the enormous moment of inertia of the turret made it difficult to accelerate and brake with existing electrical motors. Instead, a low-speed, high-torque radial piston hydraulic motor[588] was created by TsNII-73.

The switch from an electrical motor to a hydraulic system did, however, come with a penalty in cost and complexity, and it introduced an additional fire hazard. A rupture in a hydraulic line from combat damage most often leads to a high-pressure mist spray of hydraulic fluid, typically ignited soon after by whatever penetrated the tank. Ordinarily, a low-order explosion ensues, but with luck, there would "only" be a fireball.

This risk was partly mitigated in the 2E28 system by its turret traverse drive being located in the hull rather than in the turret, avoiding the need for extensive pipework running across large areas of the fighting compartment (e.g. in Leopard 1, M60A1, etc.), and by operating on an unusually small quantity of hydraulic fluid - only 10 liters. The elevation drive in the turret was a much larger risk, holding 17 liters of fluid, 6 liters of which was in a reservoir on the turret ceiling.

Both the traverse and elevation drives were constant-capacity systems, otherwise known as constant-flow systems, as opposed to the more common constant-pressure system. In a constant-capacity system, the fluid flow rate is constant but the pressure could fluctuate depending on the load. In a constant-pressure system, a stable operating pressure is maintained regardless of load, typically through a gas-charged accumulator – a vessel with a nitrogen-filled chamber exerting pressure on a large volume of hydraulic fluid[589]. The lack of a hydraulic accumulator in the 2E28 system meant that the traverse drive, and elevation drive to a lesser extent, lost their pressure quickly once ruptured.

The control system for the traverse drive was open-looped, featuring pressure feedback sensors in the pistons of the motor. If, for example, the motor received no signal to traverse the turret and the hull began to turn, the turret would remain fixed in azimuth on its own accord under inertia. This creates negative pressure in the pistons, which were perceived by the feedback sensors, letting the motor be driven in reverse. It was for this reason that the turret could maintain its orientation regardless of the turning rate of the hull. When

---

[587] The torque needed to accelerate the T-72 turret reached 40 kN.m, whereas for Western tank turrets it was just 1–1.5 kN.m.
[588] Fixed stator type, with 11 pistons.
[589] The M60 and Leopard 1 used a floating-piston gas-charged hydraulic accumulator.

performing a pivot turn in the 1ˢᵗ gear, for instance, the hull would be turning a rate of >40°/s, which is around twice the powered traverse speed, but the turret remains fixed in traverse nonetheless.

The vertical drive had a closed-loop control system. Controlled elevation was possible to a speed of 3.5°/s, equal to the elevating speed of the TPD-2-49 or TPD-K1 primary sight. The gun was elevated by controlling the pump to create a pressure differential between the upper and lower chambers of the double-ended drive piston. If, however, the hull drove over a bump or a dip in the terrain, the gun itself supplied additional hydraulic pressure with its own moment of inertia. This raised the pressure differential, in turn raising the elevating (or lowering) speed, up to a technical limit of 7.0-8.5°/s, and this allowed the gun to remain on target when the tank pitched up and down at a rate quicker than the gun could elevate through the gunner's control handles.

Hydraulic braking zones were incorporated into the control system to slow down the gun if it approached the hard stops at angular speed of 8.5°/s[590], and if the gun slammed into these hard stops, it would be briefly hydrolocked to prevent it from rebounding off in the opposite direction. This way, the gun returned to its initial orientation more quickly once the tank rode back up from a dip, or pitched back down after a bump[591].

A linear accelerometer was fitted to detect and compensate for momentary disturbances that arose from the imbalance of the turret, such as when it was subjected to complex rolling and yawing accelerations over rough terrain. This was a simplified method of negative feedback inertial compensation, which is normally provided by a rate gyro in more modern systems.

Benchmark tests at a speed of 35 km/h established that the system gave a stabilization precision[592] of 0.8 mils in the vertical plane and 2.0 mils in the horizontal plane[593]. These figures were subsequently used for the official tactical-technical characteristics of the stabilizer. In actual operation, the degree of precision could differ noticeably. A long-term technical study[594] recorded that in 1977-1978 and in 1978-1979, the median stabilization error (aggregate of both axes) was 0.777 and 1.01 mils respectively. By 1981, the median error improved to just 0.59 mils. The tank speeds reached during testing ranged from 22-33 km/h.

---

[590] Министерство Обороны СССР (1979). *Стабилизаторы Танкового Вооружения 2Э28М (2Э28М-2). Техническое Описание.* Москва: Воениздат. p. 42.
[591] Ibid. p. 99.
[592] This is equivalent to the root mean square error (RMS error).
[593] Stabilization error refers to the angular size of the misalignment between the true bore axis and the indicated axis.
[594] Шамарин, О. В. (1985). 'Электромеханические Стабилизаторы Танкового Вооружения', *Вестник Бронетанковой Техники* 1985, сборник 1. pp. 18-19.

2E42 *Zhasmin*

The 2E42 stabilizer belonged to the third generation of Soviet tank gun stabilizers, featuring the use of semiconductor electronics and a new control scheme. The controls, working procedures and other operational aspects were identical to the previous stabilizer, but stabilization quality improved and the 2E42 took up less space in the tank.

The highlight of the new stabilizer was the return to electric turret traverse. Previously, a conventional electric motor running at high speed (geared down by 1500:1 at the turret ring) generated a substantial moment of inertia in its rotor and gears, contributing an additional 50% to the moment of inertia of a turret[595]. With the new EDM-16U low-speed, low-inertia motor[596], it became possible to surpass hydraulic motors in high-torque performance at a smaller volume and with a lower power consumption.

Compared to the 2E28M, the 2E42 stabilizer enjoyed a longer service life, higher reliability, lower maintenance requirements, and required less labor in production. Forced air cooling enabled long continuous operation, and regenerative braking kept power consumption under control. The braking time for a moving turret was also diminished to as little as a third of that from the 2E28M stabilizer, and the damping strength was also increased.

The fire hazard posed by high-pressure hydraulic systems to combat damage was demonstrated during the 1973 Arab-Israeli war[597], especially in the M48 and M60A1, which were found to have suffered an alarming frequency of hydraulic detonations. Compensatory measures of various types were implemented in various countries.

In the U.S, FRH fluid[598], or Fire Resistant Hydraulic fluid, began replacing OHT[599]. Compared to an ordinary hydraulic fluid, FRH and similar fluids did not fundamentally alter the propensity of hydraulic systems to detonate or ignite, but they could shorten the combustion time of a fireball and lower the likelihood of sustained fires by certain ignition sources[600].

West Germany, also working with Israeli specialists, concluded for the Leopard 2 that it was more worthwhile to isolate the hydraulic pump in a turret

---

[595] With a low-speed hydraulic system, it was approximately 1%.
[596] НПО Электромашина (2022). *Электродвигатель* ЭДМ-16У. Available at: https://web.archive.org/web/20211029070132/https://www.npoelm.ru/product/spetsproduktsiya/elektrodvigate edm-16u/ (Archived).
[597] Brocklin, C. V. (1989). *Single Hydraulic Fluid for Army Ground Combat and Tactical Vehicles and Equipment*. Fort Belvoir: U.S. Army Belvoir RD&E Center. p. 1.
[598] MIL-48600
[599] MIL-6083. The designation OHT was a military symbol and not an acronym like FRH.
[600] Department of Defense (1995). *Design of Combat Vehicles for Fire Survivability*, MIL-HDBK-684. pp. 3-13, 3-15.

bustle blow-out compartment [601]. The additional length of hydraulic lines between the pump to the gun control drives at the front of the turret was accepted as a calculated risk.

Switching from a hydraulic drive to an all-electric drive would have been a decisive solution to this issue, but in tanks, full electrification was rare. The most notable example was the Merkava. The 2E42 stabilizer was a hybrid solution in that the gun elevation drive remained hydraulic, but in a unique system containing only 3 liters of hydraulic fluid in total, including 2.8 liters in a reservoir built into the hydraulic booster, all positioned under the gun. The fire hazard from the replenisher reservoir fitted on the turret ceiling from the 2E28M was thus eliminated.

The elevation speed did not change, but the maximum traverse speed was slightly increased. The main reason was to help shorten the time taken in the target hand-over process from the commander to the gunner, and thereby reduce engagement times.

Prototype testing of the 2E42 system on T-72 testbeds in 1981 and 1982 showed a median stabilization error of 0.4 mils and 0.377 mils respectively[602]. The trend of year-on-year improvements in precision through production refinements continued through the early 1980's. On average, in different weather conditions and at different times of the year, the recorded stabilization accuracy was 0.28 mils during testing in Ukraine and 0.34 mils during testing in the Urals, under somewhat more difficult conditions.

---

[601] Hilmes R. (1988). *Kampfpanzer. Die Entwicklungen der Nachkriegszeit.* Bonn: Mittler Report Verlag GmbH. p. 44.
[602] Шамарин, О. В. (1985). Электромеханические Стабилизаторы Танкового Вооружения, Вестник Бронетанковой Техники 1985, сборник 1. pp. 18-19.

Autoloader

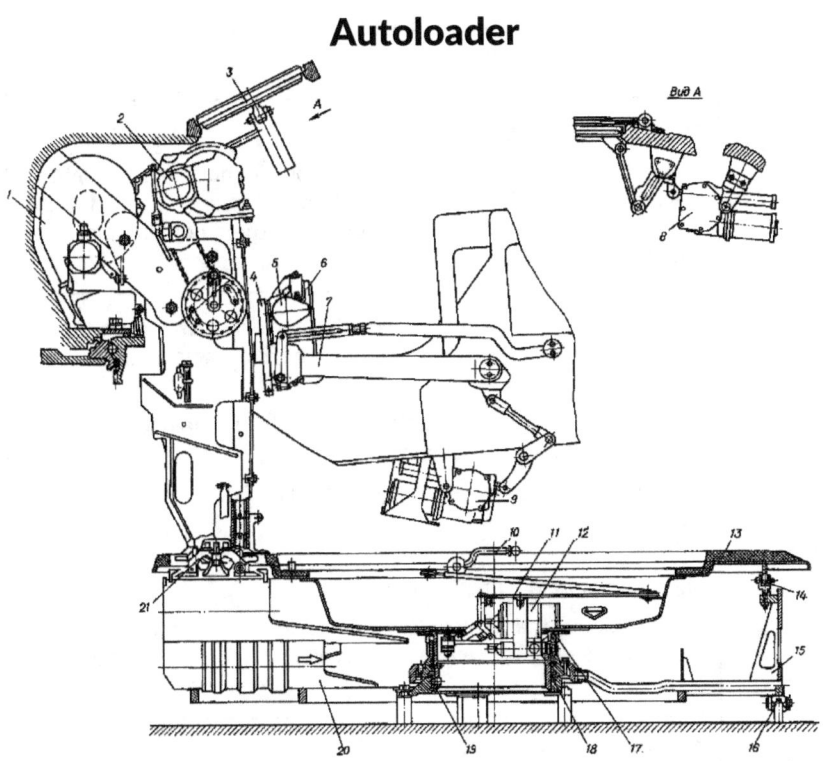

Figure 112: Overall view of T-72 autoloader. (*Soviet Ministry of Defense*)

The T-72's gun was loaded by a carousel-type autoloader with a 22-round capacity. It was entirely driven by electric motors, and was thus known as an electromechanical autoloader to distinguish it from the T-64's electrohydraulic autoloader. It was referred to simply as an autoloader, "AZ" (АЗ – автомат заряжания), while the formal term for the T-64 autoloader was a loading mechanism, or "MZ" (МЗ – механизм заряжания). All T-72 models were equipped with the AZ-172 autoloader with the exception of the T-72B, which featured the AZ-184 autoloader with the ability to carry gun-launched ATGMs.

The loading cycle was completely automated from the moment the gunner pressed the "load" switch on his control panel after choosing an ammunition type. The end of the loading cycle was the moment an ejected stub case was caught in the stub case catcher, after the freshly loaded round was fired.

Table 23: Steps of the loading cycle (Next page)

Step	Actions
1	The carousel turns. In parallel with the rotation of the carousel, the gun is brought to the loading angle by the gun elevation piston and stopped at the correct angle by an electric bolt, and then held in place by hydraulically locking the elevation piston.
2	The carousel starts braking once the correct ammunition type is reached and stops once it lines up the correct cassette to the elevator. During the braking process, the case ejector frame rises if the gun has been hydraulically locked. The cassette rises until the projectile is lined up to the breech and stops.
3	The rammer loads the projectile. While loading the projectile, the ejection hatch opens, and the spent case from the previous shot is ejected after a predetermined delay.
4	The ejection hatch closes. The cassette is lowered until the powder charge is lined up to the breech. The rammer loads the powder charge and trips the ejector levers, retracting while the gun breech closes.
5	The empty cassette and the stub catcher frame both lower to their original positions simultaneously. The gun is released from its locks and returns into alignment with the gunner's sight.
6	The gun fires and ejects the spent case stub into the stub catcher. The stub catcher signals the presence of a stub on the gunner's control panel, ending the loading cycle.

Министерство Обороны СССР (1979). *Вооружение Танка Т-72*. Москва: Военная академия бронетанковых войск. p. 76.

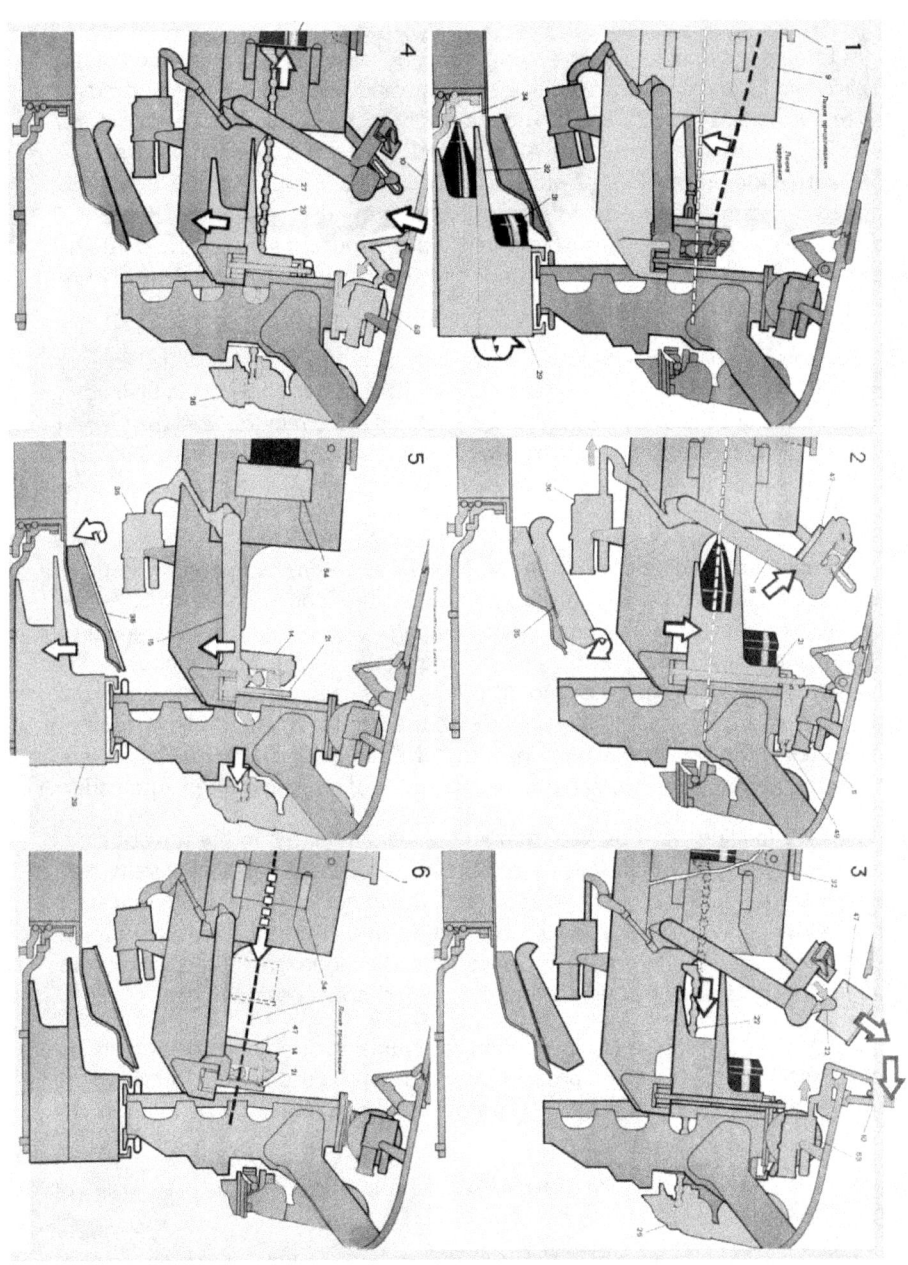

Figure 113: Autoloader cycle diagram. (*Soviet Ministry of Defense*)

In total, the complete loading and firing cycle took 7.8 seconds to complete. This could be maintained at a vehicle tilt (slope and side slope) of 15°, and a tank speed of no less than 25 km/h. The maximum technical rate of fire, therefore, was 7.7 rounds per minute. The AZ autoloader was effectively on par with the MZ autoloader in the T-64A, which achieved a total cycle time of 7.5 seconds under the same criteria of loading the third round in sequence. If the AZ autoloader took the next available round in the carousel instead of the third, the cycle time was reduced to 7.1 seconds and the maximum technical rate of fire rose to 8.4 rounds per minute[603].

For comparison, simulated manual loading of a 125 mm gun with HE-Frag rounds in a tank mockup demonstrated that the first two shots could be loaded in 1 minute, but the 5th and 6th shots were loaded in 1.5 minutes, and the 9th and 10th shots needed nearly 2 minutes to load[604]. The autoloader ensured a quick and stable loading speed, and provided a larger supply of ready rounds than what was otherwise feasible for a manually loaded tank.

The speed of loading came from the overlap between the operating steps of each individual component in the system, and the loading angle of 2.5° was optimal for minimizing the time spent bringing the gun from its firing angle to the loading angle and back, both on the move and from a standstill[605]. Nevertheless, a considerable amount of time was spent bringing the gun from the loading angle back into alignment with the gunner's sight, cycling between cassettes in the carousel, and raising a cassette to the loading path. These were fundamental limiting factors in the loading speed of a hull-mounted autoloader.

Quicker loading was possible from autoloaders mounted in the turret bustle with a conveyor belt-type feeding system. Autoloaders of this type were widely explored internationally in the 1960's, but would not begin to proliferate until the 1990's. However, a bustle autoloader or even non-mechanized ammunition stowage in the bustle greatly increased the likelihood of a direct hit to the ammunition, especially when fighting from a hull-down position.

Local shielding is inefficient when ammunition is widely dispersed or otherwise occupies a large proportion of the projected area of the tank[606]. In the case of bustle stowage, this proportion approaches 100%. Even with

---

[603] Министерство Обороны СССР (1979). *Вооружение Танка Т-72*. Москва: Военная академия бронетанковых войск. p. 75.
[604] Ibid.
[605] Беззубиков, Ю. К., Рослов, В. Б. (1987). 'Расчет быстродействия автомата заряжания', *Вестник Бронетанковой Техники* 1987, сборник 1. p. 27.
[606] Бакшинов, В. М., Комащенко, А. Г., Тимохин, В. И. (1986). 'Броневые отсеки для боекомплекта танка', *Вестник Бронетанковой Техники* 1986, сборник 1. pp. 15-16.

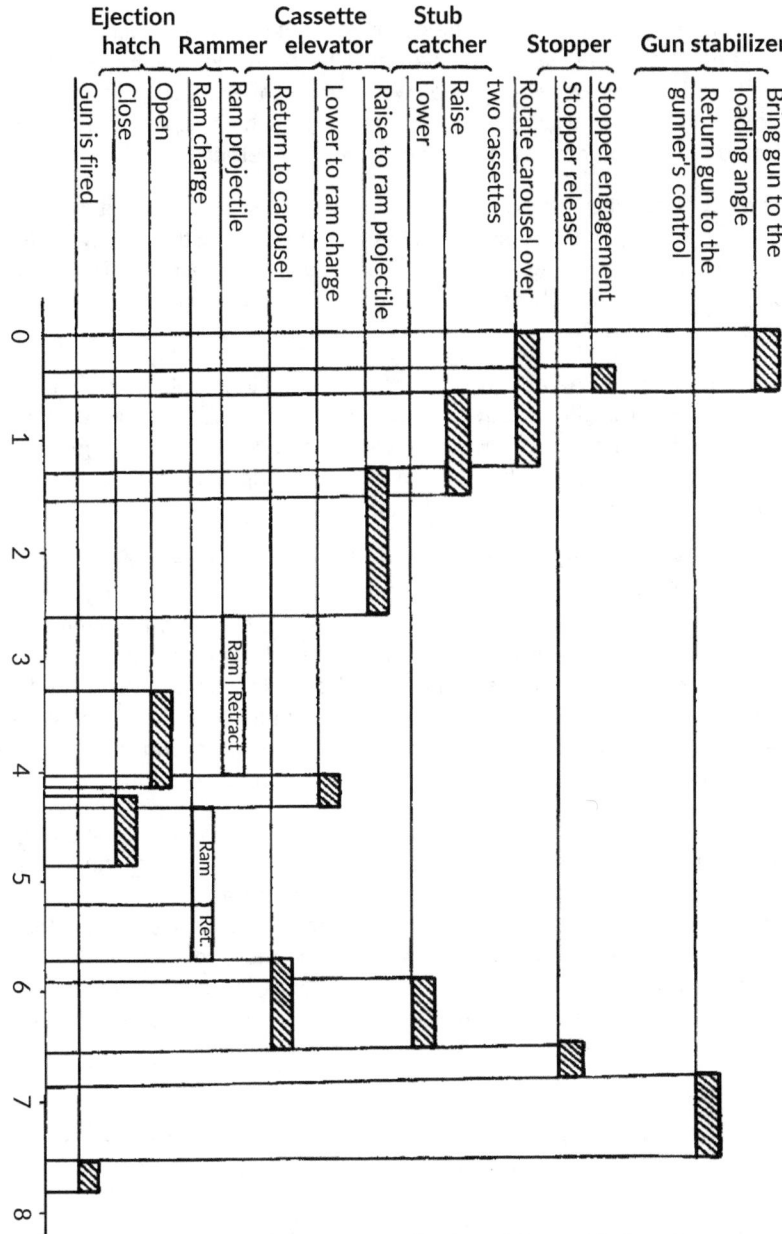

Figure 114: Autoloader cyclogram. (*Soviet Ministry of Defense*)

barriers thick enough to stop the spray of secondary fragments from a penetrating attack, the likelihood of a direct hit from the primary penetrator element (residual jet or residual rod) was very high. Protection in that case required a very large weight of additional armor, which could otherwise be dedicated to the armor to prevent a penetration altogether[607].

The most effective methods for reducing ammunition fires were to maximize the packing density of the ammunition stores and to position these stores as low in the hull as possible, reducing the likelihood of a direct hit. The T-72's autoloader carousel met these design guidelines, but the ammunition stowage scheme as a whole did not.

All 22 rounds in the AZ autoloader carousel were individually stored in two-tiered cassettes arranged radially around the carousel hub, with the turret slip ring at its center. Each cassette was a steel-walled welded structure with a wall thickness of 2.5 mm. The much heavier projectiles occupied the bottom cassette tier, keeping the center of gravity of the tank low.

The carousel was a skeleton frame made from welded steel tubes, securing the cassettes from rolling and shifting with brackets and from bouncing off their slots with the carousel cover. The carousel cover was made from 2.5 mm steel sheets, and the snag guards around the perimeter of the carousel were 3 mm thick[608]. The carousel was thus essentially equivalent to a relatively thick-walled floor stowage bin, and the semi-combustible ammunition inside was no worse protected than cased ammunition. Anti-radiation lining on both sides of the carousel cover added to the protection for the ammunition beneath.

However, rather than ammunition safety, the cabinless autoloader design mainly offered convenience. The low profile of the carousel kept the fighting compartment open along its perimeter, equivalent to a rotating turret floor instead of a turret basket. By turning the turret to various angles, the gunner or commander could reach all of the hull ammunition racks to manually load the gun in an emergency, or to restock the autoloader during short interludes in a battle, which was considered impossible for the T-64 autoloader due to its enclosed cabin design[609].

A small quantity of ammunition was stowed in the turret as a ready supply for emergency situations. On the autoloader carousel cover, there were four slots

---

[607] Ibid.
[608] Measured by calipers. The measured thickness slightly exceeded the given figure due to the paint layer.
[609] Ефремов, А., Павлов, М., Павлов, И. (2011). 'История создания первого серийного танка Т-80 с газотурбинной силовой установкой', *Техника и вооружение* 2011, (November).

behind the commander's seat and two slots behind the gunner's seat allocated for three rounds of HE-Frag and HEAT, and the turret wall behind the gunner's seat held one APFSDS projectile. In front of each seat was one powder charge apiece. In the event that the autoloader failed, it was possible to manually load the gun with up to four shots without needing to reach for ammunition in the hull. In the T-72A, this was increased to five shots. This was, in a sense, an improvement over the T-64 autoloader, which lacked any alternative means of loading apart from cranking the autoloader manually, but the safety advantages of the low-profile carousel were totally compromised.

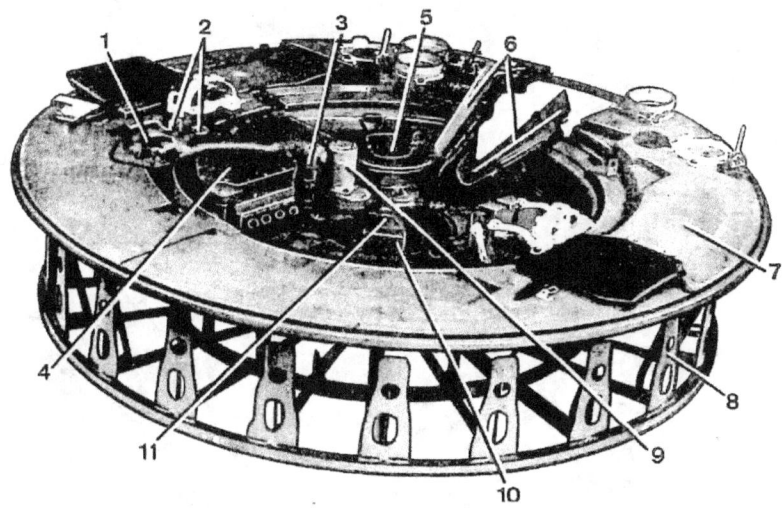

Figure 115: The carousel rotation motor (9) and its gearbox were installed atop the carousel cover, with a manual crank (3) next to it as a backup. Four rollers on the hull floor supported the not-inconsiderable weight of a full carousel.
(*Soviet Ministry of Defense*)

The T-72 had a VKU-330-4 slip ring in the hub of its autoloader carousel[610]. Command tanks had the VKU-330-1. Later, the T-72B standardized on the VKU-330-1 because of its smaller diameter, which allowed the carousel's ammunition length limit to be extended enough to load special gun-launched anti-tank guided missiles. The maximum projectile length was 680 mm[611]. In the AZ-184 autoloader for the T-72B, the slip ring was modified and a cut was made in the carousel hub to add clearance for 9M119 *Svir* guided missiles to be loaded

---

[610] Зимин, Ю., Кононов, А., Беляков, С. (1999). *Электроспецоборудование танка Т-72*. Омск: Омский Государственный Технический Университет.
[611] Потемкин, Э. К. (ed.) (1982). *Основы проектирования вооружения танка*. Том 2. Москва: Издательство МГТУ им. Н. Э. Баумана. p. 124.

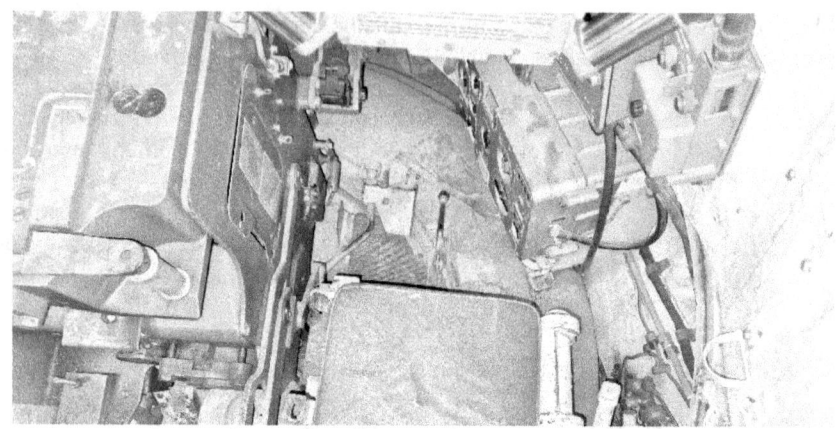

Figure 116: Carousel manual backup with the stopper release lever on the right, and the ratchet to crank the carousel on the left. (*S. Stauber, 2024*)

Figure 117: The cassette elevator manual backup was a handwheel to raise the lift through a chain. (*S. Stauber, 2024*)

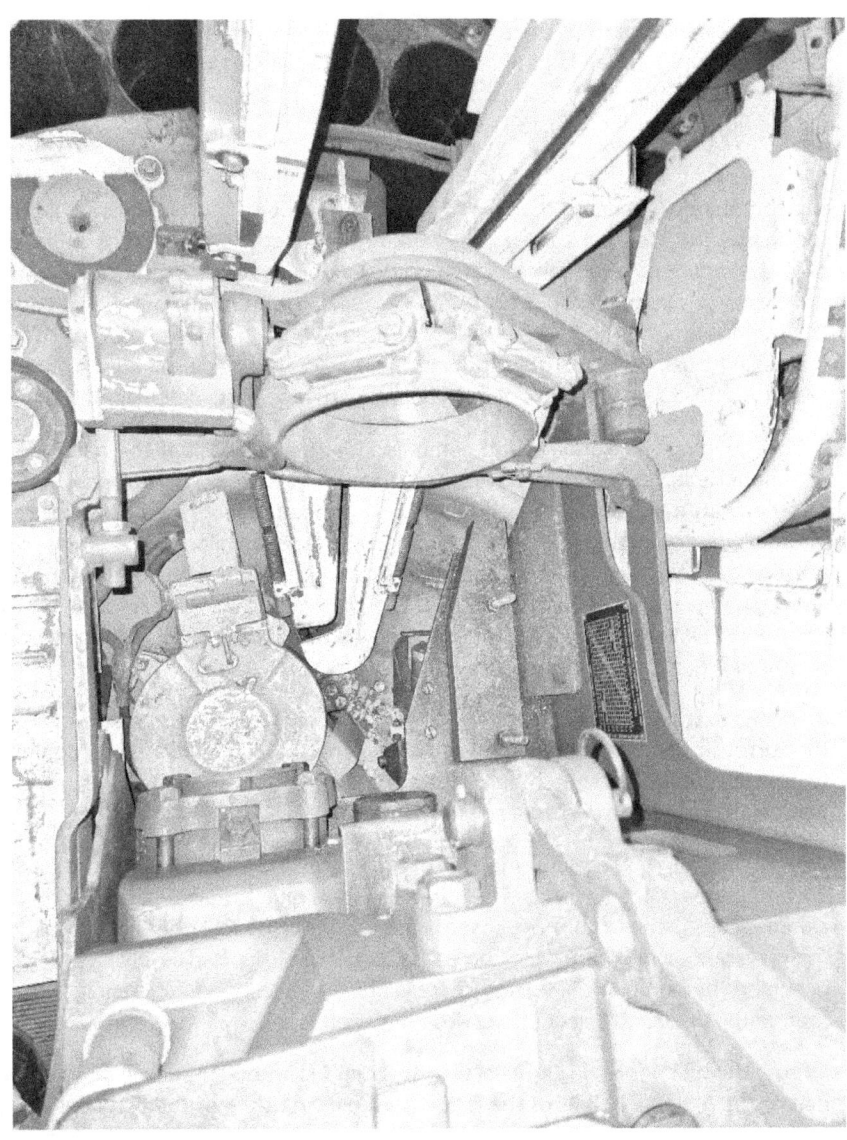

Figure 118: Top-down view of the autoloader memory unit to the left of the carousel trapdoor. While cranking the carousel manually, there was no way to see what ammunition type was positioned under the trapdoor since it was too dark, especially without power for the dome lights. (S. Stauber, 2024)

in the carousel. These missiles were 695 mm in length, the longest projectile type in the ammunition load of a T-72 through the end of the Cold War. The T-72B1 retained the AZ-172 autoloader and memory system since it lacked the missile firing capability of the T-72B.

In the event that the autoloader was inoperable due to a total power outage, the commander assumed the role of the tank's loader. The ready supply in the turret was the first to be used, followed by the reserve supply in the hull. The carousel was the last available ammunition source. By engaging the manual drive, the commander could crank the carousel until a fresh round was under the trapdoor, and then, with the elevator handwheel, raise the cassette to the restocking position. With the gunner's help releasing the projectile from the cassette, the commander extracted the two cartridge parts and loaded the gun.

Since loading the gun was the commander's responsibility if the autoloader was inoperable, an assumption was made in the tank's design that the gun would also be fired manually in this situation. For the commander's safety, a special interlock disconnected the manual firing mechanism unless the commander pressed an unlocking lever on the right side of the gun. After each shot, the interlock was reapplied by the recoil of the gun, readying the gun for a manual reload[612]. This interlock was always functioning, but had no effect when the gun was fired electrically under normal circumstances.

Thanks to the cabinless design, it was possible for the driver to escape through the turret at most turret orientation angles, and the turret occupants could in turn reach the escape hatch behind the driver's seat most of the time without needing to turn the turret to a special position.

As a general rule, maximizing the packing density of ammunition stores is advantageous. Given a certain amount of ammunition, a low packing density axiomatically means that more space is needed. This reduces the space available for crew and equipment, and increases the share of a tank's projected area occupied by ammunition, thus increasing the likelihood of damage from fragments, spall, and direct strikes from a penetrator.

From the hull floor, the carousel was 450 mm tall in total[613], occupying around half of the internal height of the hull. The skeleton frame of the carousel alone

---

[612] Министерство Обороны СССР (1974). *125-мм Танковые Пушки 2А26 (Д-81) и 2А26М2 Техническое описание и инструкция по эксплуатации (Части 1, 2, 3, 4)*. Москва: Воениздат. pp. 24-25.

[613] Бундин, В.И. et al. (2000). *Боевое Отделение Подвижной Военной Машины*. Роспатент, RU27692.

was 400 mm tall, and the cassettes in them were 380 mm tall[614]. The 22 rounds occupied a total volume of 0.9 m³. The projected side area of the ammunition was no larger than 0.6 m², half the projected area of the MZ carousel (1.18 m²)[615]. More importantly, placing the ammunition supply low in the tank reduced its exposure to direct fire threats[616].

The volumetric efficiency of the carousel was worse than a simple floor ammunition box like on a number of tanks (T-34, Centurion, etc.) and somewhat lower than the MZ autoloader of the T-64[617]. The AZ carousel was also limited to a smaller capacity than the MZ carousel due to the radial arrangement of the powder charges.

A carousel of this sort is essentially equivalent to the pan magazine of a machine gun like the Lewis gun or the USSR's own DP-27. Splitting the cartridges into two parts gave a higher packing density, but with essentially cylindrical projectiles and cylindrical charges, the large diameter of the charges was the limiting dimension to the carousel capacity[618]. The MZ carousel arrayed only the projectiles radially while laying the wider propellant charges vertically to form a ring around the fighting compartment. Here, the projectile diameter was the limiting dimension instead of the charge diameter, allowing more rounds to be stored within the same hull width. The MZ autoloader also had the advantage of loading the two cartridge parts in one ramming stroke.

However, these advantages were mitigated by the AZ autoloader's higher carousel rotating speed (70°/s against 26°/s) and quicker projectile ramming speed. To load two cartridge parts in one stroke in the MZ autoloader, the rammer would routinely move a HE-Frag projectile weighing 25 kg down the breech by pushing on its powder charge, and the risk of crushing the charge made it unacceptable to use a high ramming speed. The AZ autoloader rammed projectiles at 2.2-2.5 m/s, twice the speed of the MZ autoloader rammer[619]. The

---

[614] Вильховченко, Н. Н., Емельянов, В. Е. (1978). 'Зависимость объемно-весовых показателей автомата заряжания от числа находящихся в нем выстрелов', *Вестник бронетанковой техники* 1978, сборник 4. p. 16.

[615] Бундин, В.И. et al. (2000). *Боевое Отделение Подвижной Военной Машины*. Роспатент, RU27692.

[616] Старовойтов, В. С. (ed.) (1990). *Военные гусеничные машины*. Том 1: Устройство, Книга 1. Москва: Издательство МГТУ им. Н. Э. Баумана. p. 52.

[617] Вильховченко, Н. Н., Емельянов, В. Е. (1978). 'Зависимость объемно-весовых показателей автомата заряжания от числа находящихся в нем выстрелов', *Вестник бронетанковой техники* 1978, сборник 4. p. 17.

[618] The problem can be visualized through the circle inscribed by the polygon formed by the rounds, where the length of each side is equal to the ammunition diameter. A polygon with a large number of shorter sides approximates a circle more closely, and the packing efficiency is correspondingly higher.

[619] Even this was rather tame compared to high-speed anti-aircraft gun rammers like on the Soviet 100 mm KS-19 and 130 mm KS-30, where the ramming speed reached 7.9 m/s to maximize the cyclic firing

speed was limited only by the strength of the semi-combustible charge on APFSDS rounds. When loading the powder charges, the ramming speed was dialed back to 1.5-2 m/s[620], the same as the MZ autoloader, and then retracted at full speed. In the semi-automatic mode, pressing the "ram" button on the commander's control panel only deployed the rammer at full speed, which somewhat increased the risk of damaging a powder charge if there was additional resistance on the ramming path[621].

Control System

The autoloader was controlled by electromechanically through relay logic, the standard method of machine automation before programmable circuit boards became standard. The "load" switch on the gunner's control panel triggered the carousel and the gun stabilizer into an 'on' state, so that the carousel began rotating blindly until the selected ammunition type was reached, and the gun stabilizer raised or lowered the gun until it reached the elevation stopper. A microswitch or relay at the end of a step would then trigger the subsequent step. This type of progression repeated until the conclusion of the loading cycle. To speed up the loading process, the control logic was arranged to overlap as many steps as possible.

While replenishing the carousel, the ammunition type was recorded in the carousel memory unit by pressing one of three recorder buttons marked for HE-Frag, HEAT or KE after loading a cassette, which automatically returned the loaded cassette into the carousel and located the next empty cassette.

The center of the electromechanical memory unit was a spoked hub, each spoke corresponding to the position of a cassette in the carousel. A sliding electrical contact on the end of each of the twenty-two spokes rode on one of four conduit rings on a fixed disc. Depending on which ring the spoke rode on, it could be indexed as Empty (outermost ring), HE-Frag, HEAT or KE (innermost ring) in the memory unit. Selecting HE on the gunner's control panel prompted the autoloader to accept a signal only from the second outermost conduit ring and pressing the "load" button initiated the carousel motor. The carousel would spin counter-clockwise until a sliding contact entered the braking sector of the second ring [**Fig. 119**, ПК5], causing the carousel to slow down, and then align itself with the trapdoor and stop after the contact enters the stopping sector

---

rate. See: Королева, А. А., Кучерова, В. Г. (eds.) (2002). *Физические Основы Устройства и Функционирования Стрелково-Пушечного, Артиллерийского и Ракетного Оружия - Часть 1*. Волгоград: Волгоградский Государственный Технический Университет. p. 238.

[620] Звонков, Ю. В. (1978). 'О сокращении времени заряжания танковой пушки', *Вестник Бронетанковой Техники* 1978, сборник 3. p. 11.

[621] Balla, J. et al. (2017). 'Inserting cartridges using electrically powered ramming devices', *2017 International Conference on Military Technologies (ICMT)*. Brno: University of Defence.

[**Fig. 119**, ПК4]. The largest sector of each ring [**Fig. 119**, ПК3, 6, 9, 12] was wired to the ammunition counter, which was a milliammeter. With a constant voltage passing through the sector, the current arriving at the counter depended on the number of sliding contacts resting on the sector, each contact adding slightly more resistance to the circuit.

Pushing one of the three recorder buttons on the memory unit mechanically switched the position of a spoke when replenishing a cassette, and the system automatically reset a spoke to Empty once a loading cycle concluded. In this way, the memory unit not only stored information without relying on magnetic or electrical data storage elements but was also an integral part of the autoloader's control system.

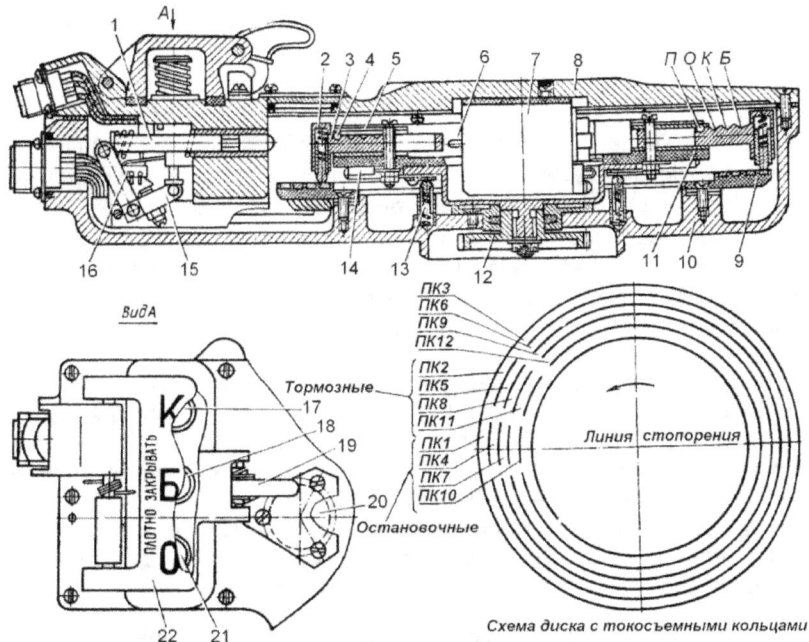

Figure 119: AZ-172 memory unit. (*Soviet Ministry of Defense*)

The AZ-184 autoloader included a new memory unit with an increased storage capacity to record ATGMs as the fourth ammunition type. The data disc was expanded to five conduit rings instead of four, and a new circular input mechanism with a rotary dial was used instead of a strip of three buttons. To reduce wear from dust abrasion, the new memory unit sealed its ammunition selector knob under a screw-on cap.

To optimize the loading time for frequent changes in ammunition type, an alternating load order was adopted as standard practice. Taking into account the standard authorized ratio of 40% KE, 40% HE-Frag and 20% HEAT[622], a typical loadout for the 22-round carousel would contain four of the following sets of five rounds, plus another KE and another HE to fill out the last two cassettes:

KE – HE– KE – HE – HEAT

Special interlocks were built into the relay logic for safety and for self-unjamming operations. The safety interlocks included the need for the gun breech to be open before the loading cycle could start. A resistance interlock in the rammer stopped the chain rammer if an unusually high resistance stopped the chain from reaching full extension. After 0.3 seconds, the chain was retracted, and then rammed forward again. This contingency was included in case of a misfeed from misalignment or abnormally high friction on the sliding parts.

The strictly sequential nature of the control logic enabled the commander to interrupt the loading cycle to set the fuze on an HE-Frag round when alternative fuze settings were needed for special targets. By turning off the chain rammer on the right circuit breaker, the autoloader would run as normal until it reached the ramming step, where it would freeze. The commander could safely set the fuze on the projectile. Once the commander was done, he turned on the rammer again, and the loading cycle continued as normal.

Control over the autoloader followed the "Rule of Two" safety principle, in that deliberate action was required from both from the commander and gunner for the autoloader to function. While preparing for combat, the commander was responsible for unblocking the gun's firing circuit and turning on the autoloader by switching on the circuit breakers for its constituent parts. The gunner was responsible for switching on the stabilizer, without which the autoloader could not function, and he started a loading cycle when explicitly assigned a target or ordered by the commander to engage a set of targets.

Recoil guards, present in all tanks, prevent injury to the gunner and commander as the gun recoiled after a shot, but guards were doubly important for stabilized guns on uneven terrain because from the perspective of the crew, the gun was constantly moving up and down at a rate that was controlled not by the gunner or commander, but by the driver, who had no way of noticing or reacting to accidents in the fighting compartment. The autoloader elevator and

---

[622] Звонков, Ю. В., Ралдугин, И. В. (1980). 'Влияние раскладки выстрелов на цикл автомата заряжания', Вестник Бронетанковой Техники 1980, сборник 3. pp. 10-11.
For the T-72 AZ autoloader, this meant 9x KE, 9x HE-Frag and 4x HEAT.

ramming mechanisms occupied the space enclosed by the recoil guards, so the safety considerations for working with the T-72 autoloader did not differ from the existing norms for basic tank operation.

In combat, placing any body parts beyond the recoil guards was strictly prohibited. The D-81 recoiled at a velocity of 11.0-12.5 m/s[623], and was quite capable of breaking anything in its path. During the loading cycle, the raising of the stub catcher frame also served as a sort of barrier blocking the gunner and commander from reaching into the ramming path. Even so, probably the most unusual detail associated with the AZ autoloader is the inexplicable spread of a rumor in the U.S. Army about the autoloader having a tendency to amputate the limbs of the crew. This became fairly common knowledge in USAREUR (United States Army Europe) Armor personnel[624] by the early 1980's, migrating into literature as early as 1985[625]. Without making undue claims to the safety of the T-72's autoloader, the rumor, at least, is apocryphal.

The ammunition stowage scheme, on the other hand, was fundamentally flawed. Apart from the vulnerability of the exposed powder charges to combat damage, they were also not weatherproofed in any of the stowage points in the T-72, and most of the charges were only indirectly sheltered from rain and snow coming in through the open hatches. The charges stowed on top of the carousel were protected by slip-on plastic sleeves. In many low-pressure gun systems, combustible charges suffer from grossly incomplete combustion when saturated with moisture. This changed the ballistics of the shot, and, more importantly, left incandescent solid particles in the chamber[626], requiring special measures to clean out the chamber after each shot. Moisture uptake issues of this type were largely mitigated by the high service pressures of the D-81.

---

[623] Жартовский, Г. С., Куртц, Д. В., Усов, О. А. (2016). *Защита оборудования и экипажа военных гусеничных машин от механоакустических и климатических воздействий: Монография.* Санкт-Петербург: Издательство «Лань». p. 12.
This range of velocities was not unusual for tank guns. The L11A5 recoiled at 10.2 m/s when firing APDS. See: Блинов, В. П., Личковах, В. А., Николахин, В. М. (1983). 'Испытания Танковой Пушки', *Вопросы оборонной техники*, 5(111). p. 23.
[624] Anecdotal accounts from former Armor crewmen, compiled by the author.
[625] *Key Weapons: Modern Soviet MBTs* in Brown, A. (ed.) (1985). *War in peace: the Marshall Cavendish illustrated encyclopedia of postwar conflict.* Volume 6. p. 165.
[626] Шишковского, В. М. (ed.) (1978). *Огневая Подготовка, Часть Вторая: Основы Устройства Вооружения.* Москва: Воениздат. p. 146. Also see: U.S. Department of the Army (1979). *Engineering Design Handbook: Breech Mechanism Design.* DARCOM-P 706-253. Alexandria: U.S. Army Materiel Development and Readiness Command. p. 2-14.

Figure 120: Autoloader elevator and rigid chain rammer. (S. Stauber, 2024)

The risk-reward balance for storing ammunition inside fuel tanks essentially boils down to this: given that a certain amount of space is needed to store both fuel and ammunition in the crew compartment, and the packing efficiency of these stores is inherently limited by the gaps between the cartridge parts, filling in the gaps between cartridge parts with fuel provisions the tank with the required volume of fuel without increasing the risk of an ammunition fire.

In certain cases, the fuel barrier between the cartridge parts and the walls of the tank may absorb enough energy from low-energy spall or light fragments to prevent the ignition of powder charges[627]. Even if a high-energy fragment passed through the ullage of a partly depleted tank, diesel vapors stubbornly refuse to ignite or detonate[628], and the fuel tank is no more dangerous than a simple ammunition rack. A direct shaped charge jet impact to the fuel usually results in the bursting of the tank followed by the eruption of a fireball, but in this case, the fuel fire is inconsequential because the jet also starts an ammunition fire.

Taking into account all of the likely outcomes with conventional ammunition racks and these fuel-ammunition containers, the use of ammunition-fuel containers gave a net benefit to survivability for a tank with open ammunition stowage while simultaneously increasing fuel capacity. It is only necessary to keep in mind that the safety of this design solution, or at least the absence of increased risk, is derived entirely from a litany of mitigating factors. It cannot be compared to dedicated armored bins, and the configuration of the T-72 with its open fuel tanks and open ammunition stowage was itself anachronistic even by the 1960's.

---

[627] Антоновский, В. П. et al. (1981). 'Взрывоопасность топливных баков и боекомплекта танков', *Вестник Бронетанковой Техники* 1981, сборник 1. pp. 19-20.
[628] Ibid. p. 20.

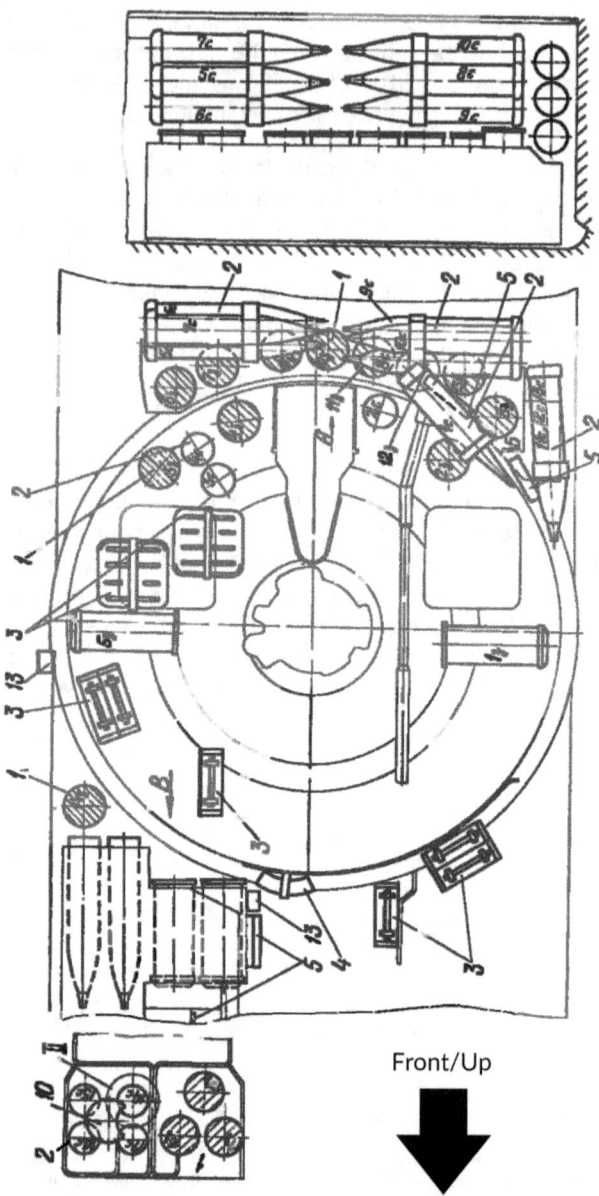

Figure 121: T-72 Ural ammunition stowage diagram (39 rounds); 1 – charges; 2 – projectiles; 3 – coaxial machine gun ammunition boxes; 4 – AK rifle magazines; 5 – F-1 hand grenades. (*Soviet Ministry of Defense*)

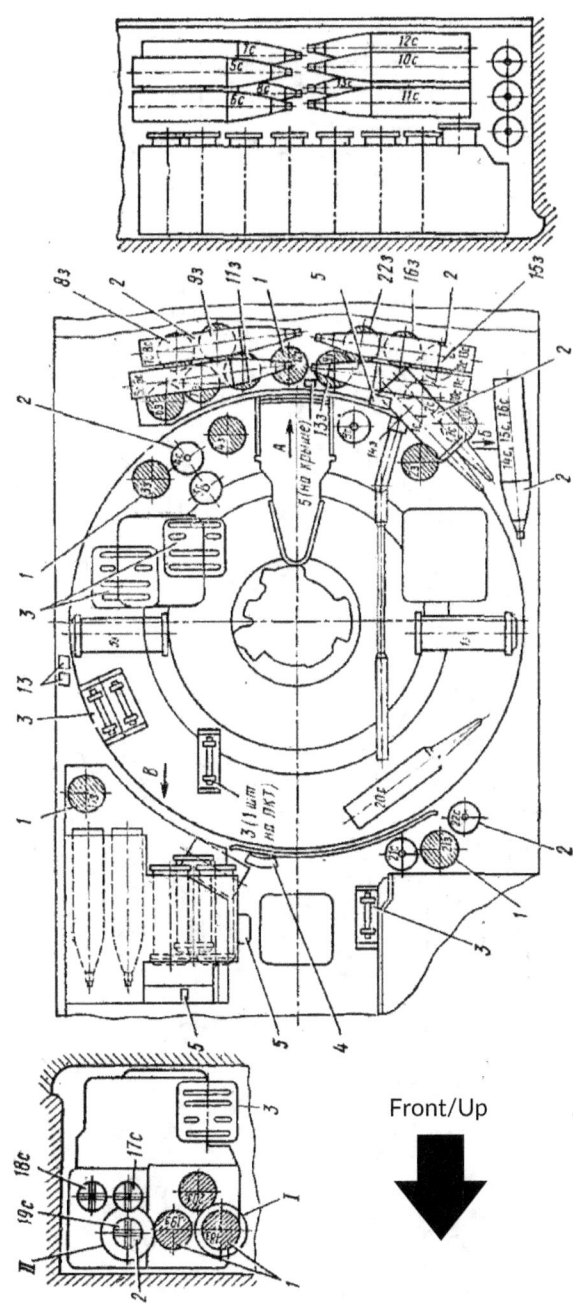

Figure 122: T-72A ammunition stowage diagram (43 rounds); 1 – charges; 2 – projectiles; 3 – coaxial machine gun ammunition boxes; 4 – AK rifle magazines; 5 – F-1 hand grenades. (Soviet Ministry of Defense)

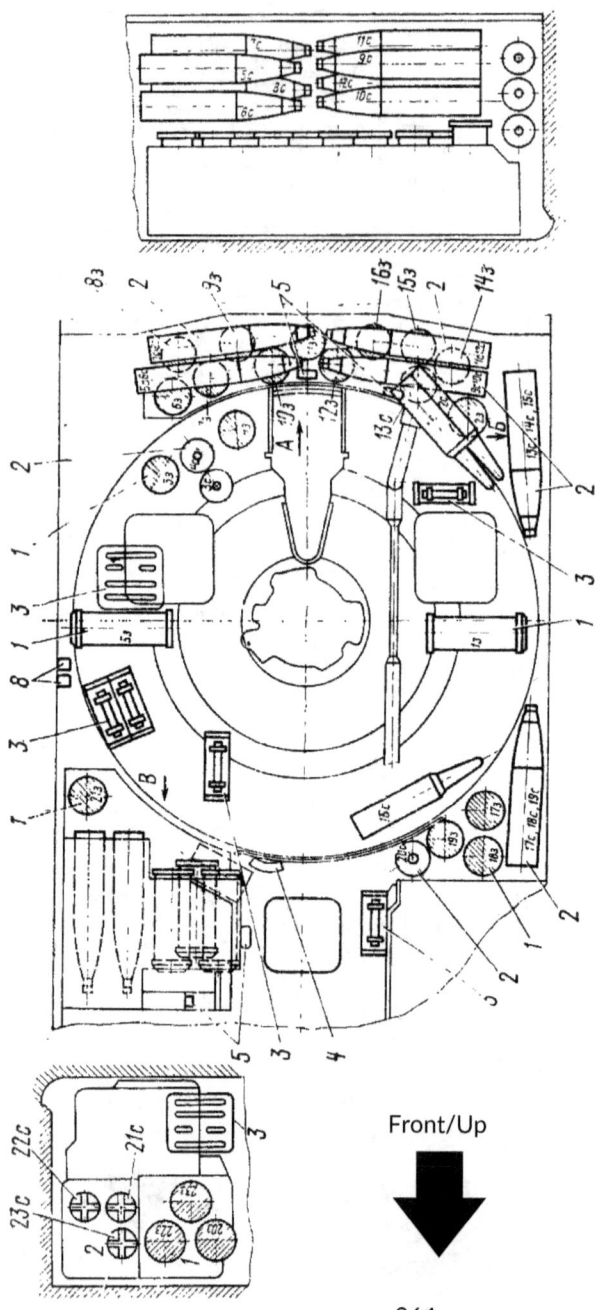

Figure 123: T-72B ammunition stowage diagram (44 rounds); 1 – charges; 2 – projectiles; 3 – coaxial machine gun ammunition boxes; 4 – AK rifle magazines; 5 – F-1 hand grenades. (*Soviet Ministry of Defense*)

D-81 Main Gun

The D-81 was a family of high-pressure, high-velocity 125 mm smoothbore tank guns created in the early 1960's. The task of designing what would become the D-81 was entrusted to the OKB-9[629] design bureau[630,631] in Sverdlovsk, headed by legendary artillery designer Fyodor Fedorovich Petrov. Artillery Plant No. 9, the factory affiliated to OKB-9, produced every model of the D-81 family for the army's needs, as did the Perm Machine-building Plant in the Urals. Each member of the D-81 family had its own internal factory index as well as a GRAU index. The first member of the D-81 family to be accepted into service was the D-81T[632] (GRAU: 2A26), fitted to the T-64A. The T-72 was equipped with the D-81TM (2A26M2), a variant adapted to the AZ autoloader. It was followed by the D-81K (2A46) in 1975, and then the D-81-3 (2A46M) in 1981[633].

The D-81 was the largest caliber high-velocity tank gun in service during the Cold War[634]. Wartime and postwar Soviet tanks typically carried a relatively large caliber gun for their size and weight, but even so, the D-81 stood out for its enormous power relative to the 41-tonne T-72. The D-81 remained competitive through the end of the Cold War through its various modifications and its growing repertoire of ammunition. The muzzle energy attained by standard service ammunition reached 10.1-10.4 MJ[635], surpassing even the 120 mm guns of the Leopard 2 and M1A1 Abrams[636] for most of the Cold War.

The single most important factor in the D-81's power was its large caliber. A large caliber is normally associated with the ability to deliver heavier high explosive shells, but the D-81 was primarily oriented towards anti-tank work[637], which reflected a widely-held belief among the major military powers that the most dangerous threat to a tank was almost invariably another tank. Modern tanks were mobile, difficult to suppress, difficult to hit, and in turn, they were

---

[629] Опытно-Конструкторское Бюро; Experimental Design Bureau.
[630] Устьянцев, С. В., Колмаков, Д. Г. (2013). *Т-72/Т-90: Опыт создания отечественных основных боевых танков*. Нижний Тагил: ОАО «Научно-производственная корпорация «Уралвагонзавод» имени Ф. Э. Дзержинского». p. 66.
[631] Лукьянов, Н. А., Лукьянов, В. Н., Близгарев, В. П. *История Создания и Совершенствования Танковой Пушки Д-81*. Available at: http://btvt.info/3attackdefensemobility/d81history.htm (Accessed: 25 November 2023).
[632] Карпенко, А. (2002). *Ракетные танки*. Библиотека журнала «Техника молодежи» №1, "Броня". Уссурийск: ООО "Восточный горизонт". Corroborated by an open-source compilation of known GRAU indices.
[633] Устьянцев, С. В., Колмаков, Д. (2004). *Боевая Машины Уралвагонзавода: Танк Т-72*. Нижний Тагил: «Медиа-Принт». p. 96.
[634] The D-81 did not possess the largest caliber; low-pressure 152 mm tank guns served in the American M60A2 and M551 tanks.
[635] Calculated with the 3BM22 APFSDS round.
[636] When compared by basic Cold War era KE ammunition - 9.8 MJ for DM23, 9.9 MJ for DM33. The American M829 round reached 10 MJ.
[637] Ефремов, А. С. (2010). *Уроки танкостроения*. Санкт-Петербург: «Гангут». p. 148.

the most potent direct-fire artillery on the battlefield. The prevalence of high-velocity guns as the weapon of choice for tanks across the world signified the priority of the anti-tank role[638], even overriding other factors like the historically higher prevalence of soft targets, light armor, or field fortifications in virtually all modern conflicts.

The impetus behind the birth of the D-81 was the appearance of the American M60 medium tank. Before the first M60 medium tanks even arrived in Europe in December 1960, Soviet analysts had already concluded that it decisively outmatched the T-55 in a tank duel[639]. The 105 mm M68 gun of the M60 was capable of knocking out a T-55 from beyond 1.5 km, and return fire from the T-55's 100 mm gun was largely impotent against the M60. In response, the Soviet Army put the T-62 into service in August 1961 as a stopgap solution.

The T-62 and its 115 mm U-5TS *Molot* smoothbore gun had been under development since late 1958, and by pure serendipity, it was powerful enough – and its APFSDS[640] ammunition potent enough – to counter the M60's thick armor. In the recommendation for service penned jointly by the Minister of Defense of the USSR and the Chairman of the GKOT[641], it was stated in no uncertain terms that the 115 mm *Molot* gave the T-62 the upper hand against the M60[642].

However, behind closed doors, industry specialists had already identified an urgent need to replace the *Molot* with something yet more powerful. In July 1961, Soviet analysts concluded that *Molot* did not give the T-62 a clear advantage in a tank duel. The 115 mm gun and its APFSDS rounds were ballistically superior to the 105 mm APDS ammunition fired by the M60 and possessed greater penetration power[643], but Soviet intelligence had determined that the M60 hull upper glacis was 120 mm thick and sloped at 65°[644]. For the existing 115 mm APFSDS ammunition to pierce this much armor, the tank had to be no more than 800-900 m away. To defeat the turret, the tank had to be no more than 2,000-2,800 m away, depending on the choice of APFSDS round.

If the T-62 had been better armored it would have eked out a slight advantage, but the GKOT noted that the T-62 could be defeated by return fire

---

[638] International experimentation on gun-launched anti-tank missiles as a tank gun option during the 1960's was also a major signifier of this belief.
[639] Устьянцев, С., Колмаков, Д. (2007). Боевые Машины Уралвагонзавод: Танки 1960-Х. Нижний Тагил: «Медиа-Принт». pp. 53-54.
[640] Armor-piercing fin-stabilized discarding sabot.
[641] State Committee of the Council of Ministers of the USSR on Defense Technology
[642] Устьянцев, С., Колмаков, Д. (2007). p. 69.
[643] U.S. Army 1973 Development Concept Paper, in US Department of Defense (1976). *Critical Considerations in the Acquisition of a New Main Battle Tank*. Report, PSAD-76-113A. p. 7.
[644] Талу, К. А. (1963). *Конструкция и Расчет Танков*. Москва: Военная академия бронетанковых войск. p. 8. Also reported in: *Вооруженные Силы Империалистических Государств* (1964).

from the M60 at the same distances[645]. Not content with parity, requirements were drawn up in the very same year for a new, more powerful gun capable of perforating 150 mm RHA at 60° from 2,000 m. To put this into perspective – at this range, the U-5TS was rated to perforate only 110 mm RHA[646]. The ammunition for the new gun would be essentially identical to the existing pattern of APFSDS ammunition, merely scaled to a larger caliber to raise its muzzle velocity. This approach led to the extraordinary 1,800 m/s muzzle velocity achieved by the basic 3VBM3 APFSDS round for the D-81.

The 3VBM3 round contained the 3BM9 projectile, which was a fairly rudimentary steel rod penetrator fletched with large steel fins. Considering that the muzzle velocity of the 3BM9 round was nearly 200 m/s higher than its direct 115 mm counterpart, and that air resistance slowed this type of projectile by 130-140 m/s for each kilometer traveled, a tank up-gunned from the U-5TS to the D-81 essentially had its effective range extended by 1,400-1,500 m. The upper glacis of the M60 would thus be defeated from 2,200-2,400 m instead of 800-900 m. Calculations by Soviet analysts in 1964 indicated that the front turret armor of the M60, M48A2, Chieftain and Leopard would be reliably penetrated at distances in excess of 3,000 m, and the upper glacis of the M60 and Chieftain would be penetrated at 2,300 m[647].

Work officially commenced at OKB-9 on August 11, 1962[648] and progressed at pace. From 1962 to 1964, working prototypes of the D-81 and its ammunition were created, and already by 1964, mounted live fire tests were being carried out on a modified T-64. In the interim, the D-68, essentially a U-5TS adapted for two-part ammunition[649] and automatic loading, was created for the T-64 (Object 432) as a stopgap.

The D-81 finally entered service in 1967 as the 2A26 for the T-64A. It was followed shortly by the 2A26M model with improved firing precision in 1969. The 2A26M2 variant entered service for the T-72 In 1973. It differed from the 2A26M only in its stub case catcher and the shape of its integral recoil guards.

---

[645] Устьянцев, С., Колмаков, Д. (2007). *Боевые Машины Уралвагонзавода: Танки 1960-Х*. Нижний Тагил: «Медиа-Принт». p. 70.
[646] Original requirement for the 115 mm gun was to perforate 100 mm RHA at 60° from 2,000 m. The gun achieved this at 2,360-2,390 m. The rated figure was hence amended to 110 mm RHA at 60° from 2,000 m
[647] Устьянцев, С., Колмаков, Д. (2007). p. 134.
[648] Павлов М. В., Павлов, И. В. (2021). *Отечественные Бронированные Машины 1945-1965 гг. - Часть I: Легкие, средние и тяжелые танки*. Кемерово: ООО "Принт". p. 626.
[649] In the technical-tactical requirements for the Object 432 issued in 1961, the tank was intended to be armed with a "U-5TS ("Molot") with a separate-loading automatic loader". See: Устьянцев, С. В., Чернышева, Е. Ю. (2020). *100 лет российского танкостроения*. Екатеринбург: Издательство ООО Универсальная Типография «Альфа Принт». p. 224.

The 2A46 was introduced in 1975 to address issues with the instability of the 2A26 recoil system after intensive fire. At the same time, the production tolerances for the barrel were tightened, again to the benefit of shot dispersion.

Towards the end of the 1970's, a major redesign of the D-81 took place to improve shot dispersion and meet new developments in NATO tank armor, resulting in the 2A46M model. The 2A46M preserved the barrel profile and chamber dimensions of its predecessors to ensure backwards compatibility in ammunition, but the strength profile of the barrel was improved to support new, high-energy subcaliber rounds[650]. It was developed jointly with new ammunition intended for fighting the new generation of NATO main battle tanks appearing in the 1980's.

Compared to earlier Soviet guns, the intrinsic shot dispersion had improved, not only in terms of ammunition dispersion, but also the technical dispersion of the gun system itself. The D-81 had more stringent standards for shot accuracy and precision than the U-5TS/D-68 and the D10-T, with significantly tighter evaluation criteria for rating shot grouping and the point of impact shift[651]. The effective range of the 2A46, defined as the range at which a 0.55 hit probability was achieved[652], was 1,800 m with standard APFSDS ammunition on a standard No. 12 tank-shaped target. This target did not correspond exactly to any NATO tank, but by using special targets designed to represent specific tank models (KST-69, KST-84)[653], known hit probabilities could be extrapolated to different tanks.

The D-81 gained several accuracy-improving accessories over the course of its service history. In 1975, a thermal sleeve was added to the 2A26M2 and 2A46 to reduce barrel bend from environmental factors by insulating and moderating temperature differences. When the 2A46M entered service, it brought with it a muzzle reference notch for the UVKV optical muzzle referencing system. This system did not affect the intrinsic mechanical precision of the gun, but it allowed the gunner to quickly boresight the gun in field conditions without leaving the tank.

---

[650] Conceptually, the 2A46M was directly analogous to the British 120 mm L30 gun in its relationship to its predecessor, the L11 gun.
[651] Потемкин, Э. К. (ed.) (1992). *Основы научной организации разработки.* Том 2. Москва: Издательство МГТУ им. Н. Э. Баумана. pp. 168, 217.
[652] Старовойтов, В. С. (ed.) (1990). *Военные гусеничные машины.* Том 1: Устройство, Книга 1. Москва: Издательство МГТУ им. Н. Э. Баумана. p. 244.
[653] KST-69: M60A1 and variants; KST-84: M1 Abrams and Leopard 2.

Table 24: D-81 primary data[a, b, c, d, e, f, g, i]

Data	2A26M2, 2A46	2A46M
Complete length (mm)	6,350	6,383
Barrel length (mm)	6,000	
Barrel rigidity (kg/cm)	335	385
Chamber length (incl. 40 mm forcing cone) (mm)	840	
Chamber volume (l)	13.1	
Loading density of basic round (KE) (kg/m^3)	510 (780)	
Maximum chamber pressure (MPa)	678.62	N/D
Average service pressure (MPa)	444.5	555
Peak average service pressure (MPa)	510	638
Barrel mass (with bore evacuator) (kg)	N/D	1,156
Gun mass (kg)	1,850	1,879
Oscillating mass (kg)	2,350*	2,443
Oscillating mass with gun mask (kg)	N/D	2,713
Normal recoil travel (mm)	270-320	260-300
Maximum recoil travel (mm)	340	310
Maximum recoil force (kN)	677	N/D
Breech ring dimensions (L x W x H) (mm)	630 x 505 x 620	

a   Министерство Обороны СССР (1974). *125-мм Танковые Пушки 2А26 (Д-81) и 2А26М2 - Техническое Описание и Инструкция по Эксплуатации. (Части 1, 2, 3, 4)*. Москва: Воениздат. p. 161.
b   Министерство Обороны СССР (1979). *125-мм Танковые пушки 2А46 и 2А46-1 - Техническое Описание и Инструкция по Эксплуатации. (Части 1 и 2)*. Москва: Воениздат.
c   Министерство Обороны СССР (1983). *125-мм Танковые Пушки 2А46М и 2А46М-1 - Техническое Описание и Инструкция по Эксплуатации*. Москва: Воениздат. pp. 4-5.
d   Розоринов, Г. Н., Хаскин, В. Ю., Лазаренко, С. В. (2013). 'Применение Лазерных Технологий Для Повышения Срока Службы Изделия КБА-3', *Збірник наукових праць*

e   СНУЯЕтаП. p. 146.
e   Устьянцев. С. В., Колмаков. Д. Г. (2013). *Т-72/Т-90. Опыт создания отечественных основных боевых танков*. Нижний Тагил: ОАО «НПК «Уралвагонзавод» имени Ф. Э. Дзержинского». p. 34.
f   ОАО НПК «Уралвагонзавод» (2021). *Танковые Пушки 2А46М, 2А46М-1, 2А46М-4, 2А46М-5*. Booklet.
g   Валеев, Г. Г., Сопин, В. Ф., Соков, Б. А. (2004). *Артиллерийские метательные заряды*. Казань: ФГУП «Государственный научно-сследовательский институт химических продуктов». p. 68.
h   Старовойтов, В. С. (ed.) (1990). *Военные гусеничные машины*. Том 1: Устройство, Книга 1. Москва: Издательство МГТУ им. Н. Э. Баумана. p. 147.
i   Исаков, П. П. (ed.) (1982). *Теория и Конструкция Танка - Том 2: Основы Проектирования Вооружения Танка*. Москва: Машиностроение. p. 54.
*   With the gun mask and stabilizer, the gun weighed 2,836 kg. See: Министерство Обороны СССР (1976). *Танк «Урал» - Руководство по Войсковому Ремонту. Книга 1: Замена и Ремонт Агрегатов и Узлов. Часть 1*. Москва: Воениздат. p. 195.

Designing a Big Gun for a Small Turret

In most respects the D-81 essentially followed contemporary standards in gun design. All of its core structural features had been well established in the Soviet artillery industry by 1967, and where size was concerned, increasing the caliber of a gun was obviously not a progressive solution if enhancing the ammunition was an option. Nevertheless, mating a 125 mm high-power gun to a diminutive two-man turret was a significant engineering achievement, made possible by a number of structural features unique to design standards set by the OKB-9 design bureau.

The main design criterion in the compactness of a gun was its swept volume[654]. On average, the swept volume accounted for approximately 8% of the total internal volume of a tank[655]. A small swept volume reduced the turret height, allowed the gun to be loaded without a turret ring of excessive size, and facilitated a functional gun elevation range.

A narrow breech ring[656] was emphasized in the design of the D-81 to fit into the same space as the U-5TS. In spite of the larger caliber and drastically increased power of the gun[657], the breech ring was unusually slim, measuring only 505 mm wide[658], essentially equal to the U-5TS and also to the 120 mm

---

[654] Turret volume that a gun "sweeps" out across its entire range of elevation under full recoil. Alternatively, some organizations define swept volume by including the ammunition load length as well as the recoil length, whichever one is greater.
[655] Horton, W. D. (1996). *Ground Vehicle System Integration (GVSI) and Design Optimization Model*. Technical Report OMI-574. Ann Arbor: OptiMetrics, Inc. p. 34.
[656] A historical and widely accepted term for the large steel housing that joins the barrel to the breech mechanism. Modern breech rings abandoned the ring or circular shape for a square profile.
[657] Hilmes R. (2007). *Kampfpanzer heute und morgen: Konzepte - System - Technologien*. Auflage: Motorbuchverlag. p. 142.
[658] Направляющая задняя 2А46М.109-62. Technical drawing.

Rh120 L/44. Only smaller caliber guns were narrower, with examples like the 105 mm M68 measuring in at 436 mm. In the T-72, the effective width as measured across the stub ejector frame on all D-81 models was 600 mm[659], and the fixed recoil guards for the gunner and commander added a few additional centimeters of clearance. This was closely comparable to the 600 mm-wide recoil guards of the Rh120 and the M68[660].

Table 25: Breech ring and recoil guard width data for various tank guns

Tank Gun	D-81	U-5TS[b]	Rh120	M68[e]	L11[f]
Breech ring width (mm)	505	500	500[c]	436	483
Recoil guard width (mm)	≤640[a]	-	630[d]	380L	368L
				292R*	251R

a   Inclusive of both fixed recoil guards for the gunner and commander.
b   Павлов, И., Павлов, М. (2023). 'Т-62: «Рабочая лошадка» советских танковых дивизий первого эшелона', *Техника и Вооружение*, (July). p. 22.
c   Technical drawing.
d   Inclusive of fixed recoil guard separating the gunner and commander from the oscillating recoil guard affixed to the gun. See: Deutsche Panzermuseum Munster (2019). Dipl-Ing. Rolf Hilmes: Wie konstruiert man einen Panzer?. [Online video]. Available from: https://youtu.be/_J1dHNtyOLI (Accessed: 21 August 2023).
e   Measured.
*   Spaulding, S., Weintraub, A., Cioch, F., Lenz, J. (1979). Parametric Engineering System Definition Model. Volume I. Main Report. Appendices A and B. Technical Report TARADCOM No. 12444. Warren: TARAD- COM. p. 99.

The inboard length of the D-81, as measured from the trunnion axis to the rear face of its breech ring[661], was just 1,240 mm (2A46M: 1,273 mm)[662]. This was only marginally longer than the U-5TS, and markedly shorter than equivalent guns like the Rh120 L/44 and the L11. When studying the possibility of installing the D-81 in a T-62 turret, its size was considered to be such a close equivalent to the U-5TS that no structural changes were needed[663], and only small modifications were needed to install it in the somewhat tighter turret of the T-64 (Object 432)[664].

---

[659] *Танк Т-72С: Каталог Деталей и Сборочных Единиц, 172М.КД-2. Книга 2: Альбом Иллюстраций.* p. 47.
[660] The M68 recoil guard was asymmetric. Its left recoil guard (loader's side) enveloped its breech opening handle and was much wider than the right recoil guard, which held a set of ballast plates to compensate for the small and light breech ring.
[661] The travel lock eye is omitted from the inboard length because it was situated low enough to be inconsequential.
[662] Based on known length of 5.11 m from muzzle to trunnion, and known total gun lengths.
[663] Колмаков, Д. Г., Устьянцев, С. В. (2017). *УКБТМ: 75 лет тагильской школе танкостроения.* Екатеринбург: Издательство ООО Универсальная. p. 68.
[664] Павлов М. В., Павлов, И. В. (2021). *Отечественные Бронированные Машины 1945-1965 гг.* - Часть I: Легкие, средние и тяжелые танки. Кемерово: ООО "Принт". p. 43.

Table 26: Inboard length data for various tank guns[a, b, c]

Tank Gun	Caliber (mm)	Inboard Length (mm)	Recoil Travel at Hard Stop (mm)	Swept Length (mm)
D-81	125	1,240/1,273*	340/310	1,560
U-5TS	115	1,222	430	1,652
M68	105	1,283/1,359**	343	1,626/1702
Rh120	120	1,375	370	1,745
L11	120	1,560	370	1,930

a   Блинов, В. П., Личковах, В. А., Николахин, В. М. (1983). 'Испытания Танковой Пушки', *Вопросы оборонной техники*, 5(111). p. 22.
b   Rheinmetall (1982). *Smooth Bore Technology 120 mm: Weapon System and Ammunition*. Düsseldorf: Rheinmetall GmbH.
c   Spaulding, S., Weintraub, A., Cioch, F., Lenz, J. (1979). *Parametric Engineering System Definition Model*. Volume I. Technical Report TARADCOM No. 12444. Warren: TARADCOM. p. 127.
*   The 2A46M breech face retained standard D-81 dimensions underneath the breech opening, to not interfere with the autoloader.
**   M68 guns in the M60A1 and M1 Abrams turrets had different inboard lengths.

The relationship between turret ring diameter and the power of a tank gun was primarily determined by inboard length, inclusive of the clearance behind the gun available to load a round[665]. In a manually-loaded tank using unitary cartridges, there was much less flexibility in how the hull and turret could be laid out, and a more compact turret with a smaller turret ring was forfeited. The T-62, for instance, had an enormous turret ring diameter of 2,245 mm for the sake of its loader, while the 1,934 mm diameter of the T-72's turret ring was entirely compatible with the much more powerful D-81.

In the T-72, the trunnion axis was 1,030 mm from the geometric center of the turret ring. With the D-81's inboard length of 1,240 mm, there was 752 mm of clearance between the breech ring and the turret ring. This was just enough for the two-part ammunition to rise from below the turret ring up to the loading position behind the breech, but even this was not particularly tight by the standards of manually loaded tanks; in a Leopard 2, the clearance was 615 mm.

In total, the D-81 occupied a total swept volume of 1.1 m³. This was actually smaller than the 1.13 m³ swept volume of the U-5TS in the T-62[666], though the

---

[665] Старовойтов, В. С. (ed.) (1990). *Военные гусеничные машины*. Том 1: Устройство, Книга 1. Москва: Издательство МГТУ им. Н. Э. Баумана. p. 51.
[666] Исаков, П. П. (ed.) (1982). *Теория и Конструкция Танка - Том 2: Основы Проектирования Вооружения Танка*. Москва: Машиностроение. p. 16.

U-5TS has a slightly larger elevation range of +16° instead of +14°. The basic reason for the small swept volume of the D-81 was its heavy breech ring[667], equal to 62% of the barrel weight[668], aided in no small part by using the stub catcher behind the breech[669] and the gun stabilizer components mounted underneath the breech as counterweights[670]. These were the hydraulic pump for the gun elevation drive, the stabilizer gyroscopes, and a control box.

Using the stabilizer as a counterweight was a particularly elegant approach to fighting compartment layout design. The clearance between the gun and the turret floor essentially did not change as compared to fitting the stabilizer components to the floor, so the gun elevation limit was unaffected. It was already common practice to utilize the space underneath the gun for stowage or for electrical and hydraulic equipment[671], so this setup did not complicate maintenance as far as accessibility was concerned. This layout gave the D-81 a profile distinct to many postwar Soviet guns, where the barrel extended from a slender gun cradle ending in a very bulky, bottom-heavy breech assembly.

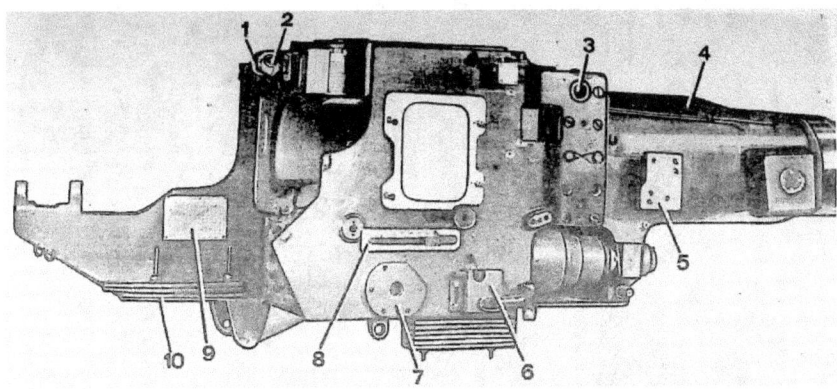

Figure 124: 2A26M2 right side view: 1 – travel lock; 2 – travel lock eye; 3 – socket for elevation stopper; 4 – gun cradle; 5 – coaxial machine gun bracket; 6 – safety blocker for percussion firing mechanism; 7 – access panel; 8 – recoil distance indicator; 9 – recoil system servicing nomogram; 10 – counterweight plate stack. (*Soviet Ministry of Defense*)

---

[667] The mathematical basis behind this concept was explored in: Zaroodny, S. J. (1972). *The heavy breech principle for tank guns*. Aberdeen: Ballistic Research Laboratories.
[668] This was slightly higher than the L11 (60%), slightly lower than the M256 (64.9%), and exactly equal to the Rh120 L/44.
[669] Королева, А. А., Кучерова, В. Г. (eds.) (2002). *Физические основы устройства и функциионирования стрелково-пушечного, артиллерийского и ракетного оружия - Часть 1*. Волгоград: Волгоградский Государственный Технический Университет. p. 307.
[670] A common Soviet practice beginning with the first postwar tank gun stabilizer project.
[671] In American and German tanks, the gun elevation piston was often installed here.

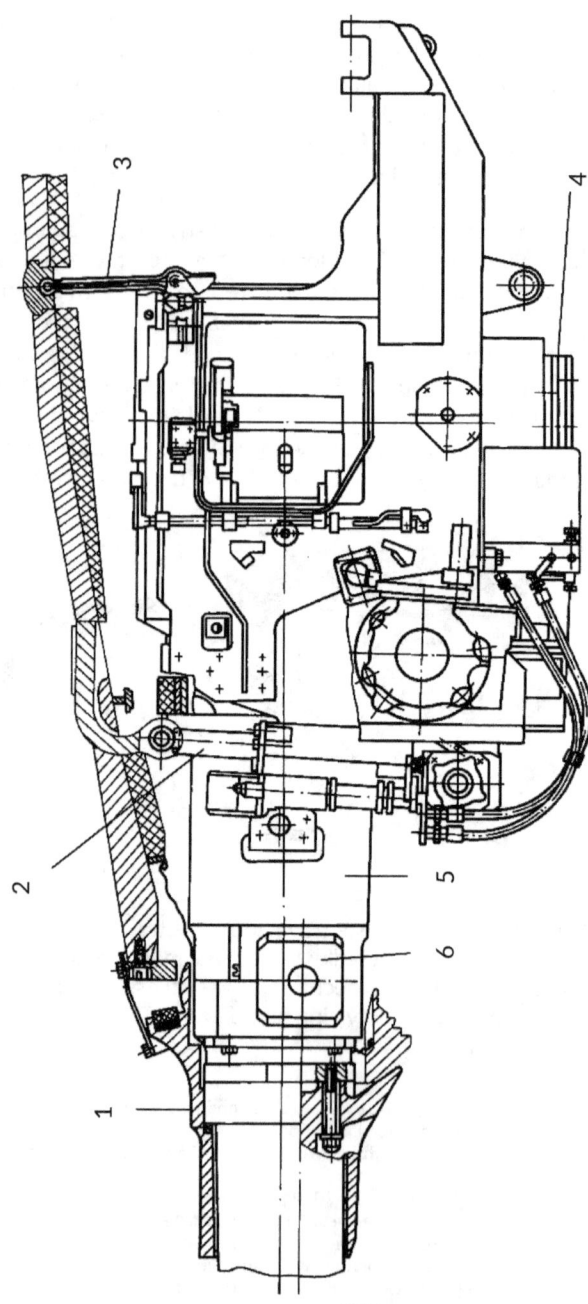

Figure 125: 2A46 gun in T-72 turret. Note the distinct profile of the gun, slender at the trunnion but bulky at the breech ring. 1 – gun mask; 2 – elevation piston; 3 – travel lock; 4 – ballast plates; 5 – gun cradle; 6 – trunnion. (*Soviet Ministry of Defense*)

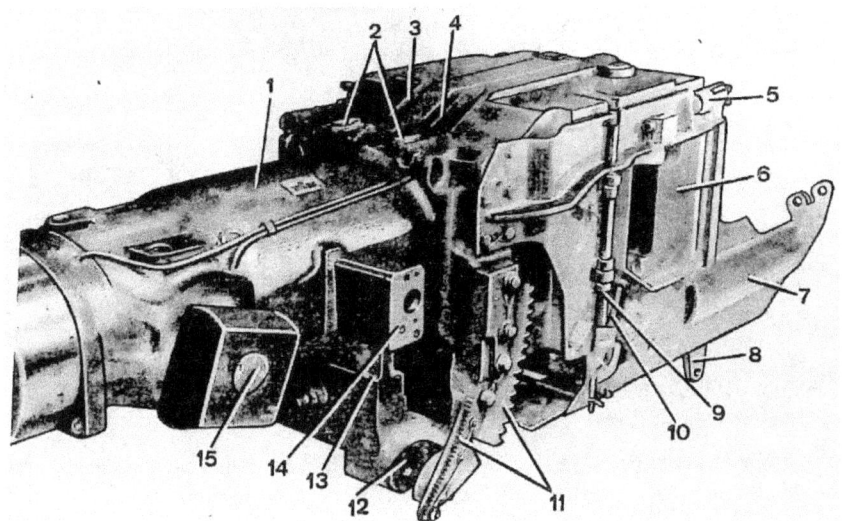

Figure 126: 2A26M2 left side view: 1 – gun cradle; 2 – shock absorbers for breech return; 3 – guide groove; 4 – guide rail; 5 – mount for gunner's quadrant; 6 – breech block in open position; 7 – left recoil guard; 8 – bracket for stub catcher frame; 9 – pin; 10 – manual firing trigger; 11 – sector gears for gun elevation (inward) and elevation angle sensor (outward); 12 – bracket for gun elevation piston; 13 – lug; 14 – bracket for day sight parallelogram linkage; 15 – trunnion. (*Soviet Ministry of Defense*)

The 2A26 and 2A46 had a conventional cage-type cradle support. A recoil guide[672] kept the gun directionally aligned during recoil. Owing to the short inboard length of the gun, the cradle was quite short, and consequently, the base support length was limited to just 890 mm. A stable support decreases the variability in the initial position of the gun after each shot, tightening the shot dispersion of a gun. Archaic models of field artillery used rail or trough-type supports[673], but the most effective type of support in terms of size, weight and rigidity[674] was the cage-type cradle, which can be found on virtually all examples of postwar gun artillery. In a cage-type cradle, the recoiling parts were supported entirely by resting the cylindrical base section of the barrel on a sleeve inside the cradle. The base section of the barrel would be unpainted or coated with a friction-reducing material, and the forward and rear ends of the

---

[672] A steel rail on the cradle, riding in a bronze trough on the top of the breech ring.
[673] Федотов, А. И. (1998). *Основы устройства артиллерийского вооружения*. Перм: Пермский государственный технический университет. pp. 102-106.
[674] A large-diameter hollow cylinder had an intrinsically high rigidity to weight ratio (specific modulus), especially when compared to bars or rails of equal length

sleeve would have panels made from a soft metal with a low sliding friction coefficient against steel, customarily brass or bronze.

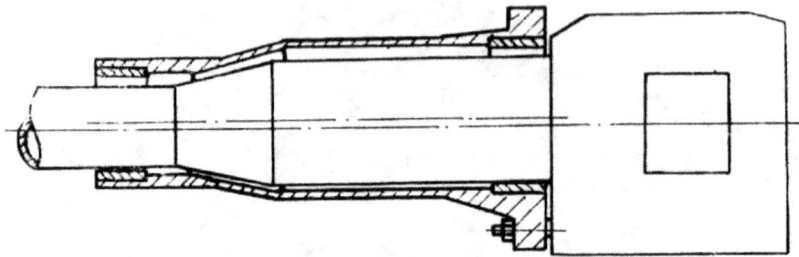

Figure 127: Diagrammatic representation of 2A26 and 2A46 cradle with its two supporting surfaces. (*Adapted from V. A. Bozhko et al., 1976*)

A longer base support length was necessary for high dynamic stability during recoil, reducing shot dispersion. Extending the cradle further forward of the trunnion was not an ideal solution because of the increased exposure to combat damage, raising the risk of jamming the gun [675]. Instead, a more creative approach had to be implemented in the 2A46M. By switching from the standard cage-type cradle to a hybrid cage-and-rail type cradle, the designers managed to extend the base support length to 1,300 mm [676]. The recoiling parts were supported by resting the barrel upon a short bronze-lined section at the forward end of the cradle, and the breech ring sat upon a pair of brass-plated support legs. These legs also functioned as recoil guide rails. In this way, the 2A46M gained a stable three-point support with a drastically longer base length [677].

Additional improvements to shot dispersion were achieved by changing the cradle sleeve to support the barrel on two points of contact instead of one, with flat bronze platform supports instead of a bronze ring. There were also two vertical preload screws [678] on the cradle, and a horizontal preload screw on the left support leg. These screws were elastic elements to restrict the gun from moving freely in its cradle during off-road driving.

---

[675] The radial gap between the gun and the cradle was small, generally averaging 0.5 mm, and was therefore rather sensitive to deformation.
[676] Лукьянов, Н. А., Лукьянов, В. Н., Близгарев, В. П. *История Создания и Совершенствования Танковой Пушки Д-81*. Available at: http://btvt.info/3attackdefensemobility/d81history.htm (Accessed: 25 November 2023).
[677] Approaching the 1,640 mm base support length of the Rh120 L/44.
[678] These were studs that controlled the vertical or horizontal vibration of the gun by spring compression from Belleville washers.

The recoil mechanism was placed under the breech ring (2A26, 2A46) or embedded inside it (2A46M), which gave the gun cradle a slender profile. The spaces on both sides of the gun cradle between the trunnions and the breech ring were very short, but space was still found for the gun elevation piston on the left and a bracket for the coaxial machine gun on the right. On most tank guns, the recoil system usually occupied the space between the trunnions and the breech ring, usually by putting one recoil brake on each side of the gun cradle[679].

On the 2A26M2 and 2A46, the recoil system had one large Schneider-type hydraulic brake and one hydropneumatic recuperator[680]. Originally introduced to Imperial Russia shortly before the First World War [681], the Schneider brake became widely used around the world for its inherent simplicity, ruggedness, and efficient recoil absorption [682]. Compared to the U-5TS and D-68, the maximum recoil impulse of the D-81 was only 26.1% higher, but to achieve a shorter recoil stroke, the peak braking resistance (peak recoil force) was nearly doubled from 343 kN to 677 kN, or 69 tonnes. This was very high for the T-72's 41-tonne weight, proportionately much higher than the Rh120's 61 tonnes (600 kN [683]) of peak recoil force for the Leopard 2's combat weight of 55 tonnes.

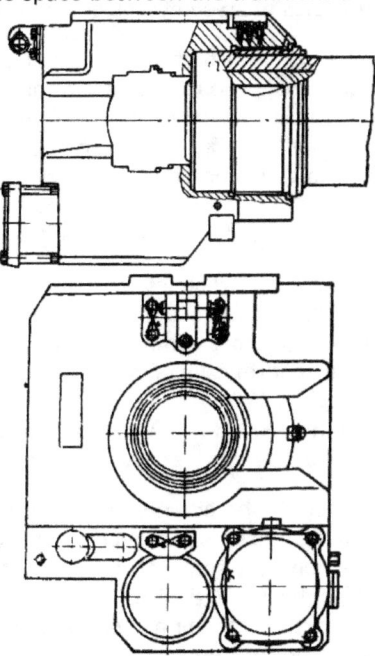

Figure 128: 2A26 and 2A46 recoil system layout. The brake is on the right. (*Soviet Ministry of Defense*)

The main shortcoming of the 2A26M and 2A46 recoil system layout was that the brake was asymmetric to the bore axis. As the gun recoiled, the hydraulic

---

[679] A cylindrical support upon which the barrel rests. The recoil system is anchored to the cradle, and the cradle is anchored to the turret via the trunnions.
[680] A device that stores recoil energy to return the gun to battery, while also providing some recoil absorption.
[681] Virtually all models of Russian and Soviet artillery featured this type of recoil system after its appearance on the Model 1910 Schneider 6-inch (152 mm) howitzer, designed by the French Schneider company under Russian commission.
[682] U.S. Army Ordnance Department (1921). *Theory and Design of Recoil Systems and Gun Carriages*. Washington, D.C.: War Department. pp. 529-534.
[683] Rheinmetall (1982). *Smooth Bore Technology 120 mm: Weapon System and Ammunition*. Düsseldorf: Rheinmetall GmbH.

brake exerted a braking force below the bore axis, without being balanced by an equal braking force on the opposite end. This created a large turning moment, tending to twist the gun upwards. This had a negative effect on shot dispersion, vertical dispersion in particular[684]. Asymmetry was very common in artillery recoil systems, but symmetric recoil systems had already begun taking over tank gun design by WW2, though not in the USSR. The 2A46M was one of the first examples of a Soviet tank gun with a symmetric recoil system.

The 2A46M featured a pair of small-diameter high-pressure brakes, which were small enough to be arranged symmetrically about the bore axis. The working cross-sectional area of the brake was reduced by 70%, raising the peak pressure by 27%, which necessitated improved seals and stronger steels[685]. A new small-diameter high-pressure pneumatic recuperator was also introduced. Peak recoil acceleration reached 300 g, imparting a maximum longitudinal acceleration of 6.5-7.0 g to the turret and 5.0-5.5 g to the hull[686].

The recoil dynamics were also completely revised in the 2A46M by reducing the braking force during the period of projectile travel (0.007 seconds) by around 10 times compared to the 2A46 and 2A26. It was not free recoil, which might theoretically eliminate any disturbances from recoil moments, but small enough (6,498 kg) to see a significant improvement shot dispersion. The 2A46M also introduced a few creature comforts, like new oil level indicators on the recoil buffers and recuperator for easier maintenance.

Since the breech ring no longer had a large brake below the breech ring, its center of mass was aligned much closer to the bore axis and the entire breech assembly lost some of its height, but also some of its weight. To add some weight back into the breech assembly to properly counterbalance the barrel, the face of the breech ring was extended by 33 mm, and in return, the recoil travel had to be shortened so that the inboard length remained unchanged.

Trunnion

Each trunnion block enclosed a set of conventional trunnion pins with needle bearings. The gun would be installed by dropping its trunnion blocks into special cavities cut into the cheek walls[687], and then a wedge would be bolted against the inclined upper surface of each trunnion. The wedges clamped the gun down, and the circular shape of the trunnion blocks kept the gun directionally aligned.

---

[684] Tvarozek, J., Gullerova, M. (2012). 'Increasing Firing Accuracy of 2A46 Tank Cannon Built-in T-72 MBT', *American International Journal of Contemporary Research*, 2(9).
[685] Голубев, В. А., Козлов, Э. П. (1985). 'Усовершенствование противооткатных устройств орудий для ВГМ', *Вестник Бронетанковой Техники* 1985, сборник 4. p. 10.
[686] Жартовский, Г. С., Куртц, Д. В., Усов, О. А. (2016). *Защита оборудования и экипажа военных гусеничных машин от механоакустических и климатических воздействий: Монография*. Санкт-Петербург: Издательство «Лань». p. 13.
[687] The cavities were approximately 170 mm tall.

The new trunnion design primarily benefited the gun installation and removal process. Like the guns of preceding serial Soviet medium tanks, the D-81 was intended to be installed and removed on rails. This method of bringing the gun in and out of the turret was first developed for the T-34 in early 1942 by the No. 112 *Krasnoe Sormovo* plant[688], and became standard for Soviet medium tanks in the postwar era. Unlike the T-34 or early T-54 models, however, the stabilizer had to be uninstalled first for the gun to be pulled out this way, which was a fairly serious oversight. This mounting system was established before the practice of mounting the gun stabilizer under the breech was established, and it simply never changed. After dismantling most of the turret internals, the turret, still with the gun in it, would be lifted and tipped forward on special brackets. The accessories attached to the gun had to be laboriously removed, including its manual elevation mechanism, and only then could the gun be moved in and out of the turret on rails resting on the hull roof.

The new embedded trunnions simplified installation by self-setting into their slots as the gun was pushed forward on rails. All that was needed afterwards was to insert and tighten the wedges[689]. This configuration placed the trunnions almost directly above the turret ring. This was a standard design solution for minimizing the clearance needed underneath the gun for a wide range of elevation. The weight of the stabilizer shifted the center of gravity of the D-81, so the gun was balanced by lowering the trunnion axis by 40 mm from the bore axis[690]. The static imbalance of the D-81, as installed in a tank with a loaded coaxial machine gun and a HE-Frag round loaded, was to be as close to zero as possible. The acceptable margin was no more than 3 kg-m towards the front[691].

A low bore axis was especially beneficial to the T-72 and other tanks with a similar ratio of recoil force to weight. Having a low center of gravity, the tank would tend to pitch more violently when firing a shot, because the recoil force had a longer moment arm to act upon the center of gravity of the tank. Given that the recoil force was tied to the gun, and a muzzle brake was considered unacceptable by the Army, lowering the bore axis was the only viable option.

---

[688] Коломиец М. (2009). *Т-34: Первая полная энциклопедия*. Москва: Эксмо. p. 202.
[689] Министерство Обороны СССР (1991). *Объект 184 - Руководство по Войсковому Ремонту, Книга первая. Замена и Ремонт Агрегатов и Узлов. Часть первая*. Москва: Воениздат. pp. 10-17.
[690] In older guns like the D-25T and D10-T, the recoil system was above the barrel, and the center of gravity was shifted upward accordingly.
[691] Жирнова, Т. А. (ed.) (2004). *Устройство, Эксплуатация, Техническое Обслуживание и Ремонт Стабилизатора Танкового Вооружения 2Э28М - Методические Указания*. Омск: Издательство ОмГТУ. p. 44.
Also see: Юрко, С. В., et al. (2016). *125-мм Танковая Пушки 2А46М: Пособие*. Минск: БИТУ. p. 103.

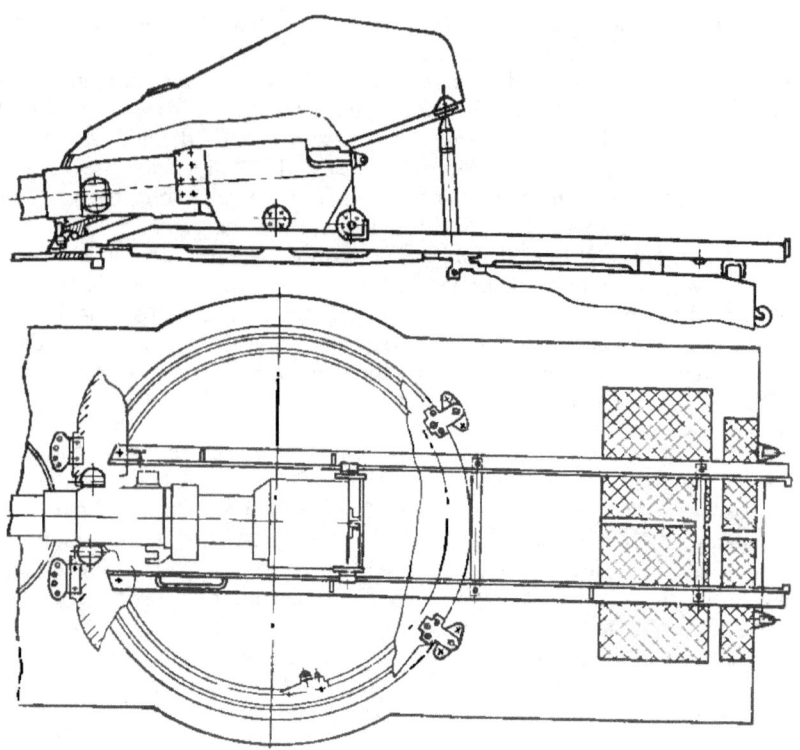

Figure 129: Gun removal procedure in T-72. (*Soviet Ministry of Defense*)

Earlier Soviet tank guns like the 122 mm D-25T and the 100 mm D10-T had hinge-type trunnions, where during installation, the gun would be aligned to a frame inside the turret, and a cross pin would be slotted through each trunnion to join the gun to the frame. The D-81 had an embedded wedge-type trunnion, inherited from the D-54. In fact, even the trunnion base width[692] was inherited from the D-54, which in turn was shared with the D10-T. Together with the U-5TS, all four guns had a trunnion base width of 410 mm[693].

The main shortcoming of the gun mount was that the round shape of the turret (in plan) required a heavy sacrifice to the thickness of the armor along the gun embrasure, particularly in front of the trunnions themselves. Thickening this part of the turret would have formed a needle-nosed shape (e.g. T-54, Chieftain), interfering with the clearance above the driver's hatch.

---

[692] The width between the ends of the trunnion pins.
[693] Исаков, П. П. (ed.) (1982). *Теория и Конструкция Танка - Том 2: Основы Проектирования Вооружения Танка*. Москва: Машиностроение. p. 54.

Nevertheless, in principle, the problem of protecting the trunnions of a tank gun can be likened to the familiar "all-or-nothing" compromise in armor design. In Soviet tank design orthodoxy, based on observations from the Great Patriotic War, the best practice to minimize the possibility of jamming the gun under shelling was to make the gun embrasure as small as possible, and the embrasure should have a fully contiguous perimeter for structural strength and rigidity[694].

Large gun mantlets with heavy armor could stop fairly powerful shells, but the momentum delivered into the mantlet must pass to the turret through the trunnions. Trunnion pin fracture or deformation jammed the gun from firing safely or elevating. Such mantlets were more likely to be struck on account of their size, and therefore suffered a higher probability of being jammed in combat. The probability of a hit on the T-72 narrow gun embrasure from a large caliber gun was small, but the probability of perforation was higher. Both types were protected from fragments and heavy machine gun fire.

Just as the "all-or-nothing" compromise posited that light armor with low exposure was a viable alternative to heavier but more exposed armor, a similar trade-off was applicable to gun embrasure design. The width of the T-72 gun mask was equal to the diameter of the barrel support on the gun cradle - just 390 mm, the same as the T-54 and T-62 gun masks despite the enormous difference in gun caliber.

This type of narrow gun embrasure was less likely to suffer a direct hit from a large caliber gun, and if a shot landed directly on the gun mask, even exceptionally heavy armor would have done little good as far as preventing the gun from jamming. Unfortunately, having such thin armor on each side of the embrasure meant that even outdated weapons could hit hard enough to disable the gun. Rather than affecting the trunnion pins through a mantlet or gun mask, striking the turret in front of the trunnions could bulge out the trunnion cavity wall, perhaps not jamming the gun outright, but extricating it would likely be impossible. Because the trunnion cavities were a part of the turret structure, repairs might not be economically feasible either.

Two large bolts secured the gun mask to a flange on the cradle. The 2A46M had a new gun mask secured by four bolts. The gun mask and gun embrasure itself were rated for non-jamming and non-perforation from 12.7 mm armor-piercing bullets only. The armor in front of the trunnion blocks ranged from 100-130 mm thick, but given that effective protection must exclude any deformation to the trunnions, the actual protection level is unknown. The gun embrasure zone was not fired upon during turret protection verification tests.

---

[694] Буров, С. С. (1973). *Конструкция и Расчет Танков*. Москва: Военная академия бронетанковых войск. pp. 78-80.

Breech Assembly

The D-81 breech ring[695] had a conventional square profile with a horizontally sliding wedge breech block. Outwardly, it had no radical differences from other tank guns, but its details deserve some interest. Most of its features were common to Soviet tank gun design practices, like the locking eye on the face of the breech ring for an internal travel lock[696], allowing the crew to transition from a march into combat without needing to exit the tank.

Despite the narrow width of the breech ring, the breech block did not protrude from its side in its open position, so the gun did not need a bulge on its recoil guard unlike most tank guns with a horizontally sliding wedge breech block (D10-T, L7, etc.). Additionally, the breech operating lever was integral to the breech ring instead of being a separate item stowed somewhere in the turret, saving space in the fighting compartment and also saving some time when performing corrective actions after a misfire.

The most interesting features of D-81 breech assembly, however, were its special adaptations for an autoloader. The structural adaptations notably came at the expense of manual loading. The breech block slid to the right to close, so the breech ring had a cutout on its right side for left-handed loading, giving the breech ring opening a C-shape like any other gun with a horizontally sliding breech block. However, when automatic loading was implemented, it turned out that the cutout introduced some risk of misfeeding when the tank was tilted to one side, so a spring-loaded guide tray was added opposite the breech wedge. This solved the problem, but the tray obstructed the commander if he were attempting to load the gun manually. In the end, the recoil guard was made to cover the entire right side of the breech ring, making the cutout entirely redundant[697].

When the entire breech end of the gun was redesigned on the 2A46M, the breech opening became oval-shaped to bring lateral symmetry to its weight distribution[698], but this of course made it much more dangerous to manually load the gun without a ramming baton, since there was absolutely no space for a loader's hand to move out of the way of the closing breech except to pull out as quickly as possible.

---

[695] A historical and widely accepted term for the large steel housing that joins the barrel to the breech mechanism. Modern breech rings abandoned the historical ring or circular shape for a square profile.
[696] Pinning the gun to the turret ceiling with a steel bar.
[697] Without the additional clearance afforded by a C-shaped breech opening, a projectile or charge had to be maneuvered more or less directly behind the opening to insert it.
[698] Зайцев, А.С. (2019). *Разработка конструкции ствола артиллерийского орудия: пособие по курсовому проектированию*. Санкт-Петербург: Балтийский государственный технический университет «Военмех». p. 25.

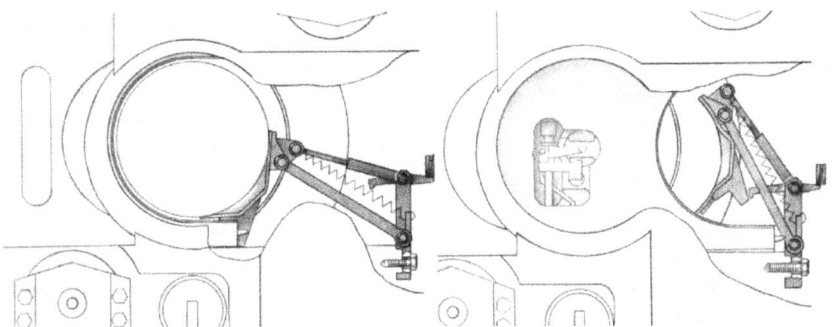

Figure 130: Loading guide tray in 2A46M, right side of breech ring cut away from drawing. (*Soviet Ministry of Defense*)

Closing the breech over a loaded round put a current-carrying firing pin in contact with the electric-percussion primer. When the firing circuit was powered, a current would be passed through the primer while simultaneously releasing the firing pin hammer. The primer would first be electrified, and then regardless of whether it went off or not, it would receive a mechanical shock from the firing pin. Electric priming was virtually instantaneous compared to percussion priming, reducing the delay between the trigger press and the firing of a shot to an absolute minimum. The percussion mechanism took 0.164 seconds to set off the primer, an eternity when firing on the move or when firing on a moving target, while the electrical firing system took just 0.034 seconds[699]. The effect was that when firing under dynamic conditions where the gunner's point of aim could change from one moment to the next, the actual firing accuracy increased.

By retaining the percussion element in the firing mechanism, however, it was still possible to fall back on an old-fashioned mechanical trigger in an emergency. This was a requirement for ensuring the operability of the D-81 in a traditional all-manual mode.

The sliding breech block had a classic cam-type operating mechanism. During the counter-recoil stroke, the breech block was pulled open by the collision of a cam in the breech ring with a hammer protruding from the cradle just before the gun returned to battery. Once the breech block cleared the case rim, it struck the extractor arms, extracting and then ejecting the spent stub case.

---

[699] Орлов, Б. В., Ларман, Э. К., Маликов, В. Г. (1976). *Устройство и проектирование стволов артиллерийских орудий*. Москва: Машиностроение. p. 332.

The breech opening speed was unusually high, to the benefit of bore evacuator efficiency and the case ejection velocity. Spent stub cases were ejected from the gun at a velocity of 14-18 m/s so that they acquired a very flat trajectory to the stub catcher. Weak or partial extraction became extremely unlikely if not completely impossible, and if the breech did not fully open, the autoloader could not begin a loading cycle. This had a special significance to safety since, in the absence of a dedicated loader, neither the gunner or commander could be relied upon to notice a problem of this nature in the heat of combat.

Of course, the gun had to retain considerable momentum during the counter-recoil stroke for this to function reliably, ruling out the possibility of firing light and undercharged rounds from the D-81. This placed some restrictions on the minimum recoil impulse of special ammunition, like guided missiles. Rather than a slow and gentle acceleration down the entire length of the bore, the launching charge for a 125 mm gun-launched guided missile had a fairly strong kick.

Interior Ballistics

The task of designing a tank gun, or any gun, was essentially to solve a parametric optimization problem. A typical goal, tempered by the constraints of fitting the gun and the ammunition in a tank with limited internal volume, was to achieve economical utilization of the powder charge by extracting as much kinetic energy as possible within a pressure limit dictated by the metallurgy of the gun barrel, while using the smallest charge-to-projectile mass ratio, and balancing charge utilization against barrel length by selecting an appropriate expansion ratio[700].

The length of the barrel could be shortened by increasing the service pressure with a large charge mass and decreasing the in-bore sectional density of the projectile with a sabot. A byproduct of the high velocity achieved with this type of ammunition was that the propellant efficiency (charge utilization)[701] fell precipitously, so a very generous charge-to-projectile mass ratio was needed. The chamber volume would therefore have to be enlarged. When a tank gun was optimized for a light projectile and an extreme muzzle velocity with a heavy charge, a design similar to the D-81 appears.

All of these relationships were derived from painstaking empirical testing and modeling of propellant behaviors at various pressures and projectile weights. By setting these relationships down as the first principles of interior ballistics

---

[700] The ratio of bore volume to chamber volume.
[701] The efficiency of converting heat into the kinetic energy of the projectile.

theory, a system of equations[702] was developed to allow ballisticians to design a gun with an extraordinary fidelity to real world conditions. In the USSR, the Serebryakov system of equations was accepted as the industry standard. New findings were applied to continuously improve the fidelity of the thermodynamic model by modifying existing expressions or introducing new coefficients to the system of equations.

Solving the problem of interior ballistics, or calculating the performance of a gun from first principles, was a relatively well-understood science even in the 1950's, and quick calculations could be completed if the propellant was restricted to simple shapes like stick powder or grain powder in the shape of spheres, flakes, strips, or chopped tubes. The complex calculations needed to derive the data in these equations were handled by nomograms and precalculated tables provided by the Soviet Main Artillery Directorate (GAU).

During this era, the direct equivalent to the Serebryakov system of equations was the Corner system of equations, which later evolved into the Baer-Frankle methodology[703], a punch card interior ballistics program coded in Fortran created by the U.S. Army Ballistics Research Laboratory (BRL). Computerization of the Serebryakov system of equations was not implemented in the USSR. The TsNIIAV[704] central scientific-research institute[705], the Soviet counterpart to the BRL, possessed a Ural-1 computer since 1955[706], but in 1959, TsNIIAV became one of many major industrial casualties to the space race when it was absorbed by the TsNII-58 research institute[707] to bolster spacefaring rocket development. Ballistics design in the USSR thus relied on paper the Serebryakov system through the end of the Cold War, and like the Baer-Frankle methodology (in the form of IBHVG2) the Serebryakov system continues to be used to this day in some parts of the world, such as China (in computerized form).

The requirements for the D-81 started with the exterior ballistics deemed necessary to defeat the M60. Knowing that an impact velocity of approximately 1,550 m/s was needed to overcome the reference thickness of armor, the penetrator had to possess a high enough muzzle velocity and low enough drag

---

[702] Chapter 10 in Серебряков, М. Е. (1949). Внутренняя баллистика ствольных систем и пороховых ракет. Москва: Государственное научно-техническое издательство.
[703] Baer, P. G., Frankle, J. M. (1962). *The Simulation of Interior Ballistic Performance of Guns by Digital Computer Program*. Report No. 1183. Aberdeen: Ballistic Research Laboratory.
[704] Central Scientific-Research Institute for Artillery Armaments.
[705] Chertok, B. (2006). Rockets and People - Volume II: Creating a Rocket Industry. Washington, D.C: National Aeronautics and Space Administration. pp. 477-489.
[706] Носкин, Г. (2011). *Первые БЦВМ космического применения и кое-что из постоянной памяти*. СПб: Реноме. p. 9.
[707] Headed by Sergey Korolev, the father of Soviet rocketry.

to be traveling no slower than 1,550 m/s at 2 km[708]. The service pressure of the gun was limited to 400 kg/cm² (392 MPa). The size of the gun had to be as close as possible to the 115 mm D-68. The powder charge had to match the dimensions of the 115 mm powder charge as closely as possible, to minimize the loss of ammunition capacity. Knowing the propellant type, propellant mass and projectile mass, this was enough to begin preliminary design work.

In the Soviet artillery design orthodoxy, this meant calculating and drawing a Serebryakov guiding diagram [709]. A guiding diagram was a contour plot calculated from unitless values that could be used to find all possible permutations of a viable gun design from a set of basic internal ballistics properties. The guiding diagram was set in a chart where the loading density was placed on the horizontal axis and the charge mass was placed on the vertical axis, these two factors being the easiest to change when adjustments were needed to the gun design.

A Serebryakov guiding diagram for a 125 mm gun system [**Fig. 131**] shows that there is a great deal of freedom in how the desired performance[710] could be achieved, but the shaded area in the shape of a guitar pick is the area where a viable solution exists. The marked border of the shaded area indicates where the powder charge completely burns out before reaching 80% of the bore length. Designing a 125 mm gun to the parameters fitting the bottom of the shaded area is optimal for charge economy but at an extreme cost to barrel length (barrel volume). Fitting a 125 mm gun within the innermost egg-shaped contour would be optimal for barrel length, but doing so with a single-base powder charge required a very large chamber volume of between 15 to 30 liters[711], and a charge mass of 12 kg to 20 kg, which was patently unacceptable.

Placing the D-81 into this diagram shows that its design was a moderate balance between a short barrel and small charge volume. The selection of the burning arch of the powder sticks (1.5 mm) was optimal for a shorter barrel, but the chamber volume (12.27 liters) and charge mass (10 kg) were suboptimal. The barrel length of the actual D-81 had to be slightly longer than the theoretical optimum due to the specific nuances in 125 mm ammunition design[712]. Given

---

[708] A standard reference range used by GRAU.
[709] Чурбанов, Е. В. (1975). *Внутренняя Баллистика*. Ленинград: Издательство «ВАОЛКА им. М. И. Калинина». pp. 174-175.
[710] 5.67 kg projectile at 1800 m/s, peak pressure of 392 MPa
[711] A modern high-velocity 152 mm howitzer like the *Msta* has a chamber volume of 23 l.
[712] Charge weight was estimated due to energy contribution of the casing of the semi-combustible charge. The powder mass equivalent cannot be determined with standard powder combustion modeling methods due to the nonstandard geometry of the case. Moreover, the charge weight must be larger than the theoretically optimal value due to gas venting in APFSDS sabots.

the strict restrictions to cartridge dimensions, this was, nevertheless, evidently the best compromise of parameters for the D-81.

Ultimately, the internal ballistics of the D-81 and all of the design compromises that came with it were decided when single-base (nitrocellulose) powder was chosen as the standard tank gun propellant, even if the D-81 was still exemplary in what it achieved with its constraints.

Of course, the combustion of the powder charge imparts kinetic energy to all of the matter expelled out the muzzle of the gun, including the combustion products of the charge itself. The pressure at the base of the chamber is an equal and opposite reaction to the acceleration of both the projectile and the combustion gasses, so naturally, increasing the charge to projectile mass ratio increases the proportion of energy wasted, as well as recoil and muzzle blast.

As an aside, the ratio of the chamber diameter to bore diameter, called the chambrage, was smaller on the D-81 than on 115 mm guns because the 125 mm cartridge retained the same base diameter. This had a positive effect on bore erosion. In heavily necked cartridges, flow constriction at the neck induces turbulence in the gas flow and in its boundary layer adjacent to the bore surface[713]. A smaller chambrage was thus experimentally associated with a longer barrel life.

more advanced powders capable of producing a greater specific force, so that the desired service pressures are obtained without an exorbitant charge volume and mass. At the same time, such powders also benefit cartridge size, but putting a powder with such lofty performance standards into service was not a trivial task. The technology of the powder charges used in the D-81 is worth its own discussion and will be examined in greater detail later.

Figure 131: Adjacent page: Serebryakov guiding diagram for 125 mm gun system accelerating 5.67 kg projectile to 1,800 m/s at 392 MPa average ballistic pressure. The D-81 is marked by a cross (*Jinpeng Zhai, 2023*)
Notation: barrel volume, liters (solid contour); thickness of the burning arch, mm (dashed contour); chamber volume, liters (dotted, diagonally fanned lines); Burnout limit (dash-dot line)

---

[713] Carlucci, D. E., Jacobson, S. S. (2007). *Ballistics: Theory and design of guns and ammunition*. Boca Raton: CRC Press. p. 104.

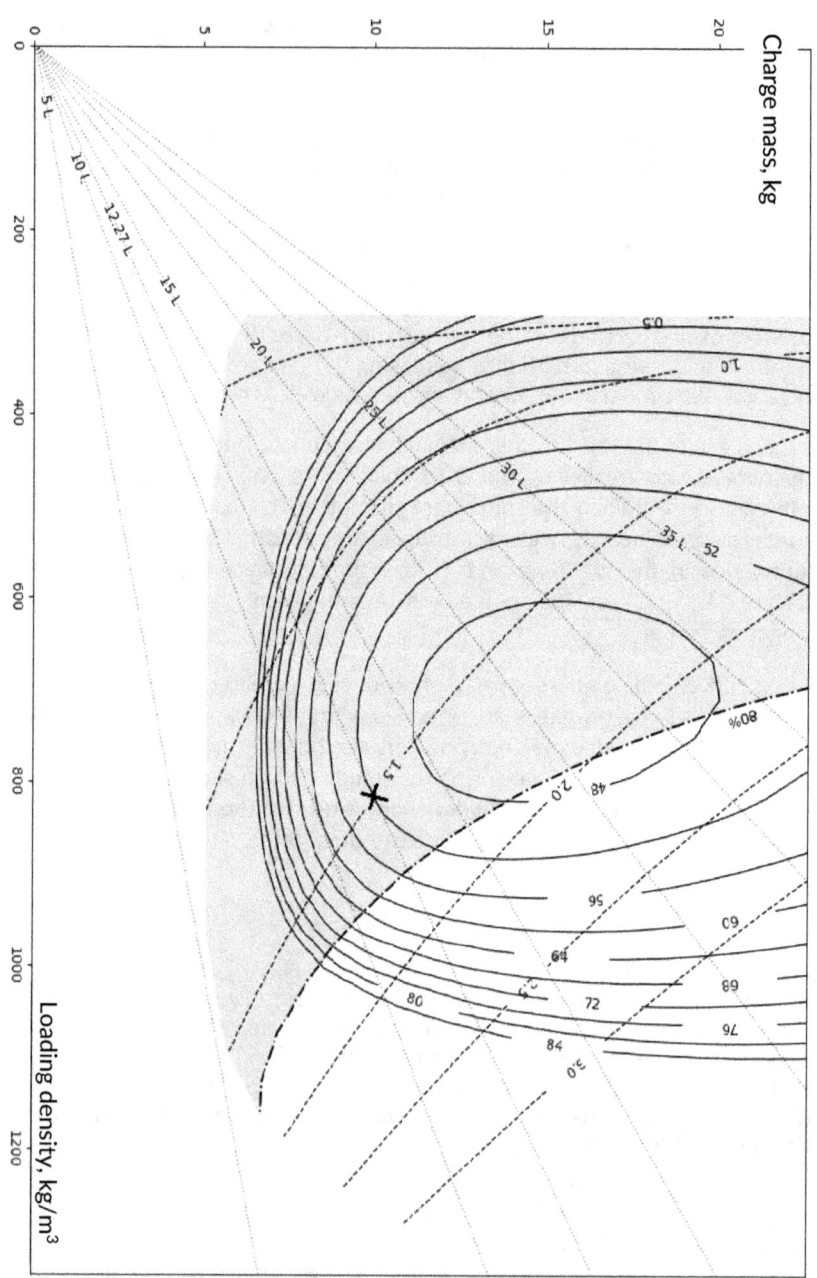

Power and Pressure

The motivation for seeking a larger caliber was simple. The force acting on the base of a projectile can be increased by raising the propellant gas pressure or by increasing the base area that perceives the gas pressure. Mechanical work is done by this propulsive force along the length of projectile travel in the barrel, and under the work-energy theorem, a greater amount of work is equal to a greater change in kinetic energy. The D-81 was reliant on the same gun steel metallurgy and propellant chemistry as its immediate predecessors when it began its development, so increasing the caliber of the gun was the most straightforward means of increasing muzzle energy.

It is well known from elementary pressure vessel theory that the tensile stress in the walls of a cylindrical pressure vessel, or hoop stress, is proportional to the pressure and the radius of the vessel, while being inversely proportional to the thickness of the walls. For a fixed thickness, the walls of a larger diameter barrel experienced a lower tensile stress from the same gas pressure, or conversely, the service pressure could be increased while maintaining the desired safety margin. For the chamber withstand the pressures that the rest of the barrel did not, while being made of the same grade of steel[714], the D-81 chamber was constructed by the so-called *frettage* method. This increased its pressure limit, at the expense of increased manufacturing complexity.

D-81 barrels were produced at the Perm plant together with barrels for various other types of artillery on an automated high-productivity radial forge[715]. After cold forging, the barrel was bored, lathed to precise external dimensions, and milled to form various details. The barrel (called the *fretée* in this case) was finished to an excess outer diameter, and an outer sleeve (called the *frette*) was made with an inner diameter smaller than this outer diameter. The final step in the barrel production process was to expand the sleeve with heat, slip it over the barrel, and then cool it[716]. As the sleeve shrunk, it placed the chamber region of the barrel under a compressive pre-stress.

Powder gasses in the chamber had to first overcome the compressive pre-stress before the tensile strength of the barrel itself came into play, and the *frettage* construction also had the effect of redistributing the stress more uniformly. A *frettage* was also used in the construction of other high-pressure

---

[714] 30KhN2MFA (CrNiMo) steel
[715] The Perm plant used an imported Austrian GFM SXP 55 high-productivity automated radial forge. See: Directorate of Intelligence (1982). *Transfer of Austrian Gun-Barrel Forging Technology to the USSR.* Available at: https://www.cia.gov/readingroom/docs/doc_0000496800.pdf (Accessed 1 September 2023).
[716] Орлов Б. В., Ларман Э. К., Маликов В. Г. (1976). *Устройство и проектирование стволов артиллерийских орудий.* Москва: Машиностроение. p. 31.

Soviet guns like the U-5TS and M62-T2S because the simpler and more effective *autofrettage* process had not been perfected in the USSR at the time.

Despite its larger caliber, the maximum outer diameter of the base of the D-81 barrel was not larger than these guns, in large part thanks to a decision to retain the maximum cartridge diameter of the 115 mm ammunition[717]. The outer diameter (inclusive of the *frettage*) was 240 mm[718], effectively equal to the 122 mm D-30 howitzer (242.6 mm), a much lower-pressure weapon with a straight-walled chamber typical of many artillery pieces, and narrower than the barrel of the 122 mm D-25 tank gun (260 mm)[719] from the IS-2 heavy tank, which had a straight-walled chamber inherited from the A-19 field gun.

Of course, such a comparison is inherently unfair because these artillery pieces were made from much more economical O-70 grade steels. The D-81 barrel was made to the O-110 grade[720], the highest standard of gun steel suitable for mass production with conventional forging methods in the USSR at the time, and an enormous achievement in metallurgy. With its slim 240 mm barrel, the 2A46 withstood the same chamber pressures as the 120 mm Rh120 L/44 and M256 guns, which were much thicker with a diameter of 310 mm[721].

The *autofrettage* process was implemented some years later in the 2A46M along with a new O-120 grade of steel. This process pre-stressed the barrel material internally[722], raising its design pressure considerably, but unlike most gun barrels produced this way, the *autofrettage* process did not simplify production by eliminating the need for an outer sleeve. A thin sleeve was retained to host the interrupted grooves of the barrel mounting system[723] without changing the 240 mm maximum diameter of the barrel.

Having a slender chamber paid dividends in breech size. A smaller chamber diameter was the most obvious way to decrease the dimensions of the breech

---

[717] 125 mm charges had a rim diameter of 172 mm, identical to 115 mm cartridges of both the metal-cased unitary and semi-combustible varieties.
[718] Орлов Б. В., Ларман Э. К., Маликов В. Г. p. 191.
[719] Andrews, W. (2023). *Tank Gun Systems: The First Thirty Years, 1916–1945: A Technical Examination.* Barnsley: Pen and Sword Military. p. 382.
[720] See: Федотов, А. И. (1998). *Основы устройства артиллерийского вооружения.* Перм: Пермский государственный технический университет. p. 56. Also see: Устьянцев, С., Колмаков, Д. p. 134. The letter "O" in the designation stands for "Орудийной" (Artillery gun), followed by the elastic strength limit (1,100 MPa).
[721] Krapke, P. -W. (2004). *Leopard 2: Sein Werden und seine Leistung.* Norderstedt: Books on Demand GmbH. p. 98.
[722] Hydraulic pressure was used to produce plastic deformation along an undersized chamber and bore. After release, the chamber and bore are permanently deformed into their final diameters and their inner surfaces come under a compressive pre-stress from the barrel walls.
[723] This was done to avoid machining stress-concentrating structures into the barrel itself.

ring [724]. The resultant increase to cartridge length was compensated by committing to designing the tank and autoloader for two-part cartridges.

The chamber had a very slight taper, while the combustible case parts were cylindrical. This prevented an air seal from forming when the charges were inserted, which could drastically increase the seating stress on the charge and lead to the rupture of the casing. Moreover, a small built-in clearance between the combustible sidewall of the semi-combustible charges and the chamber surface guaranteed a fit even for charges swelled with moisture.

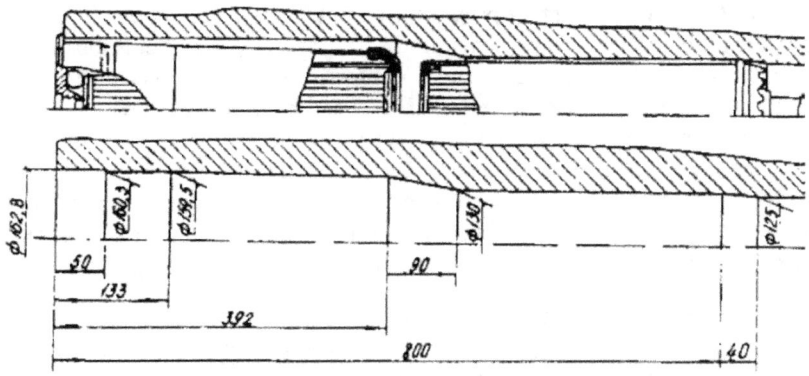

Figure 132: D-81 loaded chamber visualization (top) and dimensional specifications (bottom). (*Valeev et al.*, 2004)

Another nuance of building a chamber for two-part cartridges was that the ullage (gap between the two cartridge parts) would grow as the chamber wore out and lengthened, which created some variation in the ignition lag from the main charge to the supplementary charge. Still the projectile was engaged with the forcing cone at all times, whereas a unitary cartridge would be headspaced on the rim[725], so there would be a steadily growing gap between the projectile obturating band and the forcing cone as the chamber wore out. When a round is fired, the projectile "jumps" across the gap and slams into the forcing cone while some gas blows past, contributing to accelerated wear and increased shot dispersion[726], which would be absent in a two-part cartridge.

---

[724] U.S. Department of the Army (1964). *Engineering Design Handbook: Ammunition Series - Section 4: Design for Projection*. AMCP 706-247. Washington, D.C.: U.S. Army Materiel Command. pp. 4-117, 4-121.

[725] The consistent seating of the cartridge was controlled by its rim resting on the mouth of the chamber.

[726] Гальвиц, У., Мигрина, Б. А. (1950). *Артиллерийские пороха и заряды*. Москва: Издательство Оборонгиз. p. 155.

The total free volume of the D-81 chamber (including the forcing cone) was 12.5 liters[727]. For comparison, the total free chamber volume of the Rh120 L/44 was 10.5 liters[728], usually simplified to 10 liters in period documents. In one instance, a volume of 13.1 liters was used to approximate the D-81 chamber[729], and a volume of 12.27 liters has been cited in a presentation relating the 2A82 to the 2A46M[730]. This type of inconsistency is not atypical even if an empty chamber is specified.

Ballistically, the acceleration of the projectile is driven by the pressure gradient [**Fig. 133**] in the combusting propellant gasses between the moving projectile and the base of the chamber, where the pressure at the chamber is much higher than the pressure at the shot (projectile) base. A high pressure is desirable only at the shot base so as to perform useful work instead of spending energy to accelerate the propellant gasses, but when the combustion pressure is increased by simply increasing the propellant powder mass, the mass of particles in the gas prior to complete combustion also increases, increasing the energy lost in accelerating these particles. This, in turn, is reflected in an inflated chamber pressure for a much smaller gain in the shot base pressure.

Mathematically, the relationship between the chamber pressure and the shot base pressure was a very straightforward linear function of the charge-to-projectile mass ratio[731]. The ballistic average pressure, calculated based on these two real pressures, is also straightforward. Outside of the USSR and its spheres of influence, the chamber pressure was used as the reference figure in design and analytical processes[732], but in standard Soviet artillery design practice, the average ballistic pressure[733] was used instead to simplify calculations. Unless specified otherwise, the service pressure figures cited for Soviet guns in the literature refer to the average ballistic pressure.

---

[727] Валеев, Г. Г., Сопин, В. Ф., Соков, Б. А. (2004). *Артиллерийские метательные заряды*. Казань: ФГУП «Государственный научно-исследовательский институт химических продуктов». p. 52. The volume has been approximated to 13.1 l on page 142 of the same source.

[728] STANAG 4385. 120 mm × 570 ammunition for smooth bore tank guns, in Furmanek, W., Kijewski, J. (2021). 'Constructional Aspects for Safe Operation of 120 × 570 mm Ammunition', *Problemy Mechatroniki. Uzbrojenie, Lotnictwo, Inżynieria Bezpieczeństwa* 12, 4 (46).

[729] Ibid. p. 58.

[730] Хлопотов, А. (2019). *2А82 - супер пушка для «Арматы»*. Available at: https://dzen.ru/media/gurkhan/2a82-super-pushka-dlia-armaty-5c31c4349175d500aabd6073 (Accessed: 29 December 2023).

[731] Серебряков, М. Е. (1962). *Внутренняя баллистика ствольных систем и пороховых ракет*. Москва: Государственное научно-техническое издательство. p. 597.

[732] Ogorkiewicz, R. M. (1991). *Technology of Tanks*. Volume 1. Coulsdon: Jane's Information Group. p. 101.
Also see: *Handbook on weaponry* (1982). First English Edition. Düsseldorf: Rheinmetall GmbH. pp. 354-359.

[733] Серебряков, М. Е. (1962). *Внутренняя баллистика ствольных систем и пороховых ракет*. Москва: Государственное научно-техническое издательство. p. 597.

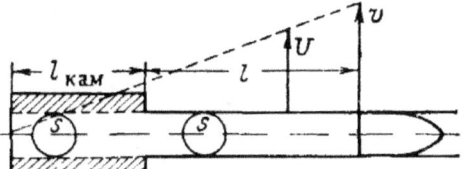

Figure 133: Velocity gradient in a gun barrel. Note that at the chamber base, net gas velocity is zero. (M. E. Serebryakov, 1962)

Neither the chamber pressure or the service pressure can be used to compare two different guns with completely different charge utilization coefficients. When comparing the D-81 to the Rh120 L/44, for instance, the 3BM15 round with its high chamber pressure of 521 MPa seems to surpass the 120 mm DM13 round (514 MPa) [734,735], but in fact it was the reverse, and by a considerable margin. The DM13 round reached and sustained a much higher peak shot base pressure, and even the 105 mm L7 (M68) had a more favorable shot base pressure curve than the D-81 [**Fig. 134**].

The low shot base pressure was exacerbated by the vents built into early 125 mm sabot petals, designed to impart a spin to the projectile before it exited the barrel. Calculations for the shot base pressure of 3BM9 to 3BM22 require a correction factor to account for the loss of pressure through gas venting[736].

The D-81 was still a formidable gun, but primarily because of its larger caliber. The larger caliber was enough to launch the 3BM15 projectile to a nominal muzzle velocity of 1,785 m/s for a gross muzzle energy of 9.4 MJ, placing it only slightly behind the 120 mm DM13 projectile (9.66 MJ). The gap in the shot base pressure curve was narrowed only in the 1980's by the introduction of new high-energy APFSDS ammunition, which, together with the 125 mm caliber, gave the 2A46M its advantage in muzzle energy.

---

[734] Colburn, J. W., Robbins, F. W. (1990). *Combustible Cartridge Case Ballistic Characterization*. Memorandum Report BRL-MR-3835. Aberdeen: Ballistic Research Laboratory. p. 8.
[735] Rocchio, J. J. (1980). *The Interior Ballistic Performance of the 120-mm Tank Gun Relative to the 105-mm Tank Gun*. Memorandum, DRDAR-BLP (30 Apr 1980). Aberdeen: Ballistic Research Laboratory.
[736] Reasonably approximated to a coefficient of 0.884.

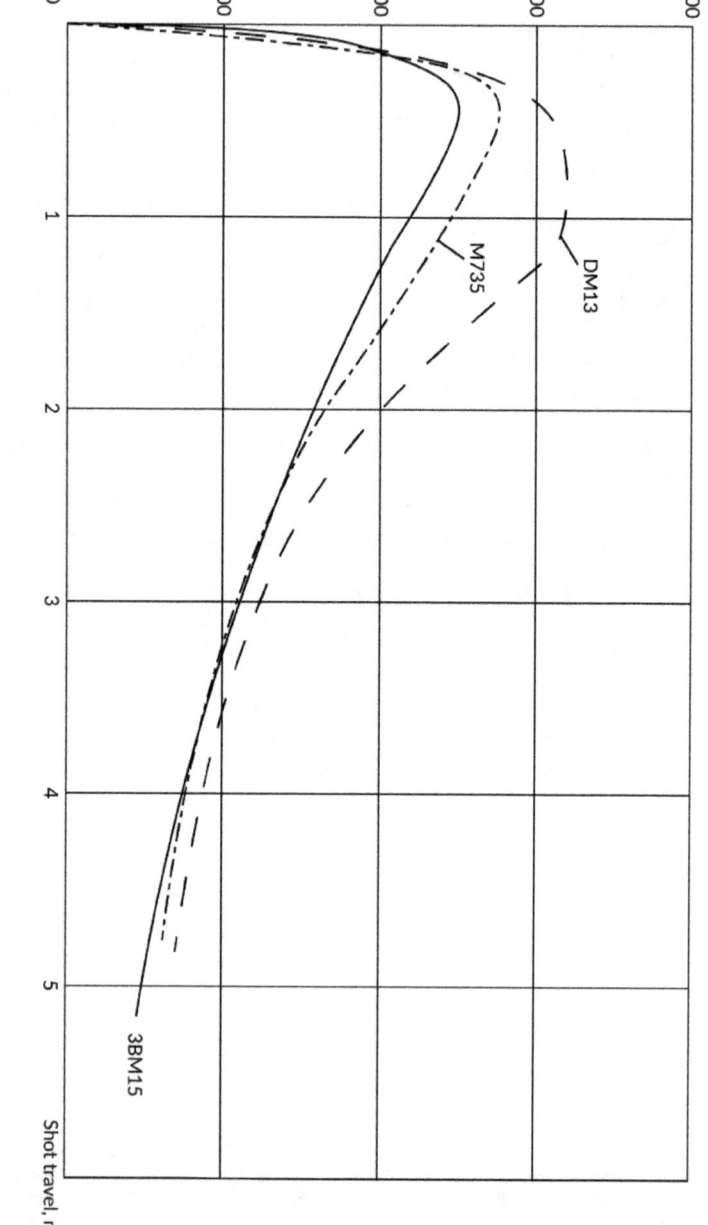

Figure 134: Shot base pressure curves for 3BM15 compared against NATO 120 mm DM13, and 105 mm M735. The curve for 3BM15 shown here is calculated. (P. Martin, 2024)

The service pressure, or more accurately, the mean peak service pressure, was the average pressure in the barrel under standard conditions[737]. For the initial reference round (3BM9) used to design the D-81, the service pressure was 392 MPa (4,000 kg/cm^2) at 15°C. Soviet artillery design regulations dictated that the normative parameters of all ammunition were standardized at 15°C[738], so the standard pressures and muzzle velocities for all ammunition reflect these values. Design regulations also stipulated that ammunition had to be functional at -40°C where the muzzle velocity would be lower, and at 40°C where it would be higher. For the 3BM9 round, a charge temperature of 40°C pushed the service pressure to 4,800 kg/cm^2 (471 MPa). The maximum chamber pressure was 588 MPa (6,000 kg/cm^2)[739].

The reference round for 2A46 was 3BM15. In the technical manual for the D-81, the maximum bore pressure at 40°C with a subcaliber round was expressed as 500 MPa (5,100 kg/cm^2), referring to 3BM15. When fired under this condition, the round would reach a muzzle velocity of 1,800 m/s[740].

The reference round for the 2A46M was the 3BM26, 3BM32 and 3BM42, all of which shared similar internal ballistics and a new mixed APTs-235P powder charge. Based on known data for 3BM42, the service pressure was 555 MPa at 15°C and the maximum service pressure was 637 MPa at 50°C[741]. Based on the known linear dependency of pressure with charge temperature in the new propellant, the service pressure at 40°C is estimated to be 614 MPa.

To put these figures into perspective, the WWII-era D10-T had a service pressure of 300 MPa at the standard charge temperature of 15°C, and its peak service pressure at 40°C was 344.33 MPa.

Assessment of the strength of the gun barrel was done at a charge temperature of 40°C, and limit testing was carried out at -40°C to 50°C to ensure safe operability at the system's absolute temperature limits[742]. During the barrel design process, a preliminary pressure curve was calculated according to the peak pressures at the temperature extremes of -40°C and 40°C, and then the safety margin was applied to the peak service pressure at 40°C[743] to obtain

---

[737] Charge temperature of 15°C, atmosphere at 760 mm Hg pressure
[738] NATO ammunition regulations standardized on a charge temperature of +21°C
[739] Устьянцев, С., Колмаков, Д. (2007). Боевые Машины Уралвагонзавода: Танки 1960-Х. Нижний Тагил: Издательский Дом «Медиа-Принт». p. 134.
[740] Валеев, Г. Г., Сопин, В. Ф., Соков, Б. А. (2004). Артиллерийские метательные заряды. Казань: ФГУП «Государственный научно-исследовательский институт химических продуктов». p. 28.
[741] Розоринов, Г. Н., Хаскин, В. Ю., Лазаренко, С. В. (2013). 'Применение Лазерных Технологий Для Повышения Срока Службы Изделия КБА-3', Збірник наукових праць СНУЯЕтаП. p. 146.
[742] Дубровин, Ф. И., Мазуренко, А. И., Морозов, Е. А., Яремчук, Л. С. (1983). 'Термостатирование Боекомплекта Танка', Вестник Бронетанковой Техники 1983, сборник 3. p. 14.
[743] Серебряков, М. Е. (1962). *Внутренняя баллистика ствольных систем и пороховых ракет.*

the design pressure[744] curve. The safety margin was enough for the gun to also fire at a charge temperature 50°C.

Throughout the postwar era of Soviet artillery design, engineers worked with a basic barrel design pressure safety margin of 1.2[745] for smoothbore guns. For rifled guns, the required safety margin factor varied from 1.2 to 1.35, increasing proportionately with the rifling thickness (groove depth) from 0.01-0.025 calibers. This margin was applied to the chamber and to the barrel along the point where the service pressure reached its theoretical peak, and the safety margin for the rest of the barrel was calculated using a simple linear expression based on the distance from the point of peak pressure, except at the muzzle. A fixed safety margin of 1.9 was applied to the muzzle[746], which gave barrels a visually obvious local thickening known as a muzzle bell.

Table 27: Estimated barrel design pressures for 2A26, 2A46 and 2A46M

Gun	2A26	2A46	2A46M
Design pressure (estimated)	≥565 MPa	≥600 MPa	≥737 MPa
Reference pressure of basic KE round at 40°C	471 MPa	500 MPa	614 MPa

For comparison, the design pressure for the Rh120 of the Leopard 2 was 707 MPa, and 681 MPa for the L30A1 gun of the Challenger 2[747].

Barrel Replacement Process

On the 2A26 and 2A46, the barrel was fitted to the breech ring by screwing it onto buttress threads. After screwing it in for ten turns, a retention pin was put in to secure the barrel from turning. On the 2A46M, the barrel mount was a set of interrupted square-profiled lugs. The barrel was first inserted, and then a special wrench would be clamped on an octagonal surface on the barrel[748] to twist it by 45°[749], mating the lugs to matching grooves in the breech ring. Then a retention pin was put in.

Москва: Государственное научно-техническое издательство. p. 614.
[744] The maximum pressure that may be reached while avoiding plastic deformation.
[745] Закаменных, Г. И., Кучерова, В. Г., Червонцев, С. Е. (eds.) (2017). *Проектирование спецмашин, часть 1, книга 1: Артиллерийские стволы*. Волгоград: ВолгГТУ. pp. 206-207.
Also see: Лаухин, А. Н. (1970). *Современная артиллерия*. Москва: Воениздат. p. 72.
[746] Due to increased stress from the release of powder gasses.
[747] U.K. Ministry of Defence (1988). Armoured warfare: A Vehicles; replacement for Chieftain; new tank for the army. DEFE 70-1890.
[748] This octagonal surface was hidden under the second thermal sleeve section.
[749] ОАО НПК «Уралвагонзавод» (2021). *Танковые Пушки 2А46М, 2А46М-1, 2А46М-4, 2А46М-5*.

Removal of the 2A46M required a screw press to be mounted on the breech opening, to press the barrel out by hand. This was laborious, but much less laborious than needing to remove the gun to replace the barrel. According to repair norms, a barrel change for the 2A46M could be done in 4 hours.

Smoothbore and Barrel Wear

Relative to the 122 mm guns fitted to domestic heavy tanks, the D-81 did not actually have a fundamentally larger bore diameter. As an example, the Soviet T-10M heavy tank's 2A17 (M62-T2S) high-velocity 122 mm gun had 2.5% rifling[750], so the bore diameter across its rifling lands was 122 mm while the diameter across the grooves was 125 mm. It would be technically quite accurate to regard a smoothbore 125 mm gun as a 122 mm rifled gun stripped of its rifling, and this came at no increase in barrel mass since the same wall thickness profile would be retained. In return, a small increase in projectile mass could be supported[751], the smoothbore barrel would be slightly lighter from the lack of rifling and be more resistant to wear.

But still, the priority behind the decision to remove rifling was the need for the barrel service life to keep up with the increase in service pressure. Rifling wasn't inherently incompatible with APFSDS ammunition; a slip-type driving band made it possible to control the spin rate to acceptable limits, even for projectiles traveling down the barrel at exceedingly high velocities. Rather, the main problem was that the engraving of the driving band into the rifling lands was the source of limiting erosion.

Bore erosion reaches its peak at the point of full rifling depth (II), followed closely by the point of rifling commencement (I). The erosion observed in this zone was strongly influenced by the fact that the engraving force was directly proportional to the bore pressure[752], which was, of course, rather problematic since increasing the service pressure was critical to improving gun performance.

Removal of the rifling changed the source of limiting erosion to bore erosion, which was not only milder, but had a uniform profile [**Fig. 135**][753]. A rifled barrel (a) wore out when its throat eroded to the prescribed limit at the initial rifled section, even if the remainder of the throat was only lightly eroded. A

---

[750] Where the rifling depth was defined as 2.5% of the caliber (0.025D); the diameter across the rifling lands, representative of the projectile diameter. The M62 and high-velocity field guns like the D-74 used 2.5% rifling, howitzers and medium-velocity field guns used 2% rifling.
[751] Compared to a 122 mm projectile, a 125 mm projectile of identical construction will be 7.5% larger in mass and volume, or a more respectable 13% if compared to a 120 mm projectile.
[752] Handbook on weaponry (1982). First English Edition. Düsseldorf: Rheinmetall GmbH. p. 575.
[753] Главное Управление Боевой Подготовки Сухопутных Войск Вооруженных Сил Российской Федерации (2004). *Учебник Сержанта Танковых Войск*. Москва: Воениздат. p. 150.

smoothbore barrel (6) eroded with much greater uniformity along the throat, making for a more economical utilization of the barrel material.

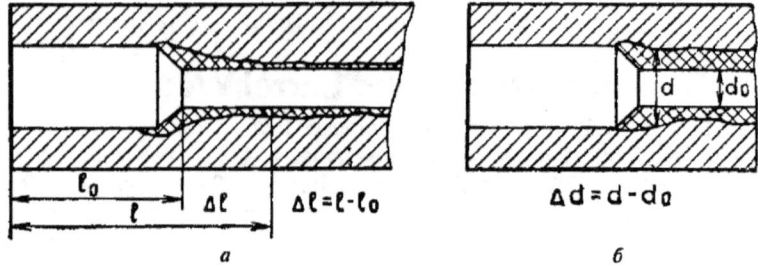

Figure 135: Nature of bore erosion in rifled (а) and smoothbore (б) barrels. (*Soviet Ministry of Defense*)

Barrel wear was measured by the lengthening of the chamber and how much the bore diameter expanded at the barrel throat, 1 m from the breech end of the bar- rel. In field conditions, this was done by inserting a PKI-26 probe down the barrel from inside the tank. From there, the difference in bore diameter from the normal 125 mm figure could be measured, and the barrel condition evaluated accordingly[754]. Categories 1-3 described the state of barrel wear, and categories 4-5 were for administrative purposes[755].

- Category 1: New and lightly used barrels (0.9 mm) with a lifespan expenditure of up to 25%.
- Category 2: Moderately worn barrels (0.9-2.6 mm) with a service life expenditure of 25% to 80%.
- Category 3: Worn barrels (2.6-3.3 mm) with a service life expenditure of 80% to 100%.
- Category 4: Dismounted from tank
- Category 5: Defective

The accuracy life of a barrel was considered to be fully exhausted once it exceeded the 100% wear threshold under Category 3. It would then be transferred to Category 5.

Each projectile type behaved differently while traveling down the barrel, vibrating in different ways and producing distinct wear patterns on the bore

---

[754] Along with a visual inspection of dents, cracks, and so on.
[755] Закаменных, Г. И., Кучерова, В. Г., Червонцев, С. Е. (eds.) (2017). *Проектирование спецмашин*, ч. 1, кн. 1: Артиллерийские стволы. Волгоград: ВолгГТУ. pp. 372-373.

surface[756]. A full-caliber round like HEAT or HE-Frag shaved off 0.0033 mm from the bore diameter at the barrel throat. The accuracy life would therefore be fully depleted after 1,000 shots, verified by field testing experience. It has also been indicated in other sources that the barrel life of the 2A46 was 900 rounds[757], perhaps due to a policy not to use barrels when they were close to fully spent.

KE ammunition was much more erosive. Firing 3BM9 removed a thickness of 0.0132 mm from the barrel bore with each shot, so it was equivalent to firing 4 full-caliber shells. The barrel service life would be 250 shots. Taking this to be 1 EFC (equivalent full charge), as compared to the full-caliber shells having only the main charge and therefore a reduced charge, it can be said that HEAT and HE-Frag rounds had an EFC rating of 0.25.

For example, given that 32.5% of a typical combat load was comprised of KE rounds, the averaged barrel life of a D-81 be calculated to be 425-450 shots[758].

Table 28: Bore erosion data for various rounds

Round Model	HEAT/ HE-Frag	3BM9	3BM15	3BM22	3BM26
Eroded diameter (mm)	0.0033	0.0132	0.0165	0.0165	0.019
EFC	0.25	1	1.25	1.25	1.44

The thickness of metal stripped away by each shot was minuscule, but with enough use, the gun would start becoming noticeably tail-heavy. To compensate, stacks of steel ballast plates were bolted behind the breech ring as barrel wear counterweights. These ballast plates were incrementally removed to maintain the balance of the gun throughout its useful service life.

It was, however, impossible to compensate for the effect of bore wear on the muzzle velocity of subcaliber (full-caliber) rounds, which varied from 0.5% (0.1%) at 0.5 mm of wear, up to 3.8% (2.9%) at 3.4 mm of wear[759]. Category 3 barrels also tended to produce increased shot dispersion and significant changes in shot trajectory from Category 1 and 2 barrels. The only type of compensation

---

[756] Jankovych, R., Beer, S. (2011). 'Wear of cannon 2A46 barrel bore'. *NAUN*.
Also see: Balla, J., Prochazka, S., Jankovych, R., Beer, S., Krist, Z., Kovarik, M. (2015). 'Technical Diagnostics of Tank Cannon Smooth Barrel Bore and Ramming Device', *Defence Science Journal*, Vol. 65, No. 5. pp. 356-362.
[757] Kotsch, S. (2011). *Die russische 125 mm Panzerkanone 2A46*. Available at: https://www.kotsch88.de/g_2a46.htm (Accessed: 1 December 2023).
[758] Гзовская, Т. В., Иванова, О. В., Поляков, В. Б. (1985). 'Обоснование требуемого уровня живучести ствола танковой пушки', *Вестник Бронетанковой Техники* 1985, сборник 2. рр. 7-9.
[759] Главное Управление Боевой Подготовки Сухопутных Войск Вооруженных Сил Российской Федерации (2004). *Учебник Сержанта Танковых Войск*. Москва: Воениздат. р. 148.

possible was to apply a simple ballistic drop correction factor to the fire control system.

Table 29: Change in vertical and horizontal shot dispersion (in mrad) from a new barrel to a totally worn barrel

Ammunition type	New barrel		Worn barrel	
	D_V	D_H	D_V	D_H
KE	0.18	0.20	0.40	0.35
HEAT	0.15	0.15	0.20	0.17
HE-Frag	0.15	0.15	0.21	0.18

Потемкин, Э. К. (ed.) (1982). *Основы проектирования вооружения танка*. Том 2. Москва: Издательство МГТУ им. Н. Э. Баумана. p. 90.

In both velocity and dispersion, subcaliber KE rounds suffered the most due to the natural inclination for the sabot to expand in a worn-out barrel, creating an unstable support base[760]. Conversely, 125 mm HE-Frag and HEAT projectiles were unusually insensitive to barrel wear, even compared to conventional rifled shells. This can be credited to the use of two obturating bands, one at the base and one at the center.

Having multiple bands was not unusual by itself, with some large caliber artillery shells having three or even four driving bands[761]. However, these were invariably a cluster of base bands, and the quantity of bands was to attain good load distribution[762] more reliably than a single band of extreme width. In unitary 100 mm and 115 mm smoothbore rounds, there was only one obturating band, located near the center of the projectile ahead of the mating point with the case. It was established long ago in ammunition design that the obturating band was best placed as far to the rear as practicable on a projectile[763], but due to the long case neck and deep seating of the projectile, the band had to be much further forward than generally considered desirable.

With 125 mm two-piece ammunition, it became possible to have a base obturating band in addition to the center obturating band, separated by a

---

[760] This was addressed in the early 1980's with a new "clamping" type sabot.
[761] Morgan, J. H., Pittman, J. (1997). *Projectile and Warhead Identification Guide - Foreign*. Charlottesville: National Ground Intelligence Center. p. 14.
[762] Fairfield, A. P. (1921). *Naval Ordnance*. Annapolis: U.S. Naval Institute. p. 507.
[763] Ibid. pp. 510-511. Also supported by much more recent sources such as: U.S. Department of the Army (1964). Engineering Design Handbook: *Ammunition Series – Section 4: Design for Projection*. AMCP 706-247. Washington, D.C.: U.S. Army Materiel Command. pp. 4-158 - 4-160.
This maximized chamber volume, placed the band at the thickest part of the shell wall, minimizing respectively the projectile yaw at muzzle exit and balloting (lateral or yawing motion) during in-bore travel. This was done by minimizing the couple from asymmetric band contact pressure, and by giving the widest possible support base with the forward bourrelet.

distance of 90-92 mm[764]. After the base obturating band passes through the forcing cone, it forms a gas seal throughout the remainder of projectile travel.

Having two widely spaced obturating bands, internal projectile yawing or balloting during the forcing process and during projectile travel was reduced, especially in worn barrels where a bourrelet lost much of its effectiveness as a supporting surface. A dual obturating band configuration thus accounted for an increase in the barrel accuracy life.

Replacing a bourrelet with an obturating band was historically uncommon. For armor-piercing projectiles, one definite reason was that a forward band introduced a stress-concentrating groove at a point of the body obligated to withstand extreme transverse loads[765]. In the D-81, the main drawback associated with double obturating bands was that the passage of the rear obturating band through the forcing cone occurred with the projectile already in motion, so that forcing of the rear band involved a strong mechanical shock.

The modest service life of the 2A26M2 and 2A46 barrels when firing KE rounds is unsurprising given its high service pressure, considering that it lacked special measures to limit erosion. Even the 2A46M was limited to only 200 rounds of standard KE (e.g. 3BM26, 3BM42). Its contemporaries, the Rh120 L/44 and M256, achieved a barrel life of up to 700 rounds firing standard KE rounds like the M829[766] thanks to chrome plating of the most heavily eroded section of the barrel, from the throat up to a point just behind the bore evacuator[767]. By the same token, it has been claimed that chroming the barrel bore can increase the service life of a 2A46M barrel to 1,200 shots[768], or with KE rounds, it can be increased to 450 shots. However, the D-81 never received a chrome lining or wear-resistant lining of any sort during its service in the Soviet Army.

---

[764] This feature was absent on the 115 mm two-piece ammunition for the D-68 gun.
[765] Fairfield. p. 501.
[766] Sopok, S., Dunn, S., O'Hara, P., Coats, D., Pflegl, G., Rickard, C. (2001). 'Cannon Coating Erosion Modeling Achievements', *10th U.S. Army Gun Dynamics Symposium*. p. 273.
[767] Oberle, W. F., White, K. J. (1991). *Electrothermal-Chemical Propulsion and Performance Limits for the 120-mm, M256 Cannon*. Technical Report BRL-TR-3264. Aberdeen: Ballistic Research Laboratory. p. 10.
[768] Yunusov, B. A. (2023). 'Steels for Tank Barrels', *International Journal of Advanced Research in Science, Engineering and Technology*, 10(8). Available at: https://www.ijarset.com/upload/2023/august/8-r-shoh- 07.PDF

Bore Evacuator and Thermal Sleeve

The purpose of a bore evacuator was to reduce powder fume contamination of the fighting compartment and thereby improve crew working conditions. Smokeless powder fumes contain asphyxiants like nitrogen oxides (NO, NO_2) and carbon oxides (CO, CO_2), along with various mild respiratory irritants. The combat effectiveness of the crew was much improved by the removal of these fumes by the suction effect created by a bore evacuator.

A bore evacuator is a special chamber to store powder gasses from the barrel and then discharge the gasses forward through angled vents when the pressure in the barrel falls. Most importantly, the bore evacuator depressurizes at a slower rate than the barrel. The gun breech is timed to open once the pressure in the barrel has fallen to atmospheric levels but before the evacuator completely empties, so that the jets of gas from the bore evacuator blow out the fumes downwind of the vents and also creates a draft to extract the fumes behind the vents by suction[769].

The D-81 bore evacuator was a thin-walled steel vessel[770] placed two-thirds down the length of the barrel. Its location was dictated by the powder burnout point in the barrel. The vent holes were drilled in a staggered layout and the barrel section hosting the vents was specially thickened to compensate for the structural weakening.

After a shot was fired and the projectile traveled passed these vents, the bore evacuator began to fill up with gasses to a small fraction of the in-bore gas pressure. A special heat-resistant nozzle embedded inside each vent accelerated the filling rate while the pressure difference between the barrel and evacuator was high. When the pressure in the barrel plummeted after the departure of the projectile, the small nozzle throat diameter instead functioned as a throttle to extend the gas discharge time.

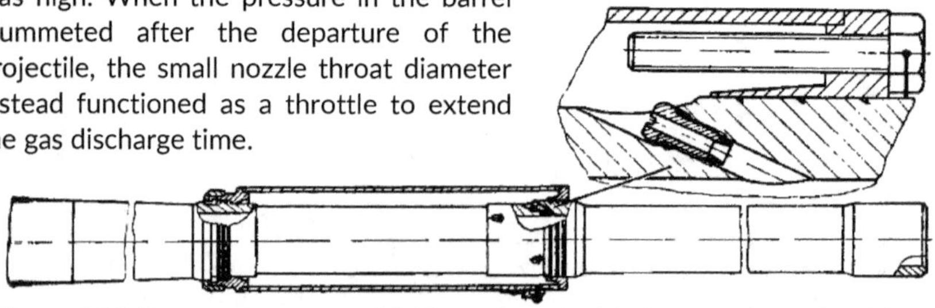

Figure 136: Bore evacuator on 2A46 gun barrel with the muzzle to the right. Note the special nozzle insert. (*Adapted from Soviet Ministry of Defense*)

---

[769] Орлов Б. В., Ларман Э. К., Маликов В. Г. (1976). *Устройство и проектирование стволов артиллерийских орудий*. Москва: Машиностроение. pp. 51-52.

[770] Besides containing the pressure of powder gasses, the bore evacuator walls had to be thick enough to resist bullets and shell fragments.

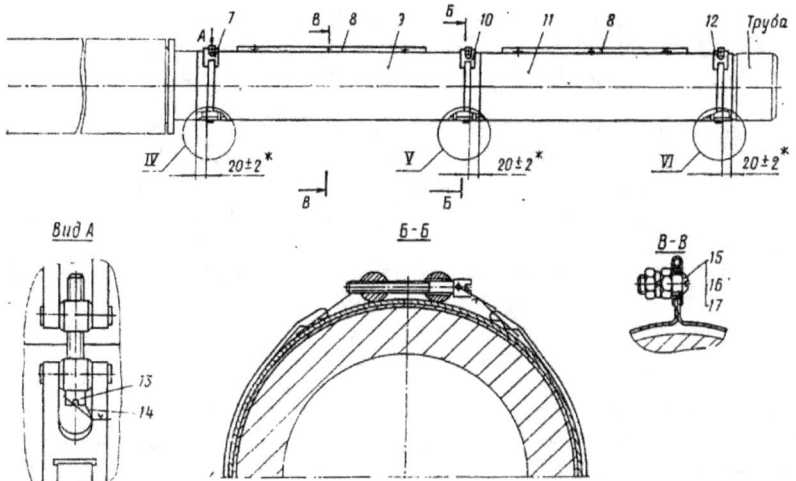

Figure 137: Thermal sleeve for 2A46. (*Soviet Ministry of Defense*)

The thermal sleeve introduced to the D-81 in 1975 was made from a set of four aluminum alloy segments. Each segment was a curved aluminum alloy sheet joined into a cylindrical shape with bolts and then clamped to the barrel on insulated rings, leaving an air gap between the thin (0.5 mm) sheet and the barrel surface. Each sleeve segment had drainage holes so that water could not accumulate inside, especially after crossing a river by deep wading. There was no sleeve segment over the concentric bore evacuator since it already functioned as one. The complete thermal sleeve set weighed a mere 5.38 kg[771].

The purpose of a thermal sleeve was to minimize the bending of the barrel from temperature gradients arising from uneven heating like from sunlight, or uneven cooling by rain, snow, and wind, including air flow from the tank's own movement while driving.

Thermal sleeves debuted on tank guns in 1966 in the form of a fiberglass fabric wrap on the 120 mm gun of the British Chieftain, known as a thermal jacket, and as a magnesium alloy sleeve on the 105 mm gun of the French AMX-30. Widespread international adoption of thermal sleeves soon followed in the early 1970's. During this period, light metal thermal sleeves[772] were universally favored over fabric-type thermal jackets for their light weight and weatherproof

---

[771] 菖万有 (1989). '身管热护套的作用原理及新结构研究', in 高膛压火炮系统论文集. 机电部兵器科学研究院. p. 155. Chang Wanyou (1989). 'Research on the principle of action and new structure of the thermal sleeve on the barrel body'. *Collection of papers on high-pressure artillery systems.*
[772] Aluminum thermal sleeves were introduced on the Leopard 1, M60A1, and T-64A.

nature[773]. Composite (fiberglass) or plastic thermal sleeves were later adopted for some guns like the Rh120 L/44 for better insulation, but at a heavy penalty to weight, so light metal sleeves understandably remain a popular choice.

Like other light metal thermal sleeves, the D-81 sleeve design worked as a thermal insulator-dissipator. The air gap between the barrel and the thermal sleeve was the main insulating element from external heating or cooling, and the inside of the sleeve was unpainted so that its bare aluminum surface, having a very low emissivity, poorly radiated heat to the barrel. The high thermal conductivity of aluminum helped the sleeve to disperse the heating or cooling effect on the barrel by ensuring a relatively uniform temperature around its circumference[774].

Tests on a 2A46M under extreme simulated conditions showed that its thermal sleeve reduced the maximum barrel bend from intense solar heating by 56%. With a barrel heated from firing, the reduction in barrel bend from a strong 10 m/s cross breeze reached 69%, and from heavy rainfall, it was 78%[775]. As expected, the sleeve had practically no effect on barrel droop on a hot barrel[776]. Droop could only be compensated by the gunner by manually adjusting the point of aim. With the UVKV optical boresighting system, it was only feasible to check for a loss of a boresight from droop, but not correct for it during combat.

Table 30: 125 mm gun barrel bend with and without thermal sleeve

Environmental Influence	Direction of barrel bend	Barrel bend (mrad)	
		w/o sleeve	w/ sleeve
Sunlight[a]	Vertical	-3.07	-1.36
Rain[b]	Vertical	1.47	0.33
Wind[b]	Horizontal	0.48	0.15

a   Strong directional heating at an intensity of 1,120 W/m2
b   Bore channel pre-heated to 300°C to simulate effect of intensive fire

---

[773] Kratzenberg, K., Kuellmer, G., Nausester, A. (1970). *Waermeschutzhuelle fuer Kanone. Wegmann and Co GmbH.* Deutsches Patentamt Patent no. DE1918422A1. Available at: https://patents.google.com/patent/DE1918422A1/en (Accessed: 23 October 2023).
[774] Богомолов, П. И., Бируля, М. А., Болотин, А. А. (2021). *Оценка Эффективности Термозащитного Кожуха Ствольной Трубы Танковой Пушки При Воздействии Солнечной Радиации.* Санкт-Петербург: Акционерное общество «Центральный научно-исследовательский институт материалов». pp. 15-16.
The same mechanism is also described in: Tauzin, M. (2008). *COMHART - Tome 9: L'armement de gros calibre.* Paris: Comité pour l'histoire de l'armement terrestre. p. 55.
[775] Лукьянов, В. Н., Крюков, И. А. (2017). 'Влияние Климатических и Внутрибаллистических Факторов на Изгиб Ствола Танковой Пушки с Различными Вариантами Термозащитных Кожухов', *Известия РАРАН*, 96. pp. 86-88.
[776] With a thermal sleeve, the maximum bend was 2.05 mrad, and without, it was 2.12 mrad.

Chapter VI

Ammunition

Overview

The T-72 could fire APFSDS, HEAT (High-Explosive Anti-Tank), and HE-Frag (High-Explosive Fragmentation) rounds. These three ammunition types formed the standard wartime repertoire of the 125 mm caliber for most of the Cold War.

Table 31: GRAU indices for 125 mm ammunition

Complete Cartridge		Projectile
HEAT		
3VBK7		3BK12M (3BK12)
3VBK10		3BK14M (3BK14)
3VBK16		3BK18M (3BK18)
HE-Frag		
3VOF22		3OF19
3VOF36		3OF26
Complete Cartridge	Projectile with Supplemental Charge	Projectile
APFSDS		
3VBM3	3BM10	3BM9
3VBM6	3BM13	3BM12
3VBM7	3BM16	3BM15
3VBM8	3BM18	3BM17
3VBM9	3BM23	3BM22
3VBM11	3BM27	3BM26
3VBM10	3BM30	3BM29

Министерство Обороны СССР (1988). *125-мм Танковые Пушки 2А26, 2А46, 2А46-1, 2А46М, 2А46М-1, 2А46-2 - Техническое Описание и Инструкция по Эксплуатации 2А46ТО1. (Части 3: Боеприпасы)*. Москва: Воениздат. pp. 12-14.

Table 32: Specifications for Soviet 125 mm ammunition[a, b, c, d, e, f]

Projectile Model	HE-Frag	HEAT		APFSDS			
	3OF19 (3OF26)	3BK12	3BK14 (3BK18)	3BM9 (3BM12)	3BM15 (3BM17)	3BM22	3BM26
Full Projectile Mass (kg)	23 (23.22)	19.05	19.05	5.67	5.92	6.55	7.05
w/o sabot (kg)				3.62	3.9	4.83	4.83
Explosive Mass (kg)	3.15 (3.401)	1.65	1.76	–	–	–	–
Muzzle Velocity (m/s)	850	905	905	1,800	1,785	1,760	1,720
Direct Fire Range on 2 m Target (m)	1,010	1,020	1,020	2,120	2,100	–	–
Penetration (1 km, 0°) (mm)	–	440	– (550)	300 (370)	425	–	–
Penetration (2 km, 0°) (mm)	–	–	–	280 (350)	400	420	–
Penetration (1 km, 60°) (mm)	–	200	250 (260)*	180 (180)	180*	–	–
Penetration (2 km, 60°) (mm)	–	–	–	150 (150)	150	170	200
Service Pressure (15°C) (MPa)	343	295	295	392	444	–	555

a	Продукция военного назначения НПК «Технологии машиностроения» (2018).
b	Оружие и технологии России. XXI век Том 12 - Боеприпасы и средства поражения (2006). Москва: Оружие и технологии. p. 466.
c	Широкорад, А. (2000). 'Пушки советских танков (1945-1970 гг.)', Техника и Вооружение, (July). p. 7.
d	Jankovych, R., Beer, S. (2011). 'Wear of cannon 2A46 barrel bore'. NAUN. p. 76.
e	Устьянцев, С., Колмаков, Д. (2007). Боевые Машины Уралвагонзавода: Танки 1960-Х. Нижний Тагил: Издательский Дом «Медиа-Принт». p. 135.
f	Селиванова, В. В. (ed.) (2016). Боеприпасы: учебник в двух томах, том 1. (2 vols). Москва: Издательство МГТУ им. Н.Э. Баумана. p. 331.
*	Calculated based on secondary information.

Inert 3P11 practice HEAT and 3P23 practice HE-Frag projectiles were available for live fire training, but no practice rounds were available for APFSDS during the Soviet era. The only requirement for practice ammunition was inertness so as not to destroy the range targets. Under this reasoning, steel APFSDS rounds like 3BM9 were used as training rounds if they were called for. The range safety issues created by the extreme velocity of APFSDS rounds were tolerated, but not ignored entirely. In the early 1990's, the 3P31 training APFSDS round was created and accepted into service. It featured a high-drag spike tip to limit its maximum range to 8 km.

Almost all 125 mm ammunition was developed by the NIMI research institute, except for the 3P31 training round, which was created by the SKB-78 factory in the 1990's. In all cases, the Kazan Research Institute of Chemical Products was involved in propellant development and production.

Production lines for 125 mm ammunition were set up at Plant No. 114 *Plastmass* in the Southern Urals, which was also the country's largest tank ammunition producer, and the Karl Liebknecht Leningrad Mechanical Plant (LMZ). In 1981, an additional production line was set up at Plant No. 790[777].

125 mm ammunition was divided into a powder charge and the projectile. APFSDS rounds have a supplementary charge packed with the projectile assembly. In effect, HE-Frag and HEAT shells were fired at a reduced charge. This was a method of solving different ballistic requirements by changing the loading density[778]. A reduced charge had a special significance for HE-Frag rounds to control the negative impacts of a high pressure and projectile velocity[779] on shell construction, filler mass, and fragmentation patterning.

---

[777] Гогин, В. В., Горчаков, В.А. (2016). 'Министр, Минмаш, НИМИ: Краткая Историография в Области Артиллерийских Боеприпасов (к 100-летнему юбилею В.В. Бахирева)', *Боеприпасы и высокоэнергетические конденсированные системы*. p. 13.

[778] Проскуров, В. А. (1972). 'Некоторые вопросы применения в танке выстрелов раздельного заряжания', *Вестник Бронетанковой Техники* 1972, сборник 2. p. 29.

[779] Проскуров, В. А., Завьялова, Г. Ф., Москвин, Г. Н., Соколов, В. Я. (1971). 'Метод Повышения Эффективности Стрельбы из Танка Осколочно-Фугасными Снарядами', *Вестник Бронетанковой Техники* 1971, сборник 4. p. 37.

Powder Charges

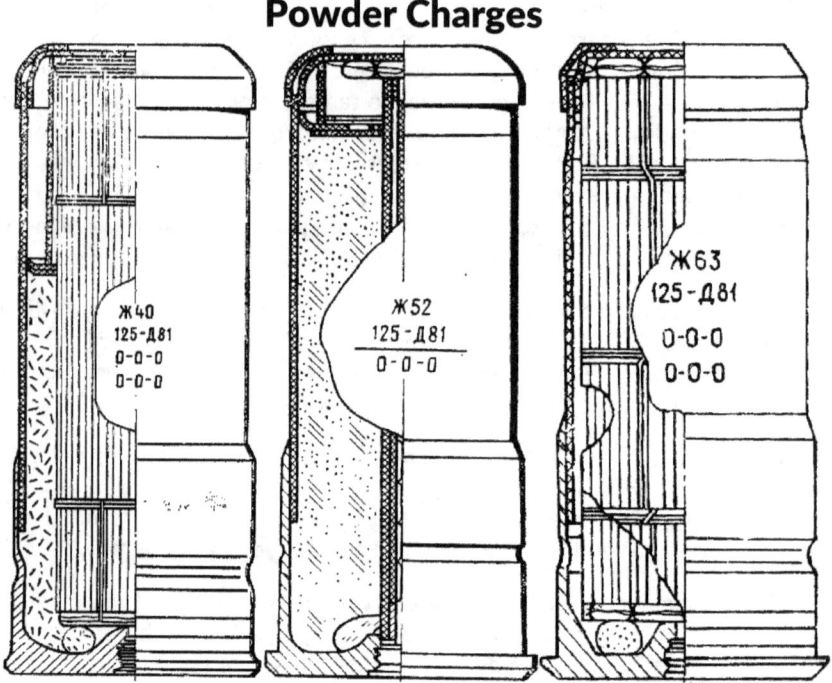

Figure 138: 4Zh40, 4Zh52 and 4Zh63 powder charges for the D-81. (*Adapted from manuals, Soviet Ministry of Defense*)

All powder charges for the D-81 were of the semi-combustible type, consisting of tubular or granular powder in a combustible casing fitted to a steel stub case. The two charge types available for the D-81 were 4Zh40 and 4Zh52. These were used for all ammunition types, and were functionally interchangeable. In the early 1980's, new high-energy KE rounds entered service with a proprietary high-energy 4Zh63 charge. The basic concept behind semi-combustible ammunition was to improve crew working conditions by reducing powder fume contamination and alleviating the clutter of spent cases, thereby improving the combat rate of fire[780].

The first examples of practical semi-combustible charges in the USSR appeared as early as 1957 for the 122 mm guns of the IS-3 and T-10 heavy tanks[781]. The abortive *Taran*, *Rezets* and *Molot-T* large caliber smoothbore gun

---

[780] Павлов, И., Павлов, М. (2023). 'Т-62: «Рабочая лошадка» советских танковых дивизий первого эшелона', *Техника и Вооружение*, (July). p. 18.
[781] Шабалин, В. А. (1970). 'О Концентрации Пороховых Газов в Зоне Дыхания Экипажа

projects (1957-1960) were also created with semi-combustible two-part ammunition in mind. When the decision was made to create a family of two-part 115 mm cartridges for the T-64 autoloader, the foundational work on semi-combustible charges had already been completed through these projects, contributing to the completion of 115 mm two-part ammunition by 1964[782]. From there, upscaling the technology to the new 125 mm caliber was a fairly straightforward endeavor.

In basic artillery design, the separation of a cartridge into two parts could be applied when a unitary cartridge would otherwise be too heavy or too cumbersome for manual loading. Two-part ammunition was also technically justified by the reduction in the minimum length of cartridge stowage points, and the higher packing efficiency that came as a result. Both factors were conducive to large-capacity hull autoloaders and gave better packing efficiency, with benefits to the tank's combat load. The short length of 125 mm charges was calculated to minimize the loss to the ammunition capacity of the Object 432 and T-62 as part of an abortive plan to up-gun these tanks when the D-81 was ready.

Table 33: 125 mm charge basic dimensions

Complete length (mm)	408
Stub case weight (kg)	3.45
Rim diameter (mm)	172
Mouth diameter (mm)	160
Combustible case diameter (mm)	156

It was also much easier to restock an autoloader from an internal reserve of ammunition when the cartridges were of a manageable size and weight; initial design drafts for the D-81 indicated that a hypothetical semi-combustible unitary 125 mm HE-Frag round would be 1,170 mm long and weigh 39 kg[783], which would have been categorically unacceptable. Restocking from an external source was also made easier, considering the size of the turret hatches.

Operationally, working with two-part ammunition was not much different from unitary ammunition. Powder charges were individually packed in hermetically sealed steel tubes, each charge paired to a projectile in a sealed wooden crate. KE rounds were packed into a long steel tube with two compartments, one for the charge and one for the projectile assembly. There was no need to manually match projectiles to the correct charge.

Бронеобъектов', *Вестник Бронетанковой Техники* 1970, сборник 4. p. 37.
[782] Немцов, А. В., Иванов, Н. А., Квашнин, В. Н. (2007). *НИМИ: Федеральное Государственное Унитарное Предприятие «Научно-Исследовательский Машиностроительный Институт» 1932-2007*. Москва: Информационно-методический центр "Арсенал образования". p. 33.
[783] Чернышев, В. Л. (2007). *Танки и люди. Дневник главного конструктора*. Available at: http://militera.lib.ru/db/morozov_aa/08.html (Accessed: 22 November 2023).

Misfires and dud rounds were also easier to deal with. Because the main charge was separate to the projectile, a misfire could be rectified by replacing the charge alone [784], which was not always possible with unitary semi-combustible cartridges because the projectile would be engaged with the forcing cone. There was always some risk that forcefully pulling on the metal rim with the projectile stuck to the forcing cone might rip the case sidewall.

The semi-combustible charge concept, sometimes referred to as a hybrid charge[785], conferred most of the advantages of a combustible charge without losing compatibility with conventional sliding wedge breeches[786], which require a metal cartridge base to seal the barrel. Moreover, replacing a large part of a metal case with a combustible material allowed the case to contribute to the total propellant energy, while also reducing manufacturing cost, metal consumption, and overall ammunition weight.

Thanks to the diminished internal surface area of the metal base, the residual powder smoke emissions were also greatly reduced[787]. Ordinarily, the smoke remaining inside the barrel after each shot was cleared by the bore evacuator, but ejected cases were unventilated and contained a significant amount of residual smoke, which was typically released into the fighting compartment unless the case was immediately ejected out of the tank. Tests on the T-10M (Object 262) heavy tank showed that, compared to metal-cased charges, semi-combustible charges gave a 77.6% and 60% reduction in carbon monoxide concentration at the commander's and loader's stations respectively[788].

The combustible case was formed out of nitrocellulose combined with plant cellulose fibers. This compound was then dipped into molten TNT. The nitrocellulose was gelatinized by TNT, turning its fibers into a gelatinous mass that functioned as a binder for the cellulose fibers. This changed its structure into a denser, mechanically stronger moldable sheet[789], and lent the combustible

---

[784] Министерство Обороны СССР (1980). *Руководство по действиям экипажа при вооружении танка Т-72*. Москва: Воениздат. р. 55.
[785] U.S. Department of the Army (1979). *Engineering Design Handbook: Breech Mechanism Design*. DARCOM-P 706-253. Alexandria: U.S. Army Materiel Development and Readiness Command. p. 2-9.
[786] Early on, this was critical to full interchangeability between metal-cased and semi-combustible charges when the technology was first applied to 122 mm tank guns.
[787] Королева, А. А., Кучерова, В. Г. (eds.) (2002). *физические основы устройства и функциионирования стрелково-пушечного, Артиллерийского и Ракетного Оружия - Часть 1*. Волгоград: Волгоградский Государственный Технический Университет. р. 309.
[788] Павлов М. В., Павлов, И. В. (2021). *Отечественные Бронированные Машины 1945-1965 гг. - Часть I: Легкие, средние и тяжелые танки*. Кемерово: ООО "Принт". pp. 40-41.
Also see: Шабалин, В. А. (1970). 'О Концентрации Пороховых Газов в Зоне Дыхания Экипажа Бронеобъектов', *Вестник Бронетанковой Техники* 1970, сборник 4. р. 37.
[789] Filipović, M. (1987). 'Dejstvo trinitrotoluena na nitrocelulozno-celulozni list', *Naučno-tehnički pregled*, Vol.XXXVII, 1987, br. 2. pp. 19-23.

casing material its characteristic orange-brown color. The finished product very closely resembled varnished fiberboard in appearance and mechanical strength. To provide the rigidity and toughness required for safe handling, the case wall was no less than 1.5 times the thickness of the steel stub case mouth wall.

The coat of TNT on the outer (and inner) surfaces of the case material served as a moisture barrier, as TNT is non-hygroscopic. The final product resisted swelling from moisture and protected the powder from mechanical damage. It was also insulative enough to prevent premature ignition ("cook-off") in a heated gun chamber for up to three minutes[790], against surface temperatures that could exceed 300°C after intensive firing.

All combustible cases in tank gun cartridges (Soviet type, British 120 mm rifled type, German 120 mm smoothbore type) contained nitrocellulose intermixed with plant cellulose fibers[791], but foreign-made cases contained an additional resin binder instead of gelatinous nitrocellulose, reducing their energy content.

The main downside of the combustible casing was that it was slightly hygroscopic owing to its nitrocellulose content. When stored in their hermetically sealed storage containers, however, the rate of degradation was no higher than unitary cased cartridges.

Alone, the casing of the main charge weighed around 0.41 kg [792] and the combustible casing of the supplementary charge on KE projectiles weighed 0.42 kg, making a sizeable contribution to the total energy of the powder charge. The main charge was sealed at the top by a reinforced cap[793] and at the bottom by a robust 3.45 kg steel stub case containing the primer. The stub case was responsible for sealing the mouth of the chamber against the breech, preventing powder gasses from bursting out of the gun. Most of its weight was from its exceptionally thick base, made to be unusually resistant to rupture because the D-81's breech block left a significant part of it unsupported when it closed over the chamber. All charges were fitted with the GUV-7 electric-percussion primer.

---

[790] Министерство Обороны СССР (1988). *125-мм Танковые Пушки 2А26, 2А46, 2А46-1, 2А46М, 2А46М-1, 2А46-2 - Техническое описание и инструкция по эксплуатации 2А46ТО1, Части 3: Боеприпасы*. Москва: Воениздат. p. 67.

[791] In various descriptions, the terms "wood pulp", "kraft", "kraft wood fiber" are used, all referring to the same thing. Kraft refers to the kraft process used to obtain cellulose from wood.

[792] Приложение: Характеристики отечественных артиллерийских выстрелов, in Королева, А. А., Кучерова, В. Г. (eds.) (2002).

[793] The reinforced cap was originally intended to ensure the integrity of the charge during the single-stage ramming cycle of the T-64 autoloader, particularly when paired with a heavy projectile like a HE-Frag shell. In the two-stage ramming cycle of the T-72 autoloader, it was largely redundant.

Figure 139: 2A46 after ejecting spent stub case. Practically all of the smoke released from the breech comes from the stub case.

Propellant Powders for 125 mm Charges

Selecting a smokeless propellant powder that meets diverse military requirements poses a formidable challenge. A delicate balance must be struck between performance, robustness, safety, propensity to produce flash, smokiness, and so on. It is generally considered desirable to reduce chamber volume and ammunition weight by utilizing powders with a larger force constant [794]. These powders must also burn at a reasonable flame temperature[795], as thermal erosion is major component of overall bore erosion. Regardless of the propellant composition, if the force and temperatures are equal, then the powders are, for all intents and purposes, also equal as far as the interior ballistics are concerned[796].

By refining conventional formulae and inventing new propellants, ballisticians sought to create powder charges capable of delivering high force at a low flame temperature, typically taking 3000 K (2727°C) as a benchmark, beyond which accelerated thermal erosion would occur[797]. In the Soviet Army, however, instead of advanced compositions like the various triple-base powders (American M30, British "NQ") deployed in the West, 125 mm cartridges relied predominantly on single-base propellant containing only nitrocellulose, the simplest form of smokeless powder.

For a period after the Great Patriotic War, high-powered large caliber gun artillery in the Soviet Army relied on two types of double-base powders. The first and most prolific type was conventional double-base powder, consisting of nitrocellulose (NC) and nitroglycerin (NG), originally developed by the Soviet Navy. These powders were designated under the NDT index[798]. For naval artillery, the low hygroscopicity[799] of NC-NG powders compared to pure NC was likely the decisive advantage. This was supplanted by a nitrocellulose and diethylene glycol dinitrate (DEGDN) powder, designated under the DG index[800] and commonly referred to simply as diglycol powder[801].

---

[794] The amount of energy transferred to the projectile by the combustion of the propellant by mass.
[795] The adiabatic flame temperature, i.e. the temperature of combustion products without heat loss or gain from external bodies.
[796] Ball, A. M. (1964). *Engineering Design Handbook: Explosives Series - Part One: Solid Propellants.* AMCP 706-175. Washington, D.C.: U.S. Army Materiel Command. p. 53.
[797] Ibid. p. 59. Also see: Teipel, U. (ed.) *Energetic Materials: Particle Processing and Characterization.* Weinherm: Wiley-VCH Verlag GmbH & Co. KGaA. p. 10.
Various Soviet-era textbooks also refer to 3000K as a limiting temperature.
[798] НДТ - Нитроглицериновый, с Дибутилфталата и Динитротолуола (Nitroglycerin, with dibutyl phtalate and dinitrotoluene).
[799] Ability or tendency to absorb moisture from the environment.
[800] ДГ – Дигликолевый (Diglycol).
[801] Гальвиц, У., Мигрина, Б. А. (1950). *Артиллерийские пороха и заряды.* Москва: Издательство Оборонгиз. pp. 171-187.
Also see: Горст, А. Г. (1949). *Пороха и взрывчатые вещества.* Москва: Государственное Издательство Оборонной Промышленности. p. 173.

NC-NG powders could contain a great deal of energy, reaching up to 1,700 kCal/kg while NC alone was limited to 950 kCal/kg and NG alone reached 1,485 kCal/kg [802]. However, this came at the expense of extreme temperatures disproportionate to the force produced. To create practical double-base powders, adjustments were made to the ratio of NG or DEGDN to NC, and organic coolants were added to temper the flame temperature. Six standard energy categories from 1 (lowest) to 6 (highest) were established for all multi-base propellants, of which only categories 3 and 4 (e.g. NDT-3, DG-4) were found to be actually suitable for service in the vast majority of guns.

The force of double-base powders in these categories, watered down as they were, did not surpass ordinary NC powder, so from a technical perspective, it was already questionable if conventional NC-NG or NC-DEGDN powders made sense for the ground forces. For a while, NC-DEGDN powders justified themselves by their low flame temperatures, but by the late 1960's, even this was eventually deemed to be of insufficient value for tank guns.

Instead, the Army returned to single-base powders for most large caliber guns. In subcaliber ammunition, high-nitrogen NC would be used instead of standard medium-nitrogen NC to deliver increased energy. Cool-burning NC-DEGDN powders continued to be produced, but primarily for indirect fire artillery (field guns, howitzers)[803], presumably because it was where the relative benefit of maximizing barrel life counted the most.

Nitrocellulose by itself had a number of advantages as a practical tank gun propellant, ranging from the high availability of cellulose sources (cotton and wood), to its easy upkeep in storage, its stable mechanical properties at extreme temperatures[804], and its low susceptibility to ignition compared to NC-NG powder, with an ignition temperature of 315°C compared to 150-160°C.

In storage, Soviet NC powder was stable throughout a standard 10-year warranty period. Beyond 10 years, chemical and physical degradation of the powder occurred at a rate closely tied to the storage conditions[805], but this was mainly the concern of ammunition stockpiling sites rather than depots that saw regular use. Officially, the issue of powder degradation in tank units was solved by rotating ammunition stocks in 5–7-year cycles[806].

---

[802] Гальвиц, У., Мигрина, Б. А. (1950). *Артиллерийские пороха и заряды*. Москва: Издательство Оборонгиз. p. 169.
This is because NC is oxygen deficient while NG contains an excess of oxygen, so NG supports the combustion of NC.
[803] As of 2024, DG-3 (NC-DEGDN) powder is still used in modern 152 mm ammunition for the Msta-S.
[804] NC-NG powders inherently suffered from cold embrittlement, but NC alone does not.
[805] Бирюков, И. Ю. (2006). 'Пороховые Заряды Длительных Сроков Хранения: Проблемы, Задачи и Пути их Решения', *Интегрированные технологии и энергосбережение*, 2. p. 54.
[806] Анипко, О. Б. (2014). 'Результаты Экспериментального Исследования Воздействия Перекиси Водорода на Нитроцеллюлозные Высокомолекулярные Соединения', *Інтегровані технології та*

Inspections on the condition of 4Zh40 charges at the 25–30-year mark showed that decomposition occurred at a rate that was typical for nitrocellulose powders[807]. After unsealing a semi-combustible charge, moisture uptake in a temperate climate dropped its shelf life to an average of three years. This was typically not a problem, because unsealed ammunition would either be expended in combat or carried in tanks belonging to high readiness units, and then expended later in training.

One of the downsides of nitrocellulose powder was its negative oxygen balance, preventing it from being a totally clean-burning powder. "Dirty" combustion leaves behind more solid particulate matter and produces more carbon monoxide instead of carbon dioxide, resulting in smokier and more noxious powder fumes.

In the late 1970's, the need arose for more powerful powders to support the increased muzzle energy of future KE ammunition, aimed at defeating tank composite armor[808]. The new projectiles were to be larger and heavier, and therefore operated at higher pressures while the volume available for additional powder diminished. The simplest solution would have been to revisit simple but extremely powerful double-base powders, leaving their shortcomings in cold weather stability and high bore erosion unresolved. In the U.S. and West Germany, triple-base powders were abandoned for the JA2 double-base powder, reflecting this "brute force" approach.

A direct equivalent to JA2 in composition and specific force could be found in the NDG-6[809] double-base powder, formulated in the 1950's as a potential option for promising new anti-tank artillery, but rather than retreading old ground, an entirely new propellant called APTs-235P was put into service. This was a novel double-base powder with a high-energy additive, possessing characteristics surpassing triple-base powders. Its main shortcoming was that the poor low-temperature stability[810] intrinsic to double-base powders remained unresolved.

It could be argued that a new powder was not strictly necessary within a short timescale, because NC powder was adequate for meeting the gun performance standards of the 1980's in the D-81 system due to its voluminous chamber.

енергозбереження, 2. p. 50.
[807] Анипко, О. Б., Хайков, В. Л. (2012). 'Анализ Методов Оценки Состояния Пороховых Зарядов Как Элемент Системы Мониторинга Артиллерийских Боеприпасов', Інтегровані технології та енергозбереження, 3.
[808] Валеев, Г. Г., Сопин, В. Ф., Соков, Б. А. (2004). Артиллерийские метательные заряды. Казань: ФГУП «Государственный научно-исследовательский институт химических продуктов». p. 51.
[809] НДГ - Нитроглицериновый и Дигликолевый. Nitroglycerin with diglycol, category 6 - formulated for maximum powder force.
[810] Cold embrittlement of powder grains forms cracks, which massively increase the burn rate, sometimes to the point of a catastrophic low-order detonation in the barrel.

Nevertheless, APTs-235P served as the new standard powder for 125 mm, 115 mm and 100 mm smoothbore KE rounds throughout the remainder of the Cold War and continued to serve in this capacity long after the collapse of the Soviet Union.

High-Nitrogen Nitrocellulose

The 12/7 V/A and 15/1 Tr V/A[811] powders that filled the 4Zh40 and 4Zh52 charges were two grades of NC powder with a high nitrogen content and a diphenylamine (DPA) stabilizer. The same powders were used in the supplemental charges for subcaliber rounds. The only exception to this was the supplemental charge to the 3BM22 projectile (3VBM9 assembly), which instead contained 16/1 TR V/A powder. The increase in web thickness slightly lowered the combustion rate, likely to support an increase in muzzle energy without incurring a major change in peak service pressure.

The force constant of nitrocellulose (NC) powder is proportional to its nitrogen content, so when a low-nitration NC powder was not needed for cost or chemical stability reasons (as a component of multi-base powders), high-nitrogen NC was preferable for volume-constrained gun systems. Aside from 125 mm charges, this type of powder was mainly used in 100 mm and 115 mm subcaliber rounds to develop the high service pressures required.

The nitrogen content of medium-nitrogen NC powders was around 12.0-12.6%. High-nitrogen NC powders exceeded 13.0%. Standard Soviet high-nitrogen NC powder contained 13.0-13.2% nitrogen[812], approximately the same as standard U.S. Army single-base powders[813]. Having a specific gas volume of 208 cm^3 NO/g[814], the 12/7 V/A and 15/1 V/A powders occupied the lower end of the high-nitrogen category[815]. The closest analogue to this grade of NC powder was the U.S. Army's M10 powder[816] found in some artillery charges.

---

[811] 14/1 - web thickness 1.4 mm, 1 channel; Tr - tubular - Трубчатые; V/A - high-nitrogen; B/A - высокоазотный.
[812] Косточко, А. В., Казбан, Б. М. (2019). *Пороха, ракетные твердые топлива и их свойства. Физико-химические свойства порохов и ракетных твердых топлив.* Москва: ИНФРА-М.
[813] U.S. Department of the Army (1964). *Engineering Design Handbook: Ammunition Series - Section 4: Design for Projection.* AMCP 706-247. Washington, D.C.: U.S. Army Materiel Command. p. 4-103.
[814] Octavian, et al. (1994). *Procedeu de obt, inere a pulberii granulare tip 12/7V/A, destinată încărcăturii de azvârlire pentru tunul de calibrul 125 mm.* Romanian Patent no. RO104346.
Also see: Octavian, et al. (1994). *Procedeu de obt, inere a pulberii tubulare granulare tip 15/1V/A, destinată încărcăturii de azvârlire pentru tunul de calibrul 125 mm.* Romanian Patent no. RO104528.
[815] Горст, А. Г. (1949). *Пороха и взрывчатые вещества.* Москва: Государственное Издательство Оборонной Промышленности. p. 156.
[816] Trebiński, R., Leciejewski, Z., Surma, Z. (2022). *Determining the Burning Rate of Fine-Grained Propellants in Closed Vessel Tests.* Energies, 15(7). pp. 2680-2694.

Table 34: Basic data on Soviet powders and their contemporaries[a, b]

Powder	12/7 V/A	15/1 Tr V/A	M10	M30A1	DG-4
Type (base)	SB	SB	SB	TB	DB
Force (J/g)	1,016	1,019	1,013	1,073.4	1,004
Temperature (K)	2970	2970	3000	3036	2668

[a] Приложение: Значения характеристик порохов, in Королева, А. А., Кучерова, В. Г. (eds.) (2002). *Физические Основы Устройства и Функциионирования Стрелково-Пушечного, Артиллерийского и Ракетного Оружия – Часть 1*. Волгоград: Волгоградский Государственный Технический Университет.

[b] Ball, A. M. (1964). *Engineering Design Handbook: Explosives Series - Part One: Solid Propellants*. AMCP 706-175. Washington, D.C.: U.S. Army Materiel Command. p. 59.

Perhaps the only notable aspect of these powders was their sheer ordinariness, but this fed into the difficulty of justifying more complex powders for their replacement. When the D-81 was under development in 1963-1964, the initial plan was to use the extremely potent NDG-6 for 125 mm subcaliber rounds (3BM9 by weight description), but it was found that its ignition behavior was unsatisfactory at -40°C[817]. Due to cold embrittlement, the powder cracked and shattered when subjected to the high-pressure gasses released by the ignition charge. The shattering of propellant powders multiplies its surface area, and the combustion rate rises at a meteoric rate. In a best-case scenario, the resultant pressure spike in NDG-6 caused excess wear or even bulged out the bore, but more often than not, the powder detonated. Needless to say, this could not remain unresolved if the ammunition was to be certified for arctic operations, which was a non-negotiable requirement.

Tests with NC powder showed that satisfactory mechanical properties were guaranteed in arctic temperatures while giving the required ballistic performance[818] at a reasonable flame temperature. Only triple-base powders gave better performance, but not by a decisive margin. It was projected that replacing NC powder with M30 triple-base powder in a 115 mm gun system could provide a 2-3% increase in muzzle velocity[819], which is small enough to be practically achievable with NC powder by altering the powder geometry.

NC powder was substantially more stable than double-base powders like JA2 in the higher-pressure Rh120 L/44 or M256. A 125 mm KE round with granular NC powder (4Zh52) had less than half the velocity change of JA2 at reduced temperatures, and a quarter of the velocity change at elevated temperatures.

---

[817] An unresolved intrinsic characteristic of double-base powders containing nitroglycerin.
[818] Валеев, Г. Г., Сопин, В. Ф., Соков, Б. А. (2004). Артиллерийские метательные заряды. Казань: ФГУП «Государственный научно-исследовательский институт химических продуктов». pp. 50-51.
[819] Ibid. p. 48.

Table 35: Comparison between NC powder in the D-81 to JA2 in the M256[a, b]

Cartridge	Charge temperature (°C)	Pressure (MPa)	Muzzle velocity (m/s)	Percent change in pressure (%P/°C)	Percent change in velocity (%V$_0$/°C)
125 mm 3BM15	-40	380	1735	-0.24	-0.05
	15	452	1785	-	-
	40	510	1800	+0.44	+0.032
120 mm M829	-51	452	1535	0.31	0.12
	21	526	1675	-	-
	63	653	1768	0.57	0.13

a Валеев, Г. Г., Сопин, В. Ф., Соков, Б. А. (2004). *Артиллерийские метательные заряды.* Казань: ФГУП «Государственный научно-исследовательский институт химических продуктов». p. 28.

b Kruczynski, D. L., Hewitt, J. R. (1991). *Temperature Compensation Techniques and Technologies - An Overview.* Technical report BRL-TR-3283. Aberdeen: Ballistic Research Laboratory. p. 2.

A complete ballistic match between 4Zh40 and 4Zh52 was maintained at all charge temperatures. The only difference was that 4Zh52 burned slightly faster and developed a slightly higher peak pressure at all temperatures. When KE rounds were paired with a 4Zh52 charge at the factory[820], the supplemental charge was packed with extra phlegmatizer paper to compensate for the increased erosion.

Although rather basic and inherently limited in growth potential, there was no serious obstacle preventing a gun from simply using more NC propellant to reach higher muzzle energies. This was perhaps most apparent in the first batch of Polish 120 mm Pz. 531 APFSDS-T[821] rounds (developed 2005-2014), which were loaded with 9/7 medium-NC powder, a pre-WW2 powder grade long established in local Polish production for the 122 mm M-30 howitzer.

Despite the age and simplicity of this propellant, the ballistic performance of Pz. 531 essentially matched the German 120 mm DM33 round, which contained a 7.2 kg charge of L5460 (a German powder made from JA2[822]). Pz. 531 had a 7.3 kg projectile assembly, equal to DM33, but a heavier 8.4 kg powder charge loaded in an effectively identical case volume. The round attained a slightly higher muzzle velocity (1,670 m/s vs 1,650 m/s) despite a lower mean peak pressure of 442

---

[820] Apparently, this was very rarely done.
[821] MESKO (2023). *Tank Ammunition - 120 mm.* Available at: https://www.mesko.com.pl/en/products/tank-ammunition/120-mm. (Accessed: 2 November 2023)
[822] Bohn, M. A., Müller, D. (2006). 'Insensitivity aspects of NC bonded and DNDA plasticizer containing gun propellants', 37th International Annual Conference of ICT. Pfinztal: Fraunhofer-Institut für Chemische Technologie (ICT). p. 47.

MPa[823] and maximum pressure of 490 MPa[824]. This was because the larger powder weight allowed propellant gasses to retain more pressure as they expanded towards the muzzle, but this came at the cost of a noticeable worsening of the muzzle blast, which came under complaint by Polish Leopard 2 tankers.

APTs-235P

Table 36: APTs-235P powder characteristics[a, b, c]

Powder	JA2	NDG-6	APTs-235P	M30A1
Force (J/g)	1,130.2	1,125.0	1,101.0	1,073.4
Temperature (K)	3410	3235	3060	3036

a   Teipel, U. (ed.) *Energetic Materials: Particle Processing and Characterization*. Weinherm: Wiley-VCH Verlag GmbH & Co. KGaA. p. 10.

b   Жуков, Б. П. (2000). *Энергетические конденсированные системы - Краткий энциклопедическийсловарь*. Москва: «Янус-К». p. 405.

c   Приложение: Значения характеристик порохов, in Королева, А. А., Кучерова, В. Г. (eds.) (2002). *Физические Основы Устройства и Функциионирования Стрелково-Пушечного, Артиллерийского и Ракетного Оружия - Часть 1*. Волгоград: Волгоградский Государственный Технический Университет.

APTs-235P was a double-base powder, consisting of low-nitrogen NC with NG with an additive of "substance Ts-2", or Tetranitrosotetraazadecalin (TNSTAD). TNSTAD is an explosive of the same category as RDX. TNSTAD is notable for the simplicity of its incorporation into a double-base propellant in mass production [825] and for the abundance of nitrogen in its chemical composition. APTs-235P grains had a black color from a deterrent coating[826] similar to the graphite coating on pistol and rifle powders[827].

IR spectroscopy of TNSTAD gaseous decomposition products reveals that carbon monoxide is not one of the main constituents[828], although of course, the other components of APTs-235P release a great deal of it, particularly nitrocellulose. TNSTAD is also nitrogen-rich: the atomic weight content of

---

[823] Kiński, A. (2016). Amunicja z MESKO S.A. do polskich Leopardów 2. Wojsko i Technika, 6. ZBiAM. p. 46.
[824] WITU.
[825] Phương, T. V., et al. (2023). 'Nghiên Cứu Tổng Hợp Hợp Chất Trans-1,4,5,8 Tetranitroso-1,4,5,8-Tetraaza Decalin (Ц-2) Ứng Dụng Làm Phụ Gia Năng Lượng Cho Thuốc Phóng АПЦ-235П Của Đạn 125 MM Trên Xe Tăng T90S', Journal of Military Sci- ence and Technology, 74(8). p. 59.
[826] A protective and insulating coating that reduces the combustion rate of the outer layer of the powder.
[827] It should be understood that despite being a surface-coated double-base powder, APTs-235P did not have the temperature insensitivity of the SCDB powder developed later in the U.S.
[828] Prabhakaran, K. V., Bhide, N. M., Kurian, E. M. (1993). 'XRD, spectroscopic and thermal analysis studies on trans-1,4,5,8-tetranitiosotetraazadecalin (TNSTAD)', *Thermochimica Acta*, 220. p. 178.

nitrogen in TNSTAD is 43.387 wt%[829], lower than nitroguanidine (53.8 wt%)[830] but much higher than high-nitrogen nitrocellulose (13.0-13.2 wt%).

Nitrogen gas products released from powder combustion function as a chemical inhibitor against carbon absorption into the barrel bore (carburization) via carbon monoxide, which otherwise creates micro-cracks in the bore surface. Nitrogen gas products also reduce bore erosion through a thermal protection mechanism known as dynamic nitriding, which had been noted since the late 1970's [831] but was only conclusively proven by the U.S. Army Research Laboratory in 2006[832]. The extent of Soviet knowledge on the topic is unknown, but the effects were there nonetheless. This relatively new understanding of the role of nitrogen in bore erosion offers a compelling explanation for high-temperature propellants exhibiting less erosion than expected, which is observed when comparing the barrel life of the 2A46M against the 2A46.

The direct counterpart to APTs-235P was the British "AX"-series of propellants that replaced triple-base NQ-series propellants for 120 mm APFSDS ammunition. Its high-energy additive was RDX, which has a relatively modest nitrogen content of 37.82 wt%[833] and releases much more carbon dioxide than the triple-based NQ propellant it replaced[834].

Table 37: Compositions of selected high-energy powders

Powder	NC	NG	TNSTAD	DEGDN	Stabilizer
APTs-235P	34.1% (low-N)	33.7%	27%	-	1.2% Centralite
JA2	59.5% (high-N)	14.9%	-	24.8%	0.7% Akardit II
NDG-6	59.5-60%	~12%	-	~25%	3% Centralite

Heil, M., Wimmer, K., Bohn, M. A. (2017). 'Characterization of gun propellants by long-term mass loss measurements', *Propellants, Explosives, Pyrotechnics*, 42(7). p. 706.

However, being a double-base powder, there were formidable shortcomings in low temperature testing during its development. It was found that APTs-235P

---

[829] Зиновьев, В. М., Куценко, Г. В., Ермилов, А. С., Болдавнин, И. И. (2011). *Высокоэнергетические наполнители твердых ракетных топлив и другихвысокоэнергетических конденсированных систем. Физико-термохимическиехарактеристики, получение.* Пермь: «Издательство ПНИПУ». p. 186.
[830] Ibid. p. 19.
[831] Conroy, P. J., Nusca, M. J., Chabalowskil, C., Anderson, W. (2001) Gun Tube Surface Kinetics and Implications', *10th U.S. Army Gun Dynamics Symposium*. p. 232.
[832] Conroy, P. J., Leveritt, C. S., Hirvonen, J. K., Demaree, J. D. (2006). The Role of Nitrogen in Gun Tube Wear and Erosion. Technical Report ARL-TR-3795. Aberdeen: Army Research Laboratory.
[833] Ibid. p. 68.
[834] Conroy, Leveritt, Hirvonen, Demaree. p. 1.

had poor flammability that, when combined with its embrittlement at low temperature, caused the powder grain to shatter under the ignition conditions of a KE supplementary charge. The ignition delay between the main charge and the supplemental charge would be long enough, even with the main charge being made from tubular powder, that the pressure accumulated in the chamber would begin breaking apart the powder in the supplemental charge before it could ignite and burn at its own pace. Detonations occurred frequently in testing, occasionally violently enough to destroy the gun.

After exploring several variations of ignition charges and testing the safety of exchanging APTs-235P main charges for NC main charges (to test an accidental mix-up of cartridge parts), Soviet research in this direction concluded with no workable solution. This led to new KE rounds paired with a 4Zh63 main charge reverting to high-nitrogen NC in their supplementary charges. Consequently, the additional energy supplied by the new powder could not be taken to its full potential, at least during the Cold War.

Charge Composition

4Zh40 was packed with bundled tubular powder and then additionally filled out with granular powder along its periphery. The tubular powder in 4Zh40 (and in most other Soviet artillery charges) was assembled by hand[835]. Propellant weight was measured and dispensed by machine, but the powder sticks were manually bundled together with twine, and then the additives (ignition charges, flame suppressant, de-coppering agent) were packed in by hand.

Table 38: Temperature dependence of 4Zh40 and 4Zh52 main charges, firing standard HE-Frag

Charge temperature (°C)	4Zh40	4Zh52
-40	835 m/s (289.39 MPa)	835 m/s (295 MPa)
15	850 m/s (343.35 MPa)	850 m/s (350 MPa)
40	860 m/s (397.30 MPa)	860 m/s (405 MPa)

Валеев, Г. Г., Сопин, В. Ф., Соков, Б. А. (2004). *Артиллерийские метательные заряды*. Казань: ФГУП «Государственный научно- исследовательский институт химических продуктов». pp. 27, 146.

---

[835] Валеев, Г. Г., Сопин, В. Ф., Соков, Б. А. (2004). Артиллерийские метательные заряды. Казань: ФГУП «Государственный научно-исследовательский институт химических продуктов». p. 18.
Also see: Горст, А. Г. (1949). Пороха и взрывчатые вещества. Москва: Государственное Издательство Оборонной Промышленности. p. 177.

4Zh52 was a more economical alternative to 4Zh40 by switching from tubular powder to machine-filled, loosely packed granular powder, leaving only the additives for manual packing. Despite only partial automation in the assembly process, this change yielded a labor intensity reduction of 48-84% with a total cost reduction of 5-16% compared to manually assembled charges[836].

Despite the economic advantages, production never definitively switched over to the 4Zh52. Both types continued to be produced concurrently throughout the Cold War. The advantage of long, unbroken sticks of tubular powder was in the greater strength and rigidity of the powder bundle, which was important for the structural integrity of semi-combustible charges. They were also ideal for consistent and uniform powder ignition, as the gasses from the ignition charge could easily travel through and around the tubes, minimizing the ignition lag at the front end of the charge. Lacking the natural flame-conveying advantages of tubular powder, 4Zh52 instead featured a combustible primer tube, made from the same material as its combustible casing[837].

4Zh63 featured a slightly modified composition devoid of loose grains. All of the powder was in the form of tubes, and the outermost layer of powder sticks was not APTs-235P, but 16/1 TR V/A nitrocellulose powder for improved ignition performance. Apart from these details, 4Zh63 was structurally identical to 4Zh40.

Additives

In the 4Zh40 charge, there were two pouches (50 g and 35 g) of DRP-2 black powder (BP), a pouch of VTKh-20 flash-suppressing NC powder (50 g), and a coil of fine lead wire for de-coppering. The BP ignition charges were held in donut- shaped pouches, made from special water-resistant cotton. Upon ignition by the primer, the 35 g bottom charge delivered hot, high-pressure gasses through the main propellant bundle and into the 50 g relay charge, which would then combust. If a subcaliber round was loaded, the gasses from the relay charge would burst through a set of special holes in the charge lid to ignite the supplementary charge.

In 4Zh52, the primer tube held 36 g of KZDP coarse-grained black powder, but the 50 g relay charge was retained, along with the flash suppressant and de-coppering wire coil. The configuration of ignition charges in 4Zh63 was completely identical to 4Zh40, and there was the same flash suppressant, but the de-coppering wire was omitted because copper obturator bands had been

---

[836] Валеев, Г. Г., Сопин, В. Ф., Соков, Б. А. (2004). *Артиллерийские метательные заряды*. Казань: ФГУП «Государственный научно-исследовательский институт химических продуктов». р. 19.
[837] Ibid. p. 129.

replaced by polyamide bands on new sabots. In all three charges, the large mass of the relay charge was to minimize variability in the ignition rate of the supplementary charge from the widening of the gap between it and the main charge (known as the "ullage") as the chamber forcing cone wore out.

Table 39: Basic data for 4Zh40, 4Zh52 and 4Zh63 charges

Charge	4Zh40	4Zh52	4Zh63
Powder types	15/1 Tr V/A + 12/7 V/A	12/7 V/A	APTs-235P 16/1 Tr + 16/1 Tr V/A
Powder weight (kg)*	5.0	5.0	5.0
Complete charge weight (kg)	10.0	10.0	9.6

a   Министерство Обороны СССР (1988). *125-мм Танковые Пушки 2А26, 2А46, 2А46-1, 2А46М, 2А46М-1, 2А46-2 - Техническое Описание и Инструкция по Эксплуатации 2А46ТО1. (Части 3: Боеприпасы)*. Москва: Воениздат.

*   All figures are approximate, and are inclusive of the 50 g VTKh-20 charge

VTKh-20 was a single-channel granular nitrocellulose powder with a 20 wt.% perchlorovinyl resin deterrent coating. Like other flash suppressants, which were usually corrosive potassium salts (potassium sulfate, potassium nitrate)[838], the compounds released from the combustion of the flash suppressant mix with the powder gasses and inhibit the secondary ignition of their flammable constituents (hydrogen, carbon monoxide) as they were released into the atmosphere. In the case of powerful tank guns, an impractical amount of flash suppressant was needed to fully eliminate muzzle flash, so instead, the flash suppressant additive in 125 mm charges served only to eliminate backfires[839].

The DRP-2 and KZDP-2 ignition charges also had some supplementary flash suppressing effect. BP is composed of 75% potassium nitrate [840], and its combustion is known to be capable of providing the flash-reducing effect of the equivalent weight of flash suppressant[841]. Incidentally, the sulfur content of these BP ignition charges gave 125 mm powder smoke (and that of many other powder charges) its characteristic sulfuric odor[842], easily recognizable to most people as the smell of fireworks.

---

[838] Горст, А. Г. (1957). *Пороха и взрывчатые вещества*. Москва: Государственное Издательство Оборонной Промышленности. p. 148.

[839] Министерство Обороны СССР (1988). *125-мм Танковые Пушки 2А26, 2А46, 2А46-1, 2А46М, 2А46М-1, 2А46-2 - Техническое описание и инструкция по эксплуатации 2А46ТО1. (Части 3: Боеприпасы)*. Москва: Воениздат. pp. 40, 94.

[840] ГОСТ 1028-79 «Пороха дымные: Общие технические условия».

[841] Ball, A. M. (1964). *Engineering Design Handbook: Explosives Series - Part One: Solid Propellants*. AMCP 706-175. Washington, D.C.: U.S. Army Materiel Command. p. 37.

[842] Wojciech Furmanek, W., Kijewski, J. (2021). 'Constructional Aspects for Safe Operation of 120 × 570 mm

APFSDS

The Soviet Army pioneered the use of armor-piercing fin-stabilized discarding sabot (APFSDS) ammunition during the late 1950's, using fin-stabilized steel rods as a direct substitute for the standard full-caliber steel AP shells of the time. These monobloc steel penetrators were supplemented by composite penetrators containing a small tungsten carbide (WC)[843] core, built to match the penetration power of a comparable APDS round on flat targets while using only a fraction of the tungsten carbide mass.

Unlike the massive WC cores of contemporary subcaliber penetrators or modern tungsten heavy alloy (WHA)[844] long rod penetrators, these steel-bodied penetrators were exceedingly economical in production, likely under the premise that ammunition produced from low-alloy steel facilitated ammunition stockpiling and helped ensure sustainable production during a major war. Despite the USSR being the world's second top producer of tungsten[845] and (occasionally) the world's largest consumer of tungsten[846], conservation was always a major factor in tank ammunition design.

These subcaliber projectiles, both the monobloc and composite types, were only 38-44 mm in diameter at their widest point. Machine-grade carbon steels[847] with differential hardening were used for all monobloc rods. Composite penetrators were made of a softer but tougher steel[848], mainly to better survive the rigid body penetration mode without premature disintegration. It had a negative impact on their performance on high obliquity armor (≥60°)[849]. VN-8[850] grade tungsten carbide was used in composite penetrator cores. VN-8 had sufficient hardness to sustain penetration through armor steel without significant deformation at velocities exceeding 1,500 m/s[851].

The advancement of tank ammunition was one of the top priorities at NII-24 (NIMI), and its successes awarded the institute several prestigious state

---

Ammunition'. *Problemy Mechatroniki. Uzbrojenie, Lotnictwo, Inżynieria Bezpieczeństwa* 12, 4 (46). p. 92.
[843] Wolfram Carbid. "Wolfram" is the scientific name for elemental tungsten.
[844] Wolfram Heavy Alloy.
[845] Rabchevsky, G. A. (1988). *The tungsten industry of the USSR*. Washington, D.C.: United States Department of the Interior.
[846] Dowding, R. J., Tauer, K. J. (1989). *Supply of tungsten in 1989*. Watertown: Army Materials Research Agency.
[847] Бабкин, А. В. et al. (2008). *Средства поражения и боеприпасы*. Москва: Издательство МГТУ им. Н.Э. Баумана. p. 587.
[848] Ермаков, Г. В., Орлов, В. Г. (1968). *Устройство и Действие Боеприпасов Артиллерии*. Пенза: Пензенское Высшее Артиллерийское Инженерное Ордена Красной Звезды Училище. p. 77.
[849] Ibid. p. 86.
[850] Tungsten carbide with 8% nickel matrix.
[851] Знаменский, Е. А. (2017). *Ударное и кумулятивное действие артиллерийских боеприпасов*. Санкт-Петербург: Балтийский государственный технический университет «Военмех». p. 27.

decorations[852]. Nevertheless, the evolution of Soviet APFSDS ammunition was extremely protracted, remaining essentially unchanged for over two decades after the first rounds entered service in 1961. This mirrored the absence of major improvements in the armor protection of NATO tanks during the same period, but as a consequence, the Soviet munitions industry would fail to progress at a competitive pace throughout the 1970's, and the T-72 was put on the back foot by the 1980's chiefly because of the Soviet Army's dated arsenal of armor-piercing ammunition.

The starting point was the 3BM9 steel rod projectile and the 3BM12 composite cored projectile. In 1972, these were followed by the 3BM15 composite cored projectile, and 3BM17 with a capped steel rod, both offering slightly improved penetration power. These were the most advanced types cleared for export together with the T-72.

The same year these two rounds entered service, the *Zakolka* development project was initiated by government directive[853] to enhance the tank-fighting capabilities of all large caliber anti-tank artillery (≥100 mm). The 3BM22 projectile, which took on the *Zakolka* name, was introduced for 125 mm guns in 1976. It featured a heavy WHA cap over its WC core, drastically improving penetration on moderately sloped armor.

It was immediately followed by the *Nadezhda* and *Nadfil*-2 projects, which entered service in 1985[854] as the 3BM26 and 3BM29 respectively. These were intended for improved performance against multilayered targets[855] as a belated reaction to the appearance of composite armor in new NATO tanks. Despite still having the same small WC core and a WHA (or DU) cap, another incremental gain in penetration was achieved by relocating the WC core to the tail of the projectile, and a new aluminum single-ramp sabot improved its internal ballistics.

125 mm APFSDS rounds were guided in the barrel bore by a short sabot and bore-riding fins, eliminating the need for a bourrelet on the sabot. The long

---

[852] Немцов, А. В., Иванов, Н. А., Квашнин, В. Н. (2007). *НИМИ: Федеральное Государственное Унитарное Предприятие «Научно-Исследовательский Машиностроительный Институт» 1932-2007*. Москва: Информационно-методический центр "Арсенал образования". p. 37.
Also see: Русаков, С. (2006). 'Создание боеприпасов гладкоствольной танковой и противотанковой артиллерии', in *Оружие и технологии России. XXI век Том 12 - Боеприпасы и средства поражения* (2006). Москва: Оружие и технологии. pp. 452-454.
[853] Семененко, Н. П., Гогин, В. В., Коленкин, А. В., Кузьмин, В. Н. (2020). 'Научно-Исследовательский Машиностроительный Институт (НИИ-24): Для великой победы, для мирной жизни для победы в Великой Отечественной Войне', Боеприпасы и высокоэнергетические конденсированные системы. pp. 9-10.
[854] Первов, М. А. et al. (2017). *Очерки истории артиллерии государства Российского*. Москва: ООО Издательский дом «Столичная энциклопедия». p. 461.
[855] Ibid.

distance between the copper bore contact nibs on the fins and the sabot obturator band gave the projectile an excellent supporting base length[856], and having no bourrelet, the sabot could be exceptionally light. The mass of the fins was, however, larger than subcaliber fins. When large fins were paired to light projectiles, aerodynamic drag overtook sabot mass as the limiting factor to the kinetic energy delivered to a target at long range.

As long heavy alloy monobloc rods became feasible in the late 1980's, heavier sabots with two or more bourrelets were no longer avoidable, but for the early 1980's, short sabots were still a tenable option; given a starting advantage of 7.7% in gross muzzle energy over a comparable 120 mm APFSDS round, a 125 mm round with a short sabot had a 9.2% advantage in the energy on-target at 2 km[857], despite the higher drag from large-diameter fins. There was also an advantage in the flatness of the ballistic trajectory out to at least 3 km[858].

Ring Sabot

Initially, all 125 mm APFSDS rounds carried a steel ring sabot, better known as an "unclamping" type sabot (разжимного типа) in technical literature. It was a simple and fairly primitive design, characterized by a trouble-free but rather inefficient operation. Owing to the short length of the sabot-projectile interface, large grooves and lands were needed to support the in-bore acceleration stress. After sabot separation, these large lands increased energy losses in the projectile to aerodynamic drag. A pair of angled holes in each sabot petal vented powder gasses to impart an equilibrium spin to the projectile[859], reducing shot dispersion. Beveled surfaces on the fins sustained the spin at 15-20 RPS[860].

Even though the projectile and the sabot were stable when mated together, the sabot itself had only one point of contact with the bore, allowing the petals to be pried and rotated forward and off their lands, or in other words, unclamping the sabot. Rotated in this way, the periphery of the sabot butted against the bore surface. This steel-on-steel contact increased bore erosion[861], and as the bore diameter expanded, the sabots could rotate more, further

---

[856] Also known as "wheelbase".
[857] Calculated based on 3BM26 and DM23 (120), energy at muzzle calculated with sabot.
[858] Иванов, И. К., et al. (1980). 'Особенности Стрельбы Из Танка Новыми Бронебойными Подкалиберными Снарядами', *Вестник Бронетанковой Техники* 1980, сборник 5. p. 20.
[859] The projectile spins only enough to equalize asymmetries in weight and shape, evening out any potential deviation in its trajectory. This is distinct from the spin stabilization of rifle projectiles.
[860] Бабкин, А. В. et al. (2008). *Средства поражения и боеприпасы*. Москва: Издательство МГТУ имени Н. Э. Баумана. p. 588.
[861] Beer, S., Kovařik, M., Ngo, T. S. (2015). 'Bore Wear by 3BM-15 Projectiles', *International Conference on Military Technologies 2015*.

accelerating erosion[862]. A side effect of this behavior was that the petals stretched the obturator band outwards into a tight gas seal, which made it possible for 125 mm subcaliber rounds to be safely fired when the bore diameter was worn to the condemnation limit of 128.4 mm while the obturator band itself was only 128 mm in diameter, though shot dispersion suffered greatly.

The protruding knurls around the circumference of the sabot were a modification for autoloaded ammunition to ensure smooth loading by a powered rammer, which lacked the ability to positively guide projectiles into the breech opening. If a projectile tilted off the guide trough for any reason (tank on side slope, high speed off-road driving, etc.), the knurls prevented projectiles from being caught on any flat surfaces around the side of the breech opening.

Ring sabots separate from the projectile through lift separation[863] from the incoming air and muzzle blast impinging on the scoops on both ends of each petal. The sabot petals separate cleanly and easily clear the large-diameter fins.

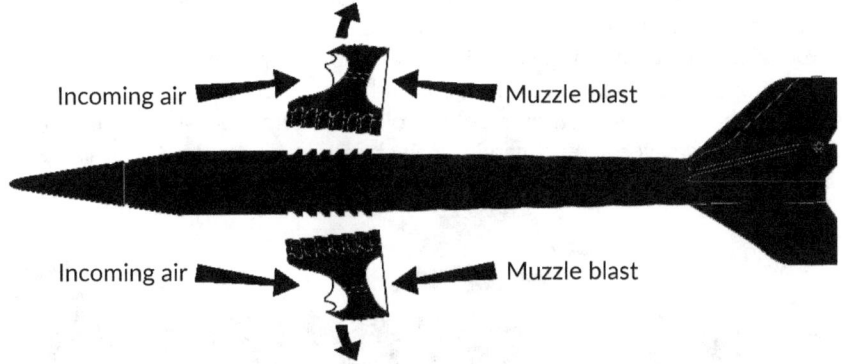

Figure 140: Ring sabot lift separation from incoming air and muzzle blast. The petals depart nearly perpendicular to the projectile. (R. A. Then, 2024)

Single-ramp Sabot

In the 1980's, NIMI moved on to a new aluminum sabot with a single-ramp profile[864] and a long projectile interface. Development was initiated by the NIMI chief designer in 1977. The new sabot was made to support more complex projectiles against lateral disturbances, and it could be mated to the projectile

---

[862] Одинцов, В. (1999). 'Танковое вооружение на пороге XXI века', *Техника и Вооружение*, (October).
[863] Cayzac, R., Carette, E., Alziary de Roquefort, T. (2001). 'Intermediate Ballistics Unsteady Sabot Separation: First Computations and Validations', 19th International Symposium of Ballistics, 7-11 May 2001. p. 297.
[864] The "ramp" or conical shape distributed shear stresses uniformly across the length of the sabot.

on fine lands, reducing aerodynamic drag[865]. Referred to as a "clamping" type (пружинного типа) sabot, the tapered body of the sabot was pressed inward by gas pressure to clamp the petals together, suppressing gas blow-by through the gaps between the petals and solving the bore erosion issue of the "unclamping" design. The new sabot also introduced a low-friction polyamide obturator band to reduce bore erosion and reduce velocity loss[866]. Polyamide notably did not suffer from the moisture and temperature sensitivity issues of conventional nylon bands[867], used in Western tank ammunition since the 1950's.

The tail of the sabot was sheathed in a thin rubber layer (petal obturator) to seal the gaps between the sabot petals, and there was a thick rubber flare (bore obturator) behind the polyamide obturator band designed expand into the gap between the sabot and the bore surface under gas pressure. Such obturators significantly reduced velocity loss from gas blow-by in highly eroded barrels[868].

Well after exiting the muzzle, residual gas pressure continued to hold the sabot petal tails under compression, so the petals rotate off the projectile after a substantial delay under the incoming air flow rather than separating perpendicularly to the projectile. This was known as drag separation. Drag-separating sabots were suitable for guns with muzzle brakes[869], but were more sensitive to slight asymmetries in petal-projectile interaction, and the departing petals had a smaller safety margin against collision with the large-diameter fins[870].

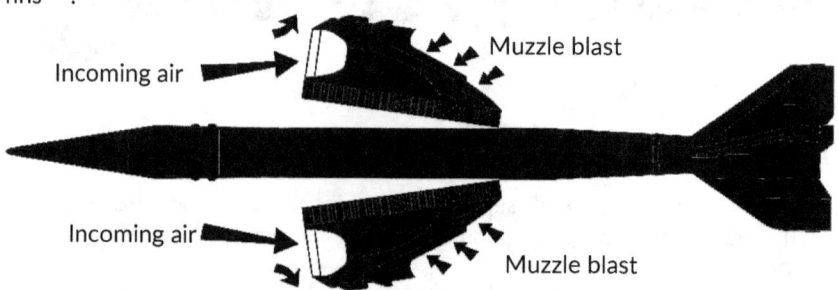

Figure 141: Single-ramp sabot drag separation. (R. A. Then, 2024)

---

[865] Бабкин, А. В. et al. (2008). *Средства поражения и боеприпасы*. Москва: Издательство МГТУ имени Н. Э. Баумана. p. 585.
[866] Polymer obturating belts tested on 125 mm Ring sabots raised the muzzle velocity by 25-30 m/s and reduced barrel wear by around 10%. See: Гашков, Ю. А. (1979). 'Исследование Механического Износа Ствола Танковой Пушки', *Вестник Бронетанковой Техники 1979*, сборник 2. p. 18.
[867] Cold weather embrittlement below -30°C led to increased shot dispersion.
[868] Ward, J. R., May, I. W. (1979). *Muzzle Velocity Drop in Wear-Limited Army Guns*. Memorandum Report ARBRL-MR-02952. Aberdeen: Ballistic Research Laboratory. p. 11.
[869] This was important for the experimental 125 mm 2A66 Anker gun created in the 1980's.
[870] Платонов, А. А. et al. (2003). *Ведущее Устройство Пружинного Типа*. Rospatent No. RU2206055C2. Available at: https://patenton.ru/patent/RU2206055C2 (Accessed: 18 January 2024).

3BM9, 3BM17

Long rod penetrators pierced through armor by erosion. Unlike rigid body penetration, which is dominated by the mechanisms of plastic deformation, eroding penetrators applied stresses far exceeding its own ultimate yield strength and that of the armor material. The penetrator and armor both erode, which is to say that the extreme strain rate rips the particles of both materials apart, ejecting them radially outward. Plastic deformation occurs at the border of this interaction, causing the penetrator to push the hole outward around itself.

The penetrator loses hardly any of its velocity during this process, but once the remaining mass is no longer enough for the residual rod to deliver the instantaneous pressure needed to maintain erosive penetration, the rod transitions to rigid body penetration. From there, it slows down until it is stopped, or breaks through the target. On medium hardness steel armor, the exit hole from a steel long rod penetrator tended to be approximately half the size of an exit hole from a 100 mm AP shell[871].

When impacting a steeply sloped plate, the nose of the rod ricochets, but instead of taking the entire rod with it, it would simply be torn off by the extreme stress, leaving the remaining length of rod to "bite" into the armor and transition to steady-state penetration [872]. Thanks to this behavior, steel long rods enjoyed exceptional penetration power at angles of 70° or above[873].

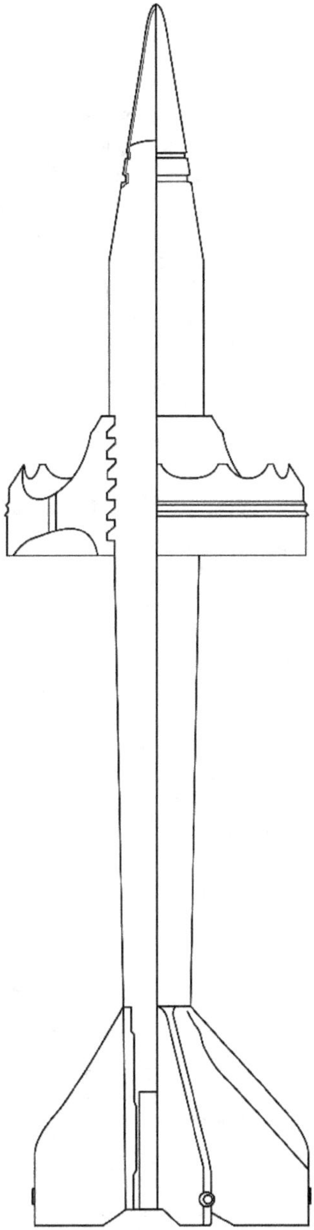

Figure 142: Right: 3BM9 (P. Martin, 2024)

---

[871] At its velocity limit on 80 mm RHA (42 SM) plate at 60°, 3BM15 exit holes ranged from 70 x 75 mm to 95 x 80 mm. 100 mm BR-412B created exit holes ranging from 145 x 95 mm to 170 x 120 mm.
[872] Ермаков, Г. В., Орлов, В. Г. (1968). *Устройство и Действие Боеприпасов Артиллерии*. Пенза: Пензенское Высшее Артиллерийское Инженерное Ордена Красной Звезды Училище. p. 86.
[873] Ibid. p. 85.

and retained some penetration power even at an armor obliquity of 80° [874].

Monobloc rods like 3BM9 showed excellent performance on spaced targets that were ordinarily quite challenging to APCR, APDS and full-caliber steel shot. With light spaced armor (10-20 mm)[875], the effect of air gaps smaller than 350 mm was to improve rather than diminish the penetration power. Thicker (30 mm) sloped spaced armor arrangements saw only a marginal (6-7%) protection increase[876], and with a large air gap (>600 mm), the loss in performance was high, but monobloc steel rods handled such targets better than conventional APDS[877].

3BM12, 3BM15

3BM12 and 3BM15 contained a VN-8 tungsten carbide in their noses, capped off with a small steel cap. The core was borrowed from the 45 mm BR-240P subcaliber shot developed during WWII. Only 20 mm in diameter and weighing only 270 grams[878], the core gave exceptional penetration power on homogeneous armor by enabling it to penetrate efficiently in the rigid body mode instead of eroding.

When traveling through armor, the WC core cut a conical flow of armor material around its pointed (ogival) tip, and in doing so, the steel body enveloping the core eroded against a sloping surface, creating a "self-sharpening" effect[879]. As

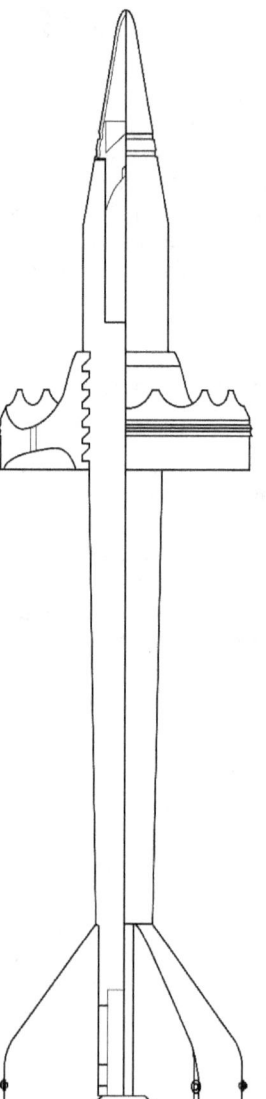

Figure 143: 3BM15 (P. Martin, 2024)

---

[874] Григорян, В. А. et al. (2006). *Частные вопросы конечной баллистики*. Москва: Издательство МГТУ имени Н. Э. Баумана. p. 225.
[875] 10-20 mm steel spaced plate over a 50-100 mm base plate, sloped at 60-65°.
[876] Relative to velocity limit. Due to the high velocity loss experienced by 125 mm APFSDS rounds, small changes in the velocity limit amounted to an insignificant difference in the distance limit.
[877] Григорян, В. А. et al. pp. 232-240.
[878] The combined mass of the cores of seven 14.5 mm BS-41 armor-piercing bullets, as used in WWII-era anti-tank rifles.
[879] 吴群彪, et al. (2019). '前置组合杆体垂直侵彻钢靶简化模型', 爆炸与冲击, 39(1). (Wu, Q. B. et al. (2019). 'Simplified Model of Nose-Composite Rod Vertically Penetrating Steel Target', *Explosion and Shock Waves*, 39(1).) Available at: https://pubs.cstam.org.cn/data/article/bzycj/preview/pdf/bzycj-39-1-013302-1.pdf (Accessed: 2 January 2024).

the penetrator progressed through the target and slowed down, penetration efficiency increased, and the core ceased to produce a cavity wider than itself in diameter as the steel body eroded against the cavity wall until it resembled the shaft to an arrowhead[880]. From then on, penetration occurred solely in the rigid body mode. Most of the projectile mass is retained[881] to devastating effect if armor perforation is achieved.

During the penetration process, the core retains its structural integrity in solid targets due to uniform compression on the ogive nose of the core. Disrupting the uniformity of this compressive load in any way could reduce the penetration power by either changing the trajectory of the penetrator or by separating the core from the body, which occurs at impact angles larger than 15-20°. The WC core would be deflected almost immediately after impact, reverting the penetrator to a steel long rod.

Flat or oblique light spaced armor was effective at neutering the core since WC was brittle. On sloped armor, a larger part of the projectile nose was lost on impact compared to a monobloc steel rod due to its complex shape and relatively low lateral strength. At an equal impact velocity, a monobloc steel rod surpassed the penetration of a cored rod at 60-80° by an average of 10%.

3BM22

To preserve the advantages of rigid body penetration on more challenging targets, a large (1.5 kg), blunt-nosed VNZh-90MT [882] tungsten

Figure 144: Right: 3BM22 (P. Martin, 2024)

---

[880] In this mode, the steel rod functioned solely as a carrier of momentum driving the core through armor.
[881] Бабкин, А. В. et al. (2008). *Средства поражения и боеприпасы*. Москва: Издательство МГТУ имени Н. Э. Баумана. pp. 586-587.
[882] ВНЖ – Вольфрам-Никель-Железо. Tungsten-nickel-iron with 90% tungsten (90W-7Ni-3Fe). Produced by sintering.

heavy alloy (WHA) armor-piercing cap was fitted over the existing WC core to create the 3BM22 projectile. VNZh-90 was directly analogous to the Teledyne X27 alloy[883].

The cap was large enough to preserve the nose of the projectile on moderately sloped armor (<60°), keeping the core in place for the penetrator to function as an arrow-type penetrator like 3BM12 and 3BM15. On steeply sloped plates, the cap functioned as a more efficient penetrator during the impact phase, giving a sizeable but relatively modest improvement in penetration power. Overall armor penetration increased by 25-30% and the range of tank destruction increased by 1.5-2 times[884].

The projectile assembly differed from previous subcaliber rounds in using 16/1 TR V/A powder instead of 15/1 TR V/A powder. The increase in web thickness slightly lowered the combustion rate. This was likely done to support an increase in muzzle energy while maintaining the same peak service pressure of 500 MPa[885].

3BM26, 3BM29

The 3BM26 projectile consisted of a steel rod with a WHA armor-piercing cap analogous to 3BM22. A WC core was placed in the tail of the projectile, resting inside a cylindrical cavity followed by a steel follow-through plug and the tracer. A hollow spacer was fitted in the gap ahead of the core to fix it in place during flight,

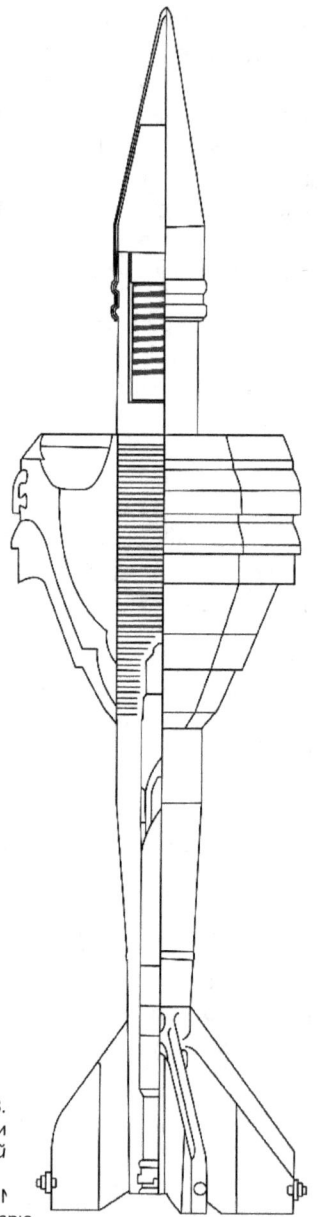

Figure 145: 3BM26 (P. Martin, 2024)

---

[883] Балакин, С. М., Белков, П. А., Данилов, П. Н., Ломов, С. В. (1987). 'Механические свойства сплава ВНЖ-90 при повышенной скорости деформации', *Вестник Бронетанковой Техники* 1987, сборник 8. pp. 59-61.

[884] Правительство Новгородской области (2018). АО «НИМ им. В.В. Бахирева» отметило свой 85-летний юбилей. Available at: https://kuginov.ru/raznoe/nimi-ao.html (Accessed: 3 October 2023).

[885] *Iran Defence Products 2013-2016*. Catalogue. p. 23.

but give way by expanding as the core passed through it once the projectile slows down after overcoming the first layers of a spaced or complex multilayer target. The gap also likely played an important role in balancing the projectile's weight distribution so that the dense tungsten carbide core in its tail did not worsen its aerodynamic stability.

When attacking heterogeneous targets, the WHA cap and the steel body of the projectile was expected to clear a path through the front plate and any other disruptive elements in the armor array to the back plate of the target, without necessarily remaining intact. The tail assembly then delivered the core to the back plate in a relatively undisturbed trajectory, where, owing to its extremely high velocity and efficient shape, it was theoretically capable of perforating a great thickness of RHA steel in the rigid body mode.

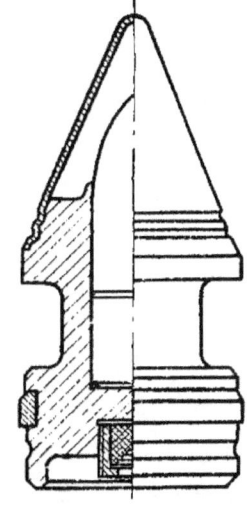

The follow-through plug behind the core augmented its penetrating power by supplying it with additional momentum, essentially similar to the 3BM15 and 3BM22 penetrator concept. In essence, it was a resurrection of the penetrator design of the late-production variant of the Soviet 76.2 mm BR-354P subcaliber APCR (armor-piercing composite rigid) shot of WWII vintage [**Fig. 146**].

The technical details of the 3BM29 projectile are largely unknown, only that it strongly resembled 3BM26, that it had the same penetration performance of 200 mm RHA at 60° at 2 km[886], and that it featured depleted uranium-nickel-zinc (UNTs) alloy. It is likely that the concept behind *Nadfil-2* was to provide an economical alternative to 3BM26 by replacing the tungsten heavy alloy armor-piercing cap with one made from depleted uranium[887] - a relatively cheap nuclear waste product.

Figure 146: Late-war variant of BR-354P subcaliber shot, used in tanks like the T-34. (*Soviet Ministry of Defense*)

Penetration testing of 3BM26 and 3BM29 took place in 1977-1978, most likely with the old steel "ring" sabot[888]. Repeated failures in testing[889] severely delayed its introduction, most

---

[886] Григорян, В. А. (ed.) (2006). *Частные Вопросы Конечной Баллистики*. Москва: Издательство МГТУ им. Н. Э. Баумана. pp. 555-556.
[887] Replacing a WHA cap with a DU cap was also tested on the British 120 mm L15A4 APDS shot.
[888] There are known photographs of proto-3BM26 projectiles used as a tank academy classroom aid.
[889] Гогин, В. В., Горчаков, В.А. (2016). 'Министр, Минмаш, НИМИ: Краткая Историография в Области Артиллерийских Боеприпасов (к 100-летнему юбилею В.В. Бахирева)', *Боеприпасы и*

likely because of the fickle nature of the complex penetrator design, especially when interacting with complex targets.

On homogeneous targets, the design of 3BM26 enabled it to maintain practically uniform penetration power[890] throughout the 0-60° range of impact angles. After the front section of the projectile fully erodes away, the WC core impacts the end of the penetration cavity[891]. Because the end of the penetration cavity is hemispheric regardless of the structural obliquity of the armor plate, and the tail of an eroding long rod projectile maintains its initial trajectory during steady-state penetration[892], the front half of the projectile essentially creates an ideal flat-on impact condition for the WC core. As such, the 3BM26 did not have to contend with core alignment issues on high obliquity armor. The only downside to abandoning the arrow-type penetrator configuration was that the penetration power of 3BM26 on flat targets fell below the level of 3BM15. The concept of a WC core in the tail has survived to the present day in some modern APFSDS ammunition, like the Chinese 125 mm DTC10-125 round[893].

высокоэнергетические конденсированные системы. p. 9.
[890] Потемкин, Э. К. (ed.) (1992). *Основы научной организации разработки.* Том 2. Москва: Издательство МГТУ им. Н. Э. Баумана. p. 172.
[891] 吴群彪, 沈培辉, 刘荣忠 (2014). '后置组合杆体侵彻机理研究', 兵工学报, 10. (Wu, Q. B., Shen, P. H., Liu, R. Zh. (2014). 'Research on the Penetration Mechanism of Tail-Composite Rod Body', Journal of Ordnance Industry, 10). Available at: http://www.co-journal.com/CN/10.3969/j.issn.1000-1093.2014.10.003 (Accessed: 2 January 2024).
[892] Ермаков, Г. В., Орлов, В. Г. (1968). *Устройство и Действие Боеприпасов Артиллерии.* Пенза: Пензенское Высшее Артиллерийское Инженерное Ордена Красной Звезды Училище. p. 85.
[893] Instead of a steel rod with a WHA cap, the entire body was a monobloc WHA rod.

HEAT

With the high ballistic performance of 125 mm subcaliber ammunition, HEAT (High-Explosive Anti-Tank) ammunition had relatively little tactical significance to the T-72 compared to older tanks[894] like the T-54, which depended on HEAT rounds to overcome the armor of tanks like the M60A1 and Chieftain. The T-72's HEAT shells traveled at a muzzle velocity of only 905 m/s, half that of its KE rounds, but the penetration power of shaped charges was both higher (on steel armor) and almost completely independent from impact velocity.

However, due to the enormous difference in muzzle velocity and the high drag of the shell body[895], the practical accuracy of 125 mm HEAT rounds was actually 1.2-1.4 times lower than 125 mm KE rounds. Against a tank like the M60A1, which had no serious protection from either ammunition type, the calculated probability of kill was 1.4-1.9 times lower with HEAT[896], making KE ammunition the obvious choice for anti-tank work, even at long range. This was reflected in the small quantity of HEAT rounds in the authorized combat loads for T-72.

For the most part, 125 mm HEAT shells instead served as multi-purpose ammunition. Its strongest niche was its high lethality against fortifications and lightly armored vehicles. The 3BK14M shell produced around 500 fragments in a 38-47° forward spray, 50% of which could perforate 5 mm of aluminum, and 10% could perforate 60 mm of aluminum[897], with devastating post-penetration effects in an APC or field shelter. For reference, the lethality criteria for a fragment was to perforate a 2 mm duralumin sheet for a man in summer clothes or a 3 mm sheet for a man in winter clothes[898]. Each perforation produced its own secondary fragmentation, or, if the barrier was thin enough, the fragment spray could cut a circular hole, through which the blast products enter the armored enclosure.

On tanks and other targets with significant armor, the fragmentation damage is low because fragments from shaped charge jets are invariably small, fast, and often lack the energy to ignite diesel fuel and gun propellant[899]. Nevertheless, with a sufficient degree of armor overmatch, the post-penetration effect was

---

[894] Лаврищев, Б. П., Соколов, В. Я., Степанов, В. В., Сушков, А. А. (1987). 'Исследование Рационального Боекомплекта Танка', *Вестник Бронетанковой Техники* 1987, сборник 1. p. 17.
[895] The spike tip, lack of a boattail, and straight fins all contributed to a relatively high speed loss.
[896] Лаврищев, Б. П. et al. pp. 18-19.
[897] Михеев, Ю. А. (1987). 'Осколочное Действие Кумулятивных И Осколочно-Фугасных Снарядов При Взрыве На Броне Танка', *Вестник Бронетанковой Техники* 1987, сборник 4. p. 24.
[898] Знаменский, Е. А. (2016). *Фугасное и осколочное действие артиллерийских боеприпасов*. Санкт-Петербург: Балтийский государственный технический университет «Военмех». p. 53.
[899] Hill, F. I. (1951). 'The Damage Effectiveness of Shaped Charges Against Tanks', *Transactions of Symposium on Shaped Charges*, Aberdeen Proving Grounds, November 13-16. Aberdeen: Ballistic Research Laboratories. pp. 364-365.

still significant[900]. As such, despite having more than enough penetration to overcome any tank with all-steel armor, improving the penetration power was still a worthy pursuit.

Tank-fired HEAT shells were limited in standoff distance, and as such, saw relatively more benefit from design measures to increase the jet velocity[901]. This was done in 125 mm HEAT shells by using explosives with a high detonation velocity, by incorporating a wave shaper[902], a variable thickness liner (4.625 mm thinning towards the cone apex), and a liner with a small cone angle of 33° [903]. The variable thickness liner also compensated for the tapered shell cavity, since the walls had to be thicker towards the base for structural reasons[904]. And of course, the large liner internal diameter of 93.5 mm[905] played its part too.

In a length-constrained warhead, the most pertinent advantage of a small cone angle was shortening the optimal standoff distance. In return, the maximum penetration potential was marginally diminished, as decreasing the cone angle also slightly decreases the maximum possible jet elongation[906]. It also had a negative effect on the volume available for the shell's explosive filler, so that in spite of the size difference between 125 mm and 120 mm HEAT shells, the mass of the explosive filler was essentially the same.

Each shell was made in steel and copper liner variants. The copper type was marked by the "M" suffix (e.g. 3BK12M). Steel liners were considered the basic type and the most suitable option for wartime production. Copper, a heavy ductile metal, could produce highly elongated jets but was considered a strategic material to be conserved where possible. The rupture of these jets occurs at relatively large distances, so copper liners had higher penetration in general, and as a rule, were more capable against spaced armor than liners made from lighter metals. For tank and artillery-fired HEAT shells, the penetration power with copper liners was approximately 10% higher than with steel liners. Conversely, steel liners delivered significantly stronger post-perforation damage[907], and were easier to produce to high precision standards.

---

[900] Hungarian live fire testing on a T-54 tank showed that 125 mm HEAT shells generated severe fragmentation damage. See: Ocskay, I. (2018). 'Kísérleti lövészet T-54-es harckocsikra 1989-ben, a „0" ponti gyakorlótéren - II. rész', *Haditechnika*, 4. p. 13.
[901] Знаменский, Е. А. (2017). *Ударное и кумулятивное действие артиллерийских боеприпасов*. Санкт-Петербург: Балтийский государственный технический университет «Военмех». p. 57.
[902] A special type of inert lens to redirect the blast waves to a more favorable incident angle on the liner.
[903] *Projectile and Warhead Identification Guide - Foreign*, NGIC-1143-782-98. (1997). p. 129.
[904] Григорян, В. А. et al. p. 48.
[905] Селиванова, В. В. (ed.) (2016). *Боеприпасы: учебник в двух томах*. Том 1. Москва: Издательство МГТУ им. Н.Э. Баумана. p. 329.
[906] Знаменский, Е. А. p. 49.
[907] Kennedy, D. R. (1951). 'Shaped Charge Damage Beyond Armor', Transactions of Symposium on Shaped Charges, Aberdeen Proving Grounds, November 13-16. Aberdeen: BRL. pp. 359-361.

3BK12

The first 125 mm HEAT shell available in service, 3BK12, was a conventional design with a spike tip for increased aerodynamic stability [908], an electro-mechanical spitback [909] fuze, and a closed-base shell cavity. The stabilizer fin assembly was screwed into a receptacle at the shell base. The charge assembly, consisting of an A-IX-1 [910] explosive filler, metal liner, and the base detonator, was pre-assembled as a single block and then loaded into the cavity through the front opening of the shell body. Screwing the nose spike onto the body secured the charge by clamping down on the lip of the liner. A small truncated cone at the base of the spike shielded the liner against spall from the shell nose and the fuze nose after striking a hard target, along with any stray particles from the spitback element [911].

The electromechanical element in the I-238 was designed for target discrimination, providing instantaneous or delayed action based on the violence of the impact. On homogeneous tank armor, where instantaneous detonation would give maximum penetration, the fuze was initiated in the superquick mode. On spaced armor screens, it gave a delayed reaction so that the shell detonated only after piercing through the screen. This was an early solution to the light spaced armor shielding systems explored during the 1950's and 1960's as a potential solution against shaped charge munitions. In principle, a fuze with a built-in delay could also permit the shell to pierce through light explosive reactive armor (ERA) to detonate only on the underlying armor.

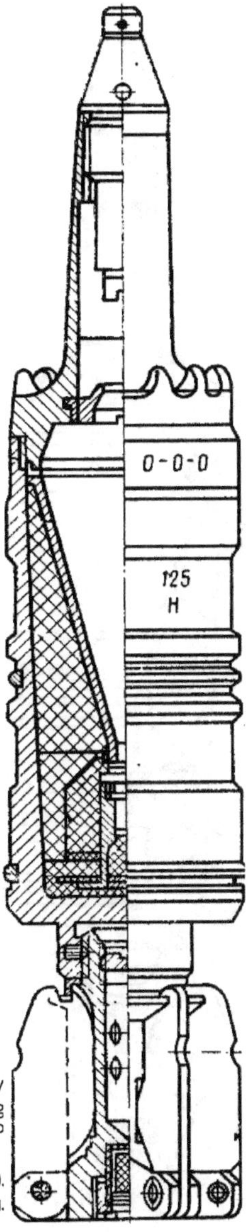

Figure 147: 3BK12 (*Soviet Ministry of Defense*)

---

[908] For the basic explanation for how the spike tip improves stability by reducing nose lift, see: U.S. Department of the Army (1966). Engineering Design Handbook: Design for Control of Projectile Flight Characteristics. AMCP 706-242. Alexandria: U.S. Army Materiel Command. p. 4-10.
The role of the spike tip is also explained in: Селиванова, В. В. (ed.) (2016). Боеприпасы: учебник в двух томах. Том 1. Москва: Издательство МГТУ им. Н.Э. Баумана. p. 328.
[909] A miniature pyrotechnic charge or shaped charge in the tip of the fuze shot down the length of the shell into a base receptable to detonate the main charge.
[910] 95% hexogen (RDX) phlegmatized with 5% ceresin wax.
[911] Лещук, О. В. (2001). Вооружение танков и БМП (Курсовая работа по огневой подготовке). Хабаровск: Хабаровский Государственный Педагогический Университет. p. 29.
This was a standard feature of Soviet HEAT shells with a spitback fuze.

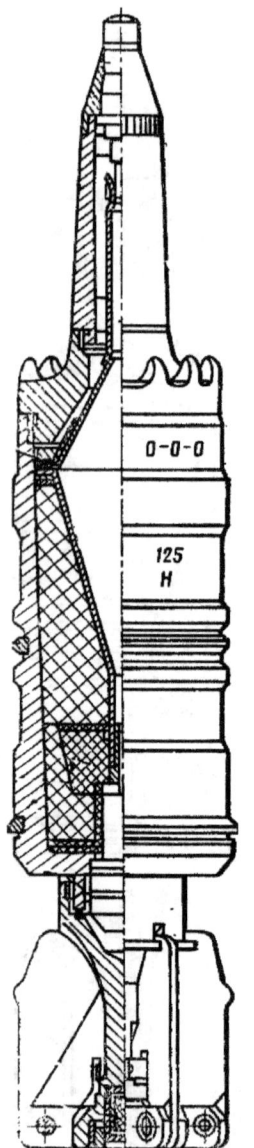

Similar to a handful of other early electromechanical fuzes, the I-238 relied a linear alternator as its source of power. As the projectile accelerated down the barrel, a magnetic core was driven through the linear alternator by setback[912], putting a voltage through a bank of capacitors. Electrical fuzing gave the shortest possible response time between sensing an impact and detonating the shaped charge warhead, which maximized its penetration performance.

When impacting terrain or thin obstacles, the fuze functioned inertially, producing a delayed HE effect. An inertial plunger in the nose was thrown forward by the momentary braking effect on the shell body, closing the ignition circuit of a pyrotechnic delay element. When operating at modest impact angles, the total delay of the inertial fuze was only 0.005 seconds. On a grazing impact, the delay could be as long as 0.25 seconds; far too long for anti-tank purposes, though perhaps still capable of some useful action on other ground targets.

On hard targets, the robust steel fuze nose was crushed against an internal contact, closing a circuit to discharge the capacitors and set off the spitback element. The strength of its nose was a prerequisite for effectively perforating spaced armor shields, but it also made for a longer delay when impacting a heavily sloped plate. The resultant loss of standoff was likely the culprit for the difference in penetration power between flat and sloped impact conditions.

3BK14, 3BK18

A new method of shell construction was adopted for 3BK14, where the liner and Okfol[913] charge assembly were fitted through the front opening of the body, but the base detonator was installed through a hole bored through the base of the shell body. A new stabilizer fin

Figure 148: 3BK18 (*Soviet Ministry of Defense*)

---

[912] For a generic example, see: *Handbook on weaponry* (1982). First English Edition. Düsseldorf: Rheinmetall GmbH. pp. 614-615, pp. 631-632.
[913] 95% octogen (HMX) phlegmatized with 5% ceresin wax.

assembly with a female-threaded base cup was then screwed over the fuze well[914]. The only difference with 3BK18 was that it had a two-piece nose spike instead of a one-piece spike.

The idea of a target-discriminating fuze was abandoned, most probably because it turned out to be, in a word, pointless. Shielded armor was not used on any tank, and even in a worst-case scenario like striking a metal side skirt at a large angle, the loss of penetration power in a large-caliber HEAT shell was not nearly enough to compensate for the thin side armor of modern tanks.

Instead, 3BK14 and 3BK18 shells were fitted with a conventional V-15 piezoelectric fuze with an all-electric fuze train. Upon impacting an obstacle, a shockwave was transmitted through the nose cap and into a piezoelectric crystal, generating a small current. An aluminum electrical conduit cone connected the positive terminal of the piezoelectric crystal to the shaped charge liner, which then connected to the detonator in the base of the shell via the spitback tube[915] formed into the liner. This avoided the need for a wire passing through the liner and charge (as found in the 105 mm M456 and 120 mm DM12 shells), which slightly disrupted jet formation and affected its penetration power. The negative terminals were linked through the shell body itself[916], thus completing the circuit between the nose and the base elements of the fuze.

The maximum angle of impact for normal fuzing action was nominally limited[917] to 70°, but reached a true maximum[918] of 77°. V-15 had no inertial mechanism for graze sensitivity, nor did it have a shoulder fuze, like on some Western tank HEAT rounds. On impact, the reaction time was no greater than 50 μs.

The new Okfol charge and improved fusing were responsible for a sizeable increase in penetration power. Tanks with early multilayered armor or homogeneous armor with heavy add-on modules (e.g. Chieftain Stillbrew) could be engaged with more confidence than the 3BK12 had offered, and conventional all-steel tanks equipped with light ERA could still be overcome by a combination of the short delay period in the V-15 fuze and the sturdy spike, which protected the jet from disruptive elements[919].

---

[914] This is a visual identification feature to distinguish between unmarked 3BK12 and 3BK14/18 shells.
[915] A cylindrical tube formed into the liner, traditionally used to funnel the flame from the charge in a spitback fuze into the base receptable to detonate the charge.
[916] An insulating insert isolated the conduit cone and the shaped charge liner from the shell body.
[917] Оружие и технологии России. XXI век Том 12 - Боеприпасы и средства поражения (2006). Москва: Оружие и технологии. pp. 398-399.
[918] Устьянцев. С. В., Колмаков. Д. Г. (2013). Т-72/Т-90. Опыт создания отечественных основных боевых танков. Нижний Тагил: ОАО «НПК «Уралвагонзавод» имени Ф. Э. Дзержинского». p. 35.
[919] Tests on T-55 and T-62 tanks showed that Kontakt-1 ERA was drastically less effective on 3BK14M compared to the lighter 100 mm 3BK3 HEAT shell, considered analogous to 105 mm HEAT. See: Тарасенко, А. (2022). Испытания ДЗ «Контакт-1» Танки Т-62 и Т-55 – Часть 3. Available at:

HE-Frag

The HE-Frag ammunition for the D-81 was born from the rather unimaginatively named "extended-range HE-Frag shell" project originally initiated in September 1963 to address the lackluster ballistics of the early 115 mm 3OF11 HE-Frag round[920]. The "extended-range HE-Frag shell" project was later expanded to cover not just 115 mm guns, but all three smoothbore calibers in service. The 125 mm variant, designated 3OF19, was ready in time for the T-64A in 1969. It was followed soon after by the 3OF26 shell, which had an enhanced blast effect. An absolute maximum range of 12,220 m could be achieved, but only by artificially increasing the elevation angle of the gun on the T-72 with a ramp.

The shell body essentially corresponded to a conventional spin-stabilized shell body, except that the nose sidewalls were thicker, and the base was much thinner, compensating slightly for the weight of the boattailed aluminum tail boom (3.31 kg). The steel body of 3OF19 was forged from S-60 low-carbon (0.6% C) structural steel[921], a grade used in most Soviet artillery shells. On 3OF26, the shell body was instead made from 45Kh1 (0.45% C, 1% Cr) structural steel with increased strength, which came at the expense of its fragmentation performance. 3OF26 had a press-filled[922] charge of A-IX-2[923]. Press-filling kept the bulk density of the charge similar to its material density, maximizing the

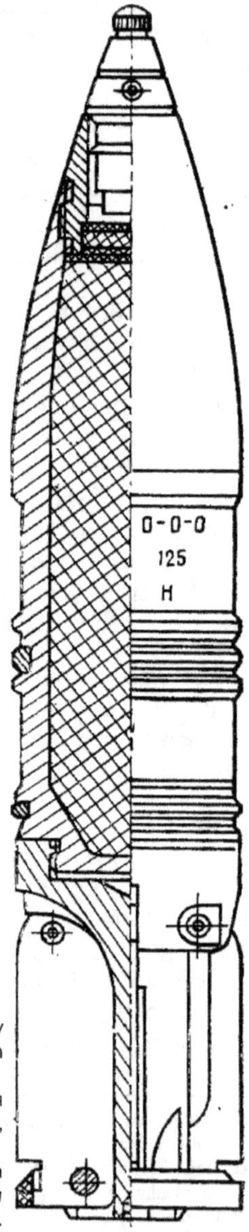

Figure 149: 3OF26 (*Soviet Ministry of Defense*)

---

http://btvt.info/3attackdefensemobility/kontakt_1_trials_pt3.htm (Accessed 20 December 2023).

[920] 3OF11 was an expedient design obtained by filling up the thin-walled cavity of the 115 mm HEAT shell with TNT, making for a usable though light shell with poor range, shot dispersion, and fragmentation effects.

[921] Саттарова, А. И., Диденко, Т. Л., Евграфова, Ю. А. (2017). 'Технология расснаряжения артиллерийских снарядов гидродинамическим методом', *Вестник технологического университета*, 20(9). p. 47.

[922] A mass of explosive filler is placed into the shell cavity, and a punch on a hydraulic press compacts the filler into the cavity. See: *Оружие и технологии России. XXI век Том 12 - Боеприпасы и средства поражения* (2006). Москва: Оружие и технологии. p. 231.

[923] A-IX-2 was aluminized A-IX-1, consisting of 73% RDX, 23% aluminum and 4% wax phlegmatizer. It was shock-insensitive enough to be handled without special precautions and for loading APHE shells, comparable to desensitized Torpex (Torpex-D1, HBX) according to its h50 value, yet had an exceptional blast effect. See: Bajić, Z., et al. (2015). 'Prediction of the Impact Sensitivity of Aluminized Explosive Mixtures Using the Response Surface Methodology', *Scientific Technical Review*, 65(3). p. 5.

charge mass for the same volume. A-IX-2 could not be melted so it was not possible to load shells by melt-pouring like TNT. Instead, it was press-filled, and the aluminum powder mixed into the compound as metallic fuel, responsible for its high blast effect, also functioned as a solid plasticizer to ease its loading[924]. The streamlined nose of the basic 3OF19 shell cavity was unsuitable for press-filling because the diameter of the punch needed to be close to that of the cavity[925], so to lessen steepness of the ogival nose without spoiling its shape, the original solid nose was switched for a screw-on type[926].

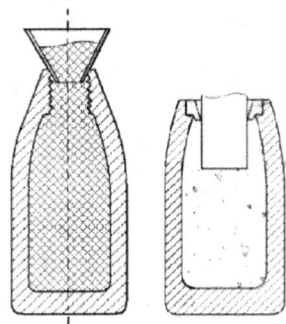

Both shells had a formidable blast and fragmentation effect, nominally comparable to a 122 mm artillery HE-Frag shell. In practice, however, the flat trajectory from direct fire severely restricts the effective fragmentation area and the cratering depth. The target effect depended a great deal on choosing the most appropriate fuze setting.

Figure 150: Loading by melt-pouring (left) and press-filling (right). (*Adapted from Andreev et al.*)

Table 40: Lethal areas of various HE shells under Soviet criteria[a, b]

Tank-fired Shell	125 mm HE-Frag (3OF19)	122 mm HE-Frag (OF-462)	115 mm HE-Frag (3OF11)	100 mm HE-Frag (OF-412)	120 mm HESH (L31A7)
50% Lethal zone (m²)	370	310	250	160	140

a   Барагин, Ю. Д., Степанов, В. В, Сушков, А. А. (1983). 'Характеристика выстрелов с бронебойными подкалиберными и бронебойно-фугасными снарядами', *Вопросы оборонной техники*, 5(111). p. 26.

b   Проскуров, В. А., Скарлато, Л. В., Соколов, В. Я. (1973). 'Об одном направлении повышения эффективности стрельбы из танков', *Вестник Бронетанковой Техники* 1973, сборник 4. p. 27.

3OF19 and 3OF26 were classified as HE-Frag shells because the explosive charge was substantial effect to produce a significant high explosive (demolition) effect, and the shell body could give a powerful fragmentation effect. Depending on the setting on the fuze and whether its nose cap was removed or not, it was also possible to obtain Frag, Delayed HE and even Ricochet effects.

---

[924] Алферов, К. Д. (1965). *Взрывчатые вещества. Часть II: инициирующие и бризантные ВВ*. Пенза: Пензенское Высшее Артиллерийское Инженерное Училище. p. 104.

[925] Андреев, В. В., Гуськов, А. В., Милевский, К. Е. (2018). *Эксплозивные вещества*. Новосибирск: Издательство НГТУ. pp. 193, 200.

[926] Министерство Обороны СССР (1974). *125-мм Танковые Пушки 2А26 (Д-81) и 2А26М2 Техническое описание и инструкция по эксплуатации (Части 1, 2, 3, 4)*. Москва: Воениздат. p. 124.

Table 41: Four possible fuze settings and their modes of functioning

Fuse setting	Cap On	Cap Removed
Setting 'O'	HE	Frag
Setting 'Z'	HE (delayed)	Ricochet

The V-429E was a conventional graze-sensitive point-detonating (PD) fuze. It was a modification of the V-429, which began development in 1947 [927] with the goal of achieving increased safety over the WWII-era RGM-series of artillery fuzes. The main innovation in the V-429 was a self-locking mechanism that kept the fuze safe even when the shell was fired from a heavily worn or clogged barrel [928]. The V-429E was a modification with a thinner cap than the basic V-429 for increased impact sensitivity [929].

The fuze was armed by setback with a relatively short arming distance of 5-7 m. This made 125 mm HE-Frag shells suitable for extremely close combat, including urban areas, forests, or even niche applications like firing over the side of the tank into trenches while driving over them. Fuzes were supplied in the HE setting from the factory and were fitted onto shells this way in the field. The fuze setting would be changed only if the tank commander determined that the HE mode was unsuitable or when a previous shot was observed to have failed to function properly. Fuze setting was done with a special key handled by the commander, and the cap was removed as needed.

In the Frag mode, the fuzing time was instantaneous (<0.001 seconds). In the HE mode, the fuzing time was up to 0.027 seconds on terrain [930], and in the Delayed HE and Ricochet modes [931], it was 0.056-0.063 seconds. The wide range of impact sensitivities and fuzing times between these modes gave a variety of unique target effects, each best suited to specific scenarios.

The Frag mode ensured that the shell detonated on contact with any surface, maximizing the distribution of shell fragments for a potent anti-personnel effect. This could even be exploited by firing shells into a tree canopy for an airburst effect. The Frag mode was often needed for reliable fuzing when firing into soft ground like fresh snow or a marsh, as the shell would otherwise detonate deep below ground or possibly fail to detonate entirely.

---

[927] Трахенберг, B. (ed.) (2000). *НИИ «Поиск»: Страницы Истории*. Санкт-Петербург: Издательство «Формика». p. 50.
[928] Ibid. p. 43. The barrel could still burst in this case, but the shell would not detonate.
[929] Цыбанев, С. А. (2022). *Справочник Сапёра*. Инженерно-Технический Отдел Омон «Зубр». p. 150. Increased-sensitivity fuzes were usually meant for low-velocity artillery to improve reliability.
[930] The fuze cap had to be crushed, which occurred only after the shell penetrated topsoil. The total 0.027-second delay included the time taken to penetrate the ground, followed by a delay of 0.005-0.01 seconds from the inertial mechanism.
[931] The only difference was in initiating the fuze inertially (Delayed HE) or by percussion (Ricochet).

The Ricochet mode was a relic of direct-fire field artillery when fighting infantry on flat terrain or in trenches overlooking flat ground. For tankers, deliberate ricochet bursts were virtually unheard of. Despite this, it was fairly easy to obtain a ricochet on flat terrain because the descent angle of a 125 mm HE-Frag shell at any direct fire range did not exceed 10°-15° [932], which practically guaranteed a ricochet[933]. Even at an impact angle of 15°-20°, the probability of ricochet was still up to 75%[934].

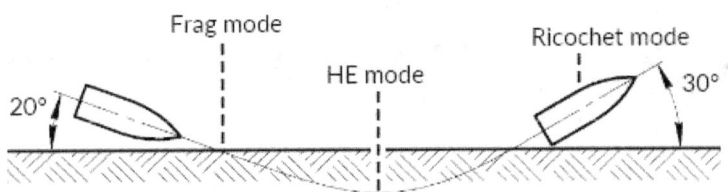

Figure 151: Trajectory and bursting points of HE shells on flat terrain. (*Adapted from E. A. Znamenskiy, 2017*)

The Delayed HE mode was intended for destroying above-ground field fortifications. Earthen parapets or log and sandbag barriers could be fired upon with a good chance of the shell bursting on the other side, to devastating effect. Houses and apartments with brick or thin concrete walls are defeated in the same way. Thick concrete walls destroy the fuze[935], but a shell can still penetrate through its kinetic energy alone, and in the process, break up into a large shower of fragments behind the wall. Keeping in mind the similarity in weight and velocity that 125 mm HE-Frag shares with 122 mm artillery shells, U.S. Army guidelines stipulate that 1.72 m (68 inches) of reinforced concrete was needed to stop a 122 mm shell set to Delayed HE[936]. On flat ground, Delayed HE in direct fire may produce inconsistent ricochet bursts.

Ultimately, however, the basic HE mode was the best choice for most circumstances. The fuze functioned inertially on sand and soil barriers for the desired HE effect on earthen field shelters, firing points with sand bag protection, and soft-skinned or lightly armored vehicle – all of the most common targets to justify a HE-Frag round on the contemporary battlefield.

---

[932] NVA (1981). *Schußtafel 125 mm Panzerkanone D-81 (ST 250/4/005)* in Kotsch, S. (2023). *Die Munition der russischen Panzerkanone D-81*. Available at: https://www.kotsch88.de/m\125\mm\d-81.htm (Accessed: 29 October 2023).
[933] Знаменский, Е. А. (2017). *Ударное и кумулятивное действие артиллерийских боеприпасов*. Санкт-Петербург: Балтийский государственный технический университет «Военмех». p. 12.
[934] Лещук, О. В. (2001). *Вооружение танков и БМП (Курсовая работа по огневой подготовке)*. Хабаровск: Хабаровский Государственный Педагогический Университет. p. 33.
[935] The solid steel fuze body protected the pyrotechnic delay element even after the destruction of its nose, but only to a limited extent. Thick walls can destroy the fuze and likely break up the shell as well.
[936] *Survivability* (1985). FM 5-103. Washington, D.C.: Department of the Army. p. 3-30.

Chapter VII

Machine guns

PKT Coaxial Machine Gun

Figure 152: PKT without its coaxial mount. (*Soviet Ministry of Defense*)

For its coaxial weapon, the T-72 was armed with a PKT long-stroke, gas-operated, belt-fed machine gun. The PKT was designed alongside the PK general purpose machine gun in 1961 by the illustrious Mikhail Kalashnikov. It fired the standard Russian 7.62x54 mmR cartridge. The main tactical niche of a tank coaxial machine gun was to suppress or eliminate infantry in the open[937], but a machine gun was a suitable weapon against soft targets of all sorts.

The PKT was mounted to the right of the main gun. Its barrel protruded from a special embrasure fitted with a sealed ball mount, lightly armored for bullet splash protection. The PKT was fed from the right[938], and its charging handle was also on the right. This arrangement precluded the need for a special charging cable, such as those typically fitted to (Western) left-feeding machine guns with charging handles on the right. It had an electric solenoid trigger connected to its trigger sear in the place of a buttstock.

The cyclic rate of the PKT was 700 RPM or 800 RPM depending on the gas setting[939]. Installed on the coaxial mount in the T-72, the actual cyclic rate was 650-750 RPM[940] due to the shock absorber spring integral to the reciprocating mount. The practical fire rate, taking into account the time for periodic reloads, aiming, and correcting fire with short bursts, was 250 RPM. The PKT had a barrel heat limit of 500 rounds when firing in continuous short bursts, and after 1,000 rounds of continuous fire, the machine gun had to be left to cool down.

---

[937] *Танки* (1973). Москва: Издательство ДОСААФ. p. 78.
[938] This long-standing tradition in Soviet gunsmithing was a holdover from the Maxim machine gun.
[939] Министерства Обороны СССР (1979). *Руководство По 7,62-мм Пулеметам Калашникова ПК, ПКМ, ПКС, ПКМС, ПКБ, ПКМБ И ПКТ*. Москва: Воениздат. p. 216.
[940] Министерство Обороны СССР (1986). *Танк Т-72А - Техническое описание и инструкция по эксплуатации: Книга Первая*. Москва: Воениздат. p. 20.

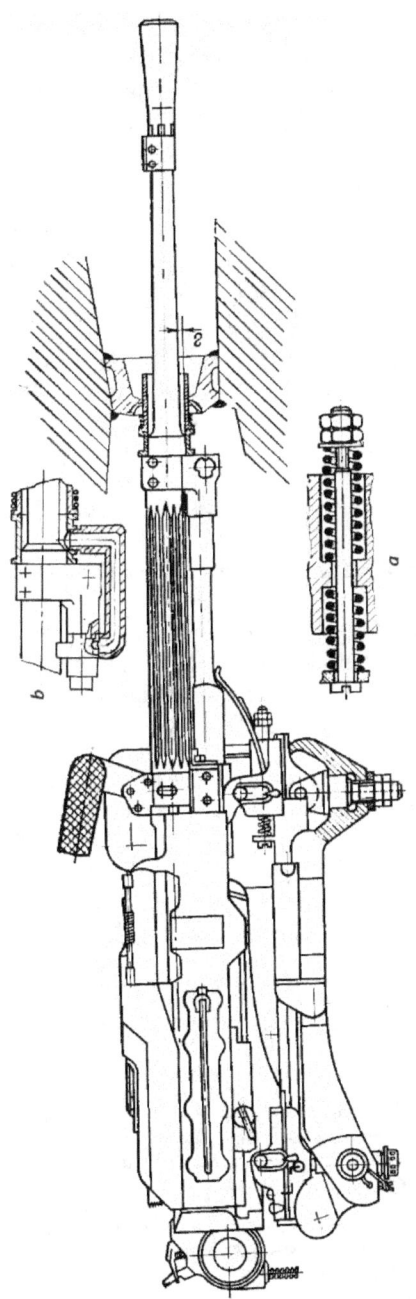

The combat load of the T-72 was divided into eight 250-round boxes for a total capacity of 2,000 rounds. This was less than the 2,500 rounds carried in a T-62, less than half the capacity of a Leopard 1 (4,600 rds.), around a third of any Patton series tank (5,900-5,950 rds.), Chieftain (6,000 rds.) or M60A1 (6,850 rds.), and just a fifth of what an M1 Abrams carried (10,000 rds.).

Spent cases and belt segments exiting the PKT passed through a large chute into a collection box (or bag) positioned directly underneath the machine gun. The collection box had a capacity of 500 spent casings and their belt segments. Another hundred or so casings and their belt segments could pile up in the duct if there was no time to empty the box.

The ready supply of 1,250 rounds was stowed in the commander's station [**Fig. 121, 122, 123**]. The machine gun had one box mounted on standby at all times, and three boxes were stowed on the floor beneath it. Two more were stowed under the commander's seat, one of which was later relocated behind the gunner's seat in the T-72B to make space for a voltage converter unit for the 9K120 guided missile system. The reserve supply of 750 rounds was kept in the hull. It could be retrieved with the turret turned to the left.

Figure 153: PKT coaxial installation with shock absorber (a). Later PKTs have a smooth barrel and a vented gas regulator (b). Early production PKTs had a fluted barrel like the PK, but in a tank with little air flow, its value was highly questionable. (*Soviet Ministry of Defense*)

Each belt had a 2:1 ratio of ball and tracer rounds. Every tenth ball round was substituted by an armor-piercing incendiary (AP-I) round[941]. The standard bullet types were LPS ball[942], T-46 tracer and B-32 AP-I. The point-blank range was 670-700 m on a running or standing soldier (1.5-1.7 m tall)[943], but effective direct fire was possible out to 1,000 m, which was the tracing range of the T-46 bullet. Area targets could be engaged out to 1,800 m.

The commander was responsible for ensuring the serviceability of the PKT. The task of reloading it during combat and occasionally emptying its spent cases were unwelcome additions to his workload, and like other tanks that relied on standard boxed ammunition for their coaxial machine guns (Leopard 1, Chieftain), it also took up valuable time in the heat of battle. If the tank was fighting on the move, the commander had to engage the stabilizer autoblocker to lock the gun in elevation when inserting the belt into the machine gun's feed tray[944], momentarily interrupting the firing of the main gun.

The PKT itself, on the other hand, required very little attention in combat. The reliability, maintainability, accuracy and overall design of the PKM were rated highly by the U.S. Army[945] during comparative testing, ranked just under the M219, MAG and M60E2[946], which was a remarkable outcome given that the only ammunition available for the PKM was captured Soviet and Chinese ammunition in various degraded and corroded conditions[947]. A subsequent assessment of the PKT by the Rock Island Arsenal after tests and engineering analyses was uncharacteristically effusive in its language[948]:

> The PKT has reportedly functioned so reliably and is so inherently rugged, that it has been recommended by highly placed sources as a candidate for exploitation by the US Arsenal System.

---

[941] Шишковского, В. М. (1978). *Огневая Подготовка, Часть Вторая: Основы Устройства Вооружения*. Москва: Воениздат. p. 168.
[942] Postwar universal ball round with a mild steel core, designed to replace Light Ball for rifles and Heavy Ball for machine guns while at the same time reducing the expenditure of lead.
[943] *Руководство По 7,62-мм Пулеметом Калашникова ПК, ПКМ, ПКС, ПКМС, ПКБ, ПКМБ И ПКТ*. Москва: Воениздат. p. 222.
[944] Министерство Обороны СССР (1980). *Руководство по действиям экипажа при вооружении танка Т-72*. Москва: Воениздат. p. 56.
[945] During a U.S. Army comparative study aimed at replacing the troublesome M219 coaxial machine gun.
[946] A special design adapted for the coaxial role, sharing only 63% parts commonality with the infantry M60. Interestingly enough, the M60E2 was ranked highest due to the nuances of the attribute scoring system but a later reevaluation favored the MAG instead, leading to the adoption of the MAG as the M240.
[947] Beeson, J. B., Mazza, T. N. (1975). *Attribute Analysis of the Armor Machine Gun Candidates*. AMSAR/SA/N-24. Rock Island: U.S. Army Armament Command, Systems Analysis Directorate. p. 6.
[948] Rocha, J. G. (1975). *Design Analysis of Machine Gun, 7.62MM, PKT, Soviet*. RTN-75-010. Rock Island: Rock Island Arsenal, General Thomas J. Rodman Laboratory. p. 1.

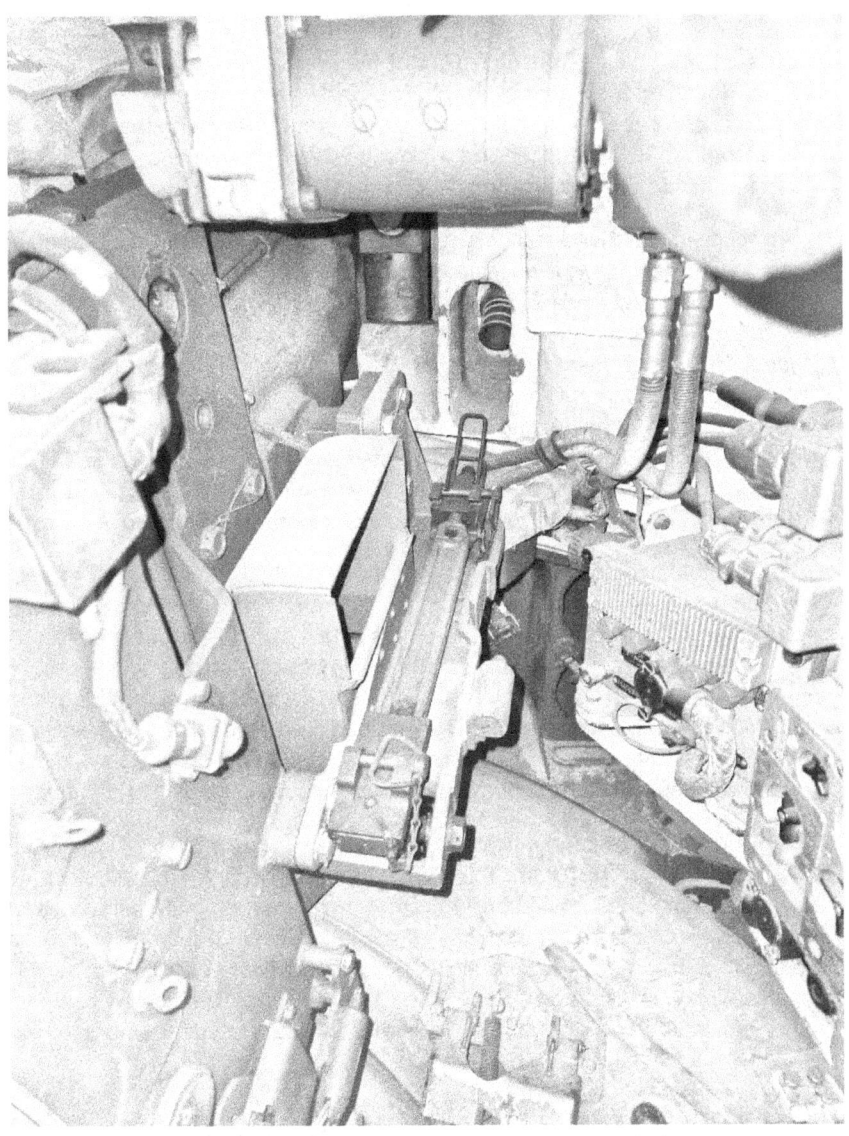

Figure 154: Bare coaxial machine gun mount, without ammunition box holder or case collection bag. The mouth of the case collector chute next to the mount is clearly seen. (*S. Stauber, 2024*)

Functionally, the PKT was still essentially the same as the PK, as the two models retained full interchangeability of the bolt carrier group (bolt) and feed system. However, the extent of the changes made to optimize the PKT to a tank mount was too deep to maintain major parts commonality with the PK. In total, only 14 out of 30 assemblies and 96 out of 176 parts were shared with the PK[949], for just 54.5% parts commonality.

The policy of adapting a standard machine gun for tank use typically kept parts commonality high[950] and reduced cost by allowing all variants to share the same production line, and for a time, the PKT had been manufactured on the same production line as the PK at the KMZ small arms plant in Kovrov[951]. In 1967, production was transferred to the ZMZ small arms plant in Zlatoust, Chelyabinsk[952], where it took over the SGM and SGMT production line. On top of that, the introduction of the PKM, a lightened and more streamlined variant of the PK, further diminished the level of interchangeability between the PKT and the standard machine guns of the infantry. By the time the T-72 began its career in the Soviet Army, the PKT was practically proprietary to armored vehicles.

Besides its solenoid firing mechanism, the PKT had a lengthened barrel, a modified trunnion and receiver, and no trigger group. The removal of the trigger group created a shorter form factor, and the new trunnion added mass around the most heat-stressed parts, making the PKT more robust in sustained fire. The lengthened barrel was not thicker than the standard PK barrel, but it gave the PKT identical ballistics to the SGMT medium machine gun used in older Soviet AFVs[953], especially when firing Heavy Ball ammunition. Heavy Ball was made obsolete by LPS ball in 1953 but was still heavily stockpiled and in circulation.

Like purpose-made tank machine guns such as the American M73/219 and the British L94A1 chain gun, the PKT was made without the possibility of dismounted fire. Even if an infantry tripod was available, compatibility was lost due to the specific configuration of the solenoid trigger mechanism and the rail mount. And despite the structural reinforcements, the service life of the PKT (receiver life) was still rated at only 30,000 rounds[954] - relatively low compared to heavier crew-served machine guns like the MAG 58, rated for 50,000 rounds.

---

[949] From Soviet instructional poster.
[950] For comparison, the tank variants of the FN MAG (MAG 60-40) and MG 3 had 95% parts commonality with their infantry versions.
[951] Пономарёв, Ю. (2011). 'От ПК к ПКМ', *Калашников. Оружие, Боеприпасы, Снаряжение*, 2011, 12.
[952] *История оружейного производства АО Златмаш* (2024). Available at: https://zmgun.ru/history (Accessed 19 December 2023).
[953] When firing the same ammunition, the effect of the higher velocity was negligible; the point blank distance to a running figure (1.5 m tall) only increased from 640 m to 670 m.
[954] Extended to 40,000 rounds. See: *История оружейного производства АО Златмаш*.

Apart from the lengthened barrel, the most important modification made to the PK for tank use was replacing its venting-type gas regulator with a non-venting three-position design[955] from the SG-43, with a sealed gas tube to reduce gas contamination in the turret[956]. Shot for shot, the amount of carbon monoxide released by a PKT was four times less than a PK in a firing port (23 mg vs 83 mg) because of the low gas blow-by in the gas tube[957].

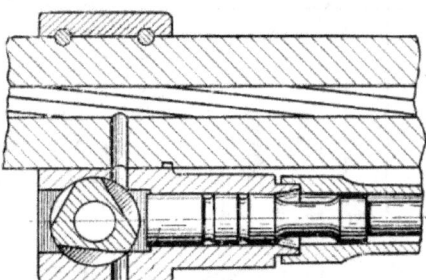

Figure 155: PKT gas regulator, adjustable to compensate for weapon age and fouling. (*Soviet Ministry of Defense*)

However, even with a tight seal and the overpressure maintained by the tank's ventilation unit, gas contamination was still high from powder fumes re-entering the turret through the barrel. This was addressed in 1988[958] by a new PKT barrel with a forward-venting gas regulator [**Fig. 153, b**]. The new regulator vented powder gasses through the coaxial machine gun port, reducing contamination in the fighting compartment by 40-60%[959] without losing compatibility with existing PKTs and tank coaxial machine gun ports.

Unlike shock absorbers in infantry and aircraft mounts, the T-72 coaxial mount was not intended for recoil reduction - the recoil of a 7.62 mm machine gun simply had no effect when it was bolted to a 125 mm gun. Rather, its purpose was to reduce wear on the receiver and mount, decrease shot dispersion by reducing receiver flex and vibration, and guarantee a consistent point of impact.

Tight shot dispersion reduced the danger to friendly troops while performing overhead fires and increased the fire density on area targets at long range (>1 km). Fire precision was also valued for efficient use of time and ammunition. At ranges where individual soldiers in a group were visible through the magnified gunner's sight, they were engaged as a collection of point targets instead of an

---

[955] Министерства Обороны СССР (1979). *Руководство по 7,62-мм Пулеметом Калашникова ПК, ПКМ, ПКС, ПКМС, ПКБ, ПКМБ И ПКТ*. Москва: Воениздат.
[956] Шишковского, В. М. (ed.) (1978). *Огневая Подготовка, Часть Вторая: Основы Устройства Вооружения*. Москва: Воениздат. p. 21.
[957] Нарбут, А. М. (1982). 'Оценка загазованности при стрельбе из БМП', *Вестник бронетанковой техники* 1982, сборник, 1. p. 22.
[958] Альбом основных конструктивных изменения, проведенных на Изд. 172М.
[959] Нарбут, А. М. , Ребриков, В. Д., Трофимов, П. В. (1987). 'Уменьшение загазованности обитаемых отделений БТТ при стрельбе', *Вестник Бронетанковой Техники* 1987, сборник 1. p. 35.

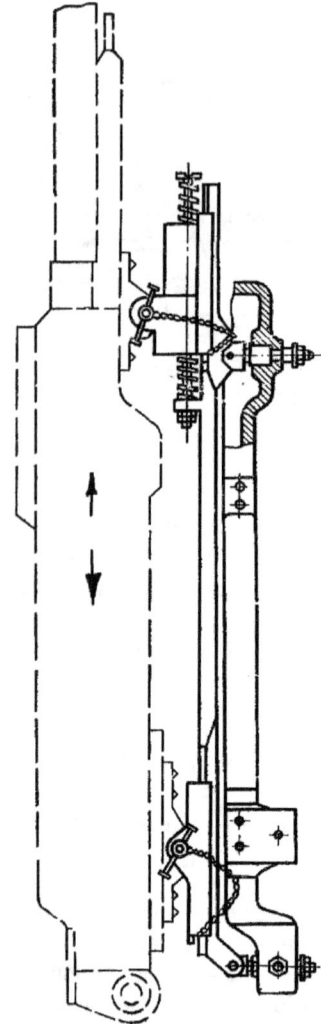

area target[960], but if area fires were needed, the gunner held a fixed elevation and raked the turret from side to side to cover the width of the target frontage.

The shock absorber functioned in a semi-cyclic mode, where the half-period of reciprocation was shorter than one cycle of the operating action of the weapon[961]. The entire machine gun reciprocated back and forth on the shock absorber under recoil, but always returned to its initial resting position at the moment of a shot[962]. Tuned this way, there was no fundamental difference in shot dynamics between bursts and single shots, giving a more consistent point of impact [963]. The cyclic rate of fire decreased because the receiver was in rearward motion during the recoil stroke of the bolt and then moved forward as the bolt ran forward again, reducing the relative velocity of the bolt and extending its cycling time. The higher cyclic rate of 700-800 RPM built into the PKT was an excess margin designed to compensate for this effect on tank mounts and return it to the intended 650-750 RPM cyclic rate of the PK system.

Figure 156: PKT on its reciprocating mount. The layout of its mounting pins made it reverse compatible with SG-43 and SGM mounts. The long wheelbase between the pins afforded it greater stability. (*Soviet Ministry of Defense*)

---

[960] Министерство Обороны СССР (1974). *Правила Стрельбы из Танков (ПСТ-74)*. Москва: Воениздат. pp. 73-84.
[961] Тер-Данилов, Р. А. (2009). 'Основы Функционирования Установок', 17.01.02 *Стрелково-пушечное, артиллерийское и ракетное оружие*. Тульский государственный университет.
[962] Balla, J. et al. (2021). 'Study Effects of Shock Absorbers Parameters to Recoil of Automatic Weapons', 2021 *International Conference on Military Technologies (ICMT)*. Brno: University of Defence.
Balla, J. et al. (2024). 'Study of relationship between motion of mechanisms in gas operated weapon and its shock absorber', Defense Technology 33. pp. 42-54.
[963] Алферов, В. В. (1977). *Конструкция и расчет автоматического оружия*. Москва: «Машиностроение». pp. 220.

The coaxial mount could be adjusted horizontally and vertically to boresight the PKT, and the amount of preload on the shock absorber springs could be adjusted by tightening or loosening the nut securing the springs to the frame. A certain amount of preload was needed to ensure proper function regardless of the weapon's elevation angle.

On the whole, as robust as the PKT was, the idiosyncrasies of the T-72's coaxial machine gun as a system, like the relatively small combat load, the 500-round capacity of the collection box[964], the 500-round heat limit of its barrel, and the use of 50-round belt segments - all of these were rooted in front line feedback from 1942[965], when the standard tank machine gun was the magazine-fed DT. Tankers lamented its small-capacity pan magazines (63 rds.), demanding a belt-fed machine gun capable of delivering 500 rounds of intensive fire. No mention was made of the ammunition capacity, which averaged 2,000 rounds (1,890-2,520) in the tanks of the time, or the absence of a provision for barrel swapping.

Having met those demands after the war in the SGMT and then the PKT, the troops were apparently satisfied, as the requirement simply never changed since. It is difficult to reconcile this against Western design practices which had also been informed by front line feedback, yet arrived at different conclusions on how much machine gun ammunition a tank really needed. Although the PK was itself an excellent weapon, perhaps even the best in its class, the circumstances surrounding the PKT, from its unusually low degree of production standardization to its limitations in the T-72 tank mount, considerably tempered the qualities of the machine gun.

Table 42: PKT data

PKT Technical Specifications	
Maximum range (m)	1,800
Complete length (mm)	1,098
Muzzle velocity (m/s)	855
Cyclic fire rate (RPM)	700-800
Barrel length (mm)	722
Barrel mass (kg)	3.23
Machine gun mass (kg)	10.5
Mass of 250-round ammo box (kg)	9.4
Single-shot Penetration Distance (m)	
Steel helmet	1,700
Soil barrier, 25-30 cm thick	1,000
Brick masonry, 10-12 cm thick	200

---

[964] This, at least, was not unusual. Even tanks like the Abrams do not have larger case collection boxes.
[965] Уланов, А. (2017). *Пулемётная драма Красной Армии. Часть 3: На замену «максиму»*. Available at: https://www.kalashnikov.ru/pulemyotnaya-drama-krasnoj-armii-3/ (Accessed: 17 December 2023).

NSV-12.7 Anti-Aircraft Machine Gun

Starting in December 1974, the T-72 began to be equipped with an NSV-12.7 large caliber anti-aircraft machine gun. It had a maximum effective range of 1,500 m against both aircraft (slant range) and ground targets. Against lightly armored vehicles, the maximum effective range was 800 m[966]. The NSV-12.7 was designed as a modular weapon lacking a trigger, bolt operating handle, and sight mount. These would be provided by specialized mounts for the machine gun. When fitted to a tripod or anti-aircraft mount, it was administratively renamed to the NSVS-12.7, and when a solenoid trigger was attached to the back of the receiver, it became the NSVT-12.7[967].

Originally developed under the TKB-084A designation, the NSV-12.7 entered service in 1972[968] to replace the DShKM as the new standard 12.7 mm machine gun family for the Soviet Army. Its relationship to the DShKM was much like what the PK had been to the SGM. As part of an ongoing effort towards improving the mobility of the infantry, a lighter, more portable heavy machine gun was created, intended to be carried on foot by small two-man teams. It had a lightened barrel with a quick-swap system to offset its thinned walls and the absence of cooling fins[969].

For the sake of complete standardization, the NSV-12.7 also replaced the DShKM on tank and anti-aircraft mounts, where its light weight and quick-swap barrel were largely meaningless since no spare barrel was carried on the T-72, and much like the change from the SGMT to the PKT, the lighter profile of the barrel had arguably been a downgrade. Nevertheless, the NSV-12.7 offered better accuracy and reliability, and its conical flash hider was a great improvement over the muzzle brake of the DShKM in low light conditions.

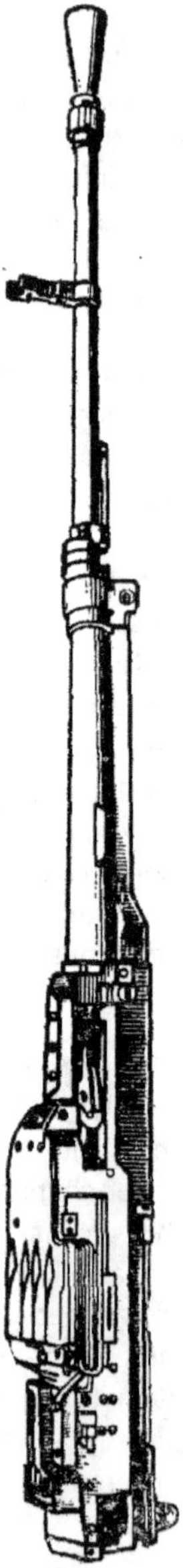

Figure 157: NSV-12.7 as seen when dismounted from the T-72's ZU-72 pintle mount. (*Soviet Ministry of Defense*)

---

[966] Министерство Обороны СССР. (1978). *Руководство по 12,7-мм Пулемету «Утес»* (НСВ-12,7). Москва: Воениздат. p. 1.
[967] Used for remote mounts such as the T-64A cupola remote machine gun.
[968] Koll, C. (2009). *Soviet Cannon*. p. 67. The *Metallist* factory was set up in Kazakhstan for the NSV-12.7.
[969] Spare barrels were provided only for infantry guns deployed in permanent firing positions.

The machine gun had a cyclic rate of fire of between 700-800 RPM, with a practical rate of fire of 80-100 RPM. This coincided with the barrel's heat limit of 100 rounds fired in continuous bursts[970]. Ammunition was fed in sixty-round boxes, feeding the machine gun from the right. Spent cartridge cases were ejected forward, and belt segments were retained in a canvas bag on the left.

Including the box mounted with the machine gun on standby at all times, the standard combat load for the NSV-12.7 was 300 rounds. For the sake of saving internal space, all of the ammunition was stowed externally. Two boxes were kept in the rear external turret bin and two more were strapped to the side of the turret next to the commander's cupola. These two boxes were the easiest to access. Considering that each box weighed 13 kg[971], ease of handling played a key part in placing all of the ammunition externally. The downside, of course, was that all of the ammunition was exposed to combat damage.

Like all firearms in the Soviet Army, ammunition stores were considered to be permanent accessories to the weapon, so the commander had to return emptied boxes and spent belts to their stowage points for later reuse. The Soviet Army retained an archaic supply system[972] of issuing loose cartridges in sealed metal tins which had to be manually loaded into reusable belts, unlike the modern NATO supply system of issuing disposable disintegrating belts prepackaged in disposable boxes[973]. Replacements for lost boxes and belts could be issued, and a spare stock of four loose (unboxed) sets of empty sixty-round belts were kept in the T-72's rear stowage bin[974].

The mix of 12.7 mm bullet types was subject to change depending on the availability of some of the more exotic types[975], but officially, the basic mix contained API and API-T in a ratio of 3:1[976] or 4:1[977] but no higher, to maintain a reasonable density of tracers for anti-aircraft work.

---

[970] *Руководство по 12,7-мм Пулемету «Утес» (НСВ-12,7).* pp. 5-6.
[971] Министерство Обороны СССР (1979). *Вооружение Танка Т-72.* Москва: Военная академия бронетанковых войск. p. 120.
[972] This system dates back to the introduction of the Maxim machine gun to the Imperial Russian army.
[973] The concept of disposable ammunition stores proliferated in NATO since the 1950's, most notably gaining traction with the introduction of the U.S. Army's M13 disintegrating link belt.
[974] Location detailed in: Запасные части, Инструмент, приспособления и Принадлежности /ЗИП/ Т-72. Poster. Quantity of belts specified in: Министерство Обороны СССР (1980). *Танк Т-72 - Каталог деталей и сборочных единиц.* Москва: Воениздат. p. 343
[975] MDZ (IAI - Instant-Action Incendiary) rounds may also be issued when available.
[976] Министерство Обороны СССР (1976). *12,7-мм Пулемет НСВ-12,7 «Утес» на Универсальном Станке: Техническое описание и инструкция по эксплуатации.* p. 101.
[977] *Руководство по 12,7-мм Пулемету «Утес» (НСВ-12,7).* p. 192.

12.7 mm Penetration Power

Table 43: Penetration limits of standard 12.7 mm ammunition[a, b]

Bullet Type	Armor Type	Range	Thickness (0°) (mm)	Thickness (30°) (mm)
B-32 (API)	RHA	0	29.5	21.8
		100	26.9	20.0
		300	21.8	16.3
	Al (5083)	0	88.6	65.8
		100	80.7	60.2
		300	65.8	49.0
		345	-	44.0
BZT-44 (API-T)	RHA	0	17.3	12.7
		100	15.2	11.4
		300	12.4	9.1
	Al (5083)	0	51.6	38.1
		100	46.0	34.0
		300	37.3	27.7

[a] Dotseth, W. D. (1971). *Survivability Design Guide for U.S. Army Aircraft. Volume 2 - Classified Data for Small-Arms Ballistic Protection.* Technical Report USAAMRDL 71-41B. Fort Eustis: USAAMRDL. pp. 20, 21.

[b] Дворянинов, В. Н. (2015). *Боевые патроны стрелкового оружия, Книга 4: Современные отечественные патроны хроники конструкторов.* Климовск: Д'Соло. p. 258.

Despite its heavy bullets and high muzzle velocity, the NSV-12.7 did not have a significantly flatter trajectory than the tank's coaxial machine gun[978], which severely limited its usefulness against targets in the open. Against long-distance targets, the lack of a magnified sight was a decisive handicap.

Its main niche was to deter or destroy low-flying aircraft, suppress and eliminate infantry in thin-walled structures and soft-skinned vehicles, and to a lesser extent, to destroy lightly armored vehicles. In practice, the merits of the NSV-12.7 as an anti-armor weapon were decidedly unconvincing, even against light armor, which should not be surprising given that the 12.7 mm threat was a basic frontal protection requirement for most armored vehicles. Armored personnel carriers were, as a rule, frontally resistant to 12.7 mm attack, and even from the side where penetration was much more likely, target destruction could only be guaranteed at relatively close ranges.

The standard 12.7 mm API (B-32) bullet was rated for a 90% probability of igniting 70-octane gasoline behind a 20 mm armor steel plate at 100 m. With it, the probability of obtaining one penetration in the front armor of an M113 (38

---

[978] Point blank range of 780-800 m on a target 1.7-1.8 m in height; only 80 m (10%) further than PKT.

mm aluminum at 45°)[979] with a seven-round burst at just 100 meters was 57%[980]. Beyond even this modest range, the probability of success fell close to 0%. Tests also showed that knocking out an M113 from its side would require an average of 27.3 shots at 300 meters and 33.5 shots at 500 meters[981].

ZU-72

The NSV-12.7 together with its mount were referred to as the ZU-72 anti-aircraft installation. The mount included a pintle ring and a reciprocating cradle with a recoil buffer spring tuned to suppress vibrations in the sights and improve shot dispersion. It also helped that the gun and cradle together had considerable mass (70 kg) to tamp the recoil. The ZU-72 could be locked to the turret facing rearward, allowing the commander's cupola to rotate independently (traveling mode) or locked to the cupola, so that the two could rotate together (combat mode)[982] for circular aiming with the commander's open hatch counterbalancing the ZU-72. The elevation range was -5° to +75°.

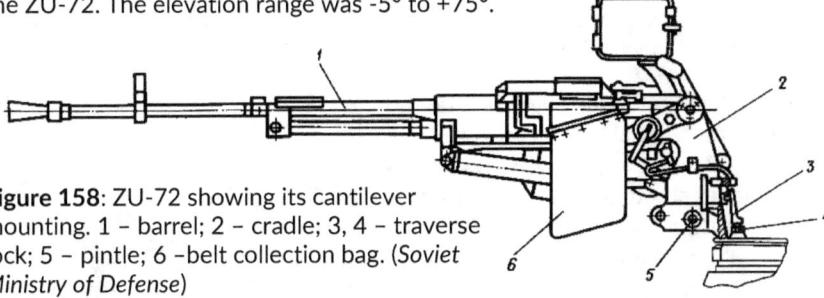

Figure 158: ZU-72 showing its cantilever mounting. 1 – barrel; 2 – cradle; 3, 4 – traverse lock; 5 – pintle; 6 –belt collection bag. (*Soviet Ministry of Defense*)

Open machine gun installations like pintle and ring mounts were the simplest way of putting a machine gun on a tank. They also exposed the operator to return fire when engaging ground targets, but for an anti-aircraft weapon, the trade-offs were not entirely unreasonable. When riding from an open hatch on a march, the tank commander could rapidly prepare the machine gun for combat upon receiving an air alert[983], or he could even have the machine gun ready on standby at all times. Moreover, the chief designer of the T-72 personally argued in favor of open mounts based on the idea that open sights gave the visibility needed for anti-aircraft work[984], but this is a decidedly less convincing factor.

---

[979] Aluminum 31 mm at 0° lower side, 43 mm at 0° upper side. ACAV gun shield – 8 mm steel at 25°.
[980] Дворянинов, В. Н. (2015). *Боевые патроны стрелкового оружия, Книга 4: Современные отечественные патроны хроники конструкторов*. Климовск: Д'Соло. p. 258.
[981] Ibid. p. 271.
[982] *Руководство по 12,7-мм Пулемету «Утес» (НСВ-12,7)*. p. 134, 136.
[983] Министерство Обороны СССР (1980). *Руководство по действиям экипажа при вооружении танка Т-72*. Москва: Воениздат. pp. 57-59.
[984] 'Власть переменилась' in Карцев, Л. Н. (2008). 'Воспоминания Главного конструктора танков', *Техника и вооружение*, (May).

Figure 159: Plan view of ZU-72, showing the elevation handwheel to the right of the gun and the traverse brake handlebar on its left. The post for the K-10T sight is shown but the sight itself is excluded. (*Soviet Ministry of Defense*)

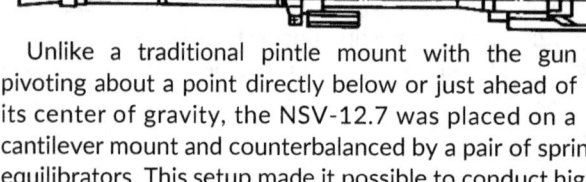

Unlike a traditional pintle mount with the gun pivoting about a point directly below or just ahead of its center of gravity, the NSV-12.7 was placed on a cantilever mount and counterbalanced by a pair of spring equilibrators. This setup made it possible to conduct high-angle fire without the long rearward overhang of the machine gun interfering with the hatch, while at the same time lowering the height of the machine gun and reducing the operator's exposure to fire. Aiming and firing was also more precise with a geared mechanism than with free (ungeared) aiming.

The K10-T reflector sight on the ZU-72 was used for both aerial targets and ground targets. It was positioned in such a way that when the commander stood up through his hatch, he would meet the K10-T at eye level. Reflector sights had replaced simple spiderweb anti-aircraft iron sights since 1944 on Soviet tanks for ease of aiming. At the time, the standard anti-aircraft weapon was the DShK outfitted with the K-8T reflector sight. The K10-T was functionally identical to the K-8T, only it had a tinted flip-up screen to reduce glare. The reticle was illuminated by sunlight collected in the sight's forward-facing collimator lens. This was satisfactory even on cloudy days, but the K10-T was, of course, totally inoperable at night[985].

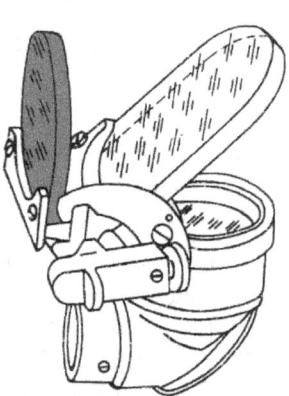

Figure 160: K10-T (*Soviet Ministry of Defense*)

With a reflector sight, the operator's eyes could be focused on the target without blurring out the reticle regardless of eye relief. The only requirement for eye relief was to remain within 155 mm and 240 mm of the reflector to keep the outer rings of the reticle in view[986]. The sight was zeroed at 400 meters, the same as the minimum range setting on the NSV's iron sights.

---

[985] The K10-T reticle was illuminated by a small plug-in lightbulb on other platforms (e.g. aircraft), but this illuminator was not included in the T-72 and there was no power outlet for it.
[986] *Техническое Описание Визира К10-Т.* p. 3.

To operate the NSV-12.7, the commander worked the elevation handwheel with his right hand and grasped the long traverse brake handlebar with his left hand, squeezing the bicycle-style firing lever on it to actuate the machine gun's trigger. The machine gun's charging handle was connected to a large lever on the right side of the mount, which provided a large mechanical advantage to overcome the rather stiff recoil spring[987].

To elevate the machine gun, the brake lever on the elevation handle was squeezed to release an internal drum brake [988] before cranking the handwheel. The full 80° elevation arc was covered by just over 2.5 turns of the handwheel, which speaks to how well the machine gun was counterbalanced. The machine gun was traversed by manually swinging the entire mount around. It could be locked in traverse by pushing down on the left handlebar, driving a wedge into a split cone ring between the mount and the turret mount, expanding the ring very much like a drum brake.

These locking mechanisms greatly improved firing accuracy over freely aimed pintle mounts. According to the technical norms, which were based on trials averaged across multiple shooters, a single five-round burst with an NSV-12.7 on a locked mount guaranteed at least one hit on an armored personnel carrier at 800 m. With the same short bursts, an average of 19 rounds were needed to score one hit on an anti-tank recoilless rifle emplacement at 1,500 m[989], and 10-15 rounds were needed to hit an anti-tank light truck (e.g. a Jeep with a recoilless rifle) at the same range. When the NSV-12.7 was fired from tripods and mounts with free aiming, the number of rounds needed to hit small point targets at long range doubled or tripled[990].

Table 44: NSV-12.7 data

NSV-12.7 Technical Specifications	
Max. effective range (m)	2,000
Rate of fire (RPM)	700-800
Muzzle velocity (m/s)	820-845*
Complete length (mm)	1,560
Barrel length (mm)	1,346
Barrel mass (kg)	9
Machine gun mass (kg)	25

* MV of 820 m/s for B-32 bullet, 845 m/s for BZT-44 bullet. The two bullets had matched trajectories.

---

[987] When the NSV-12.7 was mounted on the infantry tripod, the bolt carrier group was retracted with the aid of a cable on a pulley mechanism.
[988] The elevation brake was held in the braked position by a stiff spring until released by the lever. This gave a predictable braking force when firing and thus, a predictable recoil response.
[989] Руководство по 12,7-мм Пулемету «Утес» (НСВ-12,7). p. 221.
[990] Ibid. pp. 222-225.

Chapter VIII

Mobility

Performance

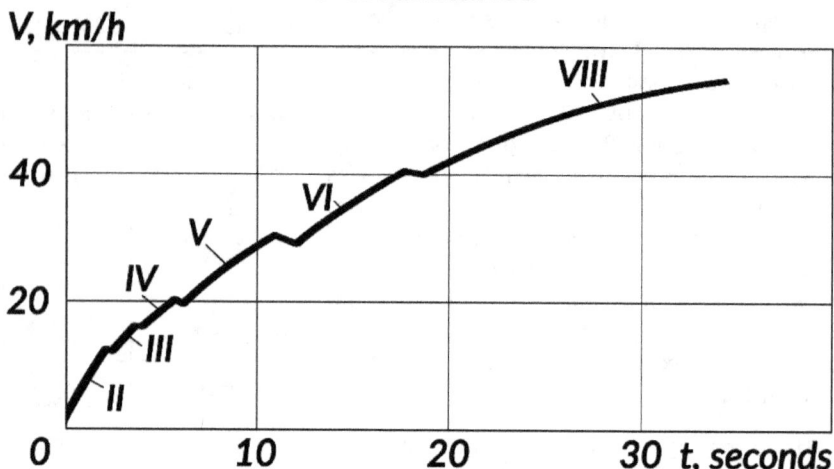

Figure 161: T-72 acceleration on concrete (top) and on dirt road (left). (O. K. Baranov et al., 1982)

Much like how the other basic technical features of the T-72 were inherited or derived from the T-64A, its drivetrain took the performance standards of the T-64A as a baseline. Equipped with the V-46 diesel multifuel V12 engine, the straight-line performance of the T-72 was equal to the T-64A, and the average speed during offroad tests was always slightly higher. Acceleration was at least as good as the T-64A. Some sources imply an acceleration time of 8 seconds to 30 km/h on concrete, while most give a time of 10 seconds[991]. Swedish tests demonstrated that the T-72M1 accelerates to 30 km/h in 10 seconds on a dry dirt road, and others give the same acceleration time of 10 seconds on a concrete, where the steel tracks on the T-72 were at a natural disadvantage due to low steel-concrete friction, and the resultant losses to track slippage.

---

[991] Баранов, О. К., Беляев, А. А., Щанкин, М. М. (1982). 'Пневмогидравлическая система управления движением танка', *Вестник Бронетанковой Техники* 1982, сборник 1. p. 32.

Table 45: Acceleration performances of various tanks (in seconds)

Speed Interval (km/h)	0-30	0-40	0-50	0-60
Concrete track				
Leopard 2[a]	5	8	12	16.5
T-80U[b]	8.4	12.6	18.4	22[c]
Leopard 1[a]	8	15	26.6	40
T-72A/B[d*]	8	-	-	27.6
T-72[d]	10	17	26	-
Strv 104[b]	13.2	25.5*	-	-
M60A1[e]	14	25	60*	-
Chieftain[f]	17	38*	-	-
Snow-covered frozen dirt road				
T-72M1[d*]	10.8	24.4	-	-
Grassy field				
Leopard 2A4[g]	8.5	-	20	24
M1A1[g]	9.0-9.5*	15-17*	33	-
Dirt road				
T-72M1	10-10.5	17	25	36

a   Krapke, P. -W. (2004). *Leopard 2: Sein Werden und seine Leistung.* Norderstedt: Books on Demand GmbH. p. 61.

       GRD: Technische Abteilung 1, Sektion 1.4 - Kampffahrzeuge (1974). *Bericht: über die technischen Versuche im Rahmen der Evaluation Panzer 68 - Kampfpanzer Leopard.* GRD report. p. 115.

b   On asphalt road. T 80 U. Delrapport 1. Framkomlighetsförsök

b*   Strv 104 top speed limited to 39.6 km/h

c   Морозов, В. (2001). *Что лучше дизеля?*. Available at: https://nvo.ng.ru/armament/2001-07-27/6_better.html (Accessed 20 April 2024).

d   Баранов, О. К., Беляев, А. А., Щанкин, М. М. (1982). 'Пневмогидравлическая система управления движением танка', *Вестник Бронетанковой Техники* 1982, сборник 1. p. 32.

d*   Unspecified T-72 model on concrete.

       Васильев, Б. И., Дорогин, С. В., Руденко, А. Г., Толмачев, Н. П. (1989). 'Теплоинерционное кратковременное форсирование мощности танкового

	двигателя и его технические возможности', *Вестник Бронетанковой Техники* 1989, сборник 2. pp. 35-36.
	Белоусов, Г. С., Бельке, А. А., Зайцев, В. А. (1987). 'Аналоговая модель движения ВГМ', *Вестник Бронетанковой Техники* 1987, сборник 6. p. 11.
d**	Edgren, S., Hansson, M. (1992). *T 72 och MTLB. Delapport 1. Trials report*. Arméns Pansarcentrum.. Snow-covered flat frozen ground, snow depth 5-15 cm, ambient air temperature -3°C.
e	Дорогин, С. В. et al. (1976). 'Ходовые испытания танка', *Вопросы Оборонной Техники*, 20(68). p. 9.
e*	M60A1 top speed limited to 48 km/h
f	Лебедев, Е. А. (1983). 'Технические Факторы, Ограничивающие Подвижность Танка', *Вопросы Оборонной Техники*, 6(112). p. 10.
f*	Chieftain top speed limited to 42 km/h
g	Swedish tests in early 1990's
g*	9.5 seconds from idle, 9.0 seconds from tac idle 15-17 seconds from idle, 14,5 seconds from tac idle

The rather inexplicable variance in acceleration performance may be explained by the fact that, despite all T-72 models having a gross engine power of 780 hp until the T-72B, the net power available at the tracks increased over time. Through refinements to the engine air cleaner, dust ejector and the cooling system in the T-72A, an additional 50-55 hp (6.4-7.0% of the gross) was freed up from the net engine power arriving at the transmission, and fuel efficiency improved by 5.5%[992]. The change in the net power-to-weight ratio of the tank was significant enough that the average speed rose from 34 km/h on the T-72 *Ural* to 37 km/h on the T-72A despite a weight increase from 41 tonnes to 41.5 tonnes. These updates were also reflected in the T-72M1. Later, the uprated V-84 engine allowed the even heavier T-72B (44.5 tonnes) to maintain the same average speed as the T-72A at 37.2 km/h[993].

In terms of acceleration, the T-72 was closely comparable to the Leopard 1, the quickest NATO main battle tank of the 1960's and 1970's, and even began surpassing it beginning with the T-72A. This was despite a 50 hp gap in gross engine power in favor of the Leopard 1, and a gross power-to-weight ratio advantage of 20.75 hp/tonne against the 19 hp/tonne of the T-72.

The T-72 also slightly surpassed the BMP-1 up to medium speeds, falling behind only at speeds above 40 km/h. Much like how the M1 Abrams and M2 Bradley represented a step forward over the previous generation of tank and

---

[992] Account given by М.Л. Наумов in Баранов, И. Н. (2010). *Главный конструктор В.Н. Венедиктов. Жизнь, отданная танкам*. Нижний Тагил: ООО «Рекламно-издательская группа «ДиАл». pp. 188-190.

[993] Иванов, В. А., Никитин, В. Т., Шпак, Ф. П., Эрнст, В. Л. (1989). 'Оценка конструкций шасси с помощью показателей Технического уровня ВГМ', *Вестник Бронетанковой Техники* 1989, сборник 2. p. 4.

APC in tactical mobility, the T-72 likewise complemented the BMP-1 in a new era for the Soviet Army.

The technical top speed of the T-72 at its rated engine speed[994] was 60 km/h, but the top speed achieved on a level road was measured at 65 km/h thanks to the relatively low rolling resistance of its suspension. For reference, the 64 km/h technical top speed of the Leopard 1 was its true top speed on level ground[995].

The angle of the suspension trailing end was lower than the leading end due to the low position of the final drives. To have the best possible obstacle crossing and trench-crossing performance, the leading end had to be tall to mount a sheer wall or the far side of a trench, and the trailing end should be low to support the tail end of the tank, usually resulting in the leading angle being larger than the trailing angle by a ratio of 1.1-1.2. When a suspension was entirely devoted to obstacle and trench crossing, this ratio could be several times larger like on the British "rhomboid" tanks of WWI.

Contemporary Soviet design guidelines stated that ideally, the leading end of the suspension can have an approach angle of up to 45°, and the trailing end could be 20° or more [996]. On the T-72, these angles were 36° and 32° respectively. This set of angles was naturally formed by placing the idler wheels on the end of the sharply pointed hull nose, and the drive sprockets inline to the planetary final drives low in the hull, balancing the tank's obstacle crossing qualities against road wheel bump travel distance and driver visibility.

Despite the low height of the idler wheels, the T-72 excelled at overcoming certain types of vertical obstacles because the idler wheels were positioned at the very end of the hull nose, with no part of the hull protruding above or ahead of the tracks. Although the T-72 was only rated for a 0.85-meter wall with its mud guards on, 1992 tests in Sweden demonstrated that flipping up the mud guards or removing them[997] allowed the tank to climb a 1.2-meter concrete wall, despite the wall almost reaching the panniers above the tracks!

The T-72 could cross a 2.6 to 2.8-meter trench. This was disproportionately wide for its hull length of 6.86 meters[998]. The M1 Abrams (7.92 m) had no advantage at all, crossing only a 2.74-meter trench, and the Leopard 2 (7.67 m) could cross a 3.0-meter trench.

---

[994] The speed at which peak gross power is achieved. For the T-72, it was 2,000 RPM.
[995] GRD: Technische Abteilung 1, Sektion 1.4 - Kampffahrzeuge (1974). Bericht: über die technischen Versuche im Rahmen der Evaluation Panzer 68 - Kampfpanzer Leopard. GRD report. p. 114.
[996] Талу, К. А. (1963). Конструкция и Расчет Танков. Москва: Военная академия бронетанковых войск. p. 45.
[997] A 20-minute task. The driver had to leave the tank.
[998] The hull length is used because the ends of the hull can support the tank while crossing a trench.

The nominal ground clearance of the T-72 was 470 mm. The ground clearance was measured from the hull belly and not the driver's "tub" which protrudes below it and is level with the suspension swing arm brackets. The minimum ground clearance as measured to the "tub" was 428 mm[999].

Drivetrain Design

The starter-generator was developed by the special electrical equipment department of the NILD engine research laboratory[1000]. The engine air cleaner was designed by VNII-100 and the engine itself was created by ChTZ in Chelyabinsk. The engine preheater was adapted from the heater of the T-64, designed jointly by KhPZ and VNII-100, and the BKP units themselves were borrowed from the T-64, designed by KhPZ. The work carried out at UKBTM was mainly aimed at working out the installation details for the drivetrain. The technical basis of its design was studied by the VNII-100 research institute with the help of prior design work by KhPZ on an ersatz T-64A, heavily modified for a transverse V-2 engine as a possible wartime "mobilization model".

The most outstanding feature of the T-72 engine compartment was its small volume, thanks to its transversely mounted engine and BKP transmission. Transversely mounted engines were not uncommon in the world of commercial automotive design, but the concept never found the same degree of popularity in tanks even towards the end of the 20th century. The concept can be traced back as early as the MS-1 light tank in 1927, the first domestically designed Soviet tank and the most mass-produced tank of its time, but it is most widely recognized from the T-44 and the T-54. The transverse engine concept made a weighty contribution to the leap in performance achieved by these tanks over the T-34 – which had a longitudinally mounted engine – with only a marginal increase in weight. Similar savings were achievable with drivetrains of other tanks without compromising other design factors[1001].

In the perpetual struggle against weight, reducing volume took priority, followed by reducing component weights. By the standards of the Soviet tank industry in the 1960's, a V-2 engine in a transverse engine layout was already seen as the bare minimum in weight reduction, and quite conservative compared to a number of more radical and more promising configurations with new engines, chief among which was the T-64 with its low-profile horizontal opposed piston engine.

---

[999] Министерство Обороны СССР (1975). *Танк «Урал». Техническое описание и инструкция по эксплуатации (172М.ТО)*. Книга Первая. Москва: Воениздат. p. 19.
[1000] Н. И. Троицкий, М. Н. Лавров. (2020). *60 лет в танковом двигателестроении*. ОАО "НИИД". Журнал «Двигатель», No. 5 (83).
[1001] For instance, an experimental transverse engine drivetrain variant developed for the Challenger 1 was claimed to reduce the engine compartment volume by 30%. See: *Transverse V-12 engine from Perkins aimed at Challenger* (1988). Jane's Defence Weekly, 2 (July). p. 1346.

The T-72 engine compartment was different from the engine compartment of the T-64A in every regard except height, which might seem strange considering that the 5TDF set itself apart from other engines by its uniquely squat design, only 580 mm tall, but because the T-64 had a conventional layout with a driver in the hull, the minimum hull height was constrained by the driver's station. Only the protruding armored radiator grilles kept the engine compartment roof from being completely flush to the rest of the hull.

In the T-64, the opposed-piston 5TDF engine was nestled between the two BKP gearboxes. The engine had two crankshafts geared together to connect the two opposing piston banks, using only the rearmost crankshaft to link the engine to the two BKPs. All engine accessories and auxiliary components were connected to the engine directly through the same set of internal gears. The drivetrain had a low profile so that the cooling system could be packaged in its entirety above the engine, thus shortening the engine compartment.

A transverse mount allowed the V-46 to be nested between the torsion bars of the suspension, which made for a much shorter engine bay than if it had to sit above them, but the engine was still 902 mm tall, and to fit it into a hull that measured only 901.5 mm internally at its peak[1002], a bulge conforming to the outline of the engine was stamped into the engine access panel. The transmission was then laid out behind the engine, with the cooling fan occupying the space between the BKP units. The absence of a conventional gearbox allowed the engine compartment to be shorter and thus smaller than the T-54/55, but nevertheless, it was still 0.5 m^3 larger than the T-64A engine compartment at a total internal volume of 3.1 m^3.

To compensate for the weight gain, the thickness of the side armor over the engine compartment was reduced from 80 mm to 70 mm, and with the added power, the power-to-weight ratio of the T-72 was brought up to the same level as the T-64A. The power density of the engine compartment remained on a competitive footing, measuring in at 252 hp/m^3 against the T-64A's power density of 269 hp/m^3, even comparing well against the Leopard 2[1003] (259 hp/m^3), only falling short against tanks with gas turbine engines like the T-80 (357 hp/m^3) and M1 Abrams (349 hp/m^3)[1004].

[1002] Specified structural height at a point between the turret ring and the engine compartment.
[1003] Based on 5.8 m^3 engine compartment volume without fuel (6.38 m^3 minus 0.58 m^3 of fuel). See: Hilmes R. (2007). *Kampfpanzer heute und morgen: Konzepte - System - Technologien*. Auflage: Motorbuchverlag. p. 130.
[1004] Engine compartment volume of 4.3 m^3 without fuel. See: Hilmes R. (1988). *Kampfpanzer. Die Entwicklungen der Nachkriegszeit*. Bonn: Mittler Report Verlag GmbH. pp. 52, 53, 56.

The V-46-4 engine was used on the T-72 *Ural* and most other T-72 variants. The V-46-6[1005] was used in the T-72A. The V-46-6 differed from the basic V-46 only in having a slightly reconfigured layout of accessories, mainly related to its lubrication system[1006]. The V-84 was installed in late production T-72A tanks and BREM-1 armored recovery vehicles since 1984, and the T-72B was fitted with the V-84-1. These were all liquid-cooled diesel engines, descendants of the legendary V-2 that powered the T-34.

Liquid cooling was considered to be categorically superior to air cooling in performance – of course, being a direct descendant of the V-2, liquid cooling was unavoidable either way. Water for radiator refills was traditionally allocated on the request of company commanders and portioned out as part of a delivery of drinking water[1007]. Given that a regular supply of clean water was needed to sustain life anyway, the burden of maintaining a liquid cooling system was generally not considered decisive in most countries. Like all of the accessories, the fan was shaft-driven through a sealed unit impervious to dust[1008], which was particularly important in an engine compartment with open air flow.

The open engine compartment configuration was exploited to reduce the complexity of the oil cooling circuit. With the engine and transmission cooled by oil radiators, the cooling fan drive, the starter-generator, the air compressor, and the transfer case were all air-cooled. The engine air intake port had a small duct with an inertial slat dust filter directing a part of the flow through the air compressor, which had a cowl to increase the air flow velocity across its fins[1009]. A bypass duct from the radiator supplied cool air to the starter-generator, also cleaned through inertial slats.

The starter-generator was connected to the electrical system by a relay-controlled regulator. In the charging mode, it was connected to the electrical network only once the engine reached 800-1,000 RPM, which restricted the engine idling speed to no less than 800 RPM. The electrical systems of NATO tanks could maintain normal charging performance in a larger engine speed range due to the use of alternators (A/C) instead of generators (D/C)[1010].

[1005] First tested in 1976 in the Object 176 experimental tank.
[1006] Министерство Обороны СССР (1983). *Двигател и В-46 и В-46-6 – Техническое Описание (ТО)*. Москва: Воениздат. pp. 125-127.
[1007] Прямицыно, В. Нашем., Чертов, В.В. (2019). 'В боевой обстановке войска зачастую готовы пить любую воду'. *Военно-Исторический Журнал* 2019, №1. p. 16.
[1008] Belted fan drives were categorically abandoned for high-powered tank engines early on in the USSR. The initial V-2 (for the BT, T-34, KV, IS) creatively combined its flywheel and cooling fan.
[1009] According to its technical specifications, an air velocity of at least 9 m/s was required.
[1010] Барсов, Ф. Ф., Усков, А. Ф. (1976). 'Аккумуляторные Батареи Зарубежной БТТ'. *Вестник Бронетанковой Техники* 1976, сборник 2. pp. 49-52.

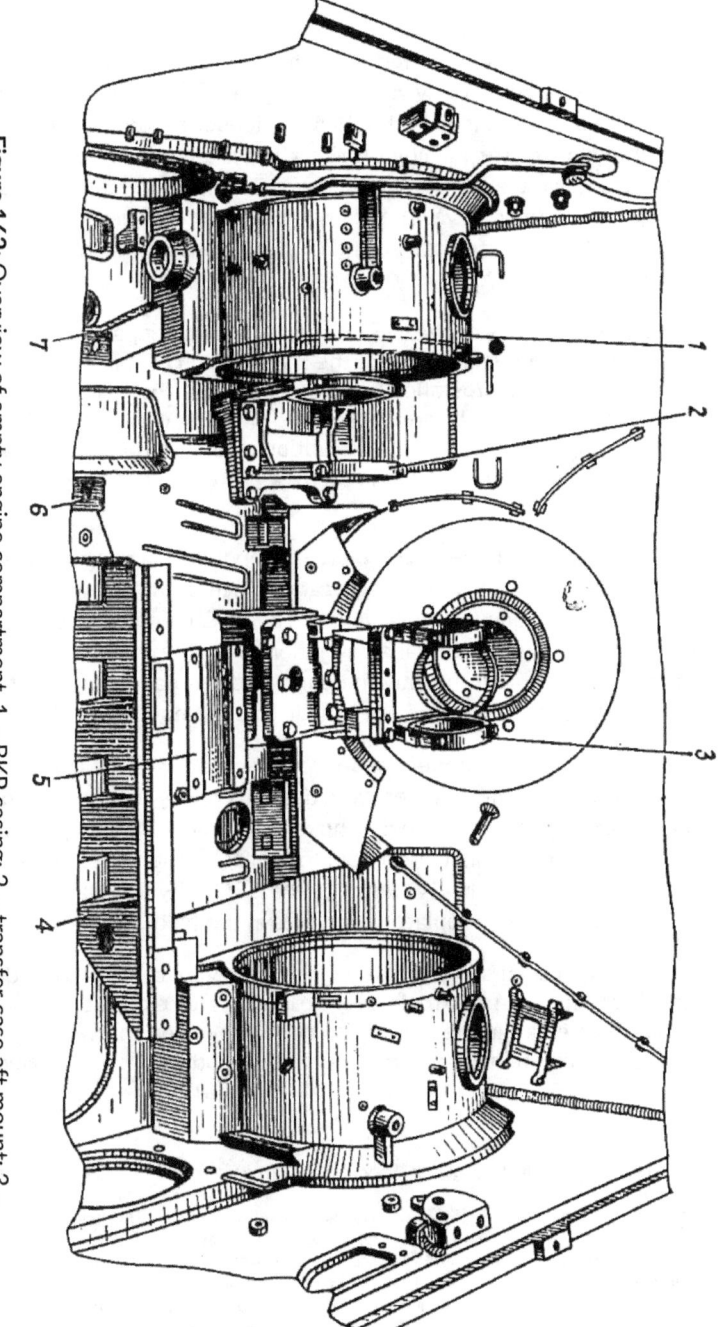

Figure 162: Overview of empty engine compartment. 1 – BKP casing; 2 – transfer case aft mount; 3 – cooling fan pedestal mount; 4 – engine mount; 5 – starter-generator mount; 6 – transfer case forward right mount; 7 – transfer case forward left mount. (*Soviet Ministry of Defense*)

The V-46 and V-84 inherited all of the basic features of the V-2 family, most of which were typical of aviation V12 engines[1011]. Its 60° bank angle was ideal for engine balance and gave the V-46 a slender silhouette, highly sought after for aerodynamic streamlining on aircraft. On tanks, a lower and boxier profile was more desirable for space efficiency. To make up for the gaps, the engine accessories were creatively positioned astride the crankcase to fill out most of the unused volume, and the engine compartment bulkhead was slanted to take up the rest, making for a very compact package overall.

The primary focus of the T-72 drivetrain design department was to increase the net engine power available at the transmission by minimizing losses through mechanical or hydrostatic inefficiencies. At the same time, maximum interchangeability with previous T-72 models was adopted as a guiding objective so as to minimize the disruption to the production line, and to enable existing tanks to be modernized to a new performance standard during scheduled overhauls[1012].

The drivetrain lacked complete component aggregation like modern powerpacks, or even partial aggregation such as the coupling of the cooling system to the engine as a single unit like in the Chieftain. The only instances that might be classified as partial aggregation were the integration of a final drive into each BKP, and in the way the engine access panel and radiator pack were fitted to the same hinge and were made to be removed as a single unit.

Even the type of fastener used to connect the radiator cores to the cooling system was dated; metal pipes were used to convey fluid throughout the entire engine compartment with rubber hoses as connectors, clamped onto the pipes with hose clamps. This was in stark contrast to the widespread use of quick-disconnect fittings in NATO tanks by the 1960's, which was driven by a desire to decrease drivetrain replacement times in the field.

Achieving a such a high degree of efficiency involved a great deal of effort, especially with respect to losses to intake and exhaust resistances and to auxiliary systems like the cooling fan drive. However, this was largely in spite of its volume-constrained engine compartment and its open (unsealed) nature.

[1011] United States Flight Standards Service (1971). *Airframe and Powerplant Mechanics Powerplant Handbook*. p. 5
Also see: Котельников, В. Р. (2010). *Отечественные авиационные поршневые моторы (1910—2009)*. Москва: Русский Фонд Содействия Образованию и Науке.
Also see: Barnes J. D., Pearce, D. M., Cantab, B. A. (1944). *Report on Russian C.I. Tank Engine Type "V2" From T-34 Cruiser Tank*. Military College of Science – School of Tank Technology.
[1012] Account given by М.Л. Наумов in Баранов, И. Н. (2010). *Главный конструктор В.Н. Венедикто:. Жизнь, отданная танкам*. Нижний Тагил: ООО «Рекламно-издательская группа «ДиАл». p. 188.

The repair concept for the T-72 followed the standard automotive repair convention, which was reflected in the design of the engine compartment layout. Like the engine bay of a car, the engine compartment of a T-72 was essentially a container, into which all of the powertrain assemblies were bolted. Each unit of the drivetrain was individually installed and then mutually aligned, using the BKPs or engine as the datum points. Then, everything else was packed into the remaining space in the compartment, but in a deliberate order. This was to ensure the possibility of replacing individual modules without touching any of the major powertrain assemblies.

As such, a certain amount of empty (wasted) space had to be left in critical zones to permit technicians to work on the drivetrain. In this way, they could perform all basic maintenance and repair tasks without removing the engine or transmission from the engine compartment, but if removal was necessary, the free space facilitated the undoing of various fasteners and connectors.

All breakdowns were to be rectified on the spot if possible, and if repairs were required, they were to be done by replacing all of the offending parts and modules in-situ, without the removal of any major assemblies.

Figure 163: Engine access panel lifted together with radiator pack. (*Soviet Ministry of Defense*)

The transverse engine mount was beneficial to accessibility in that both sides of the engine could be accessed without needing to remove the engine. The right side was accessed through removable panels on the bulkhead to the crew compartment, and the left side was open to the engine compartment. Engine removals were only carried out as part of an engine replacement process, and engine replacements were only deemed necessary if there was no other way to rectify a major malfunction. As a rule, such malfunctions would involve some form of structural damage, such as bearing failure in the supercharger, jamming of the crankshaft, or the failure of the right cylinder head gasket.

All parts likely to require servicing were laid out in a way that they could be accessed from above, leaving only essential items on the floor of the engine compartment together with the semi-permanent fixtures, i.e. the engine and transmission oil tanks. Having no moving parts, it was unlikely that they would ever need to be replaced, so they were buried deep beneath major components; extricating the transmission oil tank from the engine compartment floor required the transmission prop shaft to be removed. The cooling fan drive had to be dismantled to remove the engine oil tank, which was also on the floor, and the fan itself together with its cowling had to be removed to extricate the supplementary engine oil tank from its corner behind the right BKP.

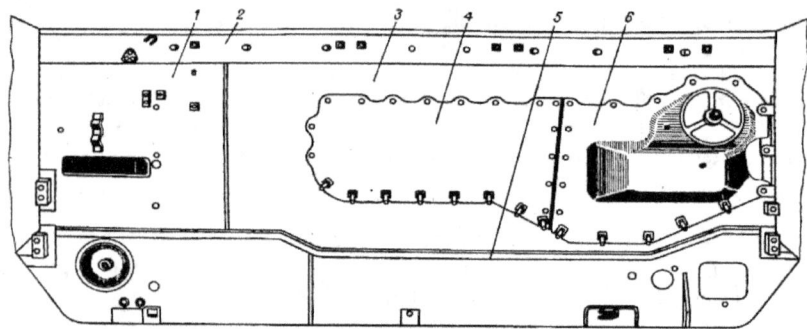

Figure 164: Engine compartment bulkhead. 1 – left bulkhead section; 2 – hull roof crossbeam; 3 – right bulkhead section; 4, 6 – removable cover; 5 – stiffening rib. (*Soviet Ministry of Defense*)

A large hatch underneath the rear end of the engine gave access to the engine oil pump and air compressor. A central access hatch allowed the engine and transmission oil primer pumps to be removed. A hatch on the right side of the engine compartment floor gave access to the service brake.

Opening the engine access panel exposed the entire top surface of the engine and its air cleaner. Opening the radiator pack gave access to the transmission hydraulic control system and the cooling fan drive. If the drivetrain was in need of more serious work, the engine access panel and radiator pack were removed as a single unit after draining the cooling system, this being a workshop-level task since a standard 1.5-tonne crane [1013] or hoist was needed. Then, the crossbar support would be removed, and with that, the entire engine compartment was exposed from above, also freeing up room to remove the starter-generator.

Serious work was generally rare for the T-72. It had a standard midlife overhaul period of 8,000 km, after which the tank had another 6,000 km to its

---

[1013] The standard TRM-A-80 mobile workshop in a Soviet tank repair battalion had a 1.5-tonne crane.

major overhaul[1014]. After the exhaustion of its service life, the next step was factory-level rebuilding. East German data gathered on T-72Ms from a variety of manufacturers showed that the tank had exceptionally high durability[1015]. Compared to the T-55, not only was less maintenance needed, it also took less time to carry out.

Table 46: Actual maintenance periods from long-term exercises in the NVA

Maintenance item	T-55A	T-72
Time required for daily inspections before use (minutes)	20-25	12-15
Routine maintenance time spent per 1,000 km (hours)	27.7	15.2
Time required for daily maintenance after use (hours)	1.9	1.3
Time spent in TO-1 maintenance (hours)	6.1	3.6
Time spent in TO-2 maintenance (hours)	12.6	8.0

Technical maintenance No. 1 (TO-1) was done at 1,000 km, and technical maintenance No. 2 (TO-2) was done at 2,000 km. Thanks to the much greater reliability of the T-72, its maintenance intervals were stretched by 1.6-1.8 times compared to the T-55, where TO-1 was carried out at 1,600-1,800 km and TO-2 was carried out at 3,300-3,500 km. These intervals were extended on the T-72B to a TO-1 of 2,000-2,200 km and TO-2 of 4,000-4,200 km.

Daily technical maintenance (ETO) included general monitoring of the technical condition of the vehicle, refueling, replenishing oil and other consumables, replenishment of ammunition (during combat operations), washing (cleaning) of the tank and checks for completeness.

[1014] Минобрнауки России (2011). *Ремонт Бронетанковой Техники*. Омск: Издательство ОмГТУ. p. 49.
[1015] Nationale Volksarmee Panzerwerkstatt (1981). *Bericht über die Weiterführung der Nutzungserprobung des Gerätes 172 M bis 14.000 Km*. Technical report.

Technical maintenance No. 1 (TO-1), was rudimentary regular maintenance carried out by the tank crews with the assistance of repair specialists after a certain operating time. Most of the volume of work consisted of basic inspections like checking if the periscope washer reservoir was filled, checking for oil leaks, checking if the marker lights were functional, and so on. The most labor-intensive operations were washing the engine oil filter and applying lubrication to select parts of the suspension and hull.

Technical maintenance No. 2 (TO-2) was carried out by the crews and technicians from the maintenance and repair units attached to the tank unit. TO-2 involved cleaning out sludge from various sumps, checking the oil level, coolant level, and lubricating the suspension.

Table 47: Reliability increase from T-54 to T-72[a, b]

Tank	Reliability coefficient*
T-54	0.62
T-55	0.77
T-62	0.74
T-72M	0.962

* The reliability coefficient was a weighted calculation from the breakdown rate and time to rectify breakdowns

a   Кос, И. И., Сидоренко, Р. В. (1968). 'Сравнительная оценка надежности средних танков (Т-54М, Т-55, Т-62)', *Вестник Бронетанковой Техники* 1968, сборник 1.

b   Nationale Volksarmee Panzerwerkstatt (1981). *Bericht über die Weiterführung der Nutzungserprobung des Gerätes 172 M bis 14.000 Km*. Technical report.

Water Obstacles

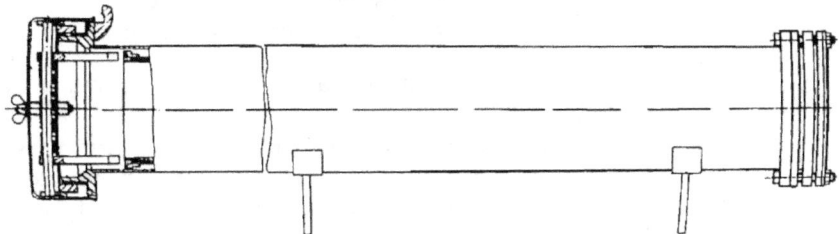

Figure 165: Snorkel telescoped and capped for stowage. (*Soviet Ministry of Defense*)

The T-72 was equipped with a set of OPVT[1016] underwater tank driving equipment to ford water obstacles and to cross rivers by deep wading to a maximum depth of 5 meters. On its own, the deep wading capability was not particularly unique. Soviet medium and heavy tanks had been equipped with the same basic capability since the 1950's, with the same wading depth limit of 5 meters, and Western tanks like the Leopard 1 could deep wade to a depth of 4 meters. The peculiarity of the deep wading capability offered by the T-72's OPVT kit was that all of the equipment needed for a river crossing operation was carried on board each individual tank, whereas on the Leopard 1, Leopard 2, and other tanks, the snorkel had to be delivered to the tanks on trucks[1017].

The T-72 could ford streams with a depth of 1.2 meters without any preparation. This was the depth limit for no water entering the exhaust pipe. At depths greater than 1.2 meters, the crew had to apply a substantial part of the deep wading preparation process to seal the tank for a depth of up to 1.8 meters. The hull roof would be submerged, flooding the engine compartment and the crew compartment ventilator, so all engine compartment roof intakes were sealed with panels. The engine had to instead draw air through the fighting compartment, so at least one turret hatch had to be left ajar, or the breather hole in the gunner's hatch had to be opened. The openings of the gun mask were sealed with ZZK-3u, a thick waterproof grease[1018]. The ventilation system had to be turned off, a valve installed in the water drainage port plug on the hull

---

[1016] "Оборудование подводного вождения танков" (ОПВТ) or "Underwater tank driving equipment" was not a specific system, but a term for a set of underwater driving equipment.
[1017] Hilmes R. (2007). *Kampfpanzer heute und morgen: Konzepte - System - Technologien*. Auflage: Motorbuchverlag. p. 152.
[1018] A thick grease containing synthetic rubber. Also used as a sealant for long-term preservation.

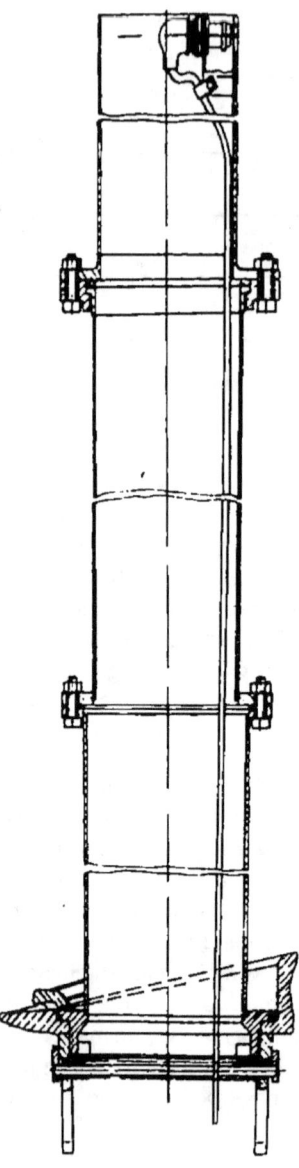

Figure 166: Fully assembled snorkel installed on breather hole in gunner's hatch. Note the signal light at the top. (*Soviet Ministry of Defense*)

belly, the driver's periscope cleaning system had to be sealed, and the engine exhaust spout replaced with a special outlet with valves.

Performing a deep-wading operation was another step up in complexity. The same preparations for wading a 1.8-meter river were applied and a snorkel was fitted. The telescoping snorkel tube was retrieved from its mount behind the turret, assembled from its three sections to the required height, and mounted onto a special adapter for the ventilation hole in the gunner's hatch.

The long telescoping snorkel, made from steel, broke down into three sections for stowage. The total length of the assembled snorkel was 3,712 mm. Fitted to the breather hole on the gunner's hatch, the total height from the ground to the tip of the snorkel was 5.9 meters, which gave a fairly large safety margin against flooding. A rear-facing signal lamp marked the position of each tank when performing a river crossing at night.

The snorkel provided ventilation for all three crew members as well as the engine. Air was sucked into the fighting compartment of the tank and into the engine through an intake hole on the engine compartment bulkhead.

The driver also had to unplug the discharge port of the bilge pump and turn it on. The pump, located in the rear left corner of the engine compartment next to the engine compartment bulkhead, sucked water from the floor of the crew compartment and discharged it through the hull roof at a rate of 100 liters per minute when operating with a back pressure of 4 meters of water, which ensured that the tank would not be flooded by a minor leak.

In total, 20 minutes were needed. Compared to the T-55, the preparation time for wading and deep wading had been halved, largely thanks to the built-in turret ring seal.

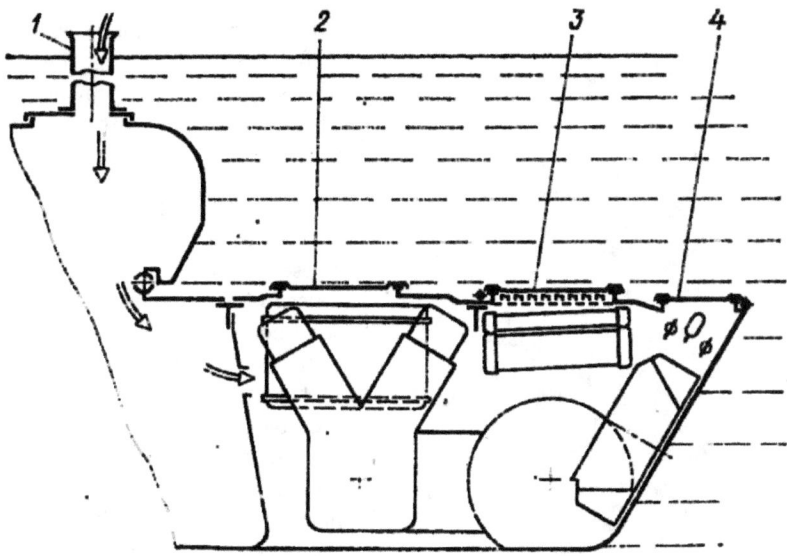

Figure 167: Air flow in sealed engine compartment while driving underwater. (V. S. Starovoytov, 1990)

The fan was left running, but with no air flow across the radiator, there was no way to remove the heat accumulating in the coolant circuit. Officially, the maximum underwater driving distance was 1,000 m, but this figure was purely nominal. The maximum traversable river width was actually determined by the coolant temperature[1019], so in practice, the driving distance depended on how hard the engine had to work to propel the tank over different river bed surfaces, and how long the engine had been left idling before the water crossing operation. According to guidelines, the coolant temperature before entering water had to be no higher than 90°C and the width of the river should be no wider than 400 meters, as driving across this distance would likely send the coolant temperature to its critical threshold of 115°C[1020].

[1019] Старовойтов, В. С. (ed.) (1990). *Военные гусеничные машины*. Том 1, Книга 2. Москва: Издательство МГТУ им. Н. Э. Баумана. p. 98.
[1020] In tanks with closed-circulation engine compartments, the radiator inlet and exhaust did not have to be sealed. The radiator cores simply became water-cooled while snorkeling or deep-wading.

The tank was to be driven in 1st gear only, and while moving underwater, it was important that the engine worked without stopping and without gear changes. If the engine stopped for some reason, it was not possible to re-start it because of the extremely high exhaust backpressure from the water depth. This was the main disadvantage of not providing the engine exhaust with its own snorkel.

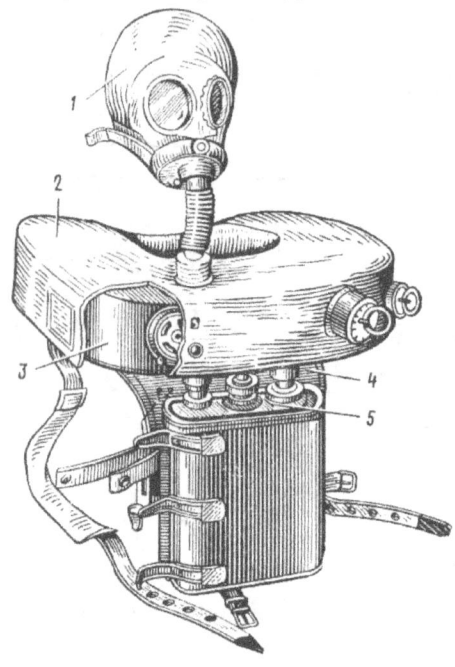

Before entering water, the crew donned their IP-5 rebreather kits. These were closed-cycle rebreathers that chemically scrubbed the carbon dioxide exhaled by the wearer and replenished the oxygen in the breathing circuit using a regenerator cartridge containing potassium superoxide and sodium peroxide. An air bladder around the wearer's neck became inflated by the regenerated air during operation, giving a slight overpressure to help the user breathe at large depths (up to 7 m), helping to moderate the temperature of the regenerated air[1021], and acting as a flotation aid.

Figure 168: IP-5 rebreather kit. 1 – ShIP-2M mask; 2 – air bladder cover; 3 – air bladder; 4 – bib; 5 – RP-5 regenerator cartridge. (*Soviet Ministry of Defense*)

The tank could open fire immediately after exiting the water without stopping. Turning the turret in either direction by 6° after exiting the water tripped on a lever on the left side of the engine compartment roof to release three spring-loaded latches to shut the engine bulkhead intake and release the sealing lids on the radiators, and fan exhaust. The turret also tripped another lever on the right side to release the cover on the engine air intake. The tank could proceed into combat in this state without any further preparation, but 1-2 minutes were

---

[1021] Rebreathers of this type released heat as part of the chemical scrubbing process, which made the oxygen released into the breathing circuit quite hot. These chemical cartridges were also infamously explosive if water leaked into them.

needed to properly prepare it for combat, and 15 minutes were needed to dismantle the external equipment and stow them away[1022].

Because of the inherent risk involved, only crews certified for diving operations and tank snorkeling in a special training course were allowed to carry out snorkeling operations in field exercises, and the actual operation involved crossing site surveillance by combat engineers, with a whole host of precautions. The river current speed could not be higher than 2 m/s, the entry and exit angles of the river banks, and the river bed could not be marshy or muddy, or have boulders taller than 30 cm, and so on[1023].

[1022] Министерство Обороны СССР (1975). *Танк «Урал». Техническое описание и инструкция по эксплуатации (172М.ТО)*. Книга Первая. Москва: Воениздат. p. 36.

[1023] Сутормин, Е. А. (2013). *Вождение боевых машин*. Екатеринбург: Издательство Уральского университета. p. 151

Engine

The V-46 was a liquid-cooled, supercharged V12 diesel-multifuel engine. It had dry-sump lubrication, double-overhead cams (DOHC), and a Bosch inline fuel injector with direct injection, and a mechanical flywheel governor. The V-84 was essentially identical save for a higher boost pressure to extract yet more power. At its core, it was a deep modification of the venerable V-2 diesel tank engine, most famous as the beating heart of the legendary T-34 medium tank. By the time the T-72 entered service in 1973, the core design of the V-2 was almost 40 years old[1024] and the V-2 family had been in service for nearly 35 years.

Work on the V-2 began in July 1931 with the explicit intention of developing a diesel tank engine under the direction of the Department of Mechanization and Motorization of the RKKA. The reasoning was simple: high-powered engines were desperately needed, but the supply of high-octane gasoline in the USSR was too limited for aircraft, let alone for the ground forces[1025]. National scientific institutions with experience in high-performance engines were closely involved in the project. It was especially noteworthy that technologies and know-how from the experimental AN-1 and AD-1 aviation diesel engines developed by TsIAM[1026] were incorporated into the V-2 through joint work.

Most of the core structural characteristics of the V-2, including its bank angle, bore diameter and stroke length[1027] were laid down as early as May 1933, and the engine essentially acquired its final form in 1934, when a master-and-articulated connecting rod layout replaced the initial fork-and-blade layout. The V-2 then spent the next 5 years in troubleshooting while actively participating in comparative assessments against the M-5 (Liberty V12) and the *Mikulin* M-17 (BMW VI) aviation gasoline engines in field tests using BT light tanks.

Equipped with the V-2, a BT-7 could drive for 500-600 km while the M-17 managed just 312 km on the same 680 liters of fuel[1028]. These results secured the fate of the V-2, and together with the successful completion of the M-17 diesel engine (unrelated to the *Mikulin* M-17) for the S-65 *Stalinets* tractor in 1937[1029], the Red Army took its first step towards dieselization.

[1024] Тарасенко, А. (2020). *Сердце «тридцатьчетвёрки» на тракторе*. Available at: https://warspot.ru/7634-serdtse-tridtsatchetvyorki-na-traktore (Accessed 21 December 2023).
[1025] Юрасов, И. В. (1975). 'К Истории Производства Отечественного Быстроходного Танкового Дизеля', *Вестник бронетанковой техники* 1975, сборник 3. pp. 48-52.
[1026] Central Institute of Aviation Engine Engineering.
[1027] М. И. Александровский (1991). 'Унифицированный Танковый Дизель', *Вопросы Оборонной Техники*, 20(120).
[1028] *Двигатели танков* (1991).
[1029] Дронг, И., Трепенёнков, И., Чухчин, Н. (1975). 'Первый Дизельный', *Историческая Серия «ТМ»*, *Техника-молодёжи* 1975, 4.

Figure 169: V-46 engine from the rear (top) and front (bottom). (*Soviet Ministry of Defense*)

The overwhelming advantage in fuel economy offered by diesel engines later inspired a postwar pivot away from gasoline tank engines among NATO countries. Clean-sheet designs like the AVDS-1790-2 of the M60A1 and the MB 838 CaM-500 of the Leopard 1 were purpose-built for tanks, and could deliver power outputs that were previously believed to be exclusive to gasoline engines. Put up against this kind of competition, the V-46 appeared oddly anachronistic, and yet, the performance of the T-72 demonstrated that the potential of the V-2 had not yet been exhausted despite some of its more outdated traits, like its low crankshaft speed and undersquare[1030] design.

A low crankshaft speed was a trait of early aviation engines. Given the range of practical propeller diameters for high performance aircraft, an engine speed of 2,000 RPM was generally considered to be the limit for direct-drive propellers[1031]. Initially, the V-2 fell just under this limit at a crankshaft speed of 1,800 RPM, and the V-2K variant had its speed raised to 2,000 RPM to extract more power for KV heavy tanks[1032]. The 2,000 RPM limit then became standard in postwar V-2 descendants, including the V-46. Further raising the crankshaft speed would erode the remaining safety margin to its maximum speed of 2,300-2,400 RPM[1033], so power growth was achieved by instead raising the mean effective pressure (MEP) of the engine.

The V-46's undersquare configuration gave it some inherent advantages in fuel efficiency[1034] and was responsible for its excellent low-end torque, but because the pistons and connecting rods had to travel a longer distance as they reciprocated, the inertial stress from rapid acceleration as they moved from their top dead center (TDC) to bottom dead center (BDC) positions was intensified[1035]. In heavy engines, the rate at which inertial stresses increase from raising the crankshaft speed outpaces the stress from increased combustion pressure, making it much more efficient to increase engine power through higher torque rather than higher speed, in contrast to square or oversquare engines[1036]. This was the biggest obstacle to power growth in the V-2 family.

[1030] Having a piston stroke length longer than the bore diameter.
[1031] United States Flight Standards Service (1971). *Airframe and Powerplant Mechanics Powerplant Handbook*. p. 26.
Also see: Johanning, A, Scholz, A. (2013). *Propeller Efficiency Calculation in Conceptual Aircraft Design*. Technical Note. Hamburg: Hamburg University of Applied Sciences.
[1032] Березкин, В. (1999). 'В-2: путь в серию', Двигатель, 6(6). p. 24.
[1033] Zero net power at 2,300 RPM, absolute maximum at 2,400 RPM. See: Двигатели танков (1991).
[1034] Pulkrabek, W. W. (1997). *Engineering Fundamentals of the Internal Combustion Engine*. Upper Saddle River: Prentice Hall. p. 40.
[1035] Peak piston acceleration in V-2 family engines was 2,025.4 m/s² (206.5 G). See: Kirasić, V. (2021). *Pogonska Jedinica V-46 TK*. Undergraduate Thesis. Zagreb: Hrvatsko vojno učilište "Dr. Franjo Tuđman".
[1036] In small-displacement automobile engines with much smaller and lighter pistons, the crankshaft speed is limited by lubrication issues at high piston velocities rather than inertial loading.

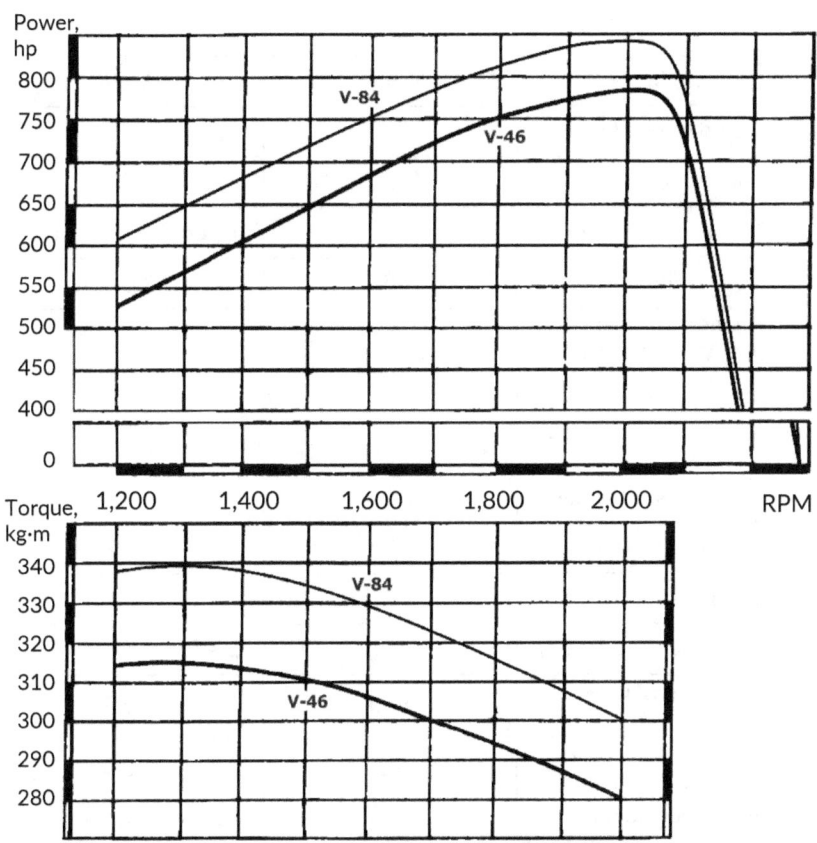

Figure 170: Power (top) and torque (bottom) of the V-46 and V-84 engines .(*Adapted from Soviet Ministry of Defense*)

Nevertheless, a peak power of 780 hp[1037] was respectable, and a high low-end torque output raised the low-end power, which had a direct and tangible effect on the mobility of the T-72. The MB 838 CaM-500 in the Leopard 1 (830 hp) was a squarer engine that surpassed the V-46 in peak power by 50 hp, but at the start of their power bands, the difference eroded to just 8 hp (572 hp against 580 hp). With the V-84, the gap in low-end power only widened. A high gross torque[1038] was also important for the T-72 due to the absence of a torque converter in its transmission.

[1037] Bench figure tested at atmospheric pressure, 20°C air and fuel temperature, 70% relative humidity.
[1038] Peak torque was developed at a crankshaft angle of 15° after TDC, with a peak combustion pressure of 95 bar at an engine speed of 1,290 RPM.

Table 48: Technical data of T-72 engines

Engine	V-46	V-84
Rated power (hp)	780	840
Maximum torque (Nm)	3,090	3,332
Bore x Stroke (mm)	150 x 180/186.7	
Dry weight (kg)	980	1,020
Engine dimensions (L x W x H) (mm)	1,480 x 896 x 902	
Engine volume (m^3)	0.835	
Performance Characteristics		
Specific power (hp/kg)	0.79	0.85
Power density (hp/m^3)	652	702
Compression ratio	14	
Mean piston speed (m/s)	12.0/12.45	
MEP at rated power (2,000 RPM) (kPa)	885	953
MEP at peak torque (1,300 RPM) (kPa)	998.7	1,078
Min. specific fuel consumption (g/hp·h)	172	171
Specific oil consumption (g/hp·h)	8	8
Supercharger		
Model	N-46	N-46-6
Impeller speed at 2,000 RPM engine speed	26,660 RPM	
Boost pressure	0.68 bar	0.88

a Министерство Обороны СССР (1983). *Двигател и В-46 и В-46-6 – Техническое Описание (ТО)*. Москва: Воениздат. pp. 3-8.
b Министерство Обороны СССР (1991). *Дизель В-84М (В-84, В-84-1) – Техническое Описание*. Москва: Воениздат. pp. 4-8, 103.
c НИИД (1978). *Анализ и систематизация материалов по уровню развития отечественных*. Технический Отчет №3443. Available at: http://btvt.info/3attackdefensemobility/niid_1978.htm (Accessed 20 August 2023).

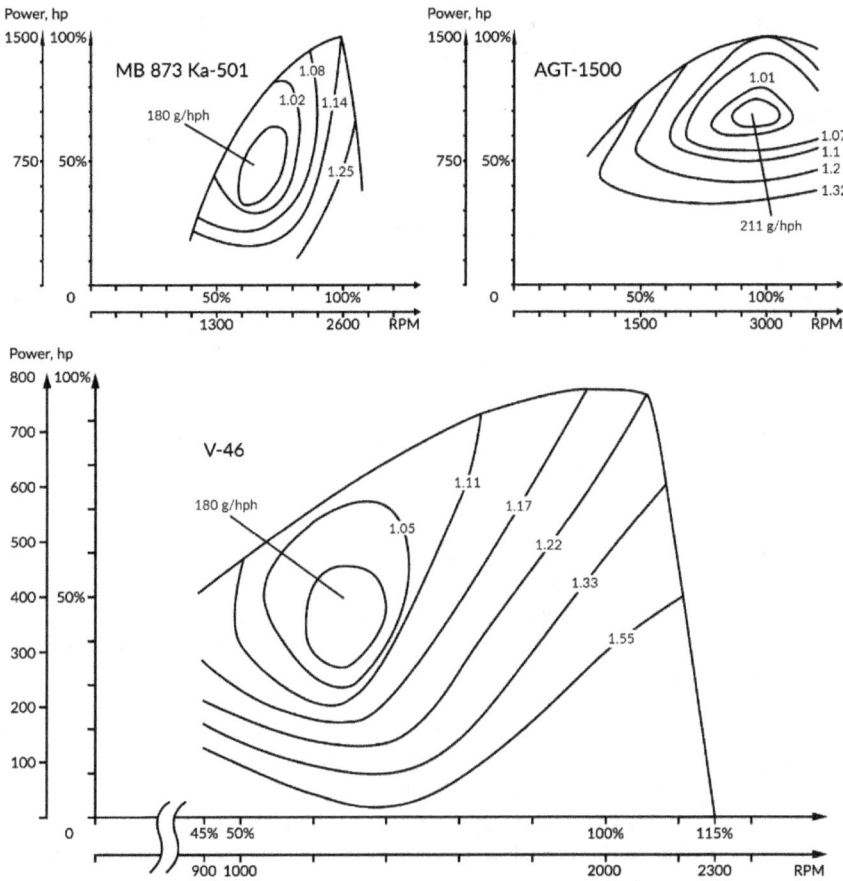

Figure 171: Specific fuel consumption (SFC) maps for the V-46, MB 873 Ka-501 and AGT-1500. Each contour is a factor of the minimum SFC. (*Adapted from V. A. Kolesov & V. A. Stepanov, 1975, W. Merhof & E.-M Hackbarth, 2015*)

The downside to the undersquare nature of the engine was that its fuel economy was good only when it could cruise in a narrow range of engine speeds. It could, however, maintain good efficiency at a wide range of loads[1039], so on long road marches, good fuel economy could still be maintained by choosing an optimal gear in the seven-speed transmission. The V-46 and V-84 otherwise compare unfavorably to the MB 873 Ka-501 of the Leopard 2, which can be

---

[1039] Derived from fuel consumption map given in: Колесов, В. А., Степанов, В. А. (1975). 'Особенности Опытной Гидромеханической Трансмиссии ГМТ-69021', *Вопросы оборонной техники*, 20(59-60). p. 6.

credited to its large displacement and low compression ratio, but they nevertheless perform very well compared to the AGT-1500 gas turbine engine of the M1 Abrams, which had acceptable fuel economy only at 65-75% load, and only when it ran at 90-100% of its maximum speed.

In general, the V-46 and V-84 were fuel-efficient engines, albeit smoky on account of their high oil consumption[1040]. On paved roads, the T-72 Ural and T-72A could travel 480-500 km on its internal fuel reserve of 705 liters. With the standard 275-liter external fuel drums fitted, the range was extended to 730. With 200-liter drums, the range was 670 km. The fuel consumption rate was 2.62-2.64 l/km on cross-country road marches[1041]. When driving off-road, the high torque of the engine enabled it to run at low crankshaft speeds with the transmission in higher gears, which gave it good fuel economy.

In the first 2,000 km, oil consumption was 2-4 liters per 100 km. By 6,000 km, oil consumption reached 8.5 liters per 100 km. The tank carried a container of reserve engine oil on top of the exhaust pipe to top up in field conditions.

The V-46 and V-84 could run on standard Soviet diesel (DL), winterized diesel (DZ, DA)[1042], kerosene (TS-1, T-1, T-2), low-octane gasoline (A-66, A-72), and even petroleum naphtha (paint thinner), or almost any arbitrary blend of these fuels. In winter, blending 20% kerosene into standard diesel to obtain winterized diesel was common practice.

The multifuel capability was made a requirement for new tank engines in the early 1960's to mirror a similar NATO requirement. Converting to other fuels took nothing more than turning an adjustment knob on the fuel injector located on top of the engine. It could be set to diesel, gasoline, or kerosene, each setting calibrated to change the injection rate based on the density of the fuel. Gasoline was the least agreeable to the engine - if it had to run on A-72 gasoline[1043], its power output plummeted and the driving range fell 20%. If possible, gasoline would be avoided entirely, or at least blended with diesel. When running on blended fuel, the injector knob was turned to the setting corresponding to the lighter fuel in the blend.

[1040] Even in new engines, high oil consumption was normal due to leaky valve stem seals, although the fitting of the piston rings was not entirely blameless. This was a hereditary trait of the V-2 family.
[1041] Маринин, В. А., Тульцев, А. А., Штейн, В. Д. (1987). 'Оценка Показателей Подвижности Модернизированных Танков на Марше', *Вестник Бронетанковой Техники* 1987, сборник 4. p. 8.
[1042] DZ winter-grade diesel was regular diesel with additives. DA arctic-grade diesel was essentially kerosene, but with additives to increase its cetane number and improve its lubrication qualities.
[1043] 72-octane gasoline.

The refueling process of a T-72 was 18 minutes in its entirety from a fully drained condition to full capacity, or 25 minutes if the external fuel drums were included[1044].

The engine produced enough power at low speed and part load to keep the tank moving at a reasonable pace, which contributed to its high reliability. A less powerful engine would have to be driven at high speed and high load to develop the same power, and when these conditions are sustained over long periods, stress-related failures inevitably arise. This was a categorical shortcoming experienced by underpowered tanks, and it was the fundamental reason why large displacement engines are used in tanks. Of course, a high power-to-weight ratio by tank standards was still low for automobiles, which is reflected by a stark contrast in reliability. Passenger car engines in particular rarely need to put out their full power, and when they do, it is only for extremely short periods[1045].

The service life of the V-46 and V-84 was 500 engine-hours, plus another 300 engine-hours after a manufacturer warranty overhaul. Long-term experience with the T-72 in the Soviet Army and in Warsaw Pact armies showed that 1,000 engine-hours was routinely exceeded in Soviet-built V-46 engines. While covered under the 500-hour warranty period, it was forbidden to tamper with the engine in any way barring basic upkeep. Lead seals covered several critical access points, and breaking them voided the warranty.

The V-46 also performed well in terms of engine elasticity, the basic metric of dynamic performance [1046]. High-power diesel engines typically had poor dynamic performance, which was a major factor in the historical reticence to leave gasoline tank engines behind for diesels in several major militaries. Dynamic performance had a strong influence on off-road acceleration within the limits of the gross engine power, and this was a notable advantage of gas turbine engines over reciprocating piston engines[1047].

A highly elastic engine, combined with a large gearing range, facilitated high average speeds [1048] on rough terrain. Tests demonstrated that the average

---

[1044] Борзов, В. К. (1986). 'Централизованная Заправка ВГМ Топливом', *Вестник Бронетанковой Техники* 1986, сборник 3. p. 30.

[1045] Зубов, Е. (1999). 'Легендарный В-2: три страницы судьбы', *Двигатель*, 4(4). p. 18.

[1046] Engine elasticity essentially describes the width and height of an engine's power band. A wide and tall powerband allows an engine to adapt easily to changing torque loads, such as those encountered on cross-country terrain. A torque converter partially compensates for low engine elasticity, but also introduces a source of large power losses

[1047] Mysłowski, J. (2014). 'Propozycja Poprawy Manewrowości Czołgu Twardy', *Zeszyty Naukowe Wsowl*, 1(171). pp. 150-151.

[1048] Маринин, В. А., Савенков, А. И., Тульцев, А. А. (1982). 'Влияние Коэффициента Приспособляемости Дизеля на Скорость Танка', *Вестник Бронетанковой Техники* 1982, сборник 5. pp. 26-30.

relative speed loss of the T-72 from its theoretical top speed through losses in traction, track flotation, control errors, and ergonomic factors reached 41%, which was significantly lower than on the T-64A (46%), and close to the level of the T-80 (39.4%)[1049]. At the same time, the increase in fuel consumption from the theoretical rate was also the lowest[1050].

The torque elasticity coefficient (E_T) is the ratio of the peak torque to the torque developed at peak power. High torque elasticity is important for negotiating terrain that imposes high fluctuating engine loads, and is therefore responsible for providing a high cross-country speed.

The speed elasticity coefficient (E_S) quantifies the width of the power band. It is the ratio of the rated speed to the peak torque speed. The wider the power band, the higher the speed elasticity. A wide power band contributes to the ease of driving in off-road terrain, especially with a mechanical transmission.

The idling speed was 800 RPM, with the option of raising it up to 2,000 RPM using a hand accelerator. Idling the engine at 1,100-1,200 RPM kept the generator at its operational speed to support power-hungry devices like the weapons stabilizer, and raising the idle speed even higher than that was sometimes necessary to speed up engine cooling during short halts[1051]. In some cases, the idle speed would have to be set to 1,300-1,600 RPM before starting the engine on a steep slope to prevent a stall[1052]. In

Table 49: Dynamic performance parameters of various tank engines

Engine	E_S	E_T
MB 873 Ka-501[a]	1.63	1.159
V-84	1.54	1.130
V-46	1.54	1.125
CV-12[b]	1.54	1.085
MB 838 CaM-500[c]	1.42	1.04
AVDS-1790-2[d]	1.27	1.041

[a] MTU. *Twelve Cylinder Diesel Engine MB 873*. Booklet.
[b] Perkins (2018). *CV12 Military Engine*. Booklet.
[c] Чернышев, В. Л. et al. «Леопард-1» - Передовая Система Немецкой Бронетанковой. Техники Середины 60-х Годов XX Века. Available at: https://btvt.narod.ru/raznoe/leo1transm/leo 1tr.htm (Accessed 2 April 2024).
[d] *Air-Cooled Diesel Tank Engines*. Booklet. Teledyne Continental Motors. p. 26.
Chavy (1962). 'Le Moteur de Char', *L'Armee*. p.30.

[1049] Шпак, Ф. П. (1986). 'Оценка Формирования Показателей Подвижности Танков по Ограничивающим Фактором', *Вестник Бронетанковой Техники* 1986, сборник 3. р. 6.
[1050] Ibid. p. 7.
[1051] This increased the coolant pumping rate and fan speed while keeping the engine at low load. The cooling rate was highest in this circumstance.
[1052] Сутормин, Е. А. (2013). *Вождение боевых машин*. Екатеринбург: Издательство Уральского университета. р. 48.

combat, the need for a high idling speed essentially served a similar function to the "tac idle" (tactical idle) setting in some U.S. Army combat vehicles. Because the engine would idle close to its powerband, acceleration from a stop could be faster than normal.

The engine could be started electrically or pneumatically. Air-starting was preferred for the sake of prolonging battery life. The air starter was a rotary distributor connected to a timing shaft, delivering sequential bursts of high-pressure air into the twelve cylinders to force the engine to turn over[1053].

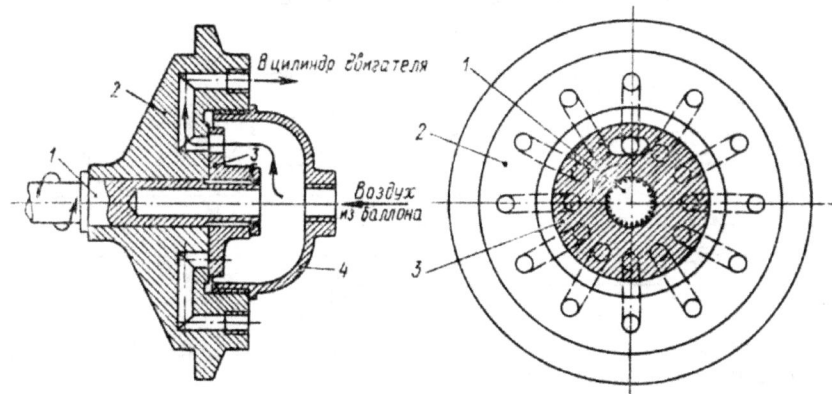

Figure 172: Air-start pneumatic distributor. (*Soviet Ministry of Defense*)

The pneumatic distributor was also a core part of the engine's integral long-term preservation system. A filler hole could be opened and a syringe of rust inhibitor oil inserted, and by turning on the air-start system (with no fuel flow), the engine cylinders were rapidly and thoroughly coated in an aerosol spray[1054].

Regardless of the starting method, cold starts were difficult in chilly weather. Preheating was needed at ambient air temperatures below 5°C, and engine starts were carried out through the so-called "combined" method, which was to run the electric starter and the air starter simultaneously. The colder the ambient temperature, the warmer the engine had to be to start reliably[1055].

At a heating rate of 10°C per 2-3 minutes[1056], the preheating period at any given temperature was already considerably shorter than most other tank diesel

---

[1053] Старовойтов, В. С. (ed.) (1990). *Военные гусеничные машины*. Том 1, Книга 2. Москва: Издательство МГТУ им. Н. Э. Баумана. p. 89.
[1054] Министерство Обороны СССР (1983). *Двигател и В-46 и В-46-6 – Техническое Описание (ТО)*. Москва: Воениздат. p. 111.
[1055] Янковский, И. Н. et al. (2020). *Устройство и эксплуатация бронетанкового вооружения*. Част 2: Эксплуатация танка Т-72Б. Минск: БНТУ. p. 400.
[1056] Гоголюк, Е. К. (1983). 'Двухрежимная Система Подогрева Дизеля', *Вестник бронетанковой*

engines, which was entirely thanks to the enormous power of the 50-kW preheater[1057]. During the preheating period, coolant flow in the cooling system was reversed, delivering hot water from the preheater boiler directly to the engine cylinder banks before flowing into the oil tanks, and then passing through the radiators last. In a mild winter weather (5°C to -10°C), the engine could start in around 7 minutes after firing up the preheater[1058], and at a temperature of -36°C, the time was doubled to 14 minutes[1059]. On short halts without a running engine, precious heat was preserved by closing all flaps on the engine compartment roof and laying heat insulating mats over the radiator flaps, and in extreme cold weather, the preheater was started every 30 minutes to maintain the minimum desirable temperature.

Table 50: Preheating temperatures

Temperature range (°C)	Engine temperature (°C)	
	Minimum desirable	Recommended
5 to -10	-	55-80
-10	60-70	80-105
-10 to -20	70-80	80-105
-20 to -30	80-90	110-115
-30 to -40	90-105	110-115

In the V-84-1, an intake heating system was introduced to reduce the preheating time to speed up emergency start-ups[1060]. Instead of glow plugs, a burner was fitted to – quite literally – shoot flames down the intake manifolds. The burner was installed just behind the supercharger. It piggybacked on the tank's primer pump to draw in fuel and it used the compressed air in the tank's pneumatic system for its fuel atomizers[1061].

Needless to say, this method of preheating was not very healthy to the engine, so the PVV system was reserved for rapid mobilization only. An ignition counter

---

техники 1983, сборник 1. p. 31.
Supported by Swedish trials in 1992. See: Edgren, S., Hansson, M. (1992). *T 72 och MTLB. Delapport 1. Trials report.* Arméns Pansarcentrum.

[1057] For comparison, this was over double the power of the Leopard 2 preheater (23.3 kW diesel burner), for a smaller engine with a much smaller total thermal mass.

[1058] Гоголюк, Е. К. p. 31.

[1059] During winter tests in the Urals in 1973, preproduction T-72 tanks were frozen to a temperature of -36°C for 43 hours. The average preheating time before engine start was 14 minutes. See: Вавилонский, Э. Б., Куракса, О. А., Неволин, В. С. (2008). *Основной боевой танк России: Откровенный разговор о проблемах танкостроения.* Нижний Тагил: Издательский Дом «Медиа-Принт». p. 70.
For comparison, a Leopard 2A4 at an ambient air temperature of -32°C with preheated, fully charged batteries can start within one hour after 50 minutes of engine preheating.

[1060] Newer engines like the MB 838 and MB 873 had prechamber ignition with glow plug heating to allow cold starting at -18°C.

[1061] Министерство Обороны СССР (1991). *Дизель В-84М (В-84, В-84-1) – Техническое Описание.* Москва: Воениздат. pp. 111-116.

on the system prevented the PVV from starting after a limit of 20 uses, but in an emergency, this counter could be bypassed with an override switch.

According to technical norms, the burner had to run for two minutes before an attempt could be made to start the engine. Additional preheating was only needed when the PVV system was used at temperatures below -20°C, in which case the minimum coolant temperature had to be 60°C plus an additional 10°C for each -10°C decrement in ambient air temperature.

Setting aside these fairly stringent preconditions, the PVV system allowed the T-72B to drastically reduce its start-up time in winter, which was an important consideration to the Soviet tank industry after performance standards for AFV diesel engines were raised in 1980 [1062], and the internal army-industry competition between the T-72 and the T-80 series ramped up.

Out of the many advantages identified in gas turbine engines for tanks, cold weather starting was among the most important, especially in northern European climates. Gas turbine engines required a great deal of battery power to spool up, but they could start reliably without preheating of any kind. As an example, the GTD-1000T gas turbine engine in the T-80 could boast of a normal start-up time of 2-3 minutes at an ambient air temperature of -18°C, slowed somewhat from its basic 50-second start-up at normal temperatures due to increased oil viscosity. At temperatures under -20°C, engine starts took 2-5 minutes and were followed by an additional 10-15 minutes spent de-icing its air cleaner[1063] and oil warming before it could start moving, for a total start-up time of 25-30 minutes[1064] at temperatures as low as -45°C[1065].

The V-84-1 was never as easy to start as a gas turbine engine, but if it was flatly impossible to wait for the preheater, the T-72 at least had the means to match the 2-3-minute start-up time of the GTD-1000T on a severe winter day, falling behind only if the weather dipped into arctic temperatures.

[1062] Гинзбург, Б. М., Иванов, В. А., Сысоев, В. Н. (1984). 'Система Подогрева Впускного Воздуха с Бесфорсуночным Факельным Подогревателем', Вестник бронетанковой техники 1984, сборник 1. p. 17. Referring to the GOST V 24480-80 standard (ГОСТ В 24480-80 «Дизели ВГМ. Общие технические требования»).
[1063] Министерство Обороны СССР (1986). *Объект 219Р – Техническое описание и инструкция по эксплуатации*. Книга вторая. Москва: Воениздат. p. 703.
[1064] Вавилонский, Э. Б., Куракса, О. А., Неволин, В. М. p. 70.
The total start-up time in an Abrams was around the same due to the waiting period for oil warming.
[1065] The GTD-1250 pushed this further to just one minute at a temperature of -30°C. The GTD-1250 was not mass-produced until the 1990's.

Engine Design

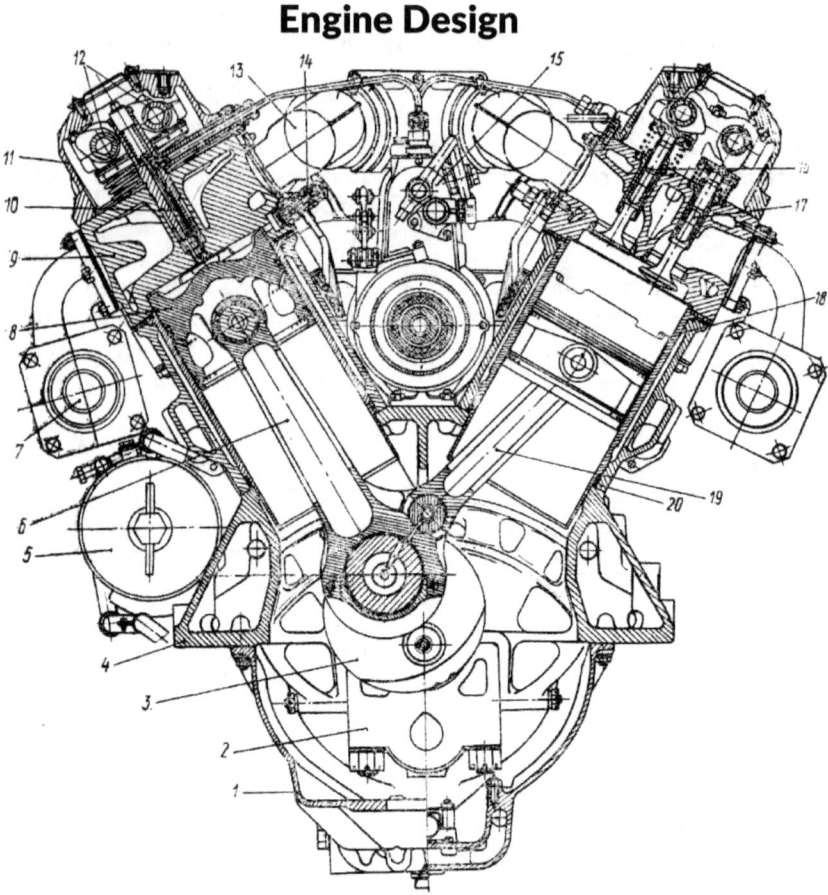

Figure 173: V-84 engine cross-section. (V. S. *Starovoytov, 1990*)

The structure of the V-46 engine crankcase was essentially the same as the original V-2. It consisted of an upper crankcase and a lower crankcase, bifurcated along the crankshaft axis. The upper crankcase bore all engine loads and the lower crankcase served as an oil pan, which greatly lightened the structure at some expense to overall crankcase rigidity. The front end (camshaft drive) was the same as a V-2, and the rear end (driveshaft)[1066] featured a rectangular housing for the supercharger gearbox. In the original V-2, the rear end was rounded, which was a vestigial tail from aviation engine design,

---

[1066] This naming convention comes from the front end of car engines facing the same direction as the car, and the rear end driving a rear prop shaft.

originally meant to help support the weight of a propeller. Eliminating this rounded shape actually made the V-46 shorter overall, despite its supercharger.

The crankcase had a circular profile. The reason for this shape was because the engine mounts were level with the crankshaft axis, so the cylinders, which must be radially aligned with the crankshaft axis, deliver a tensile force normal to the curve of the crankcase. In this way, the crankcase behaved as an arch with its two ends rooted to the tank hull. Relatively narrow main bearing caps were bolted with studs into rounded underhanging partitions and then cross-bolted again for reinforcement.

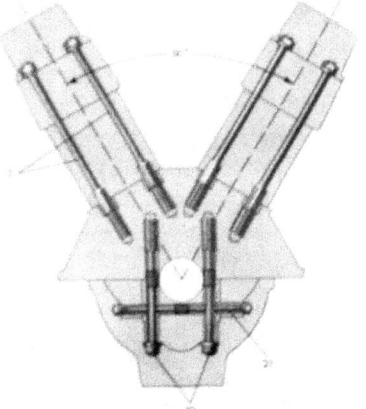

This was a completely different approach to crankcase design compared to the Soviet aviation engines available during the early 1930's, like the *Klimov* M-100 (Hispano-Suiza 12Y), the *Mikulin* M-17 (BMW VI) and also notable European aviation diesel projects from the same period, such as the Coatalen.

The engine also had unitary cylinder blocks with unitary cylinder heads, and it used hold-down studs instead of bolts or short studs to secure the cylinders to the crankcase.

Figure 174: V-46 cross section, showing the hold-down studs and main bearing cap cross-studs. *(Soviet Ministry of Defense)*

The V-2 family had this in common with some British aviation engines, most notably the Rolls-Royce Kestrel and the Merlin[1067].

Hold-down studs are long studs that pass through the entire height of the cylinder to clamp it down into the crankcase from above. Compared to bolts or short studs, hold-down studs gave a more stable and uniform clamping force between the cylinder block and crankcase[1068]. These hold-down studs were supplemented by short studs along the periphery of the cylinder block to improve load distribution. This combination of hold-down studs and short studs

---

[1067] Ellor, J. E. (1944). 'The Development of the Merlin Engine', *SAE Transactions*, Vol. 52. p. 387.
[1068] Almen, J. O. (1944). 'On the Strength of Highly Stressed, Dynamically Loaded Bolts and Studs', *SAE Transactions*, Vol. 52, pp. 151-155.
Also see: Hall, E. J. (1929). 'Reducing Transportation Cost by Means of Engine Design', *SAE Transactions*, Vol. 24. pp. 66-67.
Hold-down studs, compared to short bolts, experience a smaller change in preload for the same change in length (for the same strain energy). When warpage occurs in the cylinder block and the studs elongate or contract from their initial lengths, the total clamping force and the uniformity of the clamping force changes less than if bolts had been used in their place.

was unique to the V-2 family. It was introduced early in its development to reinforce the head gasket when problems emerged due to high pressures. These issues returned on the V-46, and failures in its bimetallic head gasket[1069] were one of the most common sources of V-46 breakdowns.

The cylinder blocks were unitary castings. This was a popular design choice in the 1930's since the cylinder blocks could contribute to the overall rigidity of the engine[1070], which was not the case with individually mounted cylinders. The cylinder block design had a major contribution towards the overall compactness of the engine. It had extremely close cylinder spacing, at just 175 mm (15 mm gap between cylinders), which was made possible by the use of wet liners with narrow water jackets taking up all of the space unoccupied by the hold-down studs. The cylinder block was therefore very light, being mostly hollow, and also very short. Thanks to these design measures, the V-46 was shorter and lighter than even engines with fewer cylinders and smaller bore diameters like the MB 838 CaM-500[1071].

The single most complex and expensive part of the entire engine was the crankshaft, which was a unitary forging lathed into shape. Pinned to it were anodized Silumin pistons[1072] with a Hesselman bowl[1073] on an articulated connecting rod layout. This was an uncommon, but not unusual layout where the left piston was connected to a shorter master rod and the right piston was connected to a longer rod, articulated on the master rod with a 67° offset.

This layout extended the stroke length of the articulated end by 6.7 mm, creating an asymmetry in the swept volumes of the left and right cylinder groups, but the effect on combustion dynamics was negligible, and the secondary imbalance from the articulated rod was entirely balanced by a simple adjustment in the counterweight[1074]. However, the offset in the axis of rotation for the articulated rods put a stronger lateral load on its pistons, which had to be taken into account when designing the piston rings. Compared to the conventional fork-and-blade connecting rod layout, the fundamental reason for

---

[1069] This type of gasket works by expanding when heated.
[1070] This refers to the structural rigidity of the engine in all planes except in torsion. Additionally, when a cylinder fires, the load is distributed through the entire cylinder block.
[1071] V10 engine with 140 mm bore diameter. Engine length of 1,562 mm.
[1072] Silicon aluminum alloy.
[1073] Бутов, В. И., Егоров, В. В. (1985). 'Повышение мощности и топливной экономичности серийного дизеля', *Вестник бронетанковой техники* 1985, сборник 1. p. 33.
[1074] A slight secondary imbalance persists, but is entirely negligible. See: Taylor, C. F. (1985). *Internal Combustion Engine in Theory and Practice. Volume 2: Combustion, Fuels, Materials, Design*. Second edition (revised). Cambridge: MIT Press. p. 278.

using this layout to begin with was to improve the rigidity of the big-ends of both connecting rods[1075], especially on the master rod[1076].

To create the V-84, the piston was modified by reshaping the bowl contour for better fuel-air mixing. The surface area of the piston bowl was reduced and the compression ratio remained unchanged, but fuel economy improved[1077]. Improved air flow was achieved by splitting the cylinder intake manifold on each cylinder bank to two separate manifolds for the 1st to 3rd cylinders, and for the 4th to 6th cylinders[1078]. The supercharger was tuned accordingly, but the boost pressure in the manifolds remained the same[1079]. All of these changes raised the engine power to 840 hp. The net power of the V-84 after intake and exhaust losses and cooling losses was 11% below its gross power[1080].

The large displacement of the V-46 and its powerful supercharger developed a peak air mass flow rate of 1.31 kg/s (60 m³/min or 2,120 CFM)[1081]. The boost pressure was 0.68 bar, but thanks to exhaust scavenging through a long valve overlap, the total cylinder overpressure was 0.804 bar.

The total overpressure rose even higher in the V-84, but like breathing air into a furnace, the cylinders burned hotter, causing a score of issues that took special design measures to solve. Among them were new cylinder valves coated in Stellite, the same substance used to line the barrels of M2HB and M60 machine guns, and new exhaust manifolds with accordion-style joints to cope with high thermal expansion. The V-84 also added a heat shield[1082] over the right exhaust manifold as a layer of insulation between it and the radiator, reducing the loss in cooling efficiency from parasitic heating. Combined with the lower heat

---

[1075] Master and slave Haas, H. H. (1957). *Design Features of the New Continental 750 Horsepower Diesel Engine and Notes On High Performance Four-Cycle Diesel Engines 600 Horsepower to 1000 Horsepower Range*. Proceedings of Meeting. New York: Society of Automotive Engineers. p. 10.
[1076] Алексеев, В. П. et al. (1990). *Двигатели внутреннего сгорания: Устройство и работа поршневых и комбинированных двигателей*. Москва: «Машиностроение».
Further supported by: Орлина, А. С. (1955). *Двигатели внутреннего сгорания. Рабочие процессы в двигателях и их агрегатах: Том II. Конструкции И Расчет*. Москва: Машиностроительной Литературы.
[1077] Бутов, В. И., Егоров, В. В. pp. 33-36.
[1078] Министерство Обороны СССР (1991). *Дизель В-84М (В-84, В-84-1) – Техническое Описание*. Москва: Воениздат. p. 102.
[1079] Ibid. p. 103.
[1080] Ефремов, А. (2002). 'Чем Выше Подвижность Танков, Тем Мобильнее Сухопутные Войска', *Двигатель*, 5(23), сентябрь-октябрь.
[1081] НИИД (1978). Анализ и систематизация материалов по уровню развития отечественных. Технический Отчет №3443. Available at: http://btvt.info/3attackdefensemobility/niid_1978.htm (Accessed 20 August 2023).
For comparison, the AVDS-1790 with comparable power had an induction air flow of 2,000 CFM. See: Haas, H. H. p. 9.
[1082] *Руководство по войсковому ремонту объект 184 кн.1 ч.2 (1992)*.

rejection from the new piston bowl geometry, the existing T-72A cooling system could cope with the V-84 when it was introduced in the T-72B.

Engine Air Cleaner

The supercharger on the engine was connected to a compact two-stage multi-cyclone air cleaner. Air was taken directly from the engine compartment through the sides of the multicyclones, but in normal conditions, nearly all of the air entering the engine came from the intake hole in the engine access panel, positioned slightly ahead of the air cleaner. The cooling fan at the tail end of the engine compartment formed a draft between the intake and the air cleaner to help feed cool air to the engine in the summer, and in winter, the intake hole was closed off with a special cap so that it took in warm air from the radiator while driving. This was done to retain as much heat as possible and to de-ice the micro-cyclones of the air cleaner[1083]. A small hole in the cap for the cooling duct intake for the AK-150SV air compressor.

The engine air cleaner was a modification of the VTI-4 air cleaner developed for V-2 engines[1084], featuring lower intake resistance, a drastically larger size[1085] and air flow capacity, and several structural adaptations for the V-46 engine supercharger intake. The T-72 air cleaner had no known name[1086].

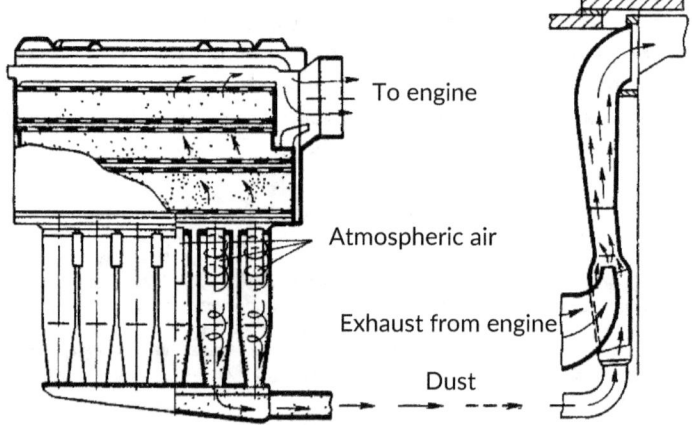

Figure 175: Air cleaner and dust ejector connection to engine exhaust. (*Soviet Ministry of Defense*)

[1083] Бершов, А. В. et al. (1981). 'Новое Воздухозаборное Устройство Для Танка'. *Вестник Бронетанковой Техники* 1981, сборник 4. p. 41.
[1084] Старовойтов, В. С. (ed.) (1990). *Военные гусеничные машины*. Том 1, Книга 2. Москва: Издательство МГТУ им. Н. Э. Баумана. p. 64.
[1085] The volume occupied by the T-72 air cleaner was 227.8 liters compared to the VTI-4's 74.5 liters.
[1086] It was referred to simply as the "172M air cleaner", referring to the T-72's Object 172M index.

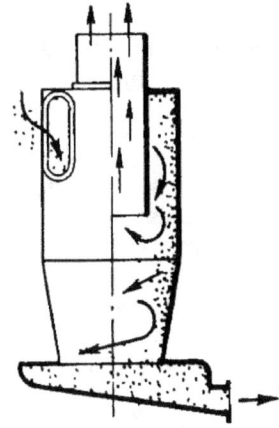

Figure 176: Air flow and dust removal in a cyclonic dust filter. (*N. A. Nosov et al., 1972*)

The first stage was an array of 96 micro-cyclones, providing 99.4% air purity and guaranteeing the removal of heavy particles and liquids like rainwater. These particles fell into a dust collector and were ejected through the engine exhaust via an exhaust ejector (scavenger) system[1087].

The second stage was a stack of three oil-wetted steel wire mesh cassettes. These trapped dust on a thin film of oil, with progressively higher mesh densities on each successive cassette, but even the last cassette was only about as dense as a steel wool kitchen scrubber, thanks to which the air cleaner supplied air at 99.8-99.9% purity[1088] with a very low intake air resistance of just 68.6 mbar, rising up to a modest 127.5 mbar at critical saturation[1089]. After reaching critical saturation, it was permitted to continue driving for an additional 5 hours in a medium-dustiness environment or 2 hours in a high-dustiness environment. The nominal time between air cleaner servicing while operating at a suspended dust level of 2.5 g/m^3 was 400 km[1090]. For reference, 1.59 g/m^3 of dust was already considered a zero-visibility condition[1091]. Otherwise, the official cleaning intervals were 1,000 km in winter and 500 km in the summer. The mesh cassettes on the air cleaner could be accessed easily through the top lid, but cleaning the cassettes could take more than two hours, most of it spent waiting for the fresh coat of oil on the mesh to drip off and leave a thin film as intended[1092].

[1087] An ejector-type dust extraction system was significantly more power-efficient than a fan-type system. See: Dziubak, T. (2017). 'Problemy Usuwania Pyłu z Multicyklonów Filtrów Powietrza Silników Terenowych Pojazdów Mechanicznych', *Archiwum Motoryzacji*, 76(2). p. 41.
For example, the electric fan dust extractor was replaced by the Vehicle Exhaust Dust Ejector System (VEDES) in the M60A1.
[1088] Министерство Обороны СССР (1986). *Танк Т-72А - Техническое описание и инструкция по эксплуатации: Книга Вторая (Часть первая)*. Москва: Воениздат. pp. 225-226.
Also see: Баранов, С. П., Никитин, В. Т., Ферштудт, А. И. (1987). 'Развитие танковых силовых установок', *Вестник бронетанковой техники 1987*, сборник 6. p. 54.
[1089] Борисюк, М. Д. et al. (2008). 'Высокоэффективная система очистки воздуха для военных гусеничных машин', *Інтегровані технології та енергозбереження*, (2) 2008. p. 19. This was nonetheless slightly higher than in the VTI-4, where intake resistance was 55 mbar up to 118 mbar at critical saturation.
[1090] Ibid. p. 20.
[1091] Overholt, L. F. (1943). 'Dust problems in military vehicle operation', *SAE Transactions* Vol. 51. p. 383.
[1092] Washing and re-oiling was a quick process, but a lot of time had to be spent waiting for the oil to drip off the mesh. This was done on special equipment and drying racks carried by the battalion maintenance platoon. For comparison, the dry mesh filters of an M1 Abrams had a working period of 20 hours before saturation and the process of cleaning them with compressed air, also provided by an

As a comparison, the Leopard 1 air cleaner, also a two-stage type, had been designed for a similar air flow requirement and was rated for the same 99.8% air purity, but it had a much higher intake resistance of 500 mbar, rising up to 700 mbar at critical saturation[1093]. The engine power loss from the intake resistances alone reached 11.44% (95 hp).

The volumetric efficiency of the air cleaner was exceptionally high, thanks to which the T-72's air cleaner was smaller than its Western counterparts for a similar air flow rate. For a given air purity, the T-72's air cleaner had a specific flow rate of 264 m³/min-m³, against 247 m³/min-m³ for an oil bath filter and 200 m³/min-m³ for a two-stage cyclone and paper filter[1094]. While investigating methods of reducing engine power losses, an optimization study found that increasing the space between the mesh cassettes and redistributing their layer thicknesses had a major impact on its air resistance[1095]. These improvements were applied to a new air cleaner introduced in the T-72A in 1979[1096]. It was designed to be completely interchangeable with the existing design, and could therefore be fitted to older tanks during scheduled overhauls.

Cooling System

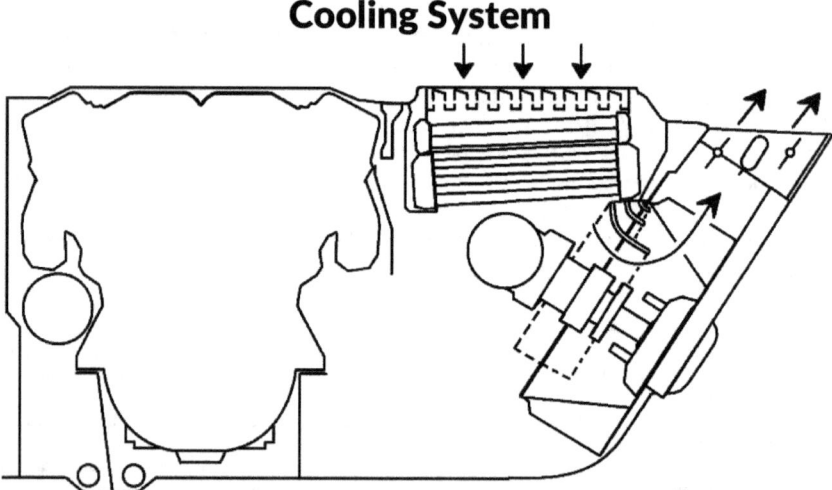

Figure 177: Air flow path in a late model T-72 cooling system. (R. A. Then, 2024)

---

attached battalion maintenance unit, took around one hour. See: Dasch, J. M., Gorish. J. D. (2013). *The TARDEC Story - 65 Years of Innovation*. Washington, D.C.: GPO. p. 86.
[1093] Spielberger, W. J. (1995). *Waffensysteme Leopard 1 und Leopard 2*. Auflage: Motorbuch Verlag. p. 389.
[1094] Millar, D. H. (1970). 'Critical Factors in the Application of Diesel Engines to Fighting Vehicles', *Proceedings of the Institution of Mechanical Engineers 1969-70*, 184(3). p. 144.
[1095] Никитин, В. Т., Ушаков, В. Я. (1980). 'Методика Расчета Двухступенчатых Танковых Воздухоочистителей', *Вестник бронетанковой техники 1980*, сборник 6. pp. 24-25.
[1096] Tested on the Object 176 and introduced under the Object 176 index.

The cooling system of the T-72 followed the basic template set by the T-54/T-55 and T-62. A large, shaft-driven centrifugal cooling fan held the engine compartment under negative pressure to draw air through the radiators, and then expelled the radiator exhaust rearward at a 30° angle. Considering that the basic configuration had remained essentially unchanged from the T-54, the performance of the cooling system was excellent.

In spite of the growth in engine power from the 520-580 hp engines of the T-54/55 and all the additional waste heat that it brought, new technical solutions allowed the designers to curb the power costs to the extent that the T-72 cooling system was both more efficient and more tolerant to a hot climate.

The T-72 could maintain stable coolant and oil temperatures at near-maximum engine load at ambient air temperatures of up to 35°C. Long marches at temperatures above 35°C was characterized by a pattern of driving at high speeds interspersed by a few minutes of relaxed speeds with the engine revving high at a lower gear to allow the coolant and oil temperatures to drop. Nevertheless, the cooling system could cope with tropical and subtropical climates where its Western counterparts generally could not, apart from special "tropicalized" tank models, while at the same time it did not demand an undue share of engine power, which kept fuel economy high and maximized useful power for combat maneuvers.

The basic layout itself was fundamentally efficient. Ambient air entered the radiators directly so that there was no additional intake heating[1097], and the exit angle of the exhaust behind the radiator ensured that waste heat did not recirculate into the intake, particularly when driving forward at speed. The choice of a single large-diameter cooling fan over smaller fans made its contribution to efficiency as well, although the most important design factor was that a centrifugal fan is best suited to generating a high static pressure[1098].

Instead of placing deep, multi-row oil and water radiators side-by-side like in a T-54/55 or T-62, the T-72 had an optimized layout with three two-row oil radiators laid on top of a pair of six-row water radiators[1099]. The leftmost oil radiator served the transmission oil circuit, and the other two were the engine

---

[1097] Unlike cooling systems where the air passes through the engine bay before reaching the radiator, like the Centurion and Chieftain tanks.
[1098] For a given air flow, a larger diameter fan requires less power, and a large-diameter centrifugal fan generates a high static pressure to overcome resistance. However, packaging constraints usually forced designers towards axial fans (Patton series, M60, Leopard 1), or mixed-flow fans (Chieftain). See: Millar, D. H. p. 143.
[1099] The T-54/55 and T-62 had a six-row oil radiator core next to a six-row water radiator, which was less efficient.

oil coolers. The face area of the radiator pack was also marginally increased thanks to the wider hull of the T-72[1100].

Running on high temperature coolant improved fuel efficiency by reducing heat loss to the combustion chamber walls, and also improved cooling efficiency by increasing heat removal from the radiators to the air. The concept of high temperature cooling was an industry standard for military liquid-cooled aviation engines during WW2 and its applications in high performance tank engines had been studied extensively in the USSR since the late 1940's, but even so, its implementation in tank engines was still a novelty in the 1970's[1101]. The T-72 system was built for a pressure limit of 2.1 kg/cm^2 (30 psi), double that of conventional automobile cooling systems and 50% higher than other tank cooling systems of the era, like on the Leopard 1 and Chieftain.

Normally, the T-72 cooling system operated well within its safety margins, typically just 70-100°C when running on diesel, petroleum naphtha or kerosene, or 80-100°C when running on gasoline. In hot weather and hard off-road driving, coolant temperatures of up to 110°C could be sustained indefinitely[1102], and a temperature of 115°C was permitted for around 10 minutes at ambient air temperatures up to 35°C[1103]. In cold weather (<5°C), the temperature limit fell to just 105°C because antifreeze had to be mixed into the coolant, which worsened its heat transfer rate compared to the basic coolant mix with the standard three-component blend of anti-corrosion and anti-scaling additives.

The absolute temperature limit was 120-121°C[1104], with a critical threshold (overheating) sensor to warn the driver at 112-118°C, and a second sensor with a threshold of 92-98°C when driving in cold weather (<5°C). The cold weather threshold was later raised to 104-109°C on the V-46-6[1105]. The sensors were connected to warning lamps on the PV-82 control panel located in the driver's station. However, the cooling system was not automatically regulated, and there was nothing to initiate an automatic engine shutdown if the critical temperatures were exceeded. All corrective actions had to be carried out by personal intervention from the driver.

[1100] Face area of water radiators: 470 x 927 mm. Face area of oil radiators: 460 x 604.6 mm.

[1101] The Leopard 2 was the first Western tank with a reciprocating piston engine to feature high temperature cooling. High-temperature cooling has also been implemented in race cars.

[1102] For comparison, in the Leopard 1, the temperature limit for indefinite movement was 93°C. Driving at 93-105°C was permitted for up to 10 minutes.

[1103] For comparison, the Leopard 1 was permitted to run for 10 minutes at 105°C. The "Tropicalized" Leopard 1 modification involved raising the coolant temperature to 115°C to allow operations under ambient air temperatures of up to 40°C.

[1104] Imposed by the boiling point of water at the 2.1 kg/cm^2 pressure limit.

[1105] Ministerium für Nationale Verteidigung (1987). *Panzer T72,T72M und T72M1 - Beschreibung*. A 051/1/130. Berlin: Militärverlag der Deutschen Demokratischen Republik. p. 157.

Temperature control was especially primitive. The fan had Neutral, Low and High settings, which could only be set by physically accessing its gearbox after opening up the radiator pack. The High gear had a 0.647 step-up ratio[1106], and the Low gear had a 0.773 step-up ratio. This type of control was known as seasonal adjustment[1107] since the Low mode was the preferred setting for practically all conditions apart from summertime or a desert climate (>25-30°C). Neutral was used to disengage the fan.

Because the fan speed could not be changed on the fly to regulate engine oil and coolant temperature, the driver instead adjusted the louvers in the cooling fan exhaust duct, which essentially functioned as large butterfly valve flaps to throttle the exhaust flow. If the engine was running too cold, particularly after starting up in winter[1108], throttling the exhaust flow curtailed the efficiency of the cooling system and worsened the ventilation inside the engine compartment so that the engine ingested more heated air from the radiator.

Of course, raising the cooling power in a volume-constrained engine compartment by increasing fan power had always been an option, but it was a deeply unattractive one. The volumetric flow rate generated by a fan is proportional to the cube of its power[1109], so doubling the air flow rate requires the fan power to be multiplied eightfold, but the cooling power rises only modestly with fan power. Doubling the flow rate of the T-72 cooling fan, for instance, would only improve the steady-state coolant temperature drop by approximately 4°C[1110]. Naturally, the designers chose to focus on improving cooling system efficiency instead by strategically leveraging and suppressing turbulent flow[1111].

As a stream of fluid flows over a solid surface, its viscosity creates a boundary layer between the fluid and the surface, where the velocity of the fluid varies from zero at the surface, essentially "sticking" to it, up to the free velocity of the flow at the end of the boundary layer. A laminar boundary layer practically does not intermix with the flow, essentially behaving as an insulating barrier where heat is transferred through conduction. However, with turbulent flow, a turbulent boundary layer is created where heat is transferred much more effectively due to the intermixing of the flow, i.e. convection.

[1106] In high gear, the T-72 cooling fan developed a flow rate of 2.7 m^3/s at an engine speed of 2,000 RPM.
[1107] Старовойтов, В. С. (ed.) (1990). *Военные гусеничные машины*. Том 1, Книга 2. Москва: Издательство МГТУ им. Н. Э. Баумана. p. 54.
[1108] Keeping the coolant temperature above the minimum 65°C threshold.
[1109] In an ideal system, volumetric flow rate is directly proportional to fan speed (First law), and fan power is proportional to the cube of the fan speed (Third law).
[1110] Venkateswaran, N., Radhakrishnan, S. R., Sathyanarayanan, P. L. (2008). 'Performance Improvement of Cooling System in T72 Bridge Layer Tank', *Defence Science Journal*, 1(58). p. 81
[1111] Summary notes for the monograph: Михайлов, Г. А. (1979). *Водяные радиаторы военных гусеницных машин*. Москва: ЦНИИ информации.

Inside the oil and water radiator cores, built-in turbulators improved coolant-to-radiator heat transfer[1112], and the entire radiator pack was positioned at a 5° offset to the intake to induce turbulence between its fins [1113], improving radiator-to-air heat transfer. The turbulence created by the ballistic grilles turned out to have a major positive impact as well. Bench tests showed that compared to open radiators with completely unrestricted air flow, the presence of ballistic grilles improved the net power efficiency of heat removal in water radiators by ~35% and in oil radiators by ~25%[1114].

At the same time, excess turbulence could not be allowed to overly restrict the flow rate, so finding an optimal balance of both characteristics in a grille design went a long way towards improving cooling efficiency. Originally, the Object 172 prototype inherited the slat-type grilles of the T-64A, but by 1974, it had been replaced by an optimized bar-type grille.

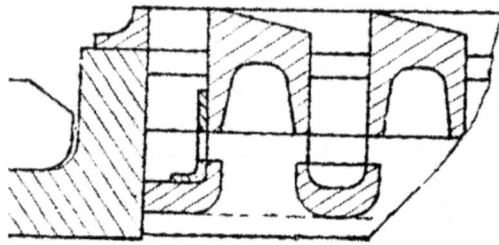

Figure 178: Cross section of T-72 radiator ballistic grilles. (*Soviet Ministry of Defense*)

Bar-type grilles were made from rows of staggered solid steel bars, and generally had a large bulk thickness, but gave good protection and had lower air resistance than other practical options[1115]. A modified bar-type ballistic grille design was created for the T-72[1116] by hollowing out the first row to reduce weight, and adding a bullet trap shape (pocket) to the second row, so that no additional rows were needed. This design was resistant to artillery fragments, bullets, and air blast, which was important for resisting a nuclear detonation at close range[1117]. The same grille design was applied to the

---

[1112] Venkateswaran, N., Radhakrishnan, S. R., Sathyanarayanan, P. L. p. 81.
[1113] Носов, Н. А., Голышев, В. Д., Волков, Ю. П., Харченко, А. П. (1972). *Расчет и Конструирование Гусеничных Машин*. Ленинград: «Машиностроение». p. 31.
A larger 10-15° radiator offset is cited in the reference as the contemporary standard in Soviet tank radiators. The T-72 radiator had a smaller structural offset, but the slanted tops of the grilles deflected the exit flow toward the front (away from the radiator), which increased the effective offset of the radiator cores relative to the incoming flow (tested in CFD simulation performed by author).
[1114] Михайлов, Г. А., Теннисон, В. Р., Фоменко, В. П. (1986). 'Повышение эффективности охлаждения масла танкового двигателя', *Вестник Бронетанковой Техники* 1986, сборник 4. p. 35.
[1115] Department of the Army (1975). *Engineering Design Handbook: Military Vehicle – Power Plant Cooling*. AMCP 706-361. Washington, D.C.: U.S. Army Materiel Command. pp. 6-3, 6-7, 6-15.
Also see: Chiou, J. P. (1975). 'Engine Cooling System of Military Combat/Tactical Vehicles', *SAE Transactions*, Vol. 84. p. 175.
[1116] The Object 172 prototypes also inherited the angled exhaust grilles of the T-54, but these were replaced in the Object 172M.
[1117] Unlike louver-type intake grilles, there was no danger of fixed grilles being jammed shut.

engine air intake[1118] beginning in 1975, and then adopted for the T-80UD in the late 1980's.

After passing through the radiator, turbulent flow was purely detrimental to power economy, so in 1979, inlet vanes were added to the cooling fan intake, and the fan impeller geometry itself was redesigned[1119]. The new fan model was less noisy and more efficient but remained interchangeable with existing fans.

Power losses to the cooling system reached a maximum of 10% in the High mode and 6% in the Low mode[1120]. This was a step forward from the T-55 cooling system (86 hp, 14.8%)[1121] and compared favorably to the cooling systems of the Leopard 1 (95 hp, 11.44%)[1122] and M60A1 (107 hp, 14.3%)[1123]. After receiving the new cooling fan and new inlet vanes, T-72 cooling losses fell to just 7.7% in the High mode and 4.9% in the Low mode[1124].

Similar efforts in the West to reduce cooling system losses can be seen in the AMX-30's centrifugal cooling fan and the novel "Ring" radiator system (with axial fans) in the Leopard 2. Apart from the potential for a comparatively slight reduction in engine compartment dimensions for the same cooling system volume, even the "Ring" solution could not qualitatively surpass the T-72 cooling system design[1125].

Transmission

The T-72 transmission consisted of a pair of side gearboxes, combining the functions of speed selection and steering. Each side gearbox, called a BKP[1126], was integral to the final drives on each side of the hull. When both side gearboxes were in the same gear, the tank would move in a straight line, and when one gearbox was downshifted, the speed difference between the two tracks put the tank into a fixed radius turn. With seven forward speeds, the

---

[1118] T-72 *Ural* tanks (1974) had a mushroom-shaped engine air intake.
[1119] Вавилонский, Э. Б., Кулюгин, А. И. (1979). 'Улучшение Характеристик Вентилятора Системы Охлаждения Танка Т-72', *Вопросы оборонной техники*, 20(83). pp. 26-29.
[1120] Грымзин, П. А., Домбровский, Ю. К., Самарин, Е. Г. (1975). 'Исследование Регулируемого Привода Вентилятора На Ходовом Макете с ГМТ', *Вопросы оборонной техники*, 20(59-60). p. 101.
[1121] Литвинов, Н. П., Мельников, Р. И. (1963). 'К вопросу о выборе вентилятора системы охлаждения среднего танка', *Вестник Бронетанковой Техники* 1963, сборник 2. pp. 40-49.
[1122] *Technische Daten Kampfpanzer LEOPARD*, p. 4 in Krauss-Mafei (1971). *Angebot über die Fertigung und Lieferung von 200 Kampfpanzer Leopard in vereinfachter Ausführung*. 17 August. Letter to the Swiss Group for Armament Services (die Schweizer Gruppe für Rüstungedienste).
[1123] Department of the Army (1975). *Engineering Design Handbook: Military Vehicle - Power Plant Cooling*. AMCP 706-361. Washington, D.C.: U.S. Army Materiel Command. p. 2-18.
[1124] Баранов, С. П., Никитин, В. Т. (1985). 'Пути Снижения Затрат Мощности в Системах Танкового Дизеля', *Вестник бронетанковой техники 1985*, сборник 2. p. 35.
[1125] Носовец, А. И., Шабашев, Л. Б. (1981). 'Система охлаждения с радиаторами на участке нагнетания вентилятора', *Вестник Бронетанковой Техники* 1981, сборник 1. pp. 22-25.
[1126] The term BKP is simply the Russian abbreviation for "side gearbox".

transmission had the flexibility to give the T-72 seven discrete turn radii, optimize its acceleration to 30 km/h, and also allow it to reach a top speed of no less than 60 km/h.

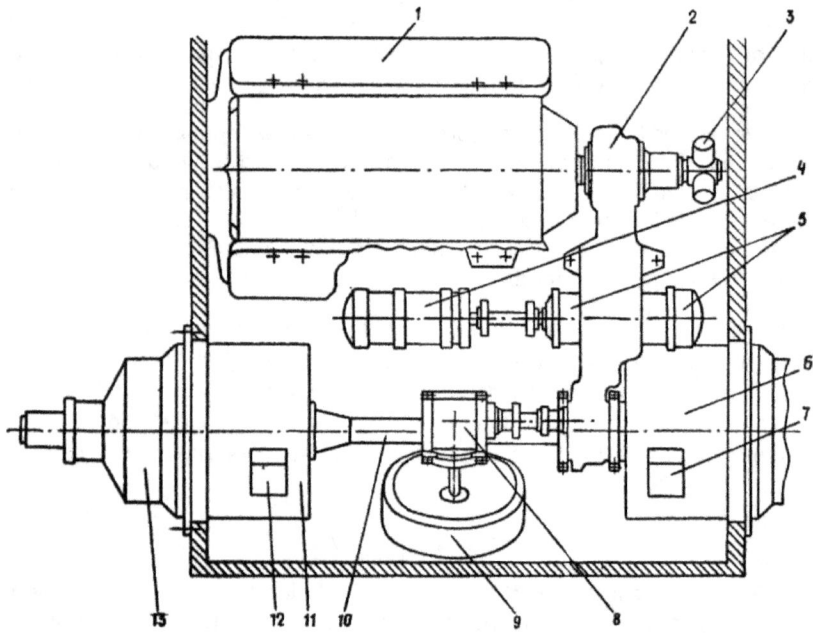

Figure 179: T-72 transmission general layout: 1 – engine; 2 – transfer case; 3 – air compressor; 4 – starter-generator; 5 – fluid coupling; 6 – right BKP; 7 – right BKP hydraulic control unit; 8 – fan drive bevel gear; 9 – cooling fan; 10 – propeller shaft; 11 – left BKP; 12 – left BKP hydraulic control unit; 13 – left final drive. (A. G. Strelkov, 2005)

The BKP side gearbox was originally conceptualized and built by KhPZ for the Object 430 in the 1950's, and was first tested in metal in 1958. It was conceived as a complement to the 5TD opposed piston engine to form an extremely low, compact drivetrain with the engine sitting in between the two BKPs. The V-46 engine was not compatible with this layout, so in the T-72, a transfer case linked the engine to the two BKPs through a common propeller shaft. After each BKP, was an integral final drive to apply an additional gear reduction before engine power arrived at the drive sprocket.

The transfer case had a step-up gearing ratio of 0.706 to the BKPs so that the V-46 matched the input speed range of the faster-running 5TDF from the T-64A. This also limited the torque load on the drivetrain units downstream of the engine, allowing thinner and lighter gears to be used, which benefitted the size and weight of the transmission[1127].

Fitted to the transfer case were the AK-150SV air compressor and ST-10-1S starter-generator. The ST-10-1S was a series-wound DC electric motor that functioned as a generator by reversing the flow of power. Because it was a generator (DC) and not an alternator (AC), push-starting or tow-starting the T-72 was possible.

The presence of an intermediate transfer case between the engine and BKPs was a source of small additional losses, previously absent in the T-64. This was compensated to some extent by using the transfer case as a power takeoff for several accessories (starter-generator, cooling fan drive, air compressor), all of which were also present in the T-64, but served by the engine through separate power takeoff channels.

The final drives did not need to be disconnected to tow the T-72 over long distances, but since its lubrication system was tied to the BKP, some preparation was still needed. Before setting off on a long journey (>5 km), the tank being towed would have both BKPs filled up with oil so that the bearings and gear teeth would be completely submerged throughout the entire journey. This could be done using the MZN engine oil primer pump (an electrical pump), or by manually draining the oil from the transmission lubrication system and transferring it all into each BKP with a bucket and a funnel[1128].

If the MZN oil primer pump was used (depending on whether the tank had functioning batteries), the starter-generator drive lubrication inflow valve could be turned to redirect the flow of oil to the BKPs, and then both BKPs would be automatically filled up from special inlets. The BKPs and final drives were thus converted from dry sump to wet sump lubrication, and the large volume of oil acted as a heat sink, massive enough for the tank to be towed indefinitely.

BKP Gearboxes

BKP gearboxes performed the functions of a main clutch, gearbox, steering mechanism (in cooperation with a parallel BKP) and a stopping brake. Each BKP

[1127] A lower torque reduces the tangential forces on gear teeth, allowing narrower and thinner gear teeth to transmit the same power. Increase in bearing wear and heating losses was negligible. See: Павлов М. В., Павлов, И. В. (2021). *Отечественные Бронированные Машины 1945-1965 гг. - Часть I: Легкие, средние и тяжелые танки.* Кемерово: ООО "Принт". p. 216.
This solution had been in use since the T-34, but was still considered an alternate transmission concept in the U.S. Army. See: Dix, D. M., Riddell, F. R. (1981). *Propulsion System Technology for Military Land Vehicles.* IDA Paper P-1578. Arlington: Institute for Defense Analyses. p. 69.
[1128] *Бронированная Ремонтно-Эвакуационная Машина БРЭМ-1: Инструкция по эксплуатации и техническое описание.* Москва: Воениздат. pp. 140-141.

was a complete planetary gearbox containing four planetary gear sets. In total, these four planetary gears gave seven forward speeds, one reverse speed, one neutral setting, and one braking setting. These ten settings were obtained by selectively engaging or locking specific pairs of multi-disc clutches[1129] using a hydraulic control system.

Table 51: T-72 transmission data

T-72 Transmission	
Complete engine compartment volume (m³)	3.1
Engine compartment volume without fuel/oil (m³)	2.96[a]
Final drive volume (m³)	0.09[c]
Transmission volume (m³)	0.43
Total BKP volume (m³)	0.33[d]
Transmission mass (kg)	1,870*
Mass of two final drives (kg)	400[b]

The T-72 BKPs were not directly interchangeable with the original T-64A BKPs [1130] due to the new propeller shaft connection, and the gears in the T-72 BKPs had been strengthened for the increased torque of the V-46 engine[1131].

The propeller shaft joining the two BKPs in the T-72 served the same purpose as the crankshaft connection given by the 5TDF engine in the T-64A. The significance of a common propeller shaft was that, in addition to complete kinematic synchronicity between the two tracks, it was also a kinetic link

a Баранов, С. П., Иванов, В. А., Никитин, В. Т. (1979). 'Направления Развития Танковых Силовых Установок с Поршневыми Двигателями', *Вопросы оборонной техники*, 20(83). p. 3.
b *Руководство по войсковому ремонту объект 184 кн.1 ч.2* (1992). p. 363.
c Баранов, А. П., Панков, Ю. Е., Степанов, В. А., Ширшов, Ю. И. (1975). 'Конструкция И Характеристики Ходового Макета с ГМТ', *Вопросы оборонной техники*, 20(59-60). p. 12.
d Солянкин, А. Г., Желтов, И. Г., Кудряшов, К. Н. (2010). *Отечественные бронированные машины. XX век - Том 3: Отечественные бронированные машины. 1946-1965 гг.* Москва: Издательство «Цейхгауз». p. 148.
* Not inclusive of cooling system or final drives. See: Сафонова, Б. С., Мураховского, В. И. (eds.) (1993). *Основные Боевые Танки*. Москва: Арсенал-Пресс. p. 158

between the two tracks[1132], allowing the torque from the engine to flow through whichever track had enough grip[1133] to make use of it. When one track had less

---
[1129] With two BKPs, there were a total of 54 friction couplings in the transmission.
[1130] Павлов М. В., Павлов, И. В. p. 223.
[1131] Чернышев, В. Л., Акиншин, А. Г. (2013). 'Оценка Нагруженности Планетарных Рядов Бортовой Коробки Передач Танка Т-64А Методом Динамического Состояния в Режиме Разгона', *збірник тез Всеукраїнської науково-практичної конференції* НУЦЗУ, 2013,
[1132] A transmission with a kinematic synchronicity between the two tracks is said to have one degree of freedom.
[1133] "Grip" in this case refers to the coefficient of adhesion, the traction available through friction and

traction than the other, the track with more traction continued to propel the tank without additional power loss[1134]. This avoided the loss of traction suffered by transmissions with certain types of types of differential-based steering systems[1135] in offroad terrain.

The friction surfaces in the clutch packs consisted of steel drive discs with MK-5 metal-ceramic (cermet) friction pads[1136] and steel driven discs, for a steel-cermet friction interface. This choice of friction material yielded a considerable improvement in durability over the steel-on-steel dry multi-disc main clutch and steering clutch packs used in the T-62, T-54/55[1137]. Despite some measures to eliminate the influence of driver skill in the operation of the clutches, the discs routinely experienced high slip loads[1138] when setting off from a standstill, during gear changes, and when entering a turn, and thin steel discs wore down relatively quickly and were sensitive to warpage under intense heating.

The temperature insensitivity of cermet pads combined with forced lubrication allowed the BKP gearboxes to tolerate long continuous slipping of its clutches, but at the expense of the fact that a wet steel-cermet clutch had a quarter of the friction coefficient of a dry steel-on-steel clutch. However, the plate pressure limit was twenty times higher[1139]. It was, therefore, abundantly clear that the only way forward was to abandon manual mechanical clutches for a powered hydraulic system. The BKP gearboxes could develop the high plate pressures needed to accommodate higher torque from more powerful engines, and the closed-circuit hydraulic system simplified the workload on the driver. Instead of prying apart clutch discs against the springs of their pressure plates with the driver's own bodily strength, the clutches were entirely controlled by turning valves in the hydraulic control units, and virtually all of the feedback resistance was artificially added by springs.

This limited the modernization potential to two directions: improving the friction pad material, and improving heat removal. in the early 1980's, improved

---

the ability of the terrain to support grousers based on its shear strength.
[1134] Забавников, Н. А. (1975). *Основы теории транспортных гусеничных машин*. Москва: «Машиностроение». pp. 231-232.
[1135] The Merritt-Brown triple differential (Centurion, Chieftain), 5SD200D (AMX-30), and Allison Cross Drive (M48, M60) notably lacked a differential lock. Several double differential steering systems (Tiger 2, Leopard) kinematically coupled the drive shaft in forward motion.
[1136] Жучков, М. Г., Фанталов, В. С. (1978). 'Повышение Долговечности Дисков Фрикционных Узлов Трансмиссии ВГМ', *Вестник Бронетанковой Техники* 1978, сборник 4. pp. 38-41.
[1137] The choice of a steel-on-steel clutch was inherited from the T-34.
[1138] Чобиток, В. А. (1984). *Конструкция и расчет*. Москва: Воениздат. p. 124.
[1139] Носов, Н. А., Голышев, В. Д., Волков, Ю. П., Харченко, А. П. (1972). *Расчет и Конструирование Гусеничных Машин*. Ленинград: «Машиностроение». p. 72.

lubrication grooves in the cermet disc pads[1140] alleviated heating in the clutch packs and compensated for the increased torque of the V-84 engine[1141].

The mechanical efficiency of the BKPs was fairly high, but being a planetary transmission, the number of meshing gears in the kinematic link between the engine and the final drives was high for a mechanical transmission, which spoiled its efficiency somewhat compared to a layshaft or parallel-shaft gearbox with simple gear pairs. In low gears (1st to 5th), the efficiency was 0.952-0.956, which translated to a power loss of 25-30 hp. In high gears (6th and 7th), the efficiency was 0.972 and 0.98 respectively[1142], which was a power loss of 50-55 hp[1143]. Reliability and durability were also high; transmission breakdowns were extremely rare, even after the exhaustion of the 6,000 km warranty period[1144].

BKP Design

Four planetary gears linked into a single compound planetary set with three degrees of freedom[1145] provided all of the functions demanded of a complete transmission. Three degrees of freedom refer to the fact that for one output to the final drive, the sun gears of the 1st, 2nd and 3rd planetary sets were three possible inputs. From these three inputs, there were a larger number of possible power flow paths to the output shaft. Three degrees of freedom provided the greatest possible economy of volume for a planetary gearbox with four to ten settings[1146]. Having a total of ten possible settings – seven forward speeds, one reverse, neutral, and brake, the T-72 BKP had no room for additional speeds.

These ten settings were obtained by selectively applying clutches and brakes to alter the gearing ratio. There were three clutch packs called clutches, F$_1$, F$_2$, and F$_3$, and three called brakes, T$_1$, T$_2$, and T$_3$. Both types used the same type of clutch pack, but clutches joined rotating elements of two adjoining planetary gears together, so the cermet discs and steel discs were both rotors. Brakes stopped an element of a planetary gear against the BKP body. The steel discs of the friction pair in this case were stators affixed to the BKP body.

[1140] Антонов, Ю. В., Жучков, М. Г., Фанталов, В. С. (1977). 'Повышение Износостойкости Фрикционных Узлов Трансмиссий ВГМ', *Вестник бронетанковой техники* 1977, сборник 4. pp. 35-38.
[1141] VNII *Transmash* p. 140.
[1142] Калинина-Иванова, Е. В. (1975). 'Анализ Топливной Экономичности Ходового Макета с ГМТ', *Вопросы оборонной техники*, 20(59-60). p. 75.
[1143] Жучков, М. Г. (1975). 'Стендовые Испытания ГМТ', *Вопросы оборонной техники*, 20(59-60). p. 39.
[1144] Георгиевский, О. Н., Дужак, Г. А., Савина, В. Ф., Ядришников, Н. И. (1987). 'Совершенствование оценки надежности при ускоренной войсковой эксплуатации танков', *Вестник Бронетанковой Техники* 1986, сборник 10. p. 20.
[1145] The simplest planetary gear has two degrees of freedom: two inputs, and one output.
[1146] Стрелков, А. Г. (2005). *Конструкция быстроходных гусеничных машин*. Москва: МГТУ «МАМИ». p. 117.

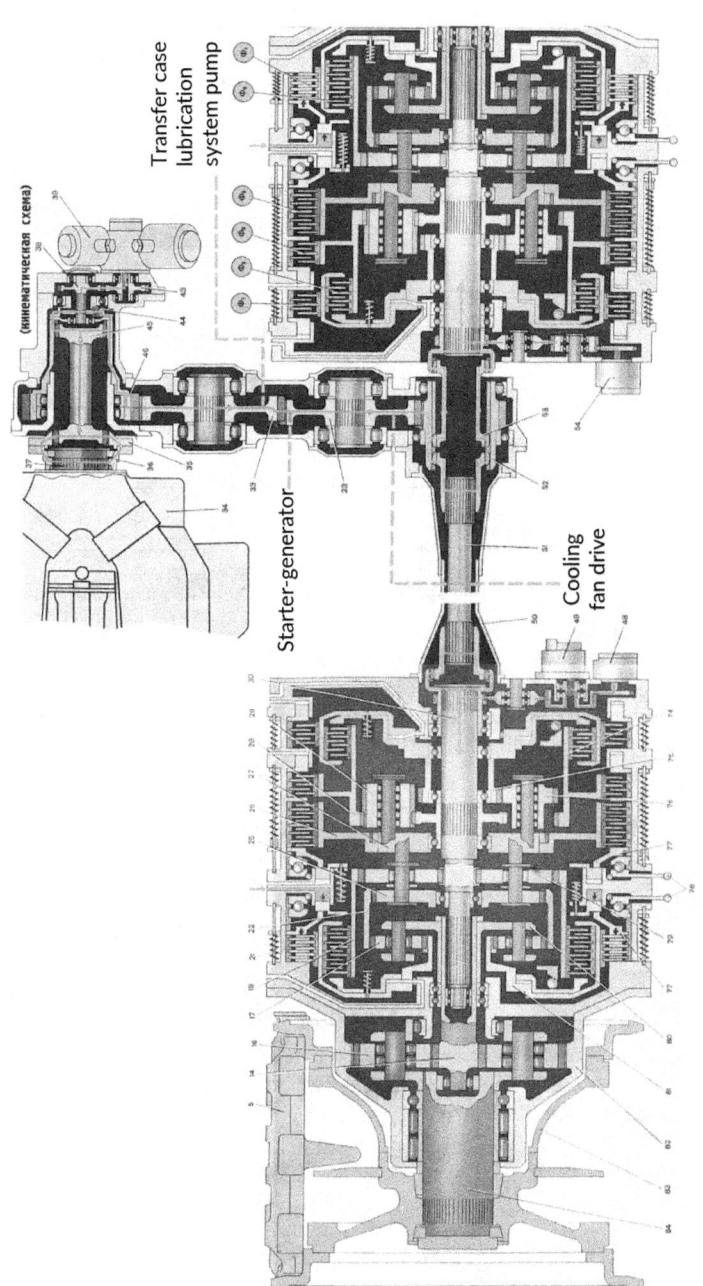

Figure 180: Cutaway of kinematic connections between the transfer case and BKP units. (*Soviet Ministry of Defense*)

405

Table 52: T-72 transmission parameters[a]

Gear	1	2	3	4	5	6	7	R
Clutches Engaged	T_4F_3	T_6T_4	T_6F_3	T_1T_4	T_1F_3	F_2T_4	F_2F_3	T_5F_3
Gear Ratio	8.173	4.400	3.485	2.787	2.027	1.467	1.000	14.353
Gear Step	1-2	2-3	3-4	4-5	5-6	6-7	-	
Step Size	1.858	1.26	1.25	1.37	1.38	1.467	-	
Transfer Case Gear Ratio	0.706							
Final Drive Gear Ratio	5.454							
Drive Sprocket Radius (m)	0.313							
Total Gear Ratio	31.47	16.94	13.42	10.73	7.80	5.64	3.85	55.06
Tank Speed (km/h)	7.3	13.8	17.4	21.8	29.8	41.2	60.5	4.2
Suspension Base Width (m)	2.79							
Min. Turn Radius* (m)	2.79	6.04	13.42	13.93	10.23	10.1	8.76	2.79
Max. Turn Rate (°/s)**	41.75							

a   Старовойтов, В. С. (ed.) (1990). *Военные гусеничные машины*. Том 1: Устройство, Книга 1. Москва: Издательство МГТУ им. Н. Э. Баумана. p. 162.
*   On dry soils, the turn radii match closely with the calculated values. On sand, mud, and other low-adhesion surfaces, the actual turn radii were 1.5-2.0 times larger due to track slip.
**   This is the maximum angular speed during a pivot turn. Due to a certain amount of time spent accelerating from a stop, the actual time to complete a full pivot turn was around 10 seconds rather than the theoretical 8-9 seconds.

All of the clutch disc packs were engaged by hydraulic pressure. Only the T_4 and T_5 brakes had an additional mechanical link (to the driver's brake pedal). Moving the gear selector lever up and down rack twisted a rotary spool valve in the hydraulic control unit mounted on top of each BKP. The position of the spool valve determined which clutch packs were pressurized. The clutch pressures were 10-11.5 kg/cm² in all gears except 1st and reverse, which were engaged at an increased pressure of 16.5-18.0 kg/cm² to ensure that the discs did not slip, since 1st gear and reverse delivered very high tractive forces[1147].

Pulling the steering lever turned the same rotary spool valve on the corresponding BKP, but only to downshift from its current gear. At the same

---

[1147] Министерство Обороны СССР (1988). *Танк Т-72А - Техническое описание и инструкция по эксплуатации. Книга Вторая, Часть Первая*. Москва: Воениздат. p. 450

time, the pressure in the BKP of the overtaking track would be raised to 16.5-18.0 kg/cm² to prevent its clutch discs from slipping due to torque overload.

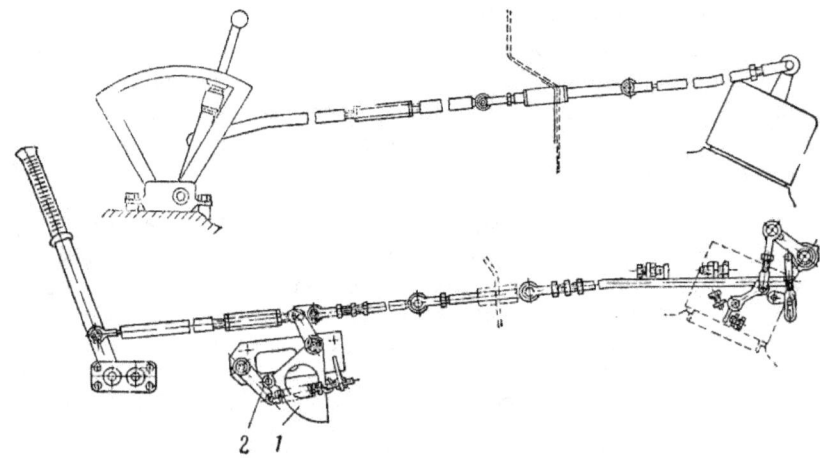

Figure 181: Gear selector (top) and steering lever (bottom). The hydraulic spool valve took very little effort to turn, so the steering linkage artificially created progressive resistance by cranking a cam (1) pushing on the roller (2) of a spring-loaded arm. The further the steering lever was pulled, the higher the resistance. (*Adapted from Soviet Ministry of Defense*)

When driving the T-72, the clutch pedal had to be pressed before shifting gears, and the shifting order was strictly sequential. The clutch pedal activated a relief valve in the hydraulic control unit to depressurize the BKP regardless of its current setting. Upon releasing the clutch pedal, smooth re-pressurization was ensured by a spring-loaded regulator valve. The driver's finesse with the clutch pedal had no significant influence, so it was imperative that he pressed and released the pedal as quickly as possible. It was possible to shift without the clutch pedal altogether, but this led to accelerated wear on the clutch discs. On the other hand, powershifting, or upshifting without releasing the accelerator pedal, was possible and even unofficially a recommended practice to reduce the speed drop while accelerating hard.

The simplest approach to calculating gearbox parameters was to carry over existing automobile gearbox design principles. Once the tank weight and suspension characteristics (traction, rolling resistance) were established, the requirements for maximum climbing slope and top speed could be used to calculate the gear ratios needed for the 1st gear and the top gear, and then a mathematical progression between them could be chosen. It could be optimized for any number of outcomes, from maximizing fuel economy to maximizing acceleration performance in a range of speeds.

However, unlike civilian automobiles, tanks operated on vastly smaller power-to-weight ratios, needed both good fuel economy and quick acceleration, and were obligated to perform in far more challenging terrain and environmental conditions. Meeting this exhaustive list of requirements demanded a number of design compromises, and various factors had to be careful weighed to choose the least disagreeable compromise.

The BKP had a geometric progression[1148] from the 2nd to 4th gears, which were on average the most commonly used gears[1149]. The 5th to 7th gears had no mathematical progression serving only to extend the tank's top speed and to enable fuel-efficient cruising at high speed on roads. When driving on dirt surfaces, the 6th gear was almost always the highest usable gear since the rolling resistances were usually too great for the 7th gear. On a hard surface, the 7th gear gave the T-72 a (recorded) true top speed of 65 km/h and a technical

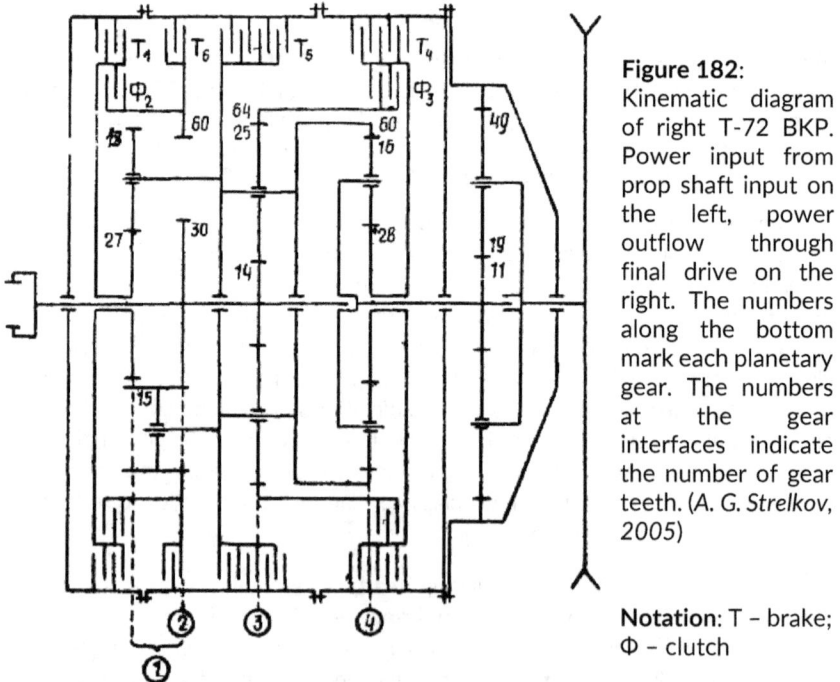

Figure 182: Kinematic diagram of right T-72 BKP. Power input from prop shaft input on the left, power outflow through final drive on the right. The numbers along the bottom mark each planetary gear. The numbers at the gear interfaces indicate the number of gear teeth. (*A. G. Strelkov, 2005*)

Notation: T – brake; Φ – clutch

[1148] Progression where the step (change in gear ratio) of two sequential gears is the same.
[1149] Behrendt, J. (2002). 'Współczynniki Wykorzystania Przełożeń Skrzyń Przekładniowych w Czołgach Rodziny T-72', *Szybkobieżne Pojazdy Gąsienicowe* 2(16). p. 6.

maximum speed of 68.8 km/h at a crankshaft speed of 2,300 RPM, likely achievable only when going downhill.

The 1st gear was a high reduction gear for circumstances demanding high tractive force. It could be used for low-speed maneuvering in restricted terrain, but it was primarily a utility gear for circumstances that demanded high tractive effort, which could be anything from towing heavy loads to overcoming obstacles, crossing trenches, and climbing hills. It gave the tank the tractive effort needed to climb a 35° slope, which it could do at 4 km/h. It was also the only gear to be used when climbing any slope lower than 35° but greater than 19°, which insured against engine stalling if the tank drove over a rut or a jutting rock while ascending a slope. The ability to creep forwards at 3-5 km/h to escort infantry at a walking pace has also been cited as a requirement[1150].

The narrow gear spacing from the 2nd to the 6th gears improved acceleration performance by ensuring that the engine was always deep in its powerband at 1,600-2,000 RPM when upshifting, so the engine was always delivering high power during hard acceleration. The gear spacing expanded after the 4th gear, but remained narrow enough that when upshifting at 1,900-2,000 RPM, the engine speed would fall to 1,300-1,400 RPM, which was still within the power band. The same could not be said for contemporary tank gearboxes with a smaller number of gears, even with a torque converter[1151].

The tank reached 30 km/h at the very end of 5th gear, beyond which the acceleration performance fell off drastically. Coincidentally, the need to spend time to upshift to the 6th gear after reaching 30 km/h meant that the T-72's acceleration time was ruined if it was rated according to the American 20 mph (32 km/h) acceleration bench mark, which became an international tank building standard in the later part of the Cold War[1152]. This was not applicable in reverse, so rating tanks at 30 km/h [**Table 45**] is fairer and does not change the significance of the acceleration bench mark.

In Neutral, the hydraulic steering control was disconnected and the engine hour clock (on the driver's instrument panel) was disconnected so as to accurately reflect the engine hours under load.

[1150] Старовойтов, В. С. (ed.) (1990). *Военные гусеничные машины*. Том 1, Книга 2. Москва: Издательство МГТУ им. Н. Э. Баумана. p. 104.

[1151] 4HP-250 gearbox in the Leopard with four speeds to the same 60 km/h top speed was a notable example. See: 'Schematisches Fahrdiagramm 4 HP-250-Leopard' in Zahnradfabrik Friedrichshafen AG. *ZF-Gangwahlschalter für das Getriebe 4 HP-250*. p. 14.

[1152] One of the major factors behind the popularity of the 32 km/h standard was that the West Germans had to abide by American testing standards while participating in comparative trials in the U.S. See: Krapke, P. -W. (2004). *Leopard 2: Sein Werden und seine Leistung*. Norderstedt: Books on Demand GmbH. pp. 28-29.

The reverse gear was required to provide the same traction as the 1st gear, most crucially enough traction to reverse up a 35° slope[1153], primarily to make sure that the tank could extricate itself from a ditch or any other obstacle it found itself stuck on.

The gear selector had an electronically controlled mechanical interlock to prevent the driver from downshifting in the 7th to 5th gears (including downshifting from 5th to 4th) unless the engine speed was below 1,500 RPM. When the interlock was active, a red warning light to the left of the driver's TNPO-168V periscope lit up.

This interlocker prevented the driver from downshifting at an engine speed higher than 1,500 RPM in the 7th to 5th gears, as that could cause the engine to overspeed – that is, exceed its governed speed of 2,300 RPM – once the lower gear engaged. Ideally, the driver should downshift at 1,300 RPM or lower. Switching off the interlocker was prohibited except in case of emergencies. There were no restrictions for the 4th to 1st gears, as the close spacing in these gears made it very hard to overspeed the engine.

Having more gears was also beneficial to the overall acceleration performance of the T-72, considering that gear changes were much quicker than in the T-54/55 and T-62, quick enough not to significantly alter the acceleration time. In principle, the greatest possible acceleration is obtained if the traction curve of the drivetrain matches the ideal traction hyperbola (ITH), a curve that represents the maximum theoretical acceleration if the engine maintained its peak power and the change in tank speed was obtained solely through gearing. The closer the traction curves of the drivetrain approximate the ITH, the higher the acceleration and the higher the average speed of the tank[1154]. For a transmission with discrete gears[1155], this essentially meant that the larger the number of gears available, the better the acceleration potential to a given speed.

As the ITH for the T-72 [**Fig. 183**] shows[1156], the T-72 transmission was inherently capable of good acceleration performance[1157], and the wide choice

---

[1153] Старовойтов, В. С. (ed.). p. 104
[1154] Сергеев, Л. В. (1973). *Теория танка*. Москва: Военная академия бронетанковых войск. pp. 147-148, 152-153.
[1155] As opposed to a continuously-variable transmission (CVT) or a transmission that uses a torque converter to cover a wide gearing range with few mechanical gear speeds (Allison Cross Drive).
[1156] Rybansky, M. et al. (2015). 'Modelling of cross-country transport in raster format', *Environmental Earth Sciences*, 74(10). p. 7054.
[1157] A torque converter may improve acceleration by improving the traction curve in the low gears (1st, 2nd). As an example, see the idealized Leopard 1 traction curve: Merhof, W., Hackbarth, E.-M. (2016). *Fahrmechanik der Kettenfahrzeuge. Überarbeitete Neuausgabe*. Neubiberg: Universität der Bundeswehr München. p. 110.

of forward gears increased its speed when climbing gentle to moderate slopes. On very steep slopes (>15°), the fully mechanical nature of the BKP transmission was a detriment, as it gave no advantages in slope climbing speed and it lacked the stall resistance and ease of stopping and starting offered by a torque converter.

Keeping in mind that the tank was steered by the difference in speed between the two tracks when the lagging side was downshifted by one gear, this acceleration-minded gear progression was somewhat less auspicious when it came to steering.

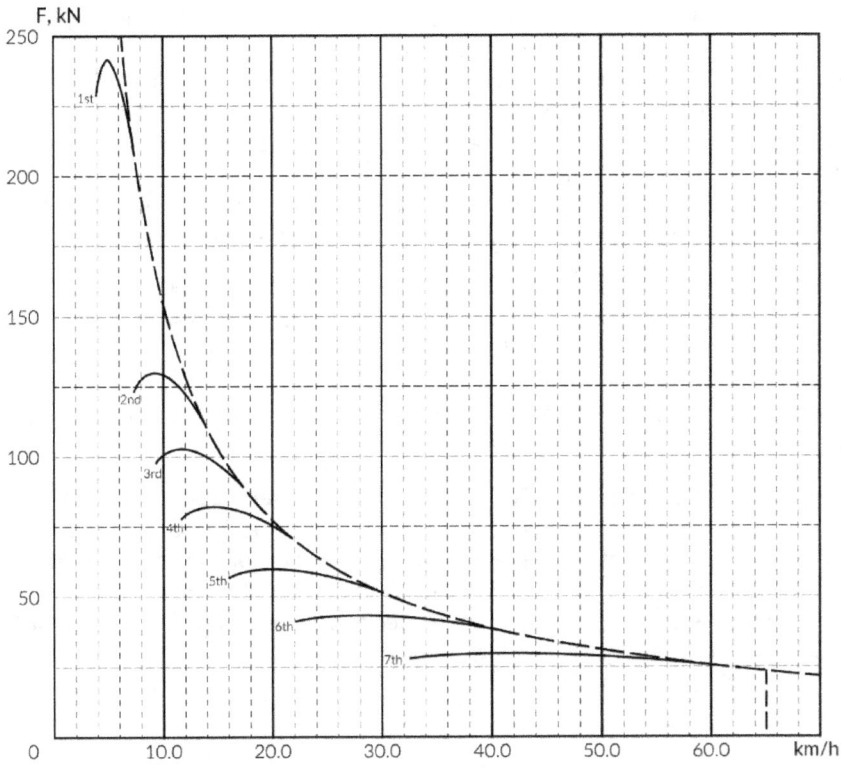

Figure 183: Tractive effort curve for T-72 (41 t). Top speed on paved road marked at 65 km/h, corresponding to 24 kN of traction. (*Adapted from M. Rybansky et al., 2015*)

Steering

The BKP transmission gave seven discrete turn radii, one in each forward gear. Steering in 1st gear was done through clutch-brake turns, and steering in all higher gears was by making geared turns. A geared turn was the primary mode of steering. It could be sustained indefinitely, since all clutches were engaged and there was no additional wear compared to straight-line driving. The secondary methods were de-clutched steering (free radius) and steering by controlled clutch slippage (variable radius). These steering modes were engaged by pulling the steering lever to different points.

For example, pulling the left steering lever while traveling in 2nd gear downshifted the left BKP to the 1st gear. The ratio of the 2nd gear to 1st gear was 0.538, and so the left track ran at nearly half the speed of the right track. This produced a tight turn radius of 6.04 meters. In the 3rd gear, the left BKP was downshifted to 2nd gear for a speed ratio of 0.792, resulting in a much wider turn of 13.42 meters. Up to the 4th gear (13.93 meters), the progression in turn radii was quite logical – the faster the tank, the wider the turn, so the risk of skidding out was low and the speed of the tank was minimally affected.

However, from the 5th to 7th gears, the progression in turn radii was reversed, and the turn radius in the 7th gear shrunk down to just 8.76 meters, which was totally unusable. The T-80 series suffered from the same issue. Despite its 4-speed BKPs[1158] giving a progressively increasing turn radius, the largest turn radius was still only 8.86 m[1159], which was grossly inadequate for the 48-70 km/h speed range provided by the 4th gear.

Because the steering system achieved a speed difference by slowing down one of the tracks, the tank always slowed down in a turn. To maintain the same speed in a turn as in straight-line motion, the average speed of the two tracks must not change. This was possible with a differential steering system, known as a Type I steering system in Soviet technical literature. The T-72's BKP transmission was a Type II steering system.

[1158] The BKPs in all T-80 tanks had two degrees of freedom with three compounded planetary gears. The BKP units were shorter and therefore more compact, but unsuitable for piston engines.
[1159] Министерство Обороны СССР (1979). *Объект 219 Техническое описание и инструкция по эксплуатации. Книга Вторая.* Москва: Воениздат. p. 203.

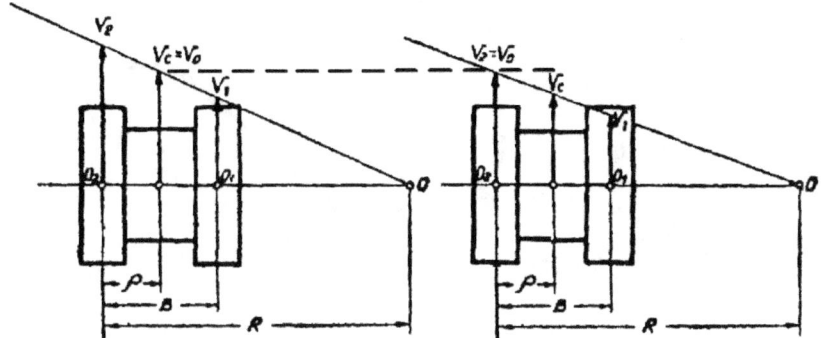

Figure 184: Type I (left) and Type II (right) steering systems, where V_0 is the speed of straight-line motion. Note the speed advantage of a Type I steering system. (*L. V. Sergeev, 1973*)

A tight turn produced the most intense slow-down, since the speed difference between the lagging and overtaking tracks was high. On low-resistance surfaces, i.e. paved roads and dry dirt roads, the T-72 was significantly slower than a tank with a Type I steering system, even if it had the upper hand in straight-line acceleration and top speed. For instance, Soviet testing found that on a track where the M60A1 saw a 15% speed loss (17.3 km/h to 14.75 km/h) from making a turn with a radius of 20 meters, a T-72 making the same turn experienced a 37.6% speed loss (17.0 km/h to 10.6 km/h)[1160].

The advantage of a Type I steering system was such that on the concrete tank course at the Kubinka proving ground, the M48A5 turned out to be faster than a T-80[1161]. The advantage eroded away in off-road driving, but for more modern tanks capable of higher off-road speeds like the M1 Abrams and Leopard 2, a Type II steering system was once again at a disadvantage[1162]. It was possible to narrow the speed gap to an extent by steering by controlled clutch slippage.

In low gears, pulling the steering lever back from its initial position by 10° dropped the hydraulic pressure in the BKP, and pulling the lever further would smoothly raise the pressure [**Fig. 185**, Zone I]. Without reaching a certain minimum pressure, the BKP never fully downshifts. Steering this way was possible only on high-resistance terrain, where the minimum gear engagement

---

[1160] Исаков, П. П. (ed.) (1985). *Теория и Конструкция Танка - Том 5: Трансмиссии Военных Гусеничных Машин*. Москва: Машиностроение. pp. 14-15.
[1161] Account given by Смолин, В. В. in Баранов, И. Н. (2010). *Главный конструктор В.Н. Венедикто: Жизнь, отданная танкам*. Нижний Тагил: ООО «Рекламно-издательская группа «ДиАл». p. 201.
[1162] Зайчиков, Ю. Н. (2011). *Трансмиссия И Ходовая Часть Танка Т-72*. Челябинск: Издательский центр ЮурГУ. pp. 14-15.

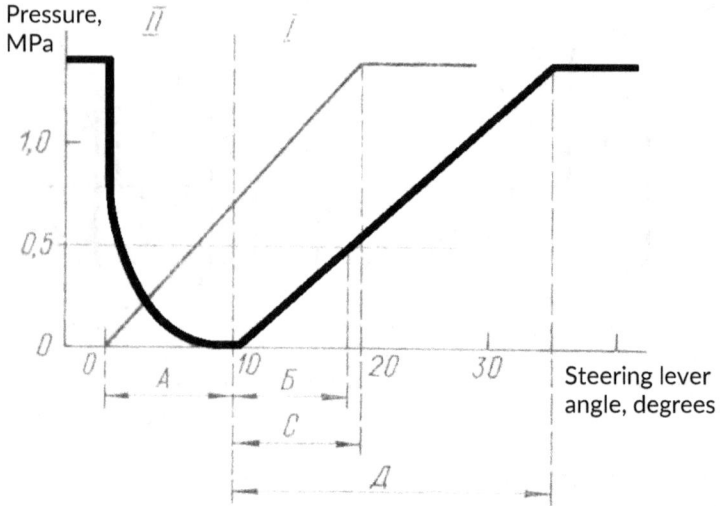

Figure 185: Change in BKP pressure with steering lever angle in low gears (Zone I, bold line) and high gears (Zone II, thin line). As a reference, 0.5 MPa (5 kg/cm²) is enough to engage 2nd gear in high-resistance terrain.
Notation: А – Idling range; Б – pressure control range; С, Д – range of full pressurization at reduced rate. (*Adapted from G. N. Gerasimov et al., 1981*)

pressure was higher due to the high traction needed to propel the tank. If the required pressure reached 0.5 MPa (5 kg/cm²), the steering lever could be positioned within the 10-19° range to enter a wide turn. At different engine speeds or loads and at different gears, the BKP could downshift at a lower pressure, narrowing the range of steering lever angles where steering by controlled clutch slippage was possible[1163]. In effect, drivers had to learn to judge the tank's reaction on different surfaces by feel.

Turning by slipping the clutches could be sustained indefinitely thanks to the use of wet clutches in the BKPs[1164], but special care had to be taken on avoiding clutch slippage when high traction was needed, like towing a heavy load, or when attempting self-extraction on a stuck tank[1165].

[1163] Герасимов, В. Н., Иванов, А. И., Седов, Б. С., Травкин, А. Д. (1981). 'Совершенствование привода управления танком с БКП', *Вестник бронетанковой техники* 1981, сборник 6. p. 28.
[1164] Obtaining a variable turn radius by slipping wet clutches with variable hydraulic pressure was a technical solution that BKPs shared with Allison Cross Drive (CD) series transmissions.
[1165] Кустылкин, В. С., Развалов, А. С. (1986). 'Анализ причин отказов при подконтрольной эксплуатации танков в 1975-1984 гг.' *Вопросы оборонной техники*, 5(126).

At higher gears, controlled clutch slippage was not possible. Positioning the steering lever at an intermediate position only delayed full pressurization of the lower gear; the further back it was pulled, the quicker the BKP would downshift. It was, however, possible to steer with a free radius and steer by pumping the steering lever intermittently, taking advantage of the pressurization delay created by the limited capacity of the pump for the hydraulic control units.

Even if the driver were to jerk the steering lever to the full-back position, there was a delay of up to 0.35 seconds until full gear engagement in the 6th and 7th gears. This was long enough to be potentially dangerous in a situation demanding an instant steering reaction, e.g. dodging a car on the road or a sudden obstacle in unfamiliar terrain. On the other hand, this intrinsic delay smoothed out the inherent jerkiness of geared steering.

The tank turned with a de-clutched lagging track at a rate proportional to terrain resistance, and as such, is known as a free radius turn. A free radius turn could be used to follow a gentle curve on a road, and a variable radius turn was applicable for small steering adjustments on all types of terrain. Making a free radius turn was most effective at high speeds, and in any case, it was less tiring. Because the steering lever was only moved a short distance, turning with a free radius took the least effort – only 5 kg[1166].

On level ground, the braking system permitted a stopping rate of 4.5 m/s^2, rated from a standard initial speed of 40 km/h[1167]. This translates to a 40-0 stopping time of 2.4-2.5 seconds, on par with contemporary tanks and tracked vehicles of the same class [1168] and approximately twice quicker than the stopping rates achieved by WWII era tanks[1169]. Engine braking[1170] was the main method of slowing down the tank when traveling downhill.

[1166] Софьин, А. П., Смолин, В. В. (1979). 'О Снижении Усилий на Органах Управления ВГМ', *Вопросы оборонной техники*, 20(84). p. 47.

[1167] Потемкин, Э. К. (ed.) (1992). *Основы научной организации разработки*. Том 2. Москва: Издательство МГТУ им. Н. Э. Баумана. p. 197.

[1168] Hilmes R. (2007). *Kampfpanzer heute und morgen: Konzepte – System – Technologien*. Auflage: Motorbuchverlag. p. 289. It is also stated by Hilmes that a stopping rate of 5 m/s^2 was targeted as a desirable figure. Heavier tanks typically fared worse, and some were unusually below par. The M60A1 had a stopping time of 30-0 of 3 seconds. See: Дорогин, С. В. et al. (1976). 'Ходовые Испытания Танка', *Вопросы Оборонной Техники*, 20(68). p. 9.

[1169] Hamparian, E. (1967). *The Evolution of Power Trains and Steering in High Speed Military Track Laying Vehicles*. SAE Technical Paper 670726. New York: Society of Automotive Engineers. p. 6.

[1170] Engine braking is usually associated with gasoline engines due to the vacuum effect with a closed throttle. In large diesel engines, high internal resistances provided engine braking. V-2 family engines had intense engine braking – for the V-45K/V-46, the maximum braking power reached 260 hp. See: Вереха, Ю. Н., Калинина-Иванова, Е. В., Корнилов, А. Н. (1976). 'О тормозных характеристиках танковых Двигателей', *Вестник бронетанковой техники* 1976, сборник 1. pp. 8-11.

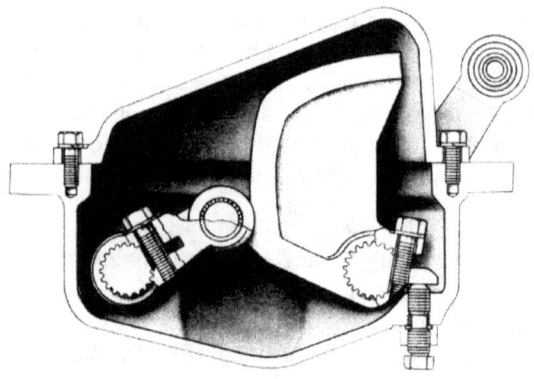

Figure 186: Brake cam and roller arm connecting to the BKPs. (*Soviet Ministry of* Defense)

The brake was entirely mechanical, with no power assistance[1171]. The driver was assisted only by a certain amount of mechanical advantage between the brake pedal to the pressure plates provided by a two-stage cam. The first stage covered the pedal play and the second stage applied pressure to the brake discs.

Another way to activate the brakes is to shift into 1st gear, and then pull both steering levers back simultaneously. This could be done to quickly control the brakes without taking a foot off the accelerator pedal while traversing obstacles at low speed, but it was also the most convenient way of stopping and starting the tank on a steep slope.

Considering the weight of the tank, the performance of the braking system was rather inexplicable, especially given that the BKP had wet clutches rather than dry clutches. Wet clutches are characterized by a much higher actuation force limit – in the case of a steel-cermet interface, by a factor of ten[1172] – but a smaller coefficient of friction, which is a combination that pairs well with an engine-powered hydraulic system, not a manual mechanical system. Based on figures for the T-64A, the brake pedal force could reach 50-100 kg[1173]. The large knee angle of the driver's seating posture at least allowed him to exert a greater pedal force[1174] to help when braking hard.

A pneumatic braking system was added beginning on January 1, 1978. It was activated by pressing and holding the brake button on the end of the left steering lever, triggering a pneumatic piston to take over for the driver's leg to push on the brake mechanism. Because of the continuous link to the brake

---

[1171] Like a go-kart.
[1172] Чобиток, В. А. (1984). *Конструкция и расчет*. Москва: Воениздат. p. 111.
[1173] Домбровский, Ю. К., Семенов, А. С. (1975). 'Исследование общемашинных показателей ходового макета, *Вопросы оборонной техники*, 20(59-60). p. 45.
[1174] *Military Standard: Human Engineering, Design Criteria for Military Systems, Equipment, And Facilities* (1989). MIL-STD-1472D. Washington, D.C.: Department of the Army. p. 119.

pedal, activating the pneumatic brake also depressed the brake pedal, which gave some physical feedback on the magnitude of the braking force.

The pneumatic braking system needed the electrical mains to be turned on for the electric control circuit, but it was otherwise operable with the engine turned off. The only limitation was the capacity of the air bottles. The T-80U had a hydraulic equivalent to this pneumatic booster for its brake, and because its hydraulic booster was powered through the same hydraulic circuit as the rest of the transmission, the engine had to be running for its system to work.

Pneumatics

The T-72 had a pair of five-liter high-pressure air bottles[1175]. The AK-150SV was an air-cooled, three-stage, reciprocating piston high-pressure air compressor, rated for a pressure of 150 kg/cm^2 (2,134 psi, or 14.71 MPa) and an output of 2.4 m^3/h (1 atm). Originally made to sustain pneumatic controls in heavy aircraft like the Tu-104, Tu-114 and Tu-95[1176], the AK-150 was adapted into the AK-150SV to support tank pneumatics. It was first used in the T-55, and from there, proliferated to nearly all Soviet armored fighting vehicles.

The compressor would charge the air bottles to a pressure of 150 kg/cm^2, and then, the outflow valve opened so that no compression occurred, acting as a power-saving idle mode. Once the pressure in the pneumatic system fell to 133-135 kg/cm^2, the valve closed and compressor began charging again[1177]. Depleting the air bottles was effectively impossible in normal operation, but if the tank was loaded with empty bottles, the full ten-liter capacity could be filled in 35-40 minutes at an engine speed of 2,000 RPM[1178]. In the previous generation of tanks (T-55, T-62), the compressor was idle most of the time (69-78%)[1179] despite a relatively heavy air usage rate.

Apart from the tank's built-in pneumatic devices, the plow lifting pistons for the KMT-6 and KMT-8 (1983) mine plows were designed to be powered by the pneumatic system. An access port next to the driver's hatch connected the air hoses for the plow lifting pistons to a distributor valve at the driver's station[1180].

[1175] *Танк Т-72М1: Техническое описание и инструкция по эксплуатации – Часть I*. p. 15.
[1176] Кожухов, Ю. В. et al. (2020). *Конструкции компрессоров объемного типа*. Санкт-Петербург: Политех-Пресс. p. 51.
[1177] Лукьянов, А. И., Бабин, Ю. С. (1966). 'Эксплуатационные характеристики компрессора АК-150 при работе его на танках', *Вестник бронетанковой техники* 1966, сборник 1. p. 50.
[1178] Ibid.
[1179] Ibid. p. 51.
[1180] Министерства Обороны СССР (1988). *Средства преодоления минно-взрывных заграждений – Минные тралы. Техническое описание и инструкция по эксплуатации*. Москва: Воениздат. pp. 14, 35.

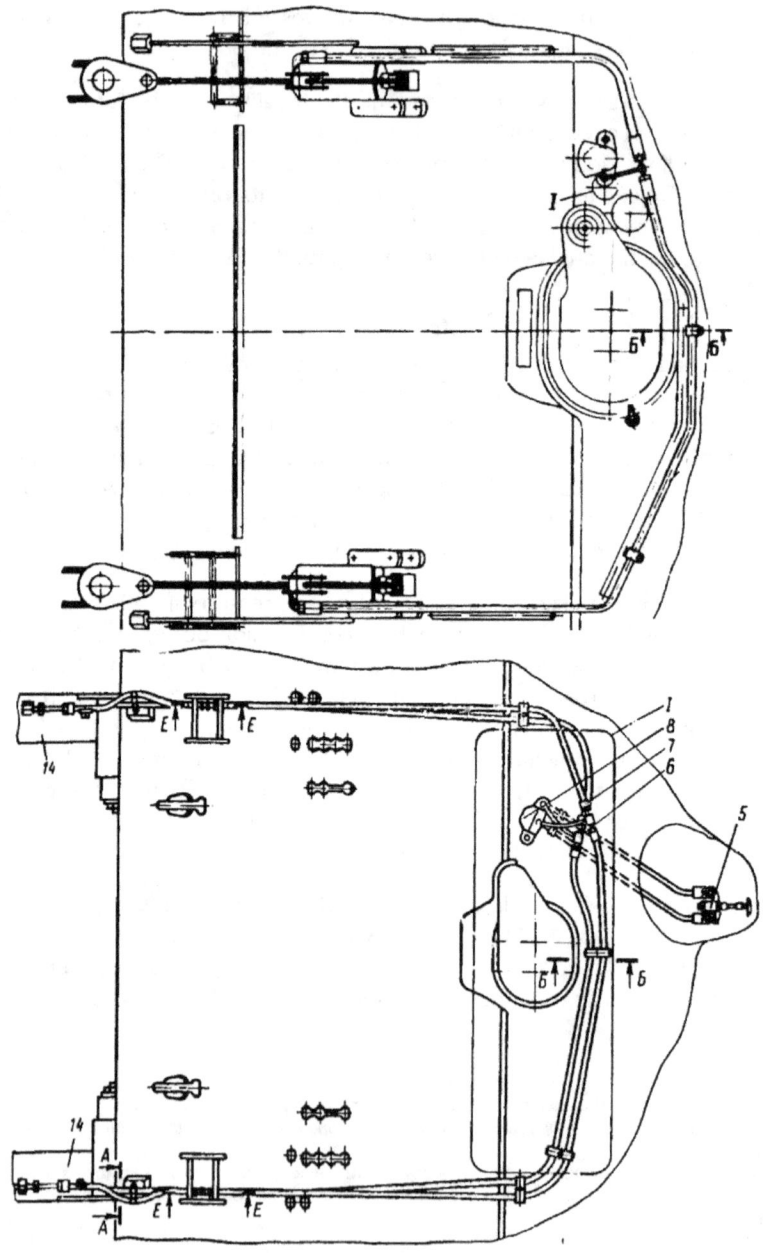

Figure 187: KMT-8 (left) and KMT-6 (right) air hose connections through access port. (*Soviet Ministry of Defense*)

Suspension

The T-72 had a torsion bar suspension with six road wheels on each side, running on metal tracks. Each suspension unit consisted of a road wheel, a swing arm, a full-width torsion bar, and two mounting brackets for each end of the torsion bar, one to anchor the bar and one for the swing arm. Hydraulic shock absorbers were installed on the 1st, 2nd and 6th suspensions, and solid bump stops were installed on the 1st, 2nd, 5th and 6th suspension units. Unlike the kilogram-counting design approach inherited from the T-64 throughout the rest of the tank, the weight of the suspension was entirely typical for medium tanks. It accounted for 19-21% of the T-72's combat weight, equal to other tanks like the M60A1 (19.76%) and T-55 (20%).

Figure 188: General view of T-72 suspension. 1 – idler wheel; 2 – road wheel; 3 – hydraulic shock absorber. (*Soviet Ministry of Defense*)

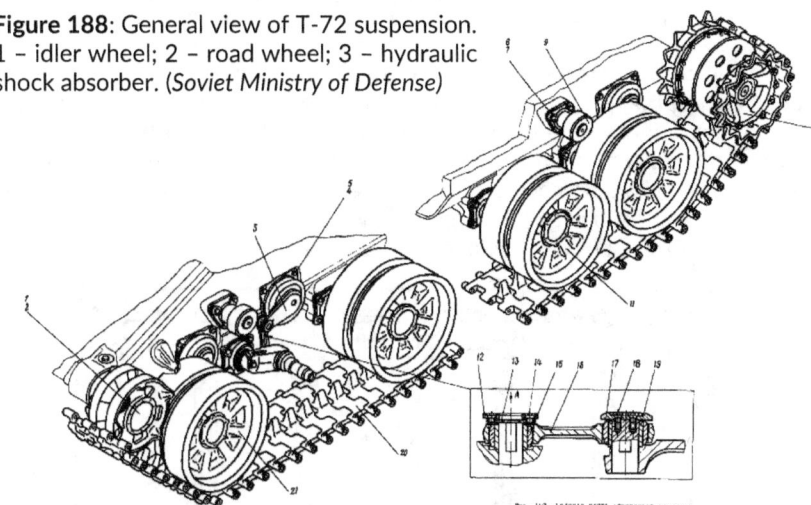

The ground contact length of 4,270 mm and suspension base width of 2,790 mm gave a steering ratio of 1.53, very close to the mathematical ideal of 1.5[1181]. Most tanks exceeded this ideal ratio, but the degree of excess varied wildly. The T-72 was similar to other tanks with a visually "squarer" appearance like the Leopard 1 (1.57) in this regard, and this steering ratio could be considered very favorable relative to long tanks like the Leopard 2 (1.775)[1182] and Chieftain (1.76)[1183]. A steering ratio close to the mathematical idea reduces steering resistance when entering and sustaining a turn, and thereby reduces engine

---

[1181] Merhof, W. (2016). *Fahrmechanik der Kettenfahrzeuge. Überarbeitete Neuausgabe*. Neubiberg: Universität der Bundeswehr München. pp. 77-78.
[1182] Spielberger, W. J. (1995). *Waffensysteme Leopard 1 und Leopard 2*. Auflage: Motorbuch Verlag. p. 373.
[1183] Hilmes R. (1988). *Kampfpanzer. Die Entwicklungen der Nachkriegszeit*. Bonn: Mittler Report Verlag GmbH. pp. 52, 53, 56.

load, with the associated benefits to fuel consumption and the load on the cooling system. However, the speed of the tank in a turn was another matter entirely, as we have seen in the nuances of the BKPs as a steering mechanism.

The suspension was, on the whole, conventional, but at the same time distinct from typical modern tank suspensions in subtle but crucial details. The use of single-pin tracks instead of parallel-pin (dual-pin) tracks was unusual but viable for a tank of the T-72's weight, and the large diameter of its wheels was a relic from the legacy of the Christie suspension.

The idler arm had a worm gear tensioning mechanism. A long steel prybar was used to wrench a nut to turn the worm gear, tipping the idler wheel forwards or backwards. The drive sprocket had a centering ring to help retain the track by blocking the center guides from moving laterally outward.

It was, in essence, the suspension of the T-54 adapted for higher off-road speeds. Conceptually, the basic difference was the change from an unsupported track configuration, where the return run of the track was left hanging or was partially on the top of the road wheels, to a supported track configuration with three support rollers on each side for the return run. The design was directly inherited from the Object 167 suspension, which in turn was a modification of the Object 140 suspension developed in the early 1950's.

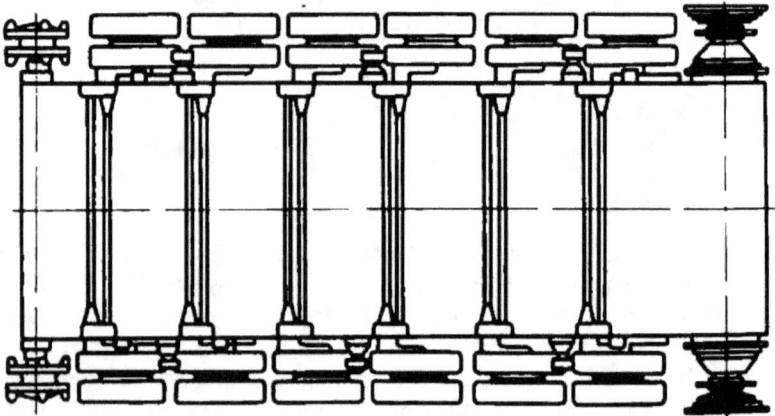

Figure 189: Suspension layout (left side facing front), showing the offset in the torsion bars and in the wheels. (*Soviet Ministry of Defense*)

A supported track suspension greatly reduced power losses from wave oscillation in the return run of the track, but more importantly, enabled new torsion bar metallurgy to be exploited to increase the bump travel of the road wheels.

The basic purpose of an off-road tank suspension is to keep the track in contact with the ground and to moderate the disturbances on the tank from terrain irregularities, chiefly in the pitching motion induced by bumps on the terrain. The critical metric for performance was minimizing vertical acceleration to an acceptable level for the crew. The foremost priority to improving off-road speed was to raise the bump travel of the suspension, and ride quality within the limits of the bump travel was maximized by having choosing an appropriate spring rate.

Fundamentally, vertical travel was limited by the torsional shear strength of the torsion bars. The higher the permissible shear strength, the higher the wheel travel. Shear stress increases linearly with the twist angle of the bar, so it is possible to reduce shear stress by using stouter and therefore stiffer torsion bars (higher spring rate) with longer swing arms to provide the desired vertical travel through a smaller twist angle[1184]. Nearly all modern tanks[1185] took this approach to suspension design. Considering the universal constraints imposed by practical considerations like torsion bar weight, acceptable floor height lost to larger torsion bars, and bending stress in long swing arms, among many others, it is not very surprising that all modern torsion bar suspensions seem to share a familial resemblance.

The T-72, on the other hand, used slender torsion bars with a soft spring rate. Each torsion bar was 2,310 mm long in total, including the splines at both ends. The working length was 2,090 mm with a 47 mm diameter[1186]. The torsion bars were about as long as those from any other tank, chiefly because similar rail transport requirements kept the hull widths fairly uniform regardless of the country of origin. 2,273 mm (89.5").

Each bar weighed 31.7 kg and was wrapped in scratch and corrosion protection tape. The bars were made from 45KhN2MFA-Sh electroslag remelted (ESR) steel[1187]. Cold rolling and grind-hardening were used to work harden the surface of the bars to HRC 56-60[1188], and elastic-plastic presetting at 145° and 105° was used to increase the shear strength[1189]. The shear strength of the T-72 torsion bars was 1,200 MPa[1190]. Despite the nature of the

---

[1184] Maclaurin, B. (2018). *High speed off-road vehicles: suspensions, tracks, wheels and dynamics*. Hoboken: John Wiley & Sons. pp. 3-4.
[1185] Stevanović, R. (2003). 'Characteristics of torsion bar suspension elasticity in MBTs and the assessment of realized solutions', *Scientific-Technical Review*, Vol. LIII, No.2, 2003. p. 69.
[1186] In the Leopard 2 and Abrams, the torsion bars were 63 mm and 62.2 mm in diameter respectively.
[1187] The ESR process was used to improve the micro-cleanliness of the steel. Impurities acted as stress concentrators between the steel grain, reducing the strength of the structure.
[1188] Исаков, П. П. (ed.) (1985). *Теория и Конструкция Танка - Том 6: Вопросы Проектирования Ходовой Части Военных Гусеничных Машин*. Москва: Машиностроение. pp. 26-27.
[1189] Presetting torsion bars was conceptually similar to a wide variety of preloading methods, including the autofrettaging of artillery barrels. Stress in the outer layers of the bar was reduced and transferred to the inner layers through residual stresses inside the material.
[1190] Бобошко, Л. С. et al. (1985). 'Результаты Контрольных Испытаний Торсионов Серийных Танков',

T-72 suspension, the torsion bar fatigue life was not worse than the T-64 and T-80 torsion bars[1191].

Further developments during the mid to late-1970's saw an increase in the shear strength of the T-72 torsion bars to 1,300 MPa[1192]. Torsion bars of this type were used in the T-72B[1193]. The vertical travel was slightly increased to make use of the improved bars, and this was complemented by a slight increase in the tank's ground clearance to reduce the likelihood of the hull bottoming out at full bump travel.

A unifying commonality of torsion bar suspensions was their linear suspension rates, which was tied to the linear spring rate of torsion bars. By intuition it seems obvious that a suspension should be soft (low spring rate), so that little force is transmitted to the tank from riding over bumps and the vertical acceleration is therefore mild. A soft suspension was also preferable for the resultant reduction in hull pitching oscillation frequency to a level most agreeable to the crew, typically considered to be between 0.5 Hz to 1 Hz.

At high speeds, a soft suspension quickly bottoms out as the tank speeds up, delivering a violent shock. A stiff suspension resists bottoming out and thereby extends the speed limit for a tolerable ride, but is acutely uncomfortable at low speeds. Due to the linear nature of torsion bars, all tanks use a very similar medium spring rate as a compromise solution, but the ideal is a progressive spring rate, which gives the smooth ride of a soft suspension at low to medium speeds with a much stronger resistance to bottoming out at high speeds[1194].

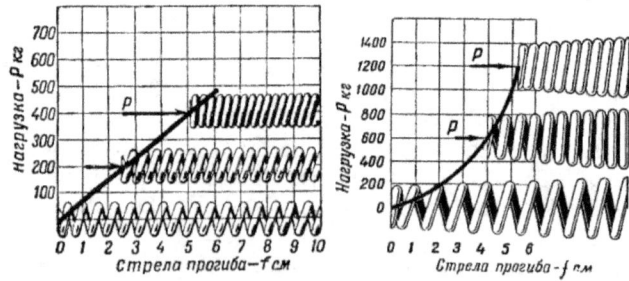

Figure 190: Characteristic spring rate curves of a linear coil spring (left) and a progressive coil spring (right). (A. S. Antonov et al., 1954)

Вестник бронетанковой техники 1985, сборник 3. p. 45.
[1191] Ibid. pp. 46-47.
[1192] Ibid. p. 45.
[1193] First wheel pair used 172.51.014sb torsion bars, the rest used 172.51.015sb torsion bars. 172.51.016-1.
[1194] Антонов, А. С., Артамонов, Б. А., Коробков, Б. М., Магидович, Е. И. (1954). *Танк*. Москва: Воениздат. pp. 462-463.

The difficulty in creating a progressive spring rate in torsion bars was recognized since the 1950's, which prompted an international exploration for alternative springing solutions. One of the most successful options was hydropneumatic springing [1195], also known as hydrogas or oleopneumatic springing, where a large bump travel with a progressive spring rate could be achieved by compressing an inert gas in a compact high-pressure chamber[1196]. It was, however, not impossible to achieve a progressive spring rate in a conventional torsion bar suspension.

On the T-72, the length of the swing arms was a geometric feature to convert the linear spring rate of the torsion bars into a progressive suspension rate[1197]. The swing arms were divided into left and right types, and into standard and reinforced types (1st, 2nd, and 6th road wheels). They were exceptionally short, only 250 mm long[1198], and were nested inside the road wheels due to the lack of clearance between the hull and the wide wheel rims. All of these details, including the swing arm length, were inherited directly from the T-54 suspension. This configuration had a few disadvantages, such as increasing the

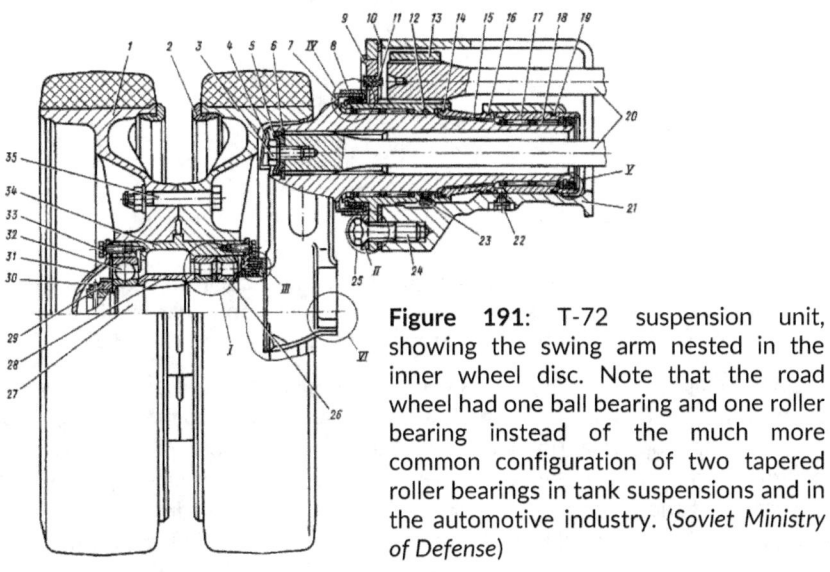

Figure 191: T-72 suspension unit, showing the swing arm nested in the inner wheel disc. Note that the road wheel had one ball bearing and one roller bearing instead of the much more common configuration of two tapered roller bearings in tank suspensions and in the automotive industry. (*Soviet Ministry of Defense*)

[1195] Ogorkiewicz, R. M. (1991). *Technology of Tanks*. Volume 1. Coulsdon: Jane's Information Group. pp. 324-327.
[1196] The progressivity of the spring rate came from the hyperbolic shape of the pressure-volume curve of a compressed gas as described under Boyle's Law.
[1197] Марецкий, П. К. (1985). 'Анализ Систем Подрессоривания Танков', *Вестник бронетанковой техники* 1985, сборник 4. p. 32.
[1198] For comparison, on the Leopard 1, the swing arms were 400 mm long. See: Krapke, P. -W. (2004). Leopard 2: Sein Werden und seine Leistung. Norderstedt: Books on Demand GmbH. p. 21.

likelihood of a deformed wheel rim jamming the wheel entirely, and adding some churning power losses from the self-cleaning action of the swing arm, which had special built-in mud scrapers.

The effect of the short swing arms on the suspension was to exaggerate the arced trajectory of the wheel during its bump travel, which in turn accentuated the geometric difference between vertical wheel travel and the twist angle of the torsion bar [**Fig. 192**]. A swing arm functions as a crank to convert vertical wheel travel into torsion bar twist. The arc of the swinging wheel reasonably approximated a straight line, where the change in twist angle with vertical travel was virtually constant. Geometrically, however, the actual relationship between these two parameters was sinusoidal, where the twist angle (x-axis) for a given vertical travel (y-axis) rose drastically towards the extreme bump position.

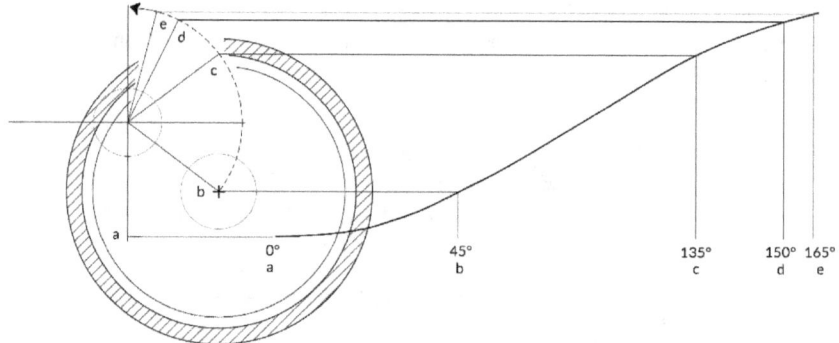

Figure 192: The curve on the right traced by the circular trajectory of the wheel has the shape of a sine wave, showing the non-linear relationship between vertical wheel travel and torsion bar twist angle. (R. A. Then, 2024)

On the majority of modern tanks with independently sprung suspensions, the swing arm deflection angle was no lower than 40° (relative to the ground) at full rebound and stopped at no more than 110° at full bump deflection, making for a fairly narrow range of deflection that, when combined with a long swing arm, ensured that the suspension rate sat at the middle of its spring rate curve[1199]. The suspension rate would not be entirely linear, but could be fairly represented as such[1200]. Only the rebound travel of some tanks might exhibit marginal non-linearity[1201]. With the short swing arms on the T-72 suspension, greater swing

---
[1199] Ogorkiewicz, R. M. (1991). *Technology of Tanks*. Volume 1. Coulsdon: Jane's Information Group. p. 325.
[1200] Сергеев, Л. В. (1973). *Теория танка*. Москва: Военная академия бронетанковых войск. pp. 406-407.
[1201] However, a flat spring rate was desirable here instead of a falling rate because it reduced the normal force to the ground during dynamic movement, thus reducing traction.

arm deflection was needed for the same vertical travel, expanding the arc wide enough to touch upon both non-linear ends of the natural sinusoidal curve.

The non-linearity of the sinusoidal curve translated into a progressive cosinusoidal suspension rate because the change in spring rate is the derivative of the sinusoidal rate of change in deflection angle with bump travel. Because the derivative of a sine function is a cosine function, the shape of the suspension rate curve became cosinusoidal[1202] [**Fig. 193**].

Additionally, the large swinging arc accentuated the shortening of the lever arm to the torsion bar towards the end of its bump travel, effectively reducing the amount of wheel load transmitted to the torsion bar as torque, and increasing the amount delivered as a normal force into the hull. This accentuated the non-linearity of the actual suspension rate curve. In the middle of its bump travel, the T-72 suspension rate was essentially similar to all other tank suspensions[1203], but towards the end, it was more closely comparable to the Challenger 1's hydrogas suspension.

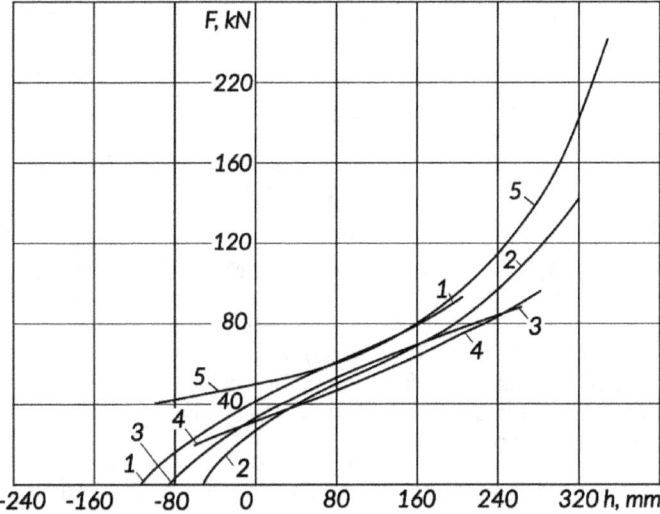

Figure 193: Suspension spring rates of the 1st road wheel on various tanks. 1 – M60A1; 2 – T-72; 3 – T-64A; 4 – Leopard 1; 5 – Challenger 1 (hydrogas). (P. Martin, 2024, adapted from P. K. Maretskiy, 1985)

[1202] The cosine curve in [**Fig. 193**] is merely the sine curve in [**Fig. 192**] after a 90° phase shift.
[1203] Not shown in [**Fig. 193**] are the Leopard 2 and Chieftain suspension rates. On the Chieftain, the suspension rate rose if both wheels in the bogie were in bump travel, and when the bump travel was high enough to engage the second coil spring.

Because of its progressive nature, the T-72 suspension was as soft as any other suspension over most of its total travel, but became drastically stiffer at high bump travel. Combined with the favorable dynamics of the V-46 engine and larger overall bump travel, the T-72 could reach noticeably higher off-road speeds than the T-64A, approaching the performance level of the T-80.

The total travel of the T-72 suspension was large[1204] in spite of the large twist angle forced upon its torsion bars, matching the travel in the T-64A and T-80 suspensions and also beating out the Leopard 1, which had the best performing Western tank suspension of its time.

Each wheel pair had a slightly different torsion bar index angle calculated against the spring rate to level out the hull and ensure the correct ground clearance. Each wheel pair was also positioned at a slightly different height corresponding to its bump travel, i.e. the first wheel was not mounted to the hull floor but on the lower glacis of the hull nose, so that the ground clearance was effectively higher.

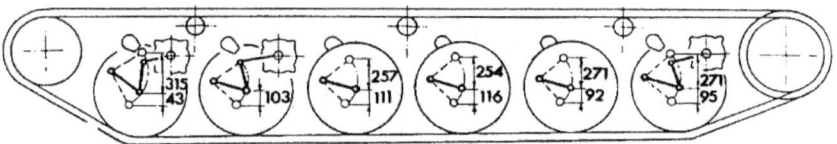

Figure 194: T-72 suspension dynamic travel distances. (*B. S. Safonov, V. I. Murakhovskiy, 1993*)

The average rebound travel was 93 mm, and the average bump travel was 272 mm, or three times larger than the rebound travel. The average total twist arc of the torsion bars was 94°. The weight distribution of the tank also made it possible to differentiate the suspension for higher bump travel on the first and last two wheel pairs. The 1st road wheel on most tanks was typically the least heavily loaded in a static condition and when driving over level ground, but it was always the most heavily stressed in dynamic loading over rough terrain due to the pitching of the hull.

The center of gravity of the T-72 was between its 3rd and 4th road wheels, but the weight distribution was heavily biased towards the rear. In both static and dynamic conditions, the 5th road wheel was the most heavily loaded. Due to the rear drive configuration, the load on the last road wheel was reduced by the increased tension in the track feeding into the drive sprocket, redistributing the wheel loads forward into the 2nd to 5th wheels[1205].

[1204] Сафонов, Б. С., Мураховский, В. И. (eds.) (1993). *Основные боевые танки*. Москва: Арсенал-пресс. p. 68.
[1205] Купцов, В. М., Тимофеев, В. Д. (1979). 'Определение вертикальных реакций грунта,

Despite the front-heavy turret and thick armor at the hull nose, the weight distribution of the T-72 was biased towards the rear like the medium tanks before it. This load distribution was considered favorable for off-road driving because the 1st road wheel pair had a very low static load – less than half that

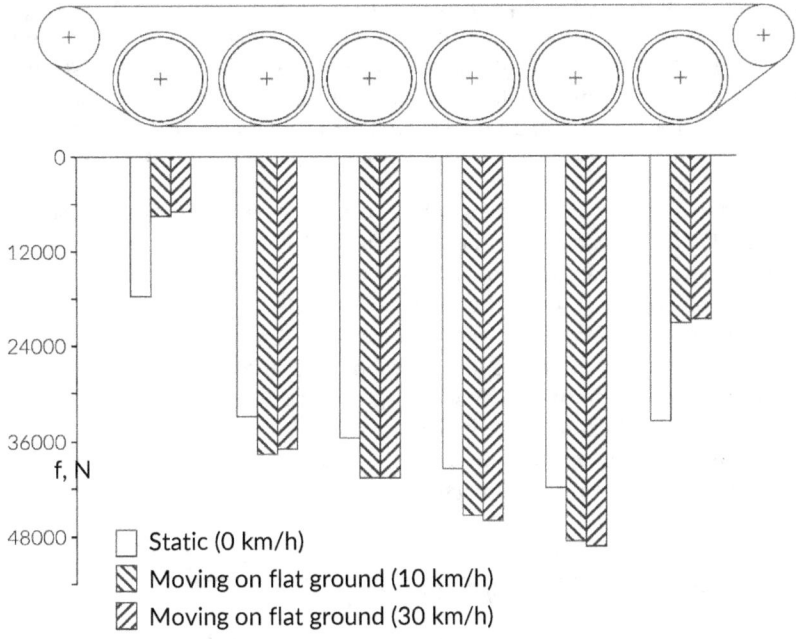

Figure 195: T-72 wheel load distribution in static and moving conditions. (V. M. Kuptsov, V. D. Timofeev, 1979)

of the 5th wheel pair, so it was easier for the wheels to handle the additional load placed upon them by the pitching motion of the tank on broken terrain. The T-72 was also capable of tolerating a great deal of added armor weight over the hull and turret front.

To withstand the increased loads from off-road driving, the 1st, 2nd and 6th swing arms had an additional roller bearing, increasing their weight to 59 kg from the basic 55 kg weight of the 3rd, 4th and 5th swing arms[1206]. Together with the swing arm and torsion bar, each complete suspension unit had a nominal weight of 265 kg, or about the same as a single T-54/55 wheel on its own.

действующих на опорные катки ВГМ при движении', *Вопросы оборонной техники*, 20(88). p. 22.
[1206] Министерство Обороны СССР (1992). *Руководство по войсковому ремонту объект 184: Книга Первая - Замена и Ремонт Агрегатов и Узлов, Часть Вторая*. Москва: Воениздат. p. 363.

When measuring the ground pressure, the nominal ground pressure is the simplest metric since it is simply the average pressure exerted by the tank's weight across the entire contact patch of the tracks. However, the actual pressure exerted by the suspension in real world conditions show significant peaks below each wheel, moderated only to a limited extent by the tracks when measured on hard ground. In effect, the mean maximum pressure (MMP) exerted by the suspension essentially corresponds to the pressure exerted by the surface area of the track link supporting each wheel, which, of course, was unfavorable for short-pitched tracks. On soft terrain, however, the effect of a large road wheel diameter and wheel spacing changed the MMP again.

If rated according to its nominal MMP, the T-72 performed at a respectable level (239 kN/m²), significantly better than the British Challenger (285 kN/m²) and slightly worse than the Leopard 1 and Leopard 2 (223 kN/m²)[1207], but even then, it is obvious that a direct comparison of MMP figures cannot accurately reflect the actual behavior of the suspension on terrain since wheel loads are almost never uniformly distributed. On the T-72, the highest wheel load was 1.25 times higher than the nominal average, which was historically middling but low enough to make a difference against some of its modern competitors. The M1 Abrams, for instance, had an MMP of 231 kN/m² but the ratio of its highest wheel load to its average load reached 1.37[1208].

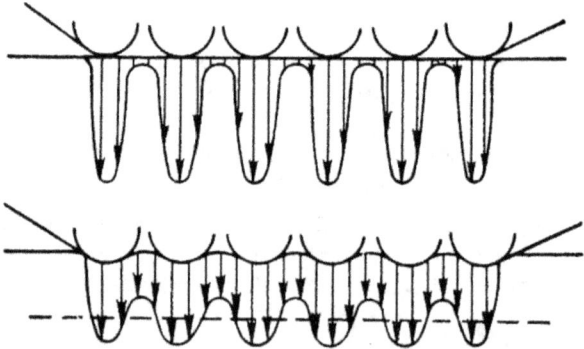

Figure 196: Mean maximum ground pressure on hard ground versus soft ground. The dashed line represents the nominal ground pressure. (*N. A. Nosov et al., 1972*)

[1207] Ogorkiewicz, R. M. (1991). *Technology of Tanks*. Volume 1. Coulsdon: Jane's Information Group. pp. 346-348.
Note that the actual MMP figures tend to be underestimated when calculated for tanks with parallel-pin tracks, especially those with rubber shoes, because the surface area is calculated from the total track pitch inclusive of the pitch of its end connectors. The supporting surface provided by a parallel-pin track link tends to be somewhat shorter than the total track pitch.
[1208] General Dynamics (1985). *M1/IPM1/M1A1 Comparison*. Briefing Book. p. 22.

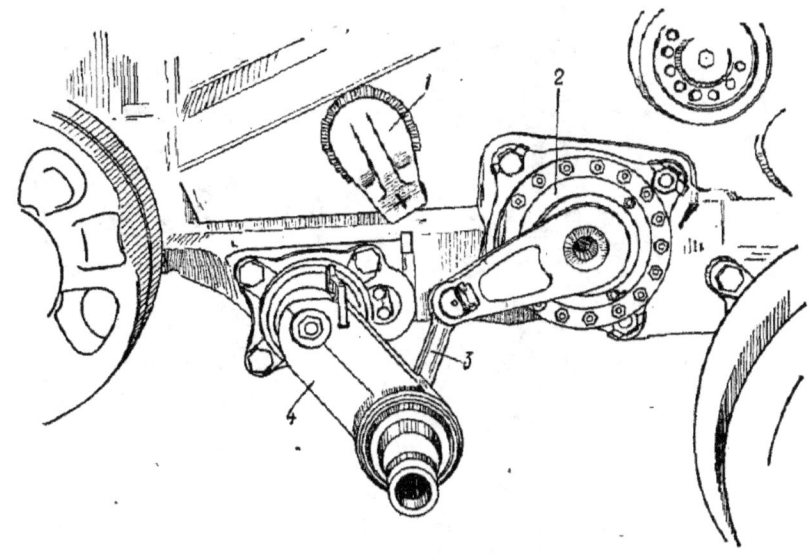

Figure 197: First suspension unit on a T-72, shown without its wheel. 1 – bump stop; 2 – hydraulic shock absorber; 3 – connecting lever; 4 – swing arm.
(*Soviet Ministry of Defense*)

The shock absorbers were of a rotary vane type, articulated by the swing arms by a connecting rod through a total operating arc of 87°. The working diameter was 250 mm, and inside it was 2.55 liters of absorber fluid. They were physically massive, which was quite appropriate for the good heat dissipation characteristics intrinsic to rotary vane shock absorbers from having a direct connection to the hull.

The shock absorbers fit into holes cut into the hull sides, so that instead of the normal 70-80 mm of steel plate, these parts of the hull were protected only by the shock absorber casings. To compensate, each shock absorber was thickly armored and as a result, very heavy; at 58 kg each, they were around 2-2.5 times heavier than contemporary tank shock absorbers. This somewhat radical design solution gave the absorbers an unusually high degree of ballistic protection without seriously compromising hull protection, and more importantly, increasing the casing thickness of the shock absorber and increasing the contact area with the hull were effective methods of dissipating heat from the absorber fluid. The tank hull itself thereby functioned as a massive inexhaustible heat sink.

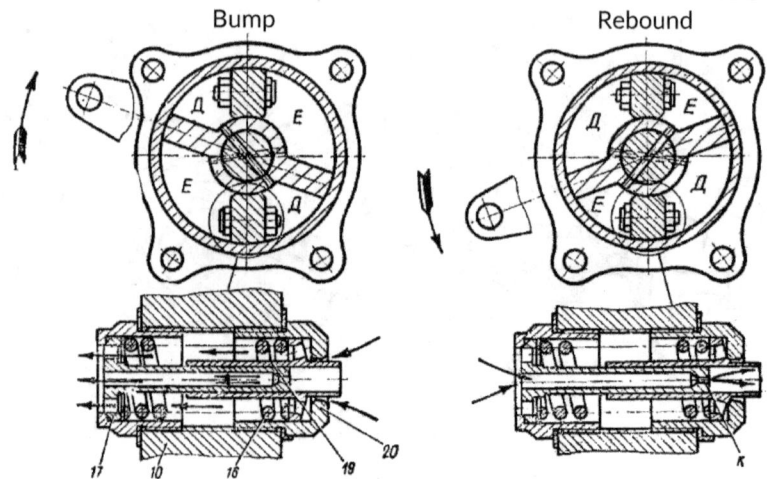

Figure 198: Action of the T-72 shock absorber. In bump travel (left), the hydraulic orifice was large, giving less resistance. In rebound travel (right), the orifice was much smaller, raising its resistance. (*Soviet Ministry of Defense*)

The exposed surface area of the casing was massive at 0.38 m², but mattered little because air cooling was difficult and largely ineffective due to the inherently poor air flow in the space between the wheels and the hull of a tank, which only worsened when side skirts were fitted.

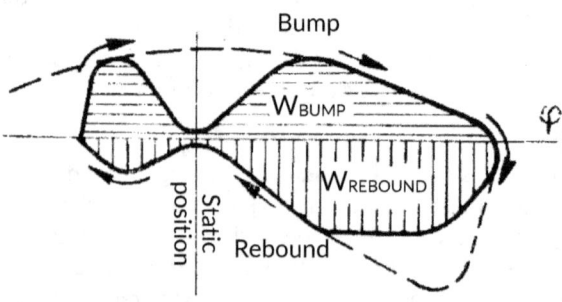

Figure 199: Energy dissipated in rotary vane shock absorber as work (W) across vane angle φ for a given forcing speed. The peak bump and rebound resistances are relatively close, but much more energy is dissipated in rebound. (*S. P. Petrakov, 1962*).

Cooling was a serious roadblock for conventional telescopic shock absorbers in modern high-speed tank suspensions with enormous work capacities. It was not without reason that in all Western modern tanks of the 1980's and 1990's, from the Leopard 2, Challenger 1 and 2, Abrams and Leclerc, none used telescopic shock absorbers. With poor cooling, a short drive could heat up the absorber fluid enough to significantly lower its viscosity, reducing absorber resistance and worsening the damping strength.

By recessing the device into the side hull plate, mud accumulation was also lessened, and the survivability of the T-72 shock absorber improved[1209]. The main downside was that rotary vane shock absorbers faced certain complications in seal design at high working pressures[1210], so the T-72 shock absorber was not as strong as the telescopic shock absorbers on the T-80 and T-64 [**Fig. 199**]. Nevertheless, the operating pressure was successfully increased for the T-72B to account for the change in weight and its effect on suspension dynamics.

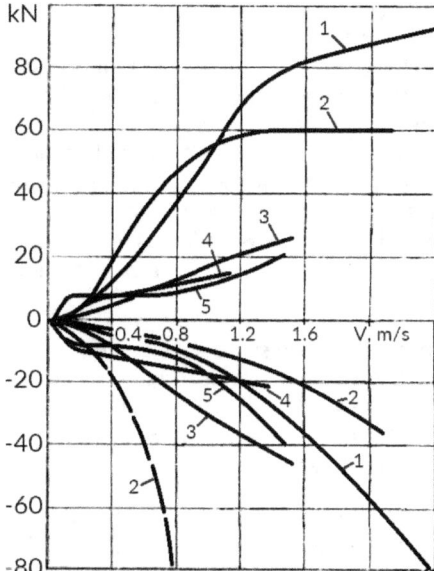

Figure 200: Shock absorber resistance against forcing speed, with positive force denoting bump travel and negative denoting rebound travel. 1 – T-80; 2 – T-64 (dashed line for 2nd and 6th road wheels); 3 – T-72; 4 – Leopard 1; 5 – M48. (*Adapted from P. K. Maretskiy, 1985*)

Unlike springs, shock absorbers dissipate kinetic energy as heat rather than storing it, and the resistance of a shock absorber is dependent on its forcing speed, not its position at any given moment. The role of shock absorbers – or dampers, to be more correct – is to damp the oscillation induced in the tank by soft suspension springs over rough ground. Powerful shock absorbers (dampers) are required to achieve good damping strengths because the oscillation frequency of a soft suspension is low, and so the travel velocity at the swing arms and thus the forcing velocity is low at any given time. But at the same time, strong shock absorbers transmit more force to the hull, stiffening the suspension and upsetting the ride quality.

Nearly all modern tank suspensions compromise by applying a moderate damping ratio and by configuring the shock absorbers for greater strength in rebound travel and less in bump travel to yield the work capacity needed[1211] while preserving the softness of the suspension as best as possible[1212].

[1209] Глухов, Г. К., Кочегаров, П. П. (1980). 'Особенности ремонта подорванных на минах танков', *Вестник бронетанковой техники* 1980, сборник, 6. p. 36.
[1210] Исаков, П. П. pp. 53-54.
[1211] Петраков, С. П. (1962). 'О создании подвесок быстроходных танков', *Вестник бронетанковой техники* 1962, сборник, 5. pp. 15-16.
[1212] The Leopard 2 was a notable exception. It used weak friction dampers that worked across the entire range of wheel travel, supplemented by telescopic hydraulic shock absorbers designed as bump

Wheel Design

The wheels ran on the same universal single-pin RMSh medium tank track as the T-54/55 and T-62, and the rubber rims matched the width of the T-54/55 wheel. This kept the wheel load very low, which had an overall positive impact on the durability of the tracks and the rubber rims, though rim width has more effects on rim durability than just wheel loading. The cantilever mounting of the suspension units gave each wheel a slight negative camber, shifting more pressure onto the inner wheel rim. This reduced the service life of road wheels by accelerating the onset of chunking [1213]. The wider the rim, the more pronounced the effect.

The T-72 wheel used new, stronger bearings in the same size category as the T-54/55 wheel, and it kept the same 85/100 axle profile [1214]. This made it possible for a T-72 to interchange wheels with a T-54/55 or T-62, although only an emergency situation might justify it [1215]. The reverse was also possible, but only if accompanied by a track tension readjustment.

At 750 mm in diameter, the wheels had a low rolling resistance on irregular terrain, but this came at the expense of weight, each wheel weighing considerably more than most contemporary types at 177 kg each. The 1st road wheel was a "strengthened" wheel containing an additional roller bearing [1216] to withstand high dynamic loading. Beginning in the early 1980's, the original eight-spoked models [1217] were replaced by a new six-spoked wheel [1218] and its own "strengthened" variant for the 1st axle. The new wheel was stronger [1219] yet lighter, weighing 164 kg each.

A massive weight reduction of 32.4% over the T-54/55 wheel [1220] had been achieved by replacing the one-piece cast steel wheel body with a pair of aluminum wheel discs, stamped from AK6 grade aluminum on a 30,000-tonne press. The discs were mated to opposite ends of a thin sleeve made from high-strength steel to form the wheel hub, and then the discs were bolted together, clamping the hub sleeve between them to form the complete road wheel. The complex shape of the disc was a means of supporting wide rims and a wide

---

stops so that they only worked in the last half of the bump travel, creating a dual-rate system.
[1213] Абрамов, А. Д., Девятовский, Ф. А., Кочегаров, П. П. (1984). 'Некоторые Причины Отказов Опорных Катков', *Вестник Бронетанковой Техники* 1984, сборник 1. pp. 27-29.
[1214] The axle profile was tapered for two sections, the outer section 85 mm in diameter and the inner section 100 mm in diameter.
[1215] Distributed over twelve wheels, the difference in weight would amount to a total weight gain of close to 1.1 tonnes.
[1216] The additional roller bearing increased the wheel weight by 4 kg.
[1217] 172.51.002sb-A standard wheel with 172.51.001sb-A strengthened wheel.
[1218] 172.50.002sb-A standard wheel with 172.50.001sb-A strengthened wheel.
[1219] It was explicitly referred to as a "reinforced wheel".
[1220] 177 kg for the T-72 wheel against 262 kg for T-55 pattern wheel

bearing surface for the hub sleeve. The need for a massive steel wheel hub was thereby eliminated.

However, modern wheel designs affix the hub to the swing arm to keep a tight seal for oil lubrication, and the two wheel discs alone constitute the only removable parts of the unit. The anachronistic T-72 wheel design was less tightly sealed, thus requiring grease lubrication (which was better supported by the larger wheel diameter anyway), and once the wheel was removed, the bearings in the hub were exposed to sand and dust carried by the wind, not to mention the contamination if a pry bar was used to lift the wheel onto the axle.

Operationally, separating the road wheel into two discs made it possible to replace each disc as needed instead of the entire wheel, and without removing its bearings[1221]. Unfortunately, with the way the discs were fitted to the hub, it was not possible to remove the inner disc without dismounting the entire wheel. This was not significant for cases of road wheel destruction by anti-tank mine[1222], since the entire wheel would invariably have to be replaced regardless of whether the mine detonated along the inner or outer edge of the track. However, if the inner disc had to be replaced due to asymmetric wear on the rubber rims, the labor involved was higher than it otherwise could have been.

The rubber rims accounted for a major share of the excess weight because of the large rim circumference, but this had its merits as well. A thicker rim had a stronger cushioning, the large wheel diameter reduced the rotating speed of the wheel (reducing wear to the axle bearings), and the large circumference helped dispersed the hysteresis heat and thus helped to prevent accelerated breakdown at high driving speeds and in hot weather.

With that in mind, it is more appropriate to compare the T-72's large-diameter wheels against a suspension with smaller diameter wheels running on rubber track pads[1223] with the same degree of cushioning. With all else being equal, larger road wheels lighten the suspension overall while reducing the rolling resistance[1224]. Inverting the steel-rubber interface to have steel wheels running on rubber track pads creates a suspension optimized for minimal total weight with even better heat dispersion, but higher rolling resistance; no production tracked vehicle uses such a configuration. Steel wheels on steel tracks is both light and efficient, but noisy, and rubber rims running on rubber track pads increases rolling resistance by 25-50%[1225], which is patently unacceptable.

[1221] Ibid. pp. 160-161.
[1222] All tank road wheels are reliably destroyed by any anti-tank mine. Sometimes, two wheels may be destroyed in one blast.
[1223] Rubber pads lining on the inner side of the track.
[1224] Merhof, W. (2016). *Fahrmechanik der Kettenfahrzeuge. Überarbeitete Neuausgabe*. Neubiberg: Universität der Bundeswehr München. p. 385.
[1225] Исаков, П. П. (ed.) (1985). *Теория и Конструкция Танка - Том 6: Вопросы Проектирования*

Tracks

The T-72 tracks were inherited from the T-55 and T-62, first introduced in 1962[1226] as a replacement for the original all-metal T-54 track. The new track had single-pin tracks cast from G13LA Hadfield steel[1227] with forged track pins and rubber pin bushings. The elasticity from the rubber bushings made it a "live" track, classified as an RMSh-type track[1228], as opposed to an all-metal "dead" track, known as the OMSh type[1229], which was equivalent to a slack chain strung through the suspension. Live tracks lose less power through friction and wave oscillation, and had inherently longer service lives because dust or sand could not get into the link-pin gap and abrade the track internally.

A set of tracks for one side of the suspension consisted of 97 links – the same number as on the T-62, and weighed around 1,723 kg. The tracks were light[1230], which was good for running efficiency and made it less burdensome to haul around during routine track maintenance. Soviet comparative tests also found the T-72 track to be noticeably easier to work with than the parallel-pin tracks of the T-64A and T-80[1231]. However, excluding a few extreme outliers, track weight made relatively little difference to critical repair benchmarks like the total track repair time after an anti-tank mine blast; in the three-way comparative test, the difference in repair time that could be positively linked to the different track weights[1232] amounted to just a few minutes out of a half-hour process. The single-pin nature of the T-72 track was the single most important characteristic that made it the easiest to work with.

Considering that the entire pitch length of each track link contributed to the supporting surface of the track, the T-72's track pitch of 137 mm was large

---

Ходовой Части Военных Гусеничных Машин. Москва: Машиностроение. p. 159.
In the case of the T-72/90 compared to the T-80 suspensions, the T-80 suspension allegedly had 50% higher rolling resistance. See: Вавилонский, Э. Б., Куракса, О. А., Неволин, В. М. (2008). *Основной боевой танк России: Откровенный разговор о проблемах танкостроения.* Нижний Тагил: Издательский Дом «Медиа-Принт». p. 108.

[1226] Солянкин, А. Г, Желтов, И. Г., Кудряшов, К. Н. (2010). *Отечественные бронированные машины. XX век - Том 3: Отечественные бронированные машины. 1946-1965 гг.* Москва: Издательство «Цейхгауз». p. 152.

[1227] Березин, И. Я., Мазепа, Г. В., Сергеев, В. Г., Шаповалов, В. В. (1984). 'Прогнозирование ресурса траков танковой гусеницы', *Вестник бронетанковой техники* 1984, сборник 6.
Hadfield steel is manganese alloy steel (Mangalloy). It was often used for tracks, and sometimes for other purposes like the U.S. Army M1 helmet.

[1228] Резино-Металлическим Шарниром (РМШ); Rubber-Metal Joint

[1229] Открытым Металлическим Шарниром (ОМШ); Open Metal Joint

[1230] The specific weight of the T-72 track was 135.4 kg per meter. For comparison, the T-142 track for the M60A1 and M1 Abrams (early) had a specific weight of 197.1 kg/m.

[1231] Глухов, Г. К., Кочегаров, П. П. (1980). 'Особенности ремонта подорванных на минах танков', *Вестник бронетанковой техники* 1980, сборник. 6. p. 36.

[1232] Hauling the track onto the drive sprocket and rewinding it through the suspension.

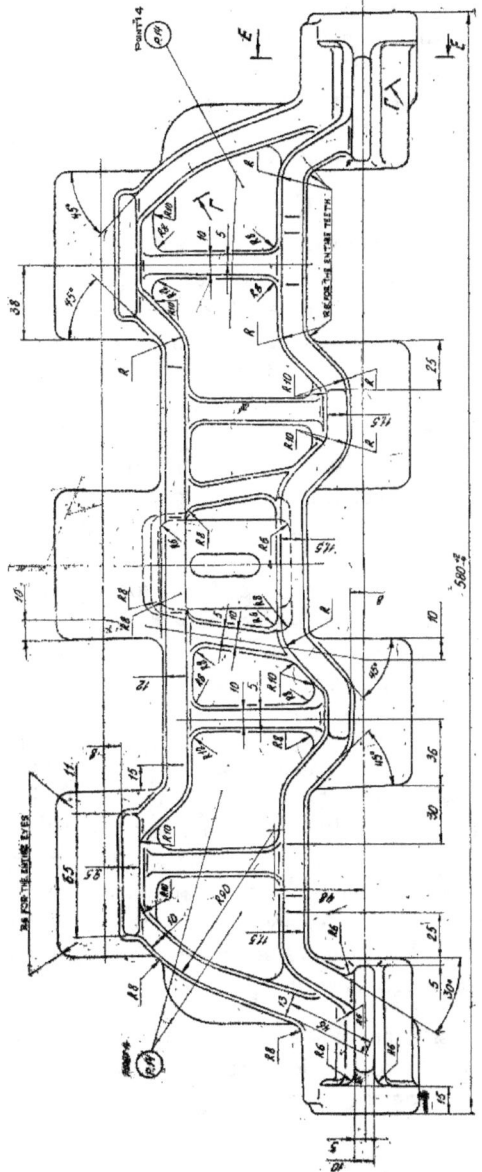

compared to practical parallel-pin tracks, but it was still modest compared to old "plate"-type tracks like on the T-34 with its pitch of 172 mm.

Smoother rolling engagement with the drive sprocket teeth from the shorter track pitch helped to lower suspension noise and contributed to a reduction in power losses [1233]. Tests showed that power losses at high speeds were noticeably lower than the parallel-pin tracks of the T-80 in cross-country terrain [1234]. When tested on T-80 test beds, the two tracks had almost the same power losses at speeds of 25-30 km/h, but at a speed 54-55 km/h, fuel economy was noticeably improved for the tanks fitted with the T-72 track.

For live tracks, a single-pin joint was inherently limited in its load-bearing capacity by the twist angle of the rubber bushing. Parallel-pin tracks, also known as dual-pin tracks, articulate through two pins instead of one, so the twist angle on each rubber bushing is halved for any given link

Figure 201: Track bottom surface. (*Soviet Ministry of Defense*)

[1233] Развалов, С., Штейн, В. Д. (1984). 'Экспериментальные данные о затратах мощности на перематывание гусениц', *Вестник бронетанковой техники* 1984, сборник 6. p. 39.
[1234] In snowy terrain, there was no significant difference.

articulation angle. The normal and tangential stresses in the rubber are reduced proportionately, allowing the track to support higher tractive force or simply to improve longevity. Load capacity was the reason for the categorical abandonment of single-pin tracks by Western tank builders, even if it was technically feasible[1235]. The T-72 track operated close to its maximum capacity, and in the 2000's, it was finally replaced by a parallel-pin track to keep up with weight gains.

Table 53: RMSh track data

T-72 RMSh track details	
Width (mm)	580
Pitch (mm)	137
Center guide height (mm)	117
Total height (mm)	195
Link weight (kg)	14.2-16
Hardness (HB)	≥170
Total length per side (mm)	13,289

The T-72 track[1236] was robust enough that it could handle loose, rocky terrain, and sand. Trafficability on soil was broadly equivalent to most other tracks regardless of type, but traction was inherently high thanks to the long, aggressive grousers and the cleats on top of the grousers. These cleats ensured that the track could penetrate into icy surfaces for traction. Of course, this also meant that the track was intensely destructive to roads and concrete, and while driving on relatively soft asphalt was not too harmful to the track, driving on concrete eroded the cleats much more intensely.

A castellated profile can be seen on the track in the transverse direction matching the knuckles (barrels) of the hinge joint. This pattern, together with the webs spanning the hinge joints, gave lateral traction, raising the skid-out speed limit of the tank and improving its climbing performance on slopes, especially icy slopes, by preventing the tank from slipping sideways. The only drawback was that lateral traction added steering resistance. Grousers with a straight profile (e.g. excavator tracks) instead of a castellated profile present nearly no lateral resistance apart from friction, which was historically quite problematic for British tanks[1237].

When driving on soils with low shear strength like mud or clay, tracks of any type deliver traction mainly through soil thrust, which is strongly dependent on the shear strength of the soil. Exceeding the shear strength causes a mud-

---

[1235] Meacham Jr., H. C., Swain, J. C., Wilcox, J. P., Doyle, G. R. (1978). Track Dynamics Program. Warren: U.S. Army TARADCOM. p. 15.
[1236] Part index 613.44.16сб-Д1.
[1237] The British Centurion tank had poor stability on slopes because of its wide, straight grousers, a legacy of outdated WW1 and interwar track design conventions.

packed track to slip, and a clogged track provides only as much traction as the coefficient of sliding friction between mud and mud allowed. In principle, the T-72 track had inherently good traction from soil thrust because of its very tall grousers (69-71 mm)[1238], but nevertheless, the suspension as a whole did not perform better than the M60A1 when tested on marshy terrain[1239], although intuition might suggest that grousers should inherently outperform rubber shoes.

Where the T-72 track decisively outperformed most other tracks was in its lifespan. Before the T-72 entered service, the lifespan of the track was no less than 6,000 km[1240] when fitted to the lighter and slower T-55 and T-62. This figure was a design goal chosen to align with the 6,000 km warranty lifespan standard set for medium tanks[1241]. During Object 172M state testing in 1972, the testing commission noticed that the track life was somewhat shorter than expected. Instead of 6,000 km, the actual service life on the Object 172M turned out to be 4,500-5,000 km[1242]. This was singled out as a point for further improvement. A goal of 6,500-7,000 km was set for the guaranteed service life[1243]. This was reached on March 1, 1978 by refining the chemical composition to cut out as much of the phosphorous content as possible when casting the track links[1244]. The improvement from this change exceeded expectations, raising the track's average lifespan to 8,000 km on the T-72[1245].

The Leopard 1 had most durable running gear of all NATO tanks throughout the 1960's and 1970's and the service life of its track was 6,800 km, and on the opposite end of the spectrum, the Chieftain's tracks, which were the least durable in NATO, had a service life of 2,000 km[1246]. By the 1980's, the track on the M1 Abrams overtook the Chieftain track with a total lifespan of just 1,370 km (850 miles), reduced even further to just 1,140 km (710 miles) when fitted

---

[1238] U.S. Department of the Army (1967). *Engineering Design Handbook: Automotive Series – Automotive Suspensions.* AMCP 706-356. Washington, D.C.: U.S. Army Materiel Command. pp. 5-10, 4-14.

[1239] Кочегаров, П. П., Лукьянов, А. И., Никулин, Е. Г. (1979). 'Результаты испытаний по оценке проходимости танков', *Вопросы оборонной техники*, 20(88). pp. 39-40.

[1240] Merhof, W., Hackbarth, E.-M. (2016). Fahrmechanik der Kettenfahrzeuge. Überarbeitete Neuausgabe. Neubiberg: Universität der Bundeswehr München. p. 398.

[1241] Развалов, А. С. (1987). 'Проблемные Вопросы Обеспечения Надежности Бронетанковой Техники', *Вестник бронетанковой техники* 1987, сборник 10. p. 3.

[1242] Устьянцев, С. В., Колмаков, Д. (2004). *Боевая Машина Уралвагонзавода: Танк Т-72.* Нижний Тагил: Издательский Дом "Медиа-Принт". p. 59.

[1243] Referring to the driving distance in which there was a 99% probability of failure-free operation within a 95% confidence interval. See: Исаков, П. П. p. 150.

[1244] Устьянцев. С. В., Колмаков Д. Г. (2013). Т-72/Т-90. Опыт создания отечественных основных боевых танков. Нижний Тагил: ОАО «НПК «Уралвагонзавод» имени Ф. Э. Дзержинского». p. 176.

[1245] Кустылкин, В. С., Развалов, А. С. (1986). 'Анализ причин отказов при подконтрольной эксплуатации танков в 1975-1984 гг.', *Вопросы оборонной техники*, 5(126).

[1246] Distance limit before complete track replacement was needed. The Chieftain's tracks were the least durable because they were "dead" tracks on a tank of considerable weight. Service life figures from Dutch trials. See: Smit, W. (2008). *De Leopard 1: Gepantserde vuist van de Koninklijke Landmacht.* Meppel: Boom. p. 30.

to the M1A1. Track replacements accounted for around half the Abrams fleet annual running costs in repair parts[1247].

When track shoe replacements are taken into account, the labor and cost savings from track maintenance make the T-72's track appear even more attractive, though this was by no means exclusive to the T-72. The absence of rubber track shoes was characteristic of all Soviet military tracked vehicles. Besides the demanding upkeep, rubber track shoes generally increase track mass by approximately 40%, increase rolling resistance, and do not necessarily provide more traction than steel tracks when driving over rough terrain. Of course, that is not to say that they do not come with any benefits. On the contrary, it was experimentally established that, in general, rubber track pads would increase traction on a concrete surface by 40% and on dry soil by 7%. When moving on roads, the average speed of a tank column could be 10-15% higher simply by eliminating track slippage from the metal-asphalt interface.

These were major gains, but considering that tracked vehicles were intended for off-road travel, the policy of using rubber track shoes ultimately boiled down to road damage prevention. In the West, rubber track shoes were strictly mandated by West German regulations on using tracked vehicles on West German roads, and all NATO partners training on German soil were beholden to this rule, regardless of their individual preferences. For instance, in 1960, the Chieftain tank, still in the middle of its development, was obligated to add rubber shoes over its original all-metal tracks to comply with this policy[1248].

Due to rapid wear, replacing track shoes often accounted for a major share of a tank's component replacement costs[1249]. Rubber shoes were invariably much cheaper overall once road reconstruction costs were factored in, but how these calculations were worked out in the USSR is somewhat unclear. Paved surfaces on military bases and training areas were made from concrete paneling, and in military testing, a "paved road" referred to concrete surfaces[1250], not asphalt.

[1247] U.S. GAO (1991). *Abrams Tank - Operating Costs More Than Expected*. Washington, D.C.: GAO/NSIAD. p. 5.
[1248] Forty, G. (1979). *Modern Combat Vehicles: 1. Chieftain*. Shepperton: Ian Allan Ltd. p. 20.
[1249] As of 1986, annual M60A3 track pad replacements cost 37.07 million U.S. dollars, making up 11.4% of the total annual component replacement costs, closely behind engine assembly replacements (16.5%).
[1250] Since the late 1960's, a concrete test track at 38NIIBT was used for "road" testing.

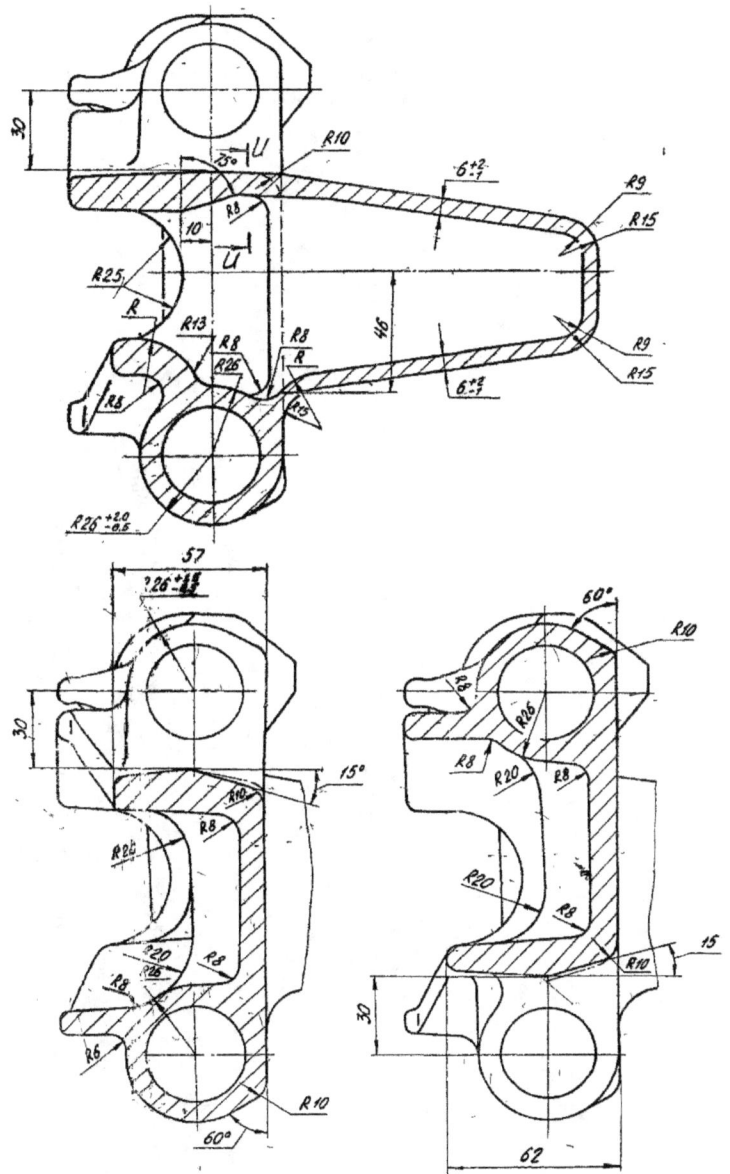

Figure 202: Cross sections of the track at three points. (Soviet Ministry of Defense)

Gallery

The following photos are of a Czechoslovakian T-72M at the Schweizerisches Militärmuseum (Full, Switzerland) by Sebastian Stauber, 2024.

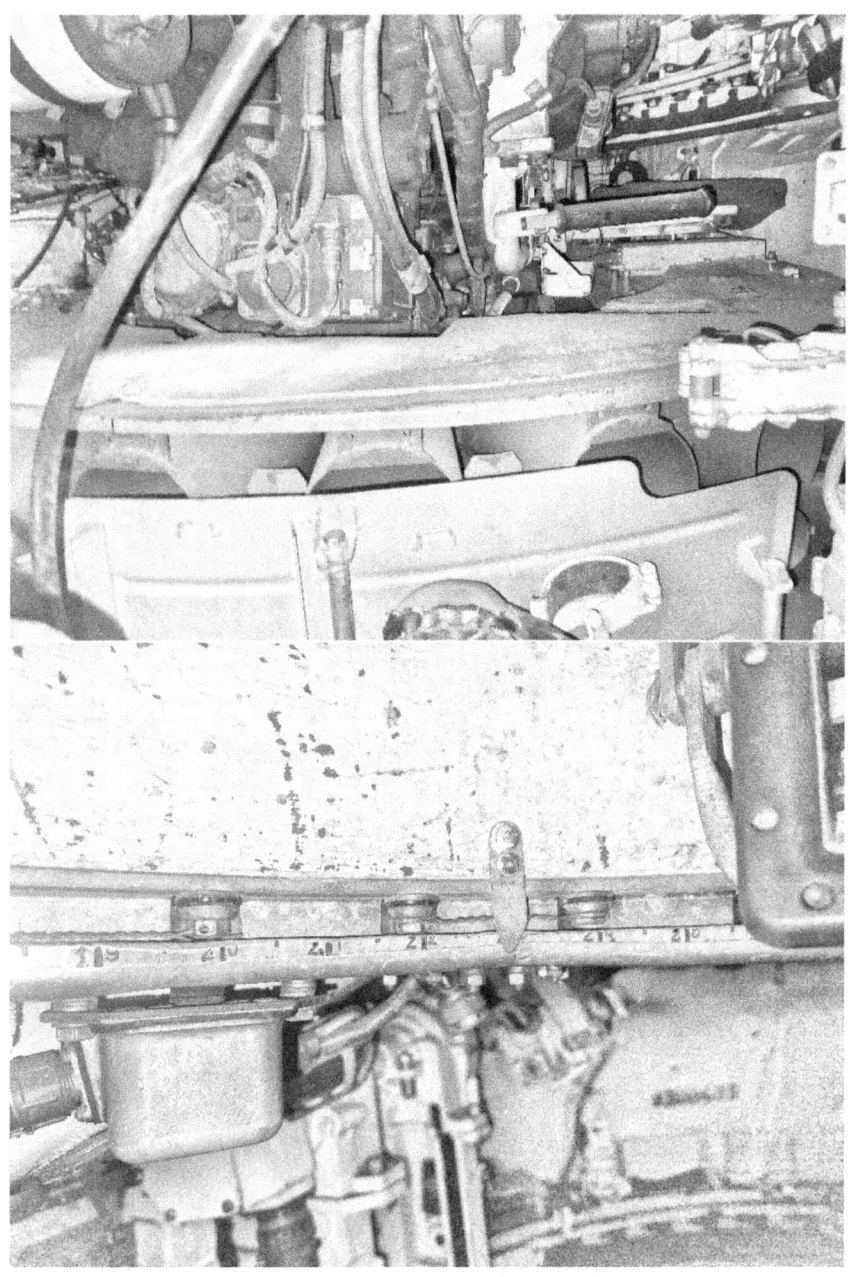

Bibliography

English Language Sources

Almen, J. O. (1944). 'On the Strength of Highly Stressed, Dynamically Loaded Bolts and Studs', SAE Transactions, Vol. 52.

Anderson, C. E., Littlefield, D. L. (1994). Pretest predictions of long-rod interactions with armor technology targets. Technical report, SwRI Project No. 07-5117. San Antonio: Southwest Research Institute.

Andrew, S. P., Caligiuri, R. D, Eiselstein, L. E. (1992). 'A review of penetration mechanisms and dynamic properties of tungsten and depleted uranium penetrators' in Computational Modeling of Dynamic Failure Mechanisms in Armor/Anti-Armor Materials. Menlo Park: Failure Analysis Associates, Inc.

Backofen, J. E. (1982). 'Armor Technology, Part 1: Armor, (May-June).

Baer, P. G., Frankle, J. M. (1962). The Simulation of Interior Ballistic Performance of Guns by Digital Computer Program. Report No. 1183. Aberdeen: Ballistic Research Laboratory.

Ball, A. M. (1964). Engineering Design Handbook: Explosives Series - Part One: Solid Propellants. AMCP 706-175. Washington, D.C.: U.S. Army Materiel Command.

Balla, J. et al. (2017). 'Inserting cartridges using electrically powered ramming devices', 2017 International Conference on Military Technologies (ICMT). Brno: University of Defence.

Balla, J. et al. (2021). 'Study Effects of Shock Absorbers Parameters to Recoil of Automatic Weapons', 2021 International Conference on Military Technologies (ICMT). Brno: University of Defence.

Barnes J. D., Pearce, D. M., Cantab, B. A. (1944). Report on Russian C.I. Tank Engine Type "V2" From T-34 Cruiser Tank. Military College of Science – School of Tank Technology.

Bajić, Z., et al. (2015). 'Prediction of the Impact Sensitivity of Aluminized Explosive Mixtures Using the Response Surface Methodology', *Scientific Technical Review*, 65(3).

Beeson, J. B., Mazza, T. N. (1975). Attribute Analysis of the Armor Machine Gun Candidates. AMSAR/SA/N-24. Rock Island: U.S. Army Armament Command, Systems Analysis Directorate.

Booz, Allen & Hamilton Inc. (2003). Final Technical Report Fires Experienced and Halon 1301 Fire Suppression Systems in Current Weapon Systems.

Brocklin, C. V. (1989). Single Hydraulic Fluid for Army Ground Combat and Tactical Vehicles and Equipment. Fort Belvoir: U.S. Army Belvoir RD&E Center.

Brown, A. (ed.) (1985). War in peace: the Marshall Cavendish illustrated encyclopedia of postwar conflict. Volume 6.

Central Weapons Laboratory (1970). Dealing with Green Camouflage Color for Vehicles. Report.

Chertok, B. (2006). Rockets and People - Volume II: Creating a Rocket Industry. Washington, D.C: National Aeronautics and Space Administration.

Chiou, J. P. (1975). 'Engine Cooling System of Military Combat/Tactical Vehicles', SAE Transactions, Vol. 84.

Colburn, J. W., Robbins, F. W. (1990). Combustible Cartridge Case Ballistic Characterization. Memorandum Report BRL-MR-3835. Aberdeen: Ballistic Research Laboratory.

Conroy, P. J., Leveritt, C. S., Hirvonen, J. K., Demaree, J. D. (2006). The Role of Nitrogen in Gun Tube Wear and Erosion. Technical Report ARL-TR-3795. Aberdeen: Army Research Laboratory.

Conroy, P. J., Nusca, M. J., Chabalowski, C., Anderson, W. (2001). Gun Tube Surface Kinetics and Implications. 10th U.S. Army Gun Dynamics Symposium.

Dakin, J. P., Brown, R. G. W. (2006). Handbook of Optoelectronics (Two-Volume Set). Boca Raton: CRC Press.

Dakin, J. P., Brown, R. G. W. (2016). Handbook of Optoelectronics: Concepts, Devices, and Techniques. Volume 1. Boca Raton: CRC Press.

Dasch, J. M., Gorish. J. D. (2013). The TARDEC Story - 65 Years of Innovation. Washington, D.C.: GPO.

Davis, G. C. (1982). 'Low Visibility Tactical Navigation", ARMOR, (January-February).

Department of Defense (1995). Design of Combat Vehicles for Fire Survivability, MIL-HDBK-684.

Department of the Army (1973). Comparison Test of Driver's Night Vision Devices, M60 Series. Fort Knox: Army Armor and Engineer Board.

Department of the Army (1975). Engineering Design Handbook: Military Vehicle – Power Plant Cooling. AMCP 706-361. Washington, D.C.: U.S. Army Materiel Command.

Dix, D. M., Riddell, F. R. (1981). Propulsion System Technology for Military Land Vehicles. IDA Paper P-1578. Arlington: Institute for Defense Analyses.

Dowding, R. J., Tauer, K. J. (1989). Supply of tungsten in 1989. Watertown: Army Materials Research Agency.

Elliott, E. C. (1983). 'Soviet land navigation', Journal of Terramechanics, 19(4).

Ellor, J. E. (1944). 'The Development of the Merlin Engine', SAE Transactions, Vol. 52.

Fairfield, A. P. (1921). Naval Ordnance. Annapolis: U.S. Naval Institute.

Forbat, J. (2012). The 'Secret' World of Vickers Guided Weapons. Stroud: The History Press.

Forty, G. (1979). Modern Combat Vehicles: 1. Chieftain. Shepperton: Ian Allan Ltd.

Furlong, R. D. M. (1980). 'Delay in Improved TOWS for Europe?', International Defense Review, 9 (September).

Gann, R. G. et al. (1990). Preliminary Screening Procedures and Criteria for Replacements for Halons 1211 and 1301. NIST Technical Note 1278. Gaithersburg: National Institute of Standards and Technology.

General Dynamics (1985). M1/IPM1/M1A1 Comparison. Briefing Book.

Glasstone, S., Dolan, P.J (eds.) (1962). The Effects of Nuclear Weapons. Second Edition. Washington, D.C.: Department of Energy.

Glasstone, S., Dolan, P.J (eds.) (1977). The Effects of Nuclear Weapons. Third Edition. Washington, D.C.: Department of Energy.

Gonda, T., et al. (2003). 'An exploration of vehicle-terrain interaction in IR synthetic scenes', The International Society for Optical Engineering.

Haas, H. H. (1957). Design Features of the New Continental 750 Horsepower Diesel Engine and Notes On High Performance Four-Cycle Diesel Engines 600 Horsepower to 1000 Horsepower Range. Proceedings of Meeting. New York: Society of Automotive Engineers.

Halbert G. A. (1983). 'Elements of Tank Design', ARMOR, (November-December).

Hall, E. J. (1929). 'Reducing Transportation Cost by Means of Engine Design', SAE Transactions, Vol. 24.

Hamparian, E. (1967). The Evolution of Power Trains and Steering in High Speed Military Track Laying Vehicles. SAE Technical Paper 670726. New York: Society of Automotive Engineers.

Handbook on weaponry (1982). First English Edition. Düsseldorf: Rheinmetall GmbH.

Harkins, T. (2003). 'Evolution of ERA: From K-1 to Relikt', Military Technology, 27(7).

Hazell, P. J. (2015). Armour - Materials, Theory, and Design. Boca Raton: CRC Press.

Held, M. (2005). 'Shaped Charge Optimisation against Bulging Targets', Propellants, Explosives, Pyrotechnics, 30(5).

Held, M. (2007). 'ERA Developments in the Post-Cold War Era', Journal of Military Ordnance, 17(1).

Held, M., Schwartz, W. (1994). 'The Importance of Jet Tip Velocity for the Performance of Shaped Charges against Explosive Reactive Armour', Propellants, Explosives, Pyrotechnics, 19(1).

Hill, F. I. (1951). 'The Damage Effectiveness of Shaped Charges Against Tanks', Transactions of Symposium on Shaped Charges, Aberdeen Proving Grounds, November 13-16. Aberdeen: Ballistic Research Laboratories.

Höhn, D. H., Büchtemann, W. (1973). 'Spectral Radiance in the S20-Range and Luminance of the Clear and Overcast Night Sky', Applied Optics 12(1).

Horton, W. D. (1996). Ground Vehicle System Integration (GVSI) and Design Optimization Model. Technical Report OMI-574. Ann Arbor: OptiMetrics, Inc.

Huber, C., Bogers, W. J. (1983). The Schuler principle: a discussion of some facts and misconceptions. EUT Report 83-E-136. Eindhoven: Technische Hogeschool Eindhoven. Available at: https://pure.tue.nl/ws/files/4325185/8407775.pdf (Accessed: 20 September 2023).

Humphrey, B.J, Smith, B.R, Skaggs, S.R. (1990). Toxicity of Halon 2402. ESL-TR-88-59. pp. 2-3.

Iran Defence Products 2013-2016. Catalogue.

Jane's (1989). 'Inside the T-72', Jane's Soviet Intelligence Review, 1(11).

Jankovych, R., Beer, S. (2011). Wear of cannon 2A46 barrel bore. NAUN.

Johanning, A, Scholz, A. (2013). Propeller Efficiency Calculation in Conceptual Aircraft Design. Technical Note. Hamburg: Hamburg University of Applied Sciences.

Kelly, R. G. (1980). Evaluation of Special Armor Technologies. Technical report. Watervliet: Benet Weapons Laboratory.

Kennedy, D. R. (1951). 'Shaped Charge Damage Beyond Armor', Transactions of Symposium on Shaped Charges, Aberdeen Proving Grounds, November 13-16. Aberdeen: Ballistic Research Laboratories.

Klein, R., Erickson, E. (1967). 'Road Test the M48', ARMOR, (January-February).

Koll, C. (2009). Soviet Cannon.

Kowalski, K. (2018). 'Automatyczne systemy przeciwpożarowe w wojskowych pojazdach bojowych', Zeszyty Naukowe SGSP 2018, 65 (2).

Krewinghaus, A. B. (1969). 'Infrared Reflectance of Paints', Applied Optics 8(4).

Kunze, H. -D., Meyer, L. W., Staskewitsch, E. (1983). ‚Dynamic strength and ductility of a tungsten-alloy for KE-penetrators in swaged and unswaged condition under various loading', Proceedings of 7th International Symposium on Ballistics 1983.

Marzloff, J. (1971). 'AMX 30 France's Main Battle Tank', International Defense Review 1971, 2.

McAlister, D. R. (2016). Neutron Shielding Materials. Illinois: PG Research Foundation. Available at: https://www.eichrom.com/wp-content/uploads/2018/02/neutron-attenuation-white-paper-by-d- m-rev-2-1.pdf (Accessed: 10 September 2023).

Military Standard: Human Engineering, Design Criteria for Military Systems, Equipment, And Facilities (1989). MIL-STD-1472D. Washington, D.C.: Department of the Army.

Millar, D. H. (1970). 'Critical Factors in the Application of Diesel Engines to Fighting Vehicles', *Proceedings of the Institution of Mechanical Engineers 1969-70*, 184(3).

Morgan, J. H., Pittman, J. (1997). Projectile and Warhead Identification Guide - Foreign. Charlottesville: National Ground Intelligence Center.

Niccols, E. H. (1976). Literature Review: Impact toughness of bainite vs. martensite. Technical report. Watervliet: Benet Weapons Laboratory. pp. 3-6.

Oberle, W. F., White, K. J. (1991). Electrothermal-Chemical Propulsion and Performance Limits for the 120-mm, M256 Cannon. Technical Report BRL-TR-3264. Aberdeen: Ballistic Research Laboratory.

Ogorkiewicz, R. M. (1991). Technology of Tanks. Volume 1. Coulsdon: Jane's Information Group.

Operator's Manual: Operator Controls and PMCS. Tank, Combat, Full Tracked, 105-MM Gun, M60A1 (RISE). Tank, Combat, Full Tracked, 105-MM Gun, M60A1 (RISE Passive) (1980). TM 9-2350-257-10-1. Washington, D.C.: Department of the Army.

Perkins aimed at Challenger (1988). Jane's Defence Weekly, 2 (July).

Pickett, C. E. (1969). 'Solar Heat Reflecting Coatings', International Automotive Engineering Congress, Detroit. January 13-17.

Prabhakaran, K. V., Bhide, N. M., Kurian, E. M. (1993). XRD, spectroscopic and thermal analysis studies on trans-1,4,5,8-tetranitiosotetraazadecalin (TNSTAD). Thermochimica Acta, 220, 178.

Projectile and Warhead Identification Guide - Foreign, NGIC-1143-782-98. (1997).

Pulkrabek, W. W. (1997). Engineering Fundamentals of the Internal Combustion Engine. Upper Saddle River: Prentice Hall.

Rabchevsky, G. A. (1988). The tungsten industry of the USSR. Washington, D.C.: United States Department of the Interior.

Raisbeck, G. et al. (1981). Design Goals for Future Camouflage Systems. Fort Belvoir: Research and Development Command.

Rocchio, J. J. (1980). The Interior Ballistic Performance of the 120-mm Tank Gun Relative to the 105-mm Tank Gun. Memorandum, DRDAR-BLP (30 Apr 1980). Aberdeen: Ballistic Research Laboratory.

Rocha, J. G. (1975). Design Analysis of Machine Gun, 7.62MM, PKT, Soviet. RTN-75-010. Rock Island: Rock Island Arsenal, General Thomas J. Rodman Laboratory.

Rosa, S. P., Lindsley, T. (1989). 'Tank Thermal Signatures: The Other Variable In the Gunnery Equation', ARMOR (September-October). p. 32.

Rosenberg, Z, Dekel, E. (2016). Terminal Ballistics. Second Edition. Heidelberg: Springer Berlin.

Rybansky, M. et al. (2015). 'Modelling of cross-country transport in raster format', Environmental Earth Sciences, 74(10). Berlin: Springer.

Schmidt, J. G. (1976). Ballisticians in War and Peace: A History of the United States Army Ballistic Research Laboratories. Volume 2, 1957-1976. Aberdeen: Aberdeen Proving Ground.

Sopok, S., Dunn, S., O'Hara, P., Coats, D., Pfl egl, G., Rickard, C. (2001). Cannon Coating Erosion Modeling Achievements. 10th U.S. Army Gun Dynamics Symposium.

STANAG 4385. 120 mm × 570 ammunition for smooth bore tank guns, in Furmanek, W., Kijewski, J. (2021). 'Constructional Aspects for Safe Operation of 120 × 570 mm Ammunition', Problemy Mechatroniki. Uzbrojenie, Lotnictwo, Inżynieria Bezpieczeństwa 12, 4 (46).

Stevanović, R. (2003). 'Characteristics of torsion bar suspension elasticity in MBTs and the assessment of realized solutions', Scientific-Technical Review, Vol. LIII, No. 2.

Survivability (1985). FM 5-103. Washington, D.C.: Department of the Army.

Tank Gunnery (1957). FM 17-12. Washington, D.C.: Department of The Army.

Tarasenko, A. (2016). Armor protection of the tanks of the second postwar generation T-64 (T-64A), Chieftain Mk5P and M60. Available at: https://btvt.info/3attackdefensemobility/432armor_eng.htm (Accessed: 2 March 2024).

Taylor, C. F. (1985). Internal Combustion Engine in Theory and Practice. Volume 2: Combustion, Fuels, Materials, Design. Second edition (revised). Cambridge: MIT Press.

Teipel, U. (ed.) Energetic Materials: Particle Processing and Characterization. Weinherm: Wiley-VCH Verlag GmbH & Co. KGaA.

Trebiński, R., Leciejewski, Z., Surma, Z. (2022). Determining the Burning Rate of Fine-Grained Propellants in Closed Vessel Tests. Energies, 15(7), 2680-2694.

U.K. Ministry of Defence (1988). Armoured warfare: A Vehicles; replacement for Chieftain; new tank for the army. DEFE 70-1890.

U.S. Department of the Army (1964). Engineering Design Handbook: Ammunition Series - Section 4: Design for Projection. AMCP 706-247. Washington, D.C.: U.S. Army Materiel Command.

U.S. Department of the Army (1966). Engineering Design Handbook: Design for Control of Projectile Flight Characteristics. AMCP 706-242. Alexandria: U.S. Army Materiel Command.

U.S. Department of the Army (1967). Engineering Design Handbook: Automotive Series – Automotive Suspensions. AMCP 706-356. Washington, D.C.: U.S. Army Materiel Command.

U.S. Department of the Army (1979). Engineering Design Handbook: Breech Mechanism Design. DARCOM-P 706-253. Alexandria: U.S. Army Materiel Development and Readiness Command.

U.S. GAO (1991). Abrams Tank - Operating Costs More Than Expected. Washington, D.C.: GAO/NSIAD.

United Nations Environment Programme (2006). Montreal Protocol On Substances that Deplete the Ozone Layer. Report.

United States Army Research Laboratory (2008). Composite Armor Performance. Technical report ARL-TR-2008. Aberdeen: ARL.

United States Flight Standards Service (1971). Airframe and Powerplant Mechanics Powerplant Handbook.

US Department of Defense (1976). Critical Considerations in the Acquisition of a New Main Battle Tank. Report, PSAD-76-113A.

Venkateswaran, N., Radhakrishnan, S. R., Sathyanarayanan, P. L. (2008). 'Performance Improvement of Cooling System in T72 Bridge Layer Tank', Defence Science Journal, 1(58).

War Office (1966). Assessment of lethality of US 105 mm HEAT shell against space-plated Centurion tank. Technical report WO 194/472. Chertsey: FVRDE.

War Office (1968). Comparative analysis of US and Soviet armor technology. Technical report WO 196/478. Chertsey: FVRDE.

Warford, J. M. (1983). 'The T95: A Gamble in High-risk Technology'. ARMOR. (September-October) 1983.

Warford, J. M. (2002). 'The Soviet T-72B Main Battle Tank: The First Look at Soviet Special Armor', Journal of Military Ordnance, 12(3).

Whitley, D. O. S. (1977). HEAT vs HESH Paper. Memorandum.

Whitley, D. O. S. (1981). Advances in Composite Armors. Memorandum.

Whitley, D. O. S. (1985). Tactical Implications of Composite Armors. Memorandum.

Wisniewski, A. Pirszel, J. (2021). 'Protection of armoured vehicles against chemical, biological and radiological contamination', Defence Technology 17(2).

Wojciech Furmanek, W., Kijewski, J. (2021). Constructional Aspects for Safe Operation of 120 × 570 mm Ammunition. Problemy Mechatroniki. Uzbrojenie, Lotnictwo, Inżynieria Bezpieczeństwa 12, 4 (46).

Xiangdong Li, Yanshi Yang, Shengtao Lv (2014). 'A numerical study on the disturbance of explosive reactive armors to jet penetration', Defence Technology, 10(1).

Yunusov, B. A. (2023). Steels for Tank Barrels. International Journal of Advanced Research in Science, Engineering and Technology, 10(8). Available at: https://www.ijarset.com/upload/2023/august/8-r-shoh-07.PDF

Russian Language Sources

Альбом основных конструктивных изменения, проведенных на Изд. 172М.

Альбом основных конструктивных изменения, проведенных на Изд. 184, 184-1 и 184К.

Абрамов, А. И., Гуменюк, Г. А., Евдокимов, В. И., Зборовский, А. А. (2015). 'Опыт оснащения бронетехники аппаратурой регистрации лазерного излучения', Известия Российской Академии Ракетных и Артиллерийских Наук, 2(87).

Абрамов, Б. А., Доронин, В. П., Лазебник, О. М., Прокуряков, В. Б. (1972). 'Некоторые Пути Повышения Прочности и Жесткости Днищ Танков', Вестник Бронетанковой Техники 1972, сборник 1. pp. 15-18.

Авдеев, В. Н. et al. (1976). 'Средства Связи', Вопросы Оборонной Техники, 20(67).

Авдеев, В. Н., Бондаренко, , В. И., Губченко, И. Н., Кузьмин, В. С. (1985). 'Танковая Взрывоустойчивая Штыревая Антенна', Вестник бронетанковой техники 1985, сборник 4.

Аксененко, М. Д., Бараночников, М. Л. (1987). Приемники Оптического Излучения. Москва: "Радио и Связь".

Александров, Ю. И. (1990). 'Бронирование современного танка', Вестник бронетанковой техники 1990, сборник 3.

Алексеев, В. П. et al. (1990). Двигатели внутреннего сгорания: Устройство и работа поршневых и комбинированных двигателей. Москва: «Машиностроение».

Алексеев, М. (2002). 'Защита современных танков', Вестник бронетанковой техники 2002, сборник 7.

Алексеев, М. et al. (2012). НИИ Стали 1942-2012. Москва: Издательство СканРус.

Алексеев, О. И., Терехин, И. И. (1976). 'О Некоторых Закономерностях, Определяющих Защитные Свойства Трехслойных Преград при Обстреле Сплошными Оперенными Бронебойно-Подкалиберными Снарядами', Вопросы оборонной техники, 20(63).

Алферов, К. Д. (1965). Взрывчатые вещества. Часть II: инициирующие и бризантные ВВ. Пенза: Пензенское Высшее Артиллерийское Инженерное Училище.

Алферов, В. В. (1977). Конструкция н расчет автоматического оружия. Москва: «Машиностроение».

Андреев, В. В., Гуськов, А. В., Милевский, К. Е. (2018). Эксплозивные вещества. Новосибирск: Издательство НГТУ.

Андреев, В. П., Горбачев, С. Н., Изосимов, Н. Г., Касьянов, В. Д. (1988). ',,Весовая" терминология в танкостроении (в порядке дискуссии)', Вестник Бронетанковой Техники 1988, Сборник 2. pp. 3-4.

Анипко, О. Б. (2014). Результаты Экспериментального Исследования Воздействия Перекиси Водорода на Нитроцеллюлозные Высокомолекулярные Соединения. Інтегровані технології та енергозбереження, 2, 50.

Анипко, О. Б., Хайков, В. Л. (2012). Анализ Методов Оценки Состояния Пороховых Зарядов Как Элемент Системы Мониторинга Артиллерийских Боеприпасов. Інтегровані технології та енергозбереження, 3.

Антонов, Ю. В., Жучков, М. Г., Фанталов, В. С. (1977). 'Повышение Износостойкости Фрикционных Узлов Трансмиссий ВГМ', Вестник бронетанковой техники 1977, сборник 4.

Антоновский, В. П. et al. (1981). 'Взрывоопасность топливных баков и боекомплекта танков', Вестник Бронетанковой Техники 1981, сборник 1.

Арбузов, В. И. (2008). Основы радиационного оптического материаловедения. Санкт-Петербург: Санкт-Петербург Государственный Университет ИТМО.

Бабкин, А. В. et al. (2008). Средства поражения и боеприпасы. Москва: Издательство МГТУ им. Н.Э. Баумана.

Бакшинов, В. М., Комащенко, А. Г., Тимохин, В. И. (1986). 'Броневые отсеки для боекомплекта танка', Вестник Бронетанковой Техники 1986, сборник 1.

Балакин, С. М., Белков, П. А., Данилов, П. Н., Ломов, С. В. (1987). 'Механические свойства сплава внж-90 при повышенной скорости деформации', Вестник Бронетанковой Техники 1987, сборник 8.

Балашов, И. В., Малофеев, А. М., Чистяков, М. В., Хазов, Н. Н. (2013). 'Противорадиационная Защита: Вчера, Сегодня, Завтра. Иллюстрации предоставлены ОАО «НИИ Стали»', Техника и Вооружение, (March).

Баннов, В. В., Чернявский, В. В., Баннова, Ю. В. (2021). 'Обитаемость Современных Танков', Транспортные Системы: Безопасность, Новые Технологии, Экология. Якутск, 16 апреля. Якутск: ФГБОУ ВО СГУВТ.

Баранов, И. Н. (2010). Главный конструктор В.Н. Венедикто: Жизнь, отданная танкам. Нижний Тагил: ООО «Рекламно-издательская группа «ДиАл».

Баранов, И. Н. (2011). Танковая броня. Москва: Яуза.

Баранов, О. К., Беляев, А. А., Щанкин, М. М. (1982). 'Пневмогидравлическая система управления движением танка', Вестник Бронетанковой Техники 1982, сборник 1.

Баранов, С. П., Никитин, В. Т. (1985). 'Пути Снижения Затрат Мощности в Системах Танкового Дизеля', Вестник бронетанковой техники 1985, сборник 2.

Баринов, Н. П., Батян, В. И., Комаров, А. В., Медов, Б. С. (1986). 'Тепловая Нестабильность Выверки Ночного Прицела с Пушкой', Вестник Бронетанковой Техники 1986, сборник 3.

Баринов, Н. П., Иванов, И. К., Комаров, А. В. (1986). 'Влияние Настройки Механизма Связи Ночного Прицела с Пушкой на Точность Стрельбы', Вестник Бронетанковой Техники 1986, сборник 2.

Барсов, Ф. Ф., Усков, А. Ф. (1976). 'Аккумуляторные Батареи Зарубежной БТТ'. Вестник Бронетанковой Техники 1976, сборник 2.

Барятинский, М. Б. (2008). Т-72. Уральская броня против НАТО. Москва: «Яуза».

Беззубиков, Ю. К., Рослов, В. Б. (1987). 'Расчет быстродействия автомата заряжания', Вестник Бронетанковой Техники 1987, сборник 1.

Беляков, С. А. (2001). Приборы Ночного Видения Бронетанковой Техники. Омск: Омский Государственный Технический Университет.

Березин, И. Я., Мазепа, Г. В., Сергеев, В. Г., Шаповалов, В. В. (1984). 'Прогнозирование ресурса траков танковой гусеницы', Вестник бронетанковой техники 1984, сборник 6.

Березкин, В. (1999). 'В-2: путь в серию', Двигатель, 6(6).

Бернштейн, Л. И. (1979). 'Улучшение Стальной Брони при Легировании Ванадием', Вестник бронетанковой техники 1979, сборник 3.

Бершов, А. В. et al. (1981). 'Новое Воздухозаборное Устройство Для Танка'. Вестник Бронетанковой Техники 1981, сборник 4.

Бирюков, И. Ю. (2006). Пороховые Заряды Длительных Сроков Хранения: Проблемы, Задачи и Пути их Решения. Интегрированные технологии и энергосбережение, 2, 54.

Блинов, В. П., Личковах, В. А., Николахин, В. М. (1983). 'Испытания Танковой Пушки', Вопросы оборонной техники, 5(111).

Богомолов, П. И., Бируля, М. А., Болотин, А. А. (2021). Оценка Эффективности Термозащитного Кожуха Ствольной Трубы Танковой Пушки При Воздействии Солнечной Радиации. Санкт-Петербург: Акционерное общество «Центральный научно-исследовательский институт материалов».

Боевой устав Сухопутных Войск. Часть III: Взвод, отделение, танк (1982). Москва: Воениздат.

Бондарь, А. И. et al. (2015). 'К вопросу совершенствования систем противопожарной защиты отечественных боевых машин', Механіка та машинобудування 2015, 1.

Борзов, В. К. (1986). 'Централизованная Заправка ВГМ Топливом', Вестник Бронетанковой Техники 1986, сборник 3.

Борисюк, М. Д., Жадина, О. В., Харланова, В. П. (1976). 'Анализ качества и совершенствование танковых систем противопожарного оборудования', Вестник Бронетанковой Техники 1976, сборник 3. pp. 14-16.

Боткина, Г. Я., Кочергин, А. К., Щелкунов, Г. М., Подрезов, В. Г. (1986). 'Применение автоматической сварки броневой стали танка Т-72', Вестник Бронетанковой Техники 1986, сборник 5.

Бронированная Ремонтно-Эвакуационная Машина БРЭМ-1 – Техническое описание и инструкция по эксплуатации. Москва: Воениздат.

Брызгов, В. Н. (1985). 'Исследование навесной динамической защиты танков израильской армии', Вестник Бронетанковой Техники 1985, сборник 3. pp. 51-53.

Брызгов, В. Н., Симаков, И. К. (1985). 'Вклад тыльного слоя высокотвердой стали в против-окумулятивную стойкость брони', Вестник Бронетанковой Техники 1985, сборник 2.

Буланкин, Н. Г., Корнилов, В. И., Михаиловский, В. Е., Погудин, Е. В. (1984). 'Сравнительные Испытания Ночных Приборов Водителя', Вестник Бронетанковой Техники 1984, сборник 1.

Бундин, В.И. et al. (2000). Боевое Отделение Подвижной Военной Машины. Роспатент, RU27692.

Буров, С. С. (1973). Конструкция и Расчет Танков. Москва: Военная академия бронетанковых войск.

Бутов, В. И., Егоров, В. В. (1985). 'Повышение мощности и топливной экономичности серийного дизеля', Вестник бронетанковой техники 1985, сборник 1.

Вавилова, С. И., Савостьяновой, М. В. (eds.) (1948). Оптика в Военном Деле - Сборник Статей (Том II). Издание Третье. Москва: Издательство Академии Наук СССР.

Вавилонский, Э. Б., Куракса, О. А., Неволин, В. М. (2008). Основной боевой танк России: Откровенный разговор о проблемах танкостроения. Нижний Тагил: Издательский Дом «Медиа-Принт».

Валеев, Г. Г., Сопин, В. Ф., Соков, Б. А. (2004). Артиллерийские метательные заряды. Казань: ФГУП «Государственный научно-исследовательский институт химических продуктов».

Вахрушева, Ю. В., Шенбергер, А. Ю., Щеглов, Е. В. (2022). 'Танковые приборы ночного видения', Специальная Техника И Технологии Транспорта 2022, 14.

Вереха, Ю. Н., Калинина-Иванова, Е. В., Корнилов, А. Н. (1976). 'О Тормозных Характеристиках Танковых Двигателей', Вестник бронетанковой техники 1976, сборник 1.

Веселов, В. Б., Диков, С. А., Харитонов, И. С. (1982) 'Анализ Поисковых Возможностей Танков', Вопросы Оборонной Техники, 20(105).

Вильховченко, Н. Н., Емельянов, В. Е. (1978). 'Зависимость объемно-весовых показателей автомата заряжания от числа находящихся в нем выстрелов', Вестник бронетанковой техники 1978, сборник 4.

Власова, И. И. et al. (1984). 'Свойства листов брони БТК-1 повышенной твердости', Вестник Бронетанковой Техники 1984, сборник 5.

'Власть переменилась' in Карцев, Л. Н. (2008). 'Воспоминания Главного конструктора танков', Техника и вооружение, (May).

Воробейчик, Г. М., Кондратьев, В. И., Потемкин, Б. В. (1972). 'О влиянии внутренних баков с топливом на уровень противорадиационной защиты экипажей танков и БМП', Вестник Бронетанковой Техники 1972, сборник 6.

Воробьева, Г. Я. (1975). Коррозионная стойкость материалов в агрессивных средах химических производств-Химия. Москва: Издательство «Химия».

Вульфельдт, Э. И., Ганчо, Ю. Г., Жуков, В. Ф., Касьянов, В. Д. (1988). 'Объемно-массовый анализ защиты серийных танков', Вестник Бронетанковой Техники 1988, Сборник 10.

Высоковский, С. Н. et al. (1983). 'Сравнение требований зарубежного и отечественных стандартов на поставку листовой противоснарядной брони', Вопросы оборонной техники, 2(108). pp. 52.

Галанова, Н. М. et al. (1983). 'исследование трещин под прибылями танковых башен', Вестник бронетанковой техники 1983, сборник 5.

Гальвиц, У., Мигрина, Б. А. (1950). Артиллерийские пороха и заряды. Москва: Издательство Оборонгиз.

Ганчо, Ю. Г. (1989). 'Конструкции и технологии повышения защищенности современных танков', Вестник бронетанковой техники 1989, сборник 5.

Гапон, В. В., Гусев, О. П., Нанава, И., Тетельбаум, Р. Д. (1987) 'Исследование Динамики Амортизированного Внутреннего Оборудования Танка', Вестник бронетанковой техники 1987, сборник 8.

Гармонова, И. В. (ed.) (1976). Синтетический Каучук. Москва: Издательство «Химия».

Гаюн, В. В. et al. (1983). 'Действие бронебойно-фугасного снаряда по броне', Вопросы оборонной техники, 5(111).

Георгиевский, О. Н., Дужак, Г. А., Савина, В. Ф., Ядришников, Н. И. (1987). 'Совершенствование оценки надежности при ускоренной войсковой эксплуатации танков', Вестник Бронетанковой Техники 1986, сборник 10.

Гзовская, Т. В., Иванова, О. В., Поляков, В. Б. (1985). Обоснование требуемого уровня живучести ствола танковой пушки. Вестник Бронетанковой Техники 1985, сборник 2.

Гинзбург, Б. М., Иванов, В. А., Сысоев, В. Н. (1984). 'Система Подогрева Впускного Воздуха с Бесфорсуночным Факельным Подогревателем', Вестник бронетанковой техники 1984, сборник 1.

Главное Управление Боевой Подготовки Сухопутных Войск Вооруженных Сил Российской Федерации (2004). Учебник Сержанта Танковых Войск. Москва: Воениздат.

Гладышев, С. А., (1982). 'Характеристики стали СБЛ-2 при изготовлении башни', Вестник бронетанковой техники 1982, сборник 1.

Глухов, Г. К., Кочегаров, П. П. (1980). 'Особенности ремонта подорванных на минах танков', Вестник бронетанковой техники 1980, сборник, 6. pp. 36-37.

Гогин, В. В., Горчаков, В. А. (2016). Министр, Минмаш, НИМИ: Краткая Историография в Области Артиллерийских Боеприпасов (к 100-летнему юбилею В.В. Бахирева). Боеприпасы и высокоэнергетические конденсированные системы.

Гоголюк, Е. К. (1983). 'Двухрежимная Система Подогрева Дизеля', Вестник бронетанковой техники 1983, сборник 1.

Голощапов, И. М. (1973). Танковые приборы ночного видения. Москва: Воениздат.

Голуб, Г. Г., Затравин, Е. И., Тютин, В. Е. (1974). 'Некоторые Статистические Характеристики Процесса Наблюдения Командира Танка', Вестник бронетанковой техники 1974, сборник 2.

Голяшов, А. В., Шамин, Б. Ф. (1983). 'Новые Танковые Средства Связи', Вестник Бронетанковой Техники 1983, сборник 5.

Горбунов, А. С., Мелихова, Т. Н., Тамбовцев, Ф. Д. (1980). 'К истории создания автоматической системы ППО', Вестник Бронетанковой Техники 1980, сборник 3.

Горбунов, А. С., Штепанек, С. М. (1984). 'Способы Оценки Аэродинамического Сопротивления Фильтров Тонкой Очистки ФВУ', Вестник Бронетанковой Техники 1984, сборник 2.

Горст, А. Г. (1949). Пороха и взрывчатые вещества. Москва: Государственное Издательство Оборонной Промышленности.

ГОСТ 9109-81. «Грунтовки ФЛ-ОЗК И ФЛ-ОЗЖ. Технические Условия».

Гребенюк, А.М., Одинцов, Л.Г., Васильев, В.А., Шеломенцев, С.В (2016). Производство взрывных работ при проведении аварийно)спасательных и других неотложных работ в различных чрезвычайных ситуациях. Москва: ФЦ ВНИИ ГОЧС.

Григорян, В. А. et al. (2006). Частные вопросы конечной баллистики. Москва: Издательство МГТУ имени Н. Э. Баумана.

Григорян, В. А., Ермаков, В. И., Мачихин, С. А., Терехин, И. И. (1987). 'Исследование стойкости комбинированной брони к воздействию бронебойных подкалиберных снарядов', Вестник бронетанковой техники 1987, сборник 8.

Гриненко, С. В., Кравченко, Ю. М., Трещевский, Н. П. (1984). 'пути совершенствования танковых систем постановки аэрозольных завес'. Вестник Бронетанковой Техники 1984, сборник 2. pp. 33.

Гуменюк, Г. А. (1986). 'Влияние внешней среды на показатели обнаружения танка по тепловому контрасту', Вестник Бронетанковой Техники 1986, сборник 5. pp. 13-15.

Гуменюк, Г. А. (1986). 'Возможность обнаружения противотанковых средств по бликам приборов', Вестник Бронетанковой Техники 1986, сборник 5.

Гуревич, Б. Г., Чепулис, Л. Л. (1985). 'Шероховатое лакокрасочное покрытие танка', Вестник бронетанковой техники 1985, сборник 1.

Двигатели танков (1991).

Дворянинов, В. Н. (2015). Боевые патроны стрелкового оружия, Книга 4: Современные отечественные патроны хроники конструкторов. Климовск: Д'Соло.

Диков, С. А. (2011). 'Танковые дальномеры', Мир измерений, 2.

Добисов, О. А., Кузьмина, Н. В. (1984). 'Исследование Приборных Комплексов Командиров Основных Танков', Вестник Бронетанковой Техники 1984, сборник 1.

Долгов, А. П. (2004). 'Разработка современных броневых материалов', Вестник бронетанковой техники 2004, сборник 6.

Драк, П. И., Касьян, В. А., Калашникова, Н. М., Калиночкина, Е. В. (1985). 'Новое защитное покрытие для ВГМ на основе двухцветной эмали ХС-5146', Вестник бронетанковой техники 1985, сборник 5. pp. 41-42.

Дрибинский, А. М., Мисюк, А. Ф., Олизаревич, Л. В. (1976). 'Броневая Защита', Вопросы Оборонной Техники, 20(67).

Дрибинский, А. М., Мисюк, А. Ф., Олизаревич, Л. В. (1983). 'Развитие керамических броневых материалов', Вестник бронетанковой техники 1983, сборник 8.

Дронг, И., Трепененков, И., Чухчин, Н. (1975). 'Первый Дизельный', Историческая Серия «ТМ», Техника-молодёжи 1975, 4.

Дубровин, Ф. И., Мазуренко, А. И., Морозов, Е. А., Яремчук, Л. С. (1983). Термостатирование Боекомплекта Танка. Вестник Бронетанковой Техники 1983, сборник 3.

Ежов, А. А., Левин, Л. С., Маслова, Ю. Н., Чикаленко, Г. А. (1981). 'Сравнительные исследования броневых сталей МБЛ-1 и СБЛ-2', Вестник бронетанковой техники 1981, сборник 1. pp. 42-43.

Ермаков, Г. В., Орлов, В. Г. (1968). Устройство и Действие Боеприпасов Артиллерии. Пенза: Пензенское Высшее Артиллерийское Инженерное Ордена Красной Звезды Училище.

Ефремов, А. (2002). 'Чем Выше Подвижность Танков, Тем Мобильнее Сухопутные Войска', Двигатель, 5(23).

Ефремов, А. С. (2010). Уроки танкостроения. Санкт-Петербург: «Гангут».

Ефремов, А., Павлов, М., Павлов, И. (2011). 'История создания первого серийного танка Т-80 с газотурбинной силовой установкой', Техника и вооружение 2011, (November).

Жартовский, Г. С., Куртц, Д. В., Усов, О. А. (2016). Защита оборудования и экипажа военных гусеничных машин от механоакустических и климатических воздействий. Санкт-Петербург: Издательство «Лань».

Желтов И. Г., Макаров А. Ю. (2017). Общее устройство танка А-20. Available at: https://t34inform.ru/publication/p01-7.html (Accessed: 12 August 2023)

Желтов И. Г., Макаров А. Ю. (2020). А.Я. Дик – начальник КБ, которое так и не было создано. Available at: https://t34inform.ru/publication/p-pers-3.html (Accessed: 12 August 2023).

Живулин, Г. А., Олихвер, А. И., Пивовар, Р. М., Снурников, А. С. (1983). 'Средства защиты органов зрения экипажей БТТ от светового излучения ядерного взрыва', Вестник Бронетанковой Техники 1983, сборник 5.

Жирнова, Т. А. (2004). Устройство, Эксплуатация, Техническое Обслуживание и Ремонт Стабилизатора Танкового Вооружения 2Э28М - Методические Указания. Омск: Издательство ОмГТУ.

Жучков, М. Г. (1975). 'Стендовые Испытания ГМТ', Вопросы оборонной техники, 20(59-60).

Жучков, М. Г., Фанталов, В. С. (1978). 'Повышение Долговечности Дисков Фрикционных Узлов Трансмиссии ВГМ', Вестник Бронетанковой Техники 1978, сборник 4.

Забавников, Н. А. (1975). Основы теории транспортных гусеничных машин. Москва: «Машиностроение».

Зайдель, И. Н., Куренков, Г. И. (1970). Электронно-оптические преобразователи. Москва: Издательство «Советское радио».

Зайцев, А.С. (2019). Разработка конструкции ствола артиллерийского орудия: пособие по курсовому проектированию. Санкт-Петербург: Балтийский государственный технический университет «Военмех».

Зайчиков, Ю. Н. (2011). Трансмиссия И Ходовая Часть Танка Т-72. Челябинск: Издательский центр ЮУрГУ.

Закаменных, Г. И., Кучерова, В. Г., Червонцев, С. Е. (eds.) (2017). Проектирование спецмашин, часть 1, книга 1: Артиллерийские стволы. Волгоград: ВолгГТУ.

Запасные части, Инструмент, приспособления и Принадлежности /ЗИП/ Т-72. Poster.

Заславский, Е. И., Погудин, Е. В. (1976). 'Особенности Формирования Лицевой Панели Пульта Управления Механика-Водителя', Вестник Бронетанковой Техники 1976, сборник 2.

Звонков, Ю. В. (1978). 'О сокращении времени заряжания танковой пушки', Вестник Бронетанковой Техники 1978, сборник 3.

Зиновьев, В. М., Куценко, Г. В., Ермилов, А. С., Болдавнин, И. И. (2011). Высокоэнергетические наполнители твердых ракетных топлив и другихвысокоэнергетических конденсированных систем. Физико-термохимические характеристики, получение. Пермь: «Издательство ПНИПУ», 186.

Знаменский, Е. А. (2017). Ударное и кумулятивное действие артиллерийских боеприпасов. Санкт-Петербург: Балтийский государственный технический университет «Военмех».

Зубарь, А. В. et al. (2018). 'Анализ Существующих Способов Выверки Нулевых Линий Визирования Прицелов Системы Управления Огнем Образцов Бронетанкового Вооружения', Научный Вестник ВВИМО, 3 (47).

Зубов, Е. (1999). 'Легендарный В-2: три страницы судьбы', Двигатель, 4(4).

Иванов, В. А., Никитин, В. Т., Шпак, Ф. П., Эрнст, В. Л. (1989). 'Оценка конструкций шасси с помощью показателей Технического уровня ВГМ', Вестник Бронетанковой Техники 1989, сборник 2.

Иванов, И. К., et al. (1980). Особенности Стрельбы Из Танка Новыми Бронебойными Подкалиберными Снарядами. Вестник Бронетанковой Техники 1980, сборник 5, 20.

Ирдынчеев, Л. А., Кудин, В. Т., Рейтблат, В. Л., Шерстюк, А. А. (1978). 'Снижение наведенной радиоактивности высокомарганцовистой стали', Вестник бронетанковой техники 1978, сборник 1.

Ирдынчеев, Л. А., Рейтблат, В. Л., Фрид, Е. С. (1977). 'Наведенная радиоактивность как фактор радиационного поражения личного состава танковых войск', Вестник Бронетанковой Техники 1977, сборник 2.

Ирдынчеев, Л. А., Фрид, Е. С. (1985). 'Расчет доз гамма-излучения наведенной радиоактивности в танках при их облучении нейтронами ядерных взрывов'. Вестник Бронетанковой Техники 1985, сборник 6.

Исаков, П. П. (ed.) (1982). Теория и Конструкция Танка - Том 2: Основы Проектирования Вооружения Танка. Москва: Машиностроение.

Исаков, П. П. (ed.) (1985). Теория и Конструкция Танка - Том 5: Трансмиссии Военных Гусеничных Машин. Москва: Машиностроение.

Исаков, П. П. (ed.) (1990). Теория и Конструкция Танка - Том 10. Кн. 2: Комплесная защита. Москва: Машиностроение.

История оружейного производства АО Златмаш (2024). Available at: https://zmgun.ru/history (Accessed 19 December 2023).

Калашников, Г. Г., Костылев, В. А. (1984). 'Развитие Танковой Радиосвязи в Послевоенный Период', Вестник бронетанковой техники 1985, сборник 4.

Калинина-Иванова, Е. В. (1975). 'Анализ Топливной Экономичности Ходового Макета с ГМТ', Вопросы оборонной техники, 20(59-60).

Калиниченко, В.И. (2017). Защита для брони. Москва: ИПО «У Никитских ворот».

Каплин, М. Е., Третьяков, В. Г., Харлашкин, С. А., Шашков, И. Н. (1984). 'Испытания Западногерманской Аппаратуры Опознавания „Свой - Чужой", Вестник Бронетанковой Техники 1984, сборник 4.

Карпенко, А. (2002). Ракетные танки. Библиотека журнала «Техника молодежи» №1, "Броня". Уссурийск: ООО "Восточный горизонт".

Карцев, Л. Н. (2008). 'Воспоминания Главного конструктора танков', Техника и вооружение, (March).

Кобылкин, И. Ф. (2016). 'Распространение Детонации в Тонких Слоях Взрывчатого Вещества с Инертными Перегородками' Физика горения и взрыва, 52(1).

Кобылкин, И. Ф., Петюков, А. В. (2015). Проявление эффекта ударно-волновой десенсибилизации при возбуждении детонации в тонких слоях взрывчатого вещества высокоскоростными ударниками. Москва: МГТУ им. Н.Э. Баумана.

Кожухов, Ю. В. et al. (2020). Конструкции компрессоров объемного типа. Санкт-Петербург: Политех-Пресс.

Козловский, Б. С. (1979). 'Исследование эффективности комбинированной брони', Вопросы оборонной техники 1979, 5(102).

Колесов, В. А., Степанов, В. А. (1975). 'Особенности Опытной Гидромеханической Трансмиссии ГМТ-69021', Вопросы оборонной техники, 20(59-60).

Колмаков, Д. Г. (2021). Тагильская школа. 80 лет в авангарде мирового танкостроения. Белгород: КОНСТАНТА-принт.

Колмаков, Д. Г., Устьянцев, С. В. (2007). Боевые Машины Уралвагонзавода: Танки 1960-Х. Нижний Тагил: «Медиа-Принт».

Колмаков, Д. Г., Устьянцев, С. В. (2017). УКБТМ. 75 лет тагильской школе танкостроения. Екатеринбург: Издательство ООО Универсальная Типография «Альфа Принт».

Коломиец М. (2009). Т-34: Первая полная энциклопедия. Москва: Эксмо.

Комяженко, А. Г. et al. (1984). 'Методический подход к выбору характеристик динамической бронезащиты танка', Вопросы оборонной техники, 20(116).

Комяженко, А. Г., Тимохин, В. И., Тренина, Н. К. (1974). 'Влияние ослабленных зон на поражение броневой защиты', Вестник бронетанковой техники 1974, сборник 6.

Королев, А. В. (2003). 'История развития динамической защиты', Вестник бронетанковой техники 2003, сборник 3.

Королёв, Г. Е., Наумик, Н. М., Трикоз, Е. И. (1979). 'Броневая Защита Американского Танка М48А3', Вестник бронетанковой техники 1979, сборник 3.

Королева, А. А., Кучерова, В. Г. (eds.) (2002). Физические Основы Устройства и Функционирования Стрелково-Пушечного, Артиллерийского и Ракетного Оружия - Часть 1. Волгоград: Волгоградский Государственный Технический Университет.

Корольков, Р. Н., Медов, Б. С. (1978). 'Исследование Работы Трансмиссии Танка с БКП в Повороте', Вестник бронетанковой техники 1978, сборник 4.

Коронин, Ю.Н., Малинин, В.В., Попов, Г.Н. (2008). 'История создания и унификации стрелковых прицелов ночного видения на примере изделий ЦКБ «Точприбор»', Интерэкспо Гео-Сибирь.

Костин, Ю. Н. et al. (2014). 'Анализ Живучести Динамической Защиты Отечественных Танков', Механіка та машинобудування, 1.

Косточко, А. В., Казбан, Б. М. (2019). Пороха, ракетные твердые топлива и их свойства. Физико- химические свойства порохов и ракетных твердых топлив. Москва: ИНФРА-М.

Косяк, Е. Г. (1986). 'Защита танков от кумулятивных боеприпасов', Вестник бронетанковой техники 1986, сборник 2.

Котельников, В. Р. (2010). Отечественные авиационные поршневые моторы (1910—2009). Москва: Русский Фонд Содействия Образованию и Науке.

Кочегаров, П. П., Лукьянов, А. И., Никулин, Е. Г. (1979). 'Результаты испытаний по оценке проходимости танков', Вопросы оборонной техники, 20(88).

Криксунов, Л. В. (1975). Приборы Ночного Видения. Киев: Издательство Техніка.

Кузнецов, Е. В., Прохорова, И. П., Файзуллина, Д. А. (1976). Производства Полимеров и Пластических Масс на их Основе. Издание Второе. Москва: Издательство «Химия».

Кузнецов, М. И., et al. (1978). Танковые навигационные системы. Москва: Воениздат.

Купцов, В. М., Тимофеев, В. Д. (1979). 'Определение вертикальных реакций грунта, действующих на опорные катки ВГМ при движении', Вопросы оборонной техники, 20(88).

Кустылкин, В. С., Развалов, А. С. (1986). 'Анализ Причин Отказов при Подконтрольной Эксплуатации Танков в 1975-1984 гг.', Вопросы оборонной техники, 5(126).

Лаврищев, Б. П., Соколов, В. Я., Степанов, В. В., Сушков, А. А. (1987). 'Исследование Рационального Боекомплекта Танка', Вестник Бронетанковой Техники 1987, сборник 1.

Лаухин, А. Н. (1970). Современная артиллерия. Москва: Воениздат.

Лепешинский, И. Ю. et al. (2009). Автоматические системы управления вооружением. Омск: Издательство ОмГТУ.

Лепешинский, И. Ю. et al. (2012). Устройство оружия и его боевое применение. Часть вторая. Омск: Издательство ОмГТУ.

Литвинов, Н. П., Мельников, Р. И. (1963). 'К вопросу о выборе вентилятора системы охлаждения среднего танка', Вестник Бронетанковой Техники 1963, сборник 2.

Лещук, О. В. (2001). Вооружение танков и БМП (Курсовая работа по огневой подготовке). Хабаровск: Хабаровский Государственный Педагогический Университет.

Лукьянов, А. И., Бабин, Ю. С. (1966). 'Эксплуатационные характеристики компрессора АК-150 при работе его на танках', Вестник бронетанковой техники 1966, сборник 1.

Лукьянов, В. Н., Крюков, И. А. (2017). Влияние Климатических и Внутрибаллистических Факторов на Изгиб Ствола Танковой Пушки с Различными Вариантами Термозащитных Кожухов. Известия РАРАН, 96.

Лукьянов, Н. А., Лукьянов, В. Н., Близгарев, В. П. История Создания и Совершенствования Танковой Пушки Д-81. Available at: http://btvt.info/3attackdefensemobility/d81history.htm (Accessed: 25 November 2023).

LZOS (2001). Оптическое стекло. Available at: http://www.lzos.ru/opglass/opgrus.htm (Defunct).

М. И. Александровский (1991). 'Унифицированный Танковый Дизель', Вопросы Оборонной Техники, 20(120).

Мазуренко, А. И., Морозов, Е. А. (1985). 'Один Из Путей Повышения Надежности Комплекса Танкового Вооружения', Вестник Бронетанковой Техники 1985, сборник 6.

Макаренко. А., Кузнецов, Ю. (1999). Электроспецоборудование танка Т-72. Омск: Военная кафедра Омский Государственный Технический Университет.

Маринин, В. А., Савенков, А. И., Тульцев, А. А. (1982). 'Влияние Коэффициента Приспособляемости Дизеля на Скорость Танка', Вестник Бронетанковой Техники 1982, сборник 5.

Маринин, В. А., Тульцев, А. А., Штейн, В. Д. (1987). 'Оценка Показателей Подвижности Модернизированных Танков на Марше', Вестник Бронетанковой Техники 1987, сборник 4.

Маркачев, Е. В., Рототаев, Д. А., Чублров, В. Д. (1991). 'Возбуждение детонации ВВ в составе динамической защиты при воздействии бронебойного подкалиберного снаряда', Вестник Бронетанковой Техники 1991, сборник 1. Available at: http://btvt.info/5library/vbtt_1991_01_dz_detonazia.htm (Accessed 15 August 2023).

Маслова, Ю. Н., Оголюк, В. И. (1976). 'Отработка оптимального режима термообработки башен танка Т-64А, изготовленных из стали СБЛ-2', Вопросы Оборонной Техники, 20(63).

Министерства Обороны СССР (1979). Руководство По 7,62-мм Пулеметам Калашникова ПК, ПКМ, ПКС, ПКМС, ПКБ, ПКМБ И ПКТ. Москва: Воениздат.

Министерство обороны РФ (2001). Танк Т-72Б: Комплекс управляемого вооружения 9К120 - Техническое описание и инструкция по эксплуатации. Москва: Воениздат.

Министерство Обороны СССР (1962). Как действовать в условиях применения ядерного, химического и бактериологического оружия - пособие солдату и матросу. Москва: Воениздат.

Министерство Обороны СССР (1972). Танковые Приборы Ночного Видения: Техническое Описание и Инструкция по Эксплуатации. Москва: Воениздат.

Министерство Обороны СССР (1974). 125-мм Танковые Пушки 2А26 (Д-81) и 2А26М2 Техническое Описание и Инструкция по Эксплуатации (Части 1, 2, 3, 4). Москва: Воениздат.

Министерство Обороны СССР (1974). Правила Стрельбы из Танков (ПСТ-74). Москва: Воениздат.

Министерство Обороны СССР (1975). Танк «Урал». Техническое описание и инструкция по эксплуатации (172М.ТО). Книга Первая. Москва: Воениздат.

Министерство Обороны СССР (1975). Танк «Урал». Техническое описание и инструкция по эксплуатации (172М.ТО). Книга Первая. Москва: Воениздат.

Министерство Обороны СССР (1975). Танк «Урал». Техническое Описание и Инструкция по Эксплуатации (172М.ТО). Книга Первая. Москва: Воениздат.

Министерство Обороны СССР (1976). 12,7-мм Пулемет НСВ-12,7 «Утес» на Универсальном Станке: Техническое описание и инструкция по эксплуатации.

Министерство Обороны СССР (1978). Руководство по 12,7-мм Пулемету «Утес» (НСВ-12,7). Москва: Воениздат.

Министерство Обороны СССР (1979). Вооружение Танка Т-72. Москва: Военная академия бронетанковых войск.

Министерство Обороны СССР (1979). Объект 219 техническое описание и инструкция по эксплуатации. Книга Вторая. Москва: Воениздат.

Министерство Обороны СССР (1979). Стабилизаторы Танкового Вооружения 2Э28М (2Э28М-2): Техническое Описание. Москва: Воениздат.

Министерство Обороны СССР (1980). Изделие 1А40 - Инструкция по эксплуатации. Москва: Воениздат.

Министерство Обороны СССР (1980). Руководство по действиям экипажа при вооружении Танка Т-72. Москва: Воениздат.

Министерство Обороны СССР (1983). Двигател и В-46 и В-46-6 – Техническое описание (ТО). Москва: Воениздат.

Министерство обороны СССР (1983). Радиостанция Р-123М – Техническое описание и инструкция по эксплуатации. Москва: Воениздат.

Министерство Обороны СССР (1983). Свинцовые Стартерные Аккумуляторные Батареи - Руководство. Москва: Воениздат.

Министерство Обороны СССР (1985). Руководство по инженерным средствам и приемам маскировки сухопутных войск. Часть I – Средства и Приемы Маскировки Войск. Москва: Воениздат.

Министерство Обороны СССР (1986). Объект 219Р – Техническое описание и инструкция по эксплуатации. Книга вторая. Москва: Воениздат.

Министерство Обороны СССР (1986). Танк Т-72А - Техническое описание и инструкция по эксплуатации: Книга Вторая (Часть первая). Москва: Воениздат.

Министерство Обороны СССР (1986). Танк Т-72А - Техническое описание и инструкция по эксплуатации: Книга Первая. Москва: Воениздат.

Министерство Обороны СССР (1987). Изделие 1А40 - Техническое Описание. Москва: Воениздат.

Министерство Обороны СССР (1988). 125-мм Танковые Пушки 2А26, 2А46, 2А46-1, 2А46М, 2А46М-1, 2А46-2 - Техническое Описание и Инструкция по Эксплуатации 2А46ТО1, Части 3: Боеприпасы. Москва: Воениздат.

Министерство Обороны СССР (1988). Танк Т-72А - Техническое описание и инструкция по эксплуатации. Книга Вторая, Часть Первая. Москва: Воениздат.

Министерства Обороны СССР (1988). Средства преодоления минно-взрывных заграждений – Минные тралы. Техническое описание и инструкция по эксплуатации. Москва: Воениздат.

Министерство Обороны СССР (1991). Дизель В-84М (В-84, В-84-1) – Техническое Описание. Москва: Воениздат.

Министерство Обороны СССР (1991). Объект 184 - Руководство по Войсковому Ремонту, Книга первая. Замена и Ремонт Агрегатов и Узлов. Часть первая. Москва: Воениздат.

Министерство Обороны СССР (1992). Руководство по войсковому ремонту объект 184: Книга Первая - Замена и Ремонт Агрегатов и Узлов, Часть Вторая. Москва: Воениздат.

Минобрнауки России (2011). Ремонт Бронетанковой Техники. Омск: Издательство ОмГТУ.

Михайлов, Г. А., Теннисон, В. Р., Фоменко, В. П. (1986). 'Повышение эффективности охлаждения масла танкового двигателя', Вестник Бронетанковой Техники 1986, сборник 4.

Михайлов, Г. А. (1979). Водяные радиаторы военных гусеницных машин. Москва: ЦНИИ информации.

Михеев, Ю. А. (1987). 'Осколочное Действие Кумулятивных И Осколочно-Фугасных Снарядов При Взрыве На Броне Танка', Вестник Бронетанковой Техники 1987, сборник 4.

Музей Рогачевского завода «Диапроектор»: у каждого экспоната – своя история (2015). Available at: https://news.21.by/society/2015/12/23/1150337.html (Accessed: 10 March 2024).

Н. И. Троицкий, М. Н. Лавров. (2020). 60 лет в танковом двигателестроении. ОАО "НИИД". Журнал «Двигатель», No. 5 (83).

Знаменский, Е. А. (2017). Ударное и кумулятивное действие артиллерийских боеприпасов. Санкт-Петербург: Балтийский государственный технический университет «Военмех».

Направляющая задняя 2А46М.109-62. Technical drawing.

Нарбут, А. М. (1982). 'Оценка загазованности при стрельбе из БМП', Вестник бронетанковой техники 1982, сборник, 1.

Нарбут, А. М. , Ребриков, В. Д., Трофимов, П. В. (1987). 'Уменьшение загазованности обитаемых отделений БТТ при стрельбе', Вестник Бронетанковой Техники 1987, сборник 1.

Неверовский, Н. А., Овчаров, В. Г., Рещиков, И. Ф., Терехин, И. И. (1966). 'Снижение радиолокационного отражения танков за счет выбора оптимальных наружных форм', Вестник бронетанковой техники 1966, сборник 1.

Немцов, А. В., Иванов, Н. А., Квашнин, В. Н. (2007). НИМИ: Федеральное Государственное Унитарное Предприятие «Научно-Исследовательский Машиностроительный Институт» 1932-2007. Москва: Информационно-методический центр "Арсенал образования".

НИИ прикладной химии (2016). Тучи, которые защищают. Available at: http://www.niiph.com/ru/novosti/stati/24-tuchi-kotorye-zashchishchayut (Accessed 2 February 2024).

НИИД (1978). Анализ и систематизация материалов по уровню развития отечественных. Технический Отчет №3443. Available at:

http://btvt.info/3attackdefensemobility/niid_1978.htm (Accessed 20 August 2023).

Никитин, В. Т., Ушаков, В. Я. (1980). 'Методика Расчета Двухступенчатых Танковых Воздухоочистителей', Вестник бронетанковой техники 1980, сборник 6.

Носкин, Г. (2011). Первые БЦВМ космического применения и кое-что из постоянной памяти. СПб: Реноме.

Носов, Н. А., Голышев, В. Д., Волков, Ю. П., Харченко, А. П. (1972). Расчет и Конструирование Гусеничных Машин. Ленинград: «Машиностроение».

Носовец, А. И., Шабашев, Л. Б. (1981). 'Система охлаждения с радиаторами на участке нагнетания вентилятора', Вестник Бронетанковой Техники 1981, сборник 1.

НПО Еврохим. Полиизобутилен. Available at: http://ehim.spb.ru/ru/ximicheskoe-syre/ximicheskie-produkty/poliizobutilen (Accessed: 10 December 2023).

НПО Электромашина (2022). Электродвигатель ЭДМ-16У. Available at: https://web.archive.org/web/20211029070132/https://www.npoelm.ru/product/spetsproduktsiya/elektrodvigatel-edm-16u/ (Archived).

ОАО НПК «Уралвагонзавод» (2021). Танковые Пушки 2А46М, 2А46М-1, 2А46М-4, 2А46М-5.

Орлина, А. С. (1955). Двигатели внутреннего сгорания. Рабочие процессы в двигателях и их агрегатах: Том II. Конструкции И Расчет. Москва: Машиностроительной Литературы.

Орлов, Б. В., Ларман, Э. К., Маликов, В. Г. (1976). Устройство и проектирование стволов артиллерийских орудий. Москва: Машиностроение.

Оружие и технологии России. XXI век Том 12 - Боеприпасы и средства поражения (2006). Москва: Оружие и технологии.

Павлов, И., Павлов, М. (2023). 'Т-62: «Рабочая лошадка» советских танковых дивизий первого эшелона', Техника и Вооружение, (July).

Павлов, М. В., Павлов, И. В. (2021). Отечественные Бронированные Машины 1945-1965 гг. - Часть I: Легкие, средние и тяжелые танки. Кемерово: ООО "Принт".

Паластров, П. С., Мелешко, И. А., Платов, А. И., Рототаев, Д. А. (1991). 'Исследование устройств динамической защиты от бронебойных подкалиберных снарядов', Вестник бронетанковой техники, 1. Available at: https://btvt.info/5library/vbtt_1991_vdz.htm (Accessed: 2 June 2024).

Панов, В. В. et al. (2018). Высокоточное оружие зарубежных стран. Том 1. Противотанковые и многоцелевые ракетные комплексы. Second Edition. Тула: АО «Конструкторское бюро приборостроения им. академика А. Г. Шипунова».

Пашков, В. В. (1989). 'Современные проблемы защиты танков', Вестник бронетанковой техники 1989, сборник 2.

Пашолок, Ю. (2019). *Промежуточное звено*. Available at: https://warspot.ru/14603-promezhutochnoe-zveno (Accessed: 12 August 2023).

Пашолок, Ю. (2023). *Бронетанковый опыт на Дальнем Востоке*. Available at: https://dzen.ru/a/ZLWeXiO-xUbLHdpr (Accessed: 20 June 2024).

Пеньков, М. Д., Ржевский, А. И., Шишков, Ю. А. (1976). 'Особенности Конструкции И Компоновки', Вопросы Оборонной Техники, 20(67).

Первов, М. А. et al. (2017). Очерки истории артиллерии государства Российского. Москва: ООО Издательский дом «Столичная энциклопедия».

Петраков, С. П. (1962). 'О создании подвесок быстроходных танков', Вестник бронетанковой техники 1962, сборник, 5.

Пономарёв, Ю. (2011). 'От ПК к ПКМ', Калашников. Оружие, Боеприпасы, Снаряжение, 2011, 12.

Потемкин, Э. К. (ed.) (1992). Основы научной организации разработки. Том 2. Москва: Издательство МГТУ им. Н. Э. Баумана.

Потемкин, Э.К. (ed.) (1999). ВНИИтрансмаш – страницы истории. Санкт-Петербург: Издательство «Петровский фонд».

Правительство Новгородской области (2018). АО «НИМИ им. В.В. Бахирева» отметило свой 85-летний юбилей. Available at: https://kuginov.ru/raznoe/nimi-ao.html (Accessed: 3 October 2023).

Приказ Военного Министра СССР от 6 октября 1951 г. №130: О введении в действие Инструкции по медицинскому освидетельствованию граждан, призываемых на действительную военную службу, военнослужащих и военнообязанных. Приложение №4. Москва: Министр Обороны СССР.

Приказ Военного Министра СССР от 3 Сентября 1973 г., №185: О Введении В Действие Положения О Медицинском Освидетельствовании В Вооруженных Силах СССР. Приложение №3: Положения о медицинском освидетельствовании в Вооруженных Силах СССР. Москва: Министр Обороны СССР.

Приказ Военного Министра СССР от 9 Сентября 1987 г., №260: О Введении В Действие Положения О Медицинском Освидетельствовании В Вооруженных Силах СССР (на мирное и военное время). Приложение №4: Положения, введенного в действие Приказом Министра обороны СССР от 9 сентября 1987 г. №260. Москва: Министр Обороны СССР.

Приложение: Характеристики отечественных артиллерийских выстрелов, in Королева, А. А., Кучерова, В. Г. (eds.) (2002). Физические Основы Устройства и Функциионирования Стрелково-Пушечного, Артиллерийского и Ракетного Оружия - Часть 1. Волгоград: Волгоградский Государственный Технический Университет.

Проскуров, В. А. (1972). Некоторые Вопросы Применения в Танке Выстрелов Раздельного Заряжания. Вестник Бронетанковой Техники 1972, сборник 2.

Проскуров, В. А., Завьялова, Г. Ф., Москвин, Г. Н., Соколов, В. Я. (1971). Метод Повышения Эффективности Стрельбы из Танка Осколочно-Фугасными Снарядами. Вестник Бронетанковой Техники 1971, сборник 4.

Прямицыно, В. Нашем., Чертов, В.В. (2019). 'В боевой обстановке войска зачастую готовы пить любую воду'. Военно-Исторический Журнал 2019, №1.

Пугачев, Б. Л., Сысоев, В. Н. (1982). 'Влияние концентрации бора на противорадиационную защиту танка', Вестник Бронетанковой Техники 1982, сборник 1.

Пугачев, Б. Л., Фрид, Е. С., Шашкин, В. И. (1986). 'Особенности выбора материалов и толщин слоев в многослойных элементах противорадиационной защиты танков', Вестник Бронетанковой Техники 1986, Сборник 4.

Развалов, А. С. (1987). 'Проблемные Вопросы Обеспечения Надежности Бронетанковой Техники', Вестник бронетанковой техники 1987, сборник 10. Москва: Воениздат.

Развалов, С., Штейн, В. Д. (1984). 'Экспериментальные Данные о Затратах Мощности на Перематывание Гусениц', Вестник бронетанковой техники 1984, сборник 6. Москва: Воениздат.

Ребриков, В. Д. (1983). 'Расчет токсической дозы в танке при стрельбе из пушки без эжек- тора', вестник бронетанковой техники 1983, сборник 1.

Ребриков, В. Д., Ширман, Б. А., Сподак, В. В. (1988). 'Особенности зарядки баллонов системы противопожарного оборудования хладоном 13В1', Вестник Бронетанковой Техники 1988, сборник 10.

Рейтблат, В. Л., Сержантов, Е. П., Студниц, М. А., Фрид, Е. С. (1975). 'Противорадиационная защита экипажей танков', Вестник Бронетанковой Техники 1975, сборник 1.

Реукова, Т. Ф. (ed.) (1980). Учебник Сержанта Мотострелковых Подразделений. Москва: Воениздат.

Розоринов, Г. Н., Хаскин, В. Ю., Лазаренко, С. В. (2013). Применение Лазерных Технологий Для Повышения Срока Службы Изделия КБА-3. Збірник наукових праць СНУЯЕтаП.

Романов, Н. И. (ed.) (1973). Теория стрельбы из танков. Москва: Военная академия бронетанковых войск.

Руденко, И. П. (1992). 'История развития композитных броневых материалов', Вестник бронетанковой техники 1992, сборник 4.

Руководство По 7,62-мм Пулеметом Калашникова ПК, ПКМ, ПКС, ПКМС, ПКБ, ПКМБ И ПКТ. Москва: Воениздат.

Руководство по войсковому ремонту объект 184 кн.1 ч.2 (1992).

Саттарова, А. И., Диденко, Т. Л., Евграфова, Ю. А. (2017). 'Технология расснаряжения артиллерийских снарядов гидродинамическим методом', Вестник технологического университета, 20(9).

Сафонов, Б. С., Мураховский, В. И. (eds.) (1993). Основные боевые танки. Москва: Арсенал-пресс.

Сафонова, Б. С., Мураховского, В. И. (eds.) (1993). Основные Боевые Танки. Москва: Арсенал-Пресс.

Свантéссон, К. Г. (2009). 'Нар финендом пришел в Швецию', у Линдстрём, Р. О., Свантéссон, К. Г. Шведский боеприпас: 90 лет в русской танковой промышленности. Available at: http://www.ointres.se/ryska_strf_till_sverige.htm (Accessed: 27 August 2023).

Селиванова, В. В. (ed.) (2016). Боеприпасы: учебник в двух томах. Том 1. Москва: Издательство МГТУ им. Н.Э. Баумана.

Селивохин, В. М. (1962). Танк. Москва: Воениздат.

Сеннов, Н. И. (1942). Оптика на танке. Москва: Объединение государственных книжно-журнальных издательств.

Сергеев, А. (2011). 'Влияние наклона брони на её защитные свойства', Вестник бронетанковой техники 2011, сборник 5.

Серебряков, М. Е. (1949). Внутренняя баллистика ствольных систем и пороховых ракет. Москва: Государственное научно-техническое издательство.

Серебряков, М. Е. (1962). Внутренняя баллистика ствольных систем и пороховых ракет. Москва: Государственное научно-техническое издательство.

Соболев, Е. Г. (1981). 'Восстанавливаемость танков при боевых повреждениях', Вестник бронетанковой техники 1981, сборник, 3.

Соколовский, В. Д. (1995). Броневая защита танков. Москва: Воениздат.

Солянкин, А. Г., Желтов, И. Г., Кудряшов, К. Н. (2010). Отечественные бронированные машины. XX век: Том 3. Отечественные бронированные машины. 1946-1965 гг. Москва: ООО «Издательство "Цейхгауз"».

Солянкин, А. Г., Желтов, И. Г., Кудряшов, К. Н. (2010). Отечественные бронированные машины. XX век - Том 3: Отечественные бронированные машины. 1946-1965 гг. Москва: Издательство «Цейхгауз».

Софьин, А. П., Смолин, В. В. (1979). 'О Снижении Усилий на Органах Управления ВГМ', Вопросы оборонной техники, 20(84).

Старовойтов, В. С. (ed.) (1990). Военные гусеничные машины. Том 1: Устройство, Книга 1. Москва: Издательство МГТУ им. Н. Э. Баумана.

Старовойтов, В. С. (ed.) (1990). Военные гусеничные машины. Том 1, Книга 2. Москва: Издательство МГТУ им. Н. Э. Баумана.

Стрелков, А. Г. (2005). Конструкция быстроходных гусеничных машин. М.: МГТУ «МАМИ».

Суворов, С. (2004). 'Танки Т-72: вчера, сегодня, завтра', Техника и Вооружение, (July).

Сутормин, Е. А. (2013). Вождение боевых машин. Екатеринбург: Издательство Уральского университета.

Талу, К. А. (1963). Конструкция и Расчет Танков. Москва: Военная академия бронетанковых войск.

Танк Т-72М1: Техническое Описание и Инструкция по Эксплуатации – Часть I.

Танки (1973). Москва: Издательство ДОСААФ.

Танковые прицелы и системы управления огнем (СУО). Available at: https://38niii.ru/analitika/173-tankovye-pritsely-i-sistemy-upravleniya-ognem-suo.html (Accessed: 1 February 2024).

Таран, Ю. И., Фролов, Л. А. (1984). 'Новая Система Целеуказания', Вестник Бронетанковой Техники 1984, сборник 5.

Тарасенко, А. (2020). Сердце «тридцатьчетвёрки» на тракторе. Available at: https://warspot.ru/7634-serdtse-tridtsatchetvyorki-na-traktore (Accessed 21 December 2023).

Тарасенко, А. (2022). Испытания ДЗ «Контакт-1» Танки Т-62 и Т-55 – Часть 3. Available at: http://btvt.info/3attackdefensemobility/kontakt_1_trials_pt3.htm (Accessed 20 December 2023).

'Тема 3: Общее устройство системы электрооборудования. Источники электрической энер- гии', Уральского федерального университета.

Тер-Данилов, Р. А. (2009). 'Основы Функционирования Установок', 17.01.02 Стрелково-пушечное, артиллерийское и ракетное оружие. Тульский государственный университет.

Техническое Описание Визира К10-Т.

Трахенберг, В. (ed.) (2000). НИИ «Поиск»: Страницы Истории. Санкт-Петербург: Издательство «Формика».

Троицкий, В. В. (1987). 'Рабочие Места Экипажей в Отечественных и Зарубежных Танках', Вестник Бронетанковой Техники 1987, сборник 12.

У-31Б - ОД0.335.442ТУ. Available at: http://www.promvpk.ru/Catalog/Product/51a269276d0 fad80e402f044 (Accessed 20 August 2023).

Уланов, А. (2017). Пулемётная драма Красной Армии. Часть 3: На замену «максиму». Available at: https://www.kalashnikov.ru/pulemyotnaya-drama-krasnoj-armii-3/ (Accessed: 17 December 2023).

Устройство Переговорное Р-124. Техническое описание и инструкция по эксплуатации.

Устьянцев, С. В., Чернышева, Е. Ю. (2020). 100 лет российского танкостроения. Екатеринбург: Издательство ООО Универсальная Типография «Альфа Принт».

Устьянцев, С., Колмаков, Д. (2007). Боевые Машины Уралвагонзавода: Танки 1960-Х. Нижний Тагил: «Медиа-Принт».

Фастовский, А. Х. (1980). 'Исследование ошибок навигационных систем ВГМ', Вестник Бронетанковой Техники 1980, сборник 3.

Филлипович, М. (1987). Dejstvo trinitrotoluena na nitrocelulozno-celulozni list. Naučno-tehnički pregled, Vol.XXXVII, 1987, br. 2.

Хлопотов, А. (2019). 2А82 - супер пушка для «Арматы». Available at: https://dzen.ru/media/gurkhan/2a82-super-pushka-dlia-armaty-5c31c4349175d500aabd6073 (Accessed: 29 December 2023).

Ходаковский, В. Н. (1995). 'Композитные броневые материалы', Вестник бронетанковой техники 1995, сборник 4.

Цыбанев, С. А. (2022). Справочник Сапёра. Инженерно-Технический Отдел Омон «Зубр».

Чепурной, А. Д. (2000). 'Разработка и промышленное освоение производства сварнокатаной башни боевого танка'. ОАО "ГСКТИ".

Чернышев, В. Л., Акиншин, А. Г. (2013). 'Оценка Нагруженности Планетарных Рядов Бортовой Коробки Передач Танка Т-64А Методом

Динамического Состояния в Режиме Разгона', збірник тез Всеукраїнської науково-практичної конференції НУЦЗУ, 2013.

Чобиток, В. А. (1984). Конструкция и расчет. Москва: Воениздат.

Чобиток, В. В. (2022). Бортовые противокумулятивные щитки. Available at: http://armor.kiev.ua/Tanks/Modern/T64/shields/ (Accessed 1 June 2024).

Чобиток, В., Саенко, М., Тарасенко, А., Чернышев, В. (2016). Основной танк Т-64 - 50 лет в строю. Москва: Яуза-каталог.

Чугасов, А. А. (1969). Ядерное Оружие. Москва: Воениздат.

Чурбанов, Е. В. (1975). Внутренняя Баллистика. Ленинград: Издательство «ВАОЛКА им. М. И. Калинина».

Шабалин, В. А. (1970). О Концентрации Пороховых Газов в Зоне Дыхания Экипажа Бронеобъектов. Вестник Бронетанковой Техники 1970, сборник 3.

Шамарин, О. В. (1985). 'Электромеханические Стабилизаторы Танкового Вооружения', Вестник Бронетанковой Техники 1985, сборник 1.

Шаповалов, А. Б., Солунин, В. Л., Костюков, В. В. (2017). Системы управления, наведения и приводы: История создания и развития. Москва: Издательство МГТУ им. Н. Э. Баумана.

Шишковского, В. М. (1978). Огневая Подготовка, Часть Вторая: Основы Устройства Вооружения. Москва: Воениздат.

Шпак, Ф. П. (1978). 'Влияние процессов "торможение – разгон" на подвижность ВГМ при совершении марша', Вестник Бронетанковой Техники 1978, сборник 2.

Шпак, Ф. П. (1986). 'Оценка Формирования Показателей Подвижности Танков по Ограничивающим Фактором', Вестник Бронетанковой Техники 1986, сборник 3.

Юрасов, И. В. (1975). 'К Истории Производства Отечественного Быстроходного Танкового Дизеля', Вестник бронетанковой техники 1975, сборник 3.

Юрко, С. В., et al. (2016). 125-мм Танковая Пушки 2А46М: Пособие. Минск: БИТУ.

Ямпольский, П. А. (1961). Нейтроны атомного взрыва. Москва: Госатомиздат.

Янковский, И. Н. et al. (2020). Устройство и эксплуатация бронетанкового вооружения. Част 2: Эксплуатация танка Т-72Б. Минск: БНТУ.

German Language Sources

Deutsches Panzermuseum Munster (2019). Dipl-Ing. Rolf Hilmes: Wie konstruiert man einen Panzer?. [Online video]. Available from: https://youtu.be/_J1dHNtyOLI (Accessed: 21 August 2023).

Friesecke (1995). ‚Neuartige Werkstoffe und Materialien für den ballistischen Schutz', Wehrtechnisches Symposium Thema: "Aktuelle Schutztechnologien für Gepanzerte und Ungepanzerte Fahrzeuge" : 15.11. - 17.11.1995. Conference proceedings. Mannheim: Bundesakad. für Wehrverwaltung und Wehrtechnik.

GRD: Technische Abteilung 1, Sektion 1.4 - Kampffahrzeuge (1974). Bericht: über die technischen Versuche im Rahmen der Evaluation Panzer 68 - Kampfpanzer Leopard. GRD report.

Hilmes R. (1988). Kampfpanzer. Die Entwicklungen der Nachkriegszeit. Bonn: Mittler Report Verlag GmbH.

Hilmes R. (2007). Kampfpanzer heute und morgen: Konzepte – System – Technologien. Auflage: Motorbuch Verlag.

Köppen, U. in Kotsch, S. (2023). Das passive Ziel- und Beobachtungsgerät PZB 200. Available at: https://www.kotsch88.de/f_pzb200.htm (Accessed: 15 September 2023)

Kotsch, S. (2011). Die russische 125 mm Panzerkanone 2A46. Available at: https://www.kotsch88.de/g_2a46.htm (Accessed: 1 December 2023).

Krapke, P. -W. (2004). Leopard 2: Sein Werden und seine Leistung. Norderstedt: Books on Demand GmbH.

Kratzenberg, K., Kuellmer, G., Nausester, A. (1970). Waermeschutzhuelle fuer Kanone. Wegmann and Co GmbH. Deutsches Patentamt Patent no. DE1918422A1. Available at: https://patents.google.com/patent/DE1918422A1/en (Accessed: 23 October 2023).

Lobitz, F. (2022). Gesamtwerk Leopard 2. Erlangen: Tankograd Publishing.

Merhof, W., Hackbarth, E.-M. (2016). Fahrmechanik der Kettenfahrzeuge. Überarbeitete Neuausgabe. Neubiberg: Universität der Bundeswehr München.

Ministerium für Nationale Verteidigung (1979). Mittlerer Panzer T72. Taktisch-technische Angaben, Turm, Panzerbewaffnung und Panzerspezialausrüstung – Beschreibung und Nutzung. A 051/1/106. Berlin: Militärverlag der Deutschen Demokratischen Republik.

Ministerium für Nationale Verteidigung (1987). Panzer T72,T72M und T72M1 – Beschreibung. A 051/1/130. Berlin: Militärverlag der Deutschen Demokratischen Republik.

Nationale Volksarmee Panzerwerkstatt (1981). Bericht über die Weiterführung der Nutzungserprobung des Gerätes 172 M bis 14.000 Km. Technical report.

NVA (1981). Schußtafel 125 mm Panzerkanone D-81 (ST 250/4/005) in Kotsch, S. (2023). Die Munition der russischen Panzerkanone D-81. Available at: https://www.kotsch88.de/m\125\mm\d-81.htm (Accessed: 29 October 2023).

'Schematisches Fahrdiagramm 4 HP-250-Leopard' in Zahnradfabrik Friedrichshafen AG. ZF-Gangwahlschalter für das Getriebe 4 HP-250.

Spielberger, W. J. (1995). Waffensysteme Leopard 1 und Leopard 2. Auflage: Motorbuch Verlag.

'Warschauer Pakt: Die Modernisierung des Kampfpanzerbestandes'; Oesterreichische mil. Zeitschrift 1980.

Other Language Sources

Velek, M. (1989). 'T-72', ATOM 1989, (April), 16(112).

Behrendt, J. (2002). 'Współczynniki Wykorzystania Przełożeń Skrzyń Przekładniowych w Czołgach Rodziny T-72', Szybkobieżne Pojazdy Gąsienicowe 2(16).

Dziubak, T. (2017). 'Problemy Usuwania Pyłu z Multicyklonów Filtrów Powietrza Silników Terenowych Pojazdów Mechanicznych', Archiwum Motoryzacji, 76(2).

Edgren, S., Hansson, M. (1992). T 72 och MTLB. Delapport 1. Trials report. Arméns Pansarcentrum.

Filipović, M. (1987). 'Dejstvo trinitrotoluena na nitrocelulozno-celulozni list', Naučno-tehnički pregled, Vol.XXXVII, 1987, br. 2.

Kirasić, V. (2021). Pogonska Jedinica V-46 TK. Undergraduate Thesis. Zagreb: Hrvatsko vojno učilište "Dr. Franjo Tuđman".

Kozlowski, G., Pirog, J. (2002). 'Osiagi bojowe czolgu PT-91 z systemem kierowania ogniem „Savan"', Wojskowy instytut Techniki Pancernej i Samochodowej.

Magier, M (2016). 'Kierunki i możliwości modernizacji czołgów sił zbrojnych RP', PTU, 139(3).

Mysłowski, J. (2014). 'Propozycja Poprawy Manewrowości Czołgu Twardy', Zeszyty Naukowe Wsowl, 1(171).

Ocskay, I. (2018). 'Kísérleti lövészet T-54-es harckocsikra 1989-ben, a „0" ponti gyakorlótéren - II. rész', Haditechnika, 4.

Octavian et al. (1994). Procedeu de obt, inere a pulberii granulare tip 12/7V/A, destinată încărcăturii de azvârlire pentru tunul de calibrul 125 mm. Romanian Patent no. RO104346.

Octavian et al. (1994). Procedeu de obt, inere a pulberii tubulare granulare tip 15/1V/A, destinată încărcăturii de azvârlire pentru tunul de calibrul 125 mm. Romanian Patent no. RO104528.

Phương, T. V., et al. (2023). Nghiên Cứu Tổng Hợp Hợp Chất Trans-1,4,5,8 Tetranitroso-1,4,5,8-Tetraaza Decalin (Ц-2) Ứng Dụng Làm Phụ Gia Năng Lượng Cho Thuốc Phóng АПЦ-235П Của Đạn 125 MM Trên Xe Tăng T90S. Journal of Military Science and Technology, 74(8), 59.

Politechniki Śląskiej (1993). Odlewnictwo: Technologia wykonywania form i rdzeni - skrypt nr 1747. Gliwice.

Svantesson, C. G. (2009). 'När fienden kom till Sverige', in Lindström, R. O., Svantesson, C. G. Svenskt Pansar: 90 år av svensk stridsfordonsutveckling. Available at: http://www.ointres.se/ryska_strf_till_sverige.htm (Accessed: 27 August 2023).

Smit, W. (2008). De Leopard 1: Gepantserde vuist van de Koninklijke Landmacht. Meppel: Boom.

Stauff, J. E. (2008). COMHART - Tome 10: Armements Antichars Missiles Guidés Et Non Guidés. Paris: Comité pour l'histoire de l'armement terrestre.

T-72 adaptacja licencji: Prace wdrożeniowe 1978-1982' (2016). Szybkobieżne Pojazdy Gasienicowe, 3(41). Gliwice: Ośrodek Badawczo-Rozwojowy Urządzeń Mechanicznych „OBRUM" sp. z o.o.

吴群彪, et al. (2019). 前置组合杆体垂直侵彻钢靶简化模型. 爆炸与冲击, 39(1). Available at: https://pubs.cstam.org.cn/data/article/bzycj/preview/pdf/bzycj-39-1-013302-1.pdf (Accessed: 2 January 2023).

吴群彪, 沈培辉, 刘荣忠 (2014). '后置组合杆体侵彻机理研究', 兵工学报, 10. Available at: http://www.co-journal.com/CN/10.3969/j.issn.1000-1093.2014.10.003 (Accessed: 2 January 2024).

梁禾, 芮兰德 (1989). '"86式100滑高腾压反坦克炮研制的回顾', in 高膛压火炮系统论文集. 机电部兵器科学研究院.

陆祥璇 (1989). '"两炮"火力系统总体论证', in 高膛压火炮系统论文集. 机电部兵器科学研究院.

Also by
Military History Group

Footnotes or bust!

ACHTUNG TIGER!
How the Allies Defeated Germany's Heavy Tank

ACHTUNG TIGER! — How the Allies Defeated Germany's Heavy Tank
Peter Samsonov

ACHTUNG TIGER! unravels the mysteries surrounding the Tiger tank's famed survivability. From analyzing the Tiger's formidable armor protection and distinctive characteristics, over to describing its first combat operations against Soviet and Allied forces during World War II, and ultimately the test trials assessing the tank's capabilities, this comprehensive volume offers an unparalleled exploration of the Tiger. Drawing on a wealth of primary sources, this book presents a vivid and detailed narrative that caters to both military history enthusiasts and WW2 tank experts. ACHTUNG TIGER! stands as an indispensable read, providing a captivating synthesis of historical insights and technical expertise for anyone eager to unravel the secrets behind this legendary war machine.

"A fresh look at the ever-fascinating subject of the Tiger tank...from the perspective of its adversaries. Using archival sources, Samsonov charts the Allied reactions to the Tiger on battlefields in various theaters. This new book provides novel and especially valuable look at the early Soviet encounters with the Tiger tank and their evaluation of its combat effectiveness." - Steven J. Zaloga, Military Historian

TANK ASSAULT

The Combat Manual of Armored and Mechanized Forces of the Red Army

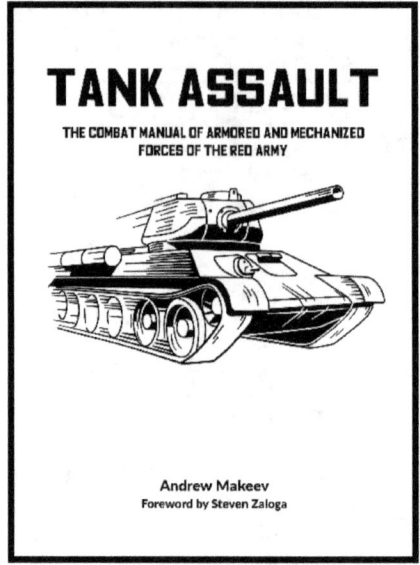

TANK ASSAULT — The Combat Manual of the Red Army
Andrew Makeev

The Combat Manual of Armored and Mechanized Forces of the Red Army is the essential manual that covers the tactics, formations and doctrine of Soviet tank platoons and companies. This document is based on the Soviet experience of fighting the German Army since 1941 and was originally published by the Soviet Union in 1944. This volume presents a faithful translation into English, offering a comprehensive insight into Soviet tank doctrine, unit structures, training and tactics. Spanning from the individual tank up to a full tank company, the information is presented in a detailed yet approachable manner. The book includes many helpful visualizations and examples, from battle formations, combined arms tactics, to the maneuver of tanks during an attack.

Translated with great care, this manual is a treasure trove for anyone interested in the Second World War. Next to the translation that provides insights into Soviet tank warfare, annotations and footnotes offer additional information, context and explanations. This manual presents a de-facto summary of Soviet experiences with tanks during WW2 and is therefore a must-have source for both the professional historian and the historically interested reader.

PANZER
The Medium Tank Company 1941

PANZER - The Medium Tank Company 1941
Bernhard Kast and Christoph Bergs

PANZER - The Medium Tank Company 1941 is a faithfully translated German World War 2 Army Regulation about the medium tank company of the German Armor Branch. This regulation was issued following the successful campaigns in Poland, the Low Countries and France and encompasses topics such as tank crew specialization, training, formations, how to engage enemy positions and tanks, as well as cooperation with other units such as the light tank company, engineers and the infantry. In addition, key information on logistical aspects is given and a breakdown of the company's force organization can be found within H. Dv. 470/7.

This translation features a side-by-side German-English translation and remains true to the original's formatting. Footnotes and supplementary information were added to provide the reader with additional context and insight into the German Army structure, the meaning of various concepts and their modern equivalents.

Available at: http://www.hdv470-7.com/

IS-2
Development, Design and Production of Stalin's Warhammer

IS-2 – Development, Design and Production of Stalin's Warhammer
Peter Samsonov

The IS-2 is the quintessential Soviet heavy tank from World War 2. Heavily armored and boasting a fearsome 122mm gun, this tank matched the German panzers on the Eastern front by more than just its fierce appearance. This tank's history is told from the beginning of the Soviet heavy tank program until the very end of World War 2, in the most detailed and complete account of its development, design and production available in English.

Supported by extensive research of Russian language sources, this publication includes a comprehensive breakdown of prototypes, the Soviet analysis of weaknesses in German tanks including the Tiger and Panther, the development of the 122mm gun, the principles of the new tank's armor layout and a wealth of technical data.

Available at: http://is-2-tank.com/

www.ingramcontent.com/pod-product-compliance
Lightning Source LLC
Chambersburg PA
CBHW052009290426
44112CB00014B/2174